Studies in Computational Intelligence 434

For further volumes:
http://www.springer.com/series/7092

El-Ghazali Talbi (Ed.)

Hybrid Metaheuristics

Editor
Prof. Dr. El-Ghazali Talbi
University of Lille 1
CNRS, INRIA
France

ISSN 1860-949X e-ISSN 1860-9503
ISBN 978-3-642-42905-7 ISBN 978-3-642-30671-6 (eBook)
DOI 10.1007/978-3-642-30671-6
Springer Heidelberg New York Dordrecht London

Softcover reprint of the hardcover 1st edition 2013

Printed on acid-free paper

Springer is part of Springer Science+Business Media (www.springer.com)

To my dear wife Keltoum. She always comforts and consoles me. I am such a lucky guy to have found such a wonderful woman.

To my two star sons Anis and Chahine. Every time that I saw your smile it lights me up inside.

To my daughter Besma, you are my sweet girl with an impressive intelligence.

To my mother Zehour for her infinite sacrifice.

To my father Ammar who continue to support me in my academic research.

Preface

Importance of This Book

Applications of optimization is countless. Every process has a potential to be optimized. There is no company which is not involved in solving optimization problems. Indeed, many challenging applications in science and industry can be formulated as optimization problems. Optimization occurs in the minimization of time, cost, risk, or the maximization of profit, quality, efficiency. For instance, there are many possible ways to design a network to optimize the cost and the quality of service; there are many ways to schedule a production to optimize the time; there are many ways to predict a 3D structure of a protein to optimize the potential energy, and so on.

A large number of real-life optimization problems in science, engineering, economics and business are complex and difficult to solve. They cannot be solved in an exact manner within a reasonable amount of time. Using hybrid algorithms is the main alternative to solve this class of problems.

Purpose of This Book

The main goal of this book is to provide a state of the art of hybrid metaheuristics. The book provides a complete background that enables readers to design and implement hybrid metaheuristics to solve complex optimization problems in a diverse range of application domains. Readers learn to solve large scale problems quickly and efficiently. Numerous real-world examples of problems and solutions demonstrate how hybrid metaheuristics are applied in such fields as telecommunication, logistics and transportation, bioinformatics, engineering design, scheduling, etc.

Audience

One of the main audience of this book is **advanced undergraduate and graduate students** in computer science, operations research, applied mathematics, control, business and management, engineering, etc. Many undergraduate courses on optimization throughout the world would be interested in the contents thanks to the introductory part of the book.

In addition, the **postgraduate** courses related to optimization and complex problem solving will be a direct target of the book. Hybrid metaheuristics are present in more and more postgraduate studies (computer science, business and management, mathematical programming, engineering, control, etc).

The intended audience is also **researchers** in different disciplines. Researchers in computer science and operations research are developing new optimization algorithms. Many researchers in different application domains are also concerned by the use of hybrid metaheuristics to solve their problems.

Many **engineers** are also dealing with optimization in their problem solving. The purpose of the book is to help engineers to use hybrid metaheuristics for solving real-world optimization problems in various domains of application. The application part of the book will deal with many important and strategic domains such as computational biology, telecommunication, engineering design, data mining and machine learning, transportation and logistics, production systems, etc.

Outline

The book is organized following 17 different chapters organized in 5 parts :

- Unified view of hybrid metaheuristics for mono and multi-objective optimization, and optimization under uncertainty.
- Combining metaheuristics with (complementary) metaheuristics.
- Combining metaheuristics with exact methods from mathematical programming approaches which are mostly used in operations research.
- Combining metaheuristics with constraint programming approaches developed in the artificial intelligence community.
- Combining metaheuristics with machine learning and data mining techniques.

Lille, Prof. Dr. El-Ghazali Talbi
March 2012 University of Lille 1, CNRS, INRIA, France

Acknowledgements

Thanks to all contributors of this book for their cooperation in bringing this book to completion.

Thanks also go to all members of my research team DOLPHIN : D. Brockoff, L. Brotcorne, F. Clautiaux, B. Derbel, C. Dhaenens, L. Jourdan, A. Liefooghe, N. Melab, S. Verel.

Finally I should like to thank the team at Springer who gave me excellent support throughout this project, and especially for their patience.

Contents

List of Contributors

Enrique Alba
Universidad de Málaga, Spain
e-mail: eat@lcc.uma.es

Filipe Alvelos
Centro Algoritmi - DPS, Universidade do Minho, 4710-057 Braga, Portugal
e-mail: falvelos@dps.uminho.pt

Sarab Al-Muhaideb
Department of Computer Science, College of Computer and Information Sciences, King Saud University, P.O. Box 51178, Riyadh 11543, Saudi Arabia
e-mail: salmuhaideb@acm.org·

Una Benlic
LERIA, University of Angers, 2 Bd Lavoisier, 49045 Angers Cedex 01, France
e-mail: benlic@info.univ-angers.fr

Christian Blum
ALBCOM Research Group, Universitat Polit'ecnica de Catalunya, Barcelona, Spain
e-mail: cblum@lsi.upc.edu

Raffaele Cipriano
Dipartimento di Matematica e Informatica, Università degli Studi di Udine via delle Scienze 208, I-33100, Udine, Italy
e-mail: cipriano@dimi.uniud.it

Pedro J. Copado-Méndez
Departament d'Enginyeria Quimica, Universitat Rovira Virgili, Tarragona, Spain
e-mail: pedrojesus.copado@urv.cat

Frederico R. B. Cruz
Departamento de Estatística - ICEx - UFMG, Av. Antônio Carlos, 6627, 31270-901 - Belo Hori-zonte - MG, Brazil
e-mail: fcruz@est.ufmg.br

Patrick De Causmaecker
Katholieke Universiteit Leuven Campus Kortrijk Etienne Sabbelaan 53 8500 Kortrijk, Belgium
e-mail: Patrick.DeCausmaecker@kuleuven-kortrijk.be

Amaro de Sousa
Instituto de Telecomunicações / DETI, Universidade de Aveiro, 3810-193 Aveiro, Portugal
e-mail: asou@ua.pt

Luca Di Gaspero
Dipartimento di Ingegneria Elettrica, Gestionale e Meccanica, Università degli Studi di Udine via delle Scienze 208, I-33100, Udine, Italy
e-mail: l.digaspero@uniud.it

Julián Domínguez
Universidad de Málaga, Spain
e-mail: julian@lcc.uma.es

Agostino Dovier
Dipartimento di Matematica e Informatica, Università degli Studi di Udine via delle Scienze 208, I-33100, Udine, Italy
e-mail: dovier@dimi.uniud.it

Jérémie Dubois-Lacoste
IRIDIA, CoDE, Université Libre de Bruxelles (ULB), Brussels, Belgium
e-mail: jeremie.dubois-lacoste@ulb.ac.be

Christophe Duhamel
Institut Charles Delaunay (LOSI) and STMR (UMR CNRS 6279), Université de Technologie de Troyes, BP 2060, 10010 Troyes Cedex, France
e-mail: christophe.duhamel@isima.fr

Paola Festa
Department of Mathematics and Applications, University of Napoli Federico II, 80126 Napoli, Italy
e-mail: paola.festa@unina.it

Christophe Gouinaud
Laboratoire d'Informatique (LIMOS, UMR CNRS 6158), Campus des Cézeaux, 63177 Aubière Cedex, France
e-mail: christophe.gouinaud@isima.fr

G. Guillén-Gosálbez
Departament d'Enginyeria Quimica, Universitat Rovira Virgili, Tarragona, Spain
e-mail: gonzalo.guillen@urv.cat

Jin-Kao Hao
LERIA, University of Angers, 2 Bd Lavoisier, 49045 Angers Cedex 01, France
e-mail: hao@info.univ-angers.fr

Nur Insani
School of Mathematical Sciences, Monash University, Victoria 3800, Australia
e-mail: nur.insani@monash.edu

Laureano Jiménez
Departament d'Enginyeria Quimica, Universitat Rovira Virgili, Tarragona, Spain
e-mail: laureano.jimenez@urv.cat

Graham Kendall
School of Computer Science, University of Nottingham, Malaysia Campus, Jalan Broga, 43500, Semenyih, Selangor Darul Ehsan, Malaysia
e-mail: graham.kendall@nottingham.edu.my

Philippe Lacomme
Laboratoire d'Informatique (LIMOS, UMR CNRS 6158), Campus des Cézeaux, 63177 Aubière Cedex, France
e-mail: placomme@isima.fr

Leo Liberti
LIX, Ecole Polytechnique, Palaiseau, France
e-mail: liberti@lix.polytechnique.fr

Andrea Lodi
DEIS, University of Bologna and IBM-UniBo Mathematical Optimization Center of Excellence, Viale Risorgimento 2 – I-40136, Bologna, Italy
e-mail: andrea.lodi@unibo.it

Leo Lopes
School of Mathematical Sciences, Monash University, Victoria 3800, Australia
e-mail: leo.lopes@monash.edu

Manuel López-Ibáñez
IRIDIA, CoDE, Université Libre de Bruxelles (ULB), Brussels, Belgium
e-mail: manuel.lopez-ibanez@ulb.ac.be

Barry McCollum
School of Electronics, Electrical Engineering and Computer Science, Queen's University Belfast, BT7 1NN, UK
e-mail: B.McCollum@qub.ac.uk

Paul McMullan
School of Electronics, Electrical Engineering and Computer Science, Queen's University Belfast, BT7 1NN, UK
e-mail: p.p.mcmullan@qub.ac.uk

Mohamed El-Bachir Menai
Department of Computer Science, College of Computer and Information Sciences, King Saud University, P.O. Box 51178, Riyadh 11543, Saudi Arabia
e-mail: menai@ksu.edu.sa

Antonio Mucherino
IRISA, University of Rennes, Rennes, France
e-mail: antonio.mucherino@irisa.fr

Caroline Prodhon
Institut Charles Delaunay (LOSI) and STMR (UMR CNRS 6279), Université de Technologie de Troyes, BP 2060, 10010 Troyes Cedex, France
e-mail: caroline.prodhon@utt.fr

Jakob Puchinger
Mobility Department, Austrian Institute of Technology, Austria
e-mail: jakob.puchinger@ait.ac

Mauricio G.C. Resende
Algorithms and Optimization Research Department, AT&T Labs Research, Florham Park, NJ 07932 USA
e-mail: mgcr@research.att.com

Ulrike Ritzinger
Mobility Department, Austrian Institute of Technology, Austria
e-mail: ulrike.ritzinger_fl@ait.ac.at

Dorabella Santos
Instituto de Telecomunicações, 3810-193 Aveiro, Portugal
e-mail: dorabella@av.it.pt

Kate Smith-Miles
School of Mathematical Sciences, Monash University, Victoria 3800, Australia
e-mail: kate.smith-miles@monash.edu

Thomas Stützle
IRIDIA, CoDE, Université Libre de Bruxelles (ULB), Brussels, Belgium
e-mail: stuetzle@ulb.ac.be

El-Ghazali Talbi
University of Lille 1, CNRS, INRIA, Lille-France
e-mail: talbi@lifl.fr

Greet Vanden Berghe
CODeS (Combinatorial Optimisation and Decision Support) Industrial Sciences KAHO St.-Lieven, Belgium
e-mail: Greet.VandenBerghe@kahosl.be

Katja Verbeeck
Vakgroep ICT, University College, Katholieke Hogeschool Sint-Lieven, Ghent (KAHO Belgium)
e-mail: Katja.Verbeeck@kahosl.be

A. Villagra
Emerging Technologies Laboratory, Universidad Nacional de la Patagonia Austral, Argentine
e-mail: avillagra@uaco.unpa.edu.ar

Tony Wauters
CODeS, KAHO Sint-Lieven, Gebroeders Desmetstraat 1, 9000 Gent, Belgium
e-mail: tony.wauters@kahosl.be

Lyndon While
School of Computer Science and Software Engineering, The University of Western Australia, Perth WA 6009, Australia
e-mail: lyndon@csse.uwa.edu.au

Brendan Wreford
School of Mathematical Sciences, Monash University, Victoria 3800, Australia
e-mail: brendan.wreford@monash.edu

Part I
Hybrid Metaheuristics for Mono and Multi-objective Optimization, and Optimization under Uncertainty

Chapter 1
A Unified Taxonomy of Hybrid Metaheuristics with Mathematical Programming, Constraint Programming and Machine Learning

El-Ghazali Talbi

Abstract. Over the last years, interest on hybrid metaheuristics has risen considerably in the field of optimization. The best results found for many real-life or classical optimization problems are obtained by hybrid algorithms. Combinations of algorithms such as metaheuristics, mathematical programming, constraint programming and machine learning techniques have provided very powerful search algorithms. Four different types of combinations are considered in this chapter:

- Combining metaheuristics with (complementary) metaheuristics.
- Combining metaheuristics with exact methods from mathematical programming approaches which are mostly used in operations research.
- Combining metaheuristics with constraint programming approaches developed in the artificial intelligence community.
- Combining metaheuristics with machine learning and data mining techniques.

1.1 Introduction

This chapter deals with the design of hybrid metaheuristics and their implementation. A taxonomy of hybrid algorithms is presented in an attempt to provide a common terminology and classification mechanisms. The goal of the general taxonomy given here is to provide a mechanism to allow comparison of hybrid algorithms in a qualitative way. In addition, it is hoped the categories and their relationships to each other have been chosen carefully enough to indicate areas in need of future work as well as to help classify future work. Among existing classifications in other domains, one can find examples of flat and hierarchical classifications schemes [74]. The taxonomy proposed here is a combination of these two schemes - hierarchical

---

El-Ghazali Talbi
University of Lille 1, CNRS, INRIA, Lille - France
e-mail: talbi@lifl.fr

E.-G. Talbi (Ed.): Hybrid Metaheuristics, SCI 434, pp. 3–76.

as long as possible in order to reduce the total number of classes, and flat when the descriptors of the algorithms may be chosen in an arbitrary order. The same classification is used for all types of combinations. For each type of hybrids, the main ideas in combining algorithms are detailed. Each class of hybrids is illustrated with some examples. A critical analysis is also carried out.

In fact, the taxonomy could usefully be employed to classify any hybrid optimization algorithm (specific heuristics, exact algorithms). The basic classification is extended by defining the space of hybrid metaheuristics as a grammar, where each sentence is a method that describes a combination of metaheuristics, mathematical programming and constraint programming. In this chapter, a "high-level" description of hybrid metaheuristics is proposed. The internal working and the algorithmic aspects of a given metaheuristic are not considered.

The chapter is organized as follows. First, in section 1.2, our concern is hybrid algorithms combining metaheuristics. The design and implementation issues of hybrid metaheuristics are detailed. A taxonomy is presented to encompass all published work up to date in the field and to provide a unifying view of it. A grammar which generalizes the basic hybridization schemes is proposed. In section 1.3, the combination of metaheuristics with mathematical programming approaches is considered. Section 1.4 deals with the combination of metaheuristics with constraint programming techniques. Then, in section 1.5 the combination of metaheuristics with machine learning and data mining algorithms is addressed. Hybrid metaheuristics for multi-objective optimization are addressed in section 1.6.

1.2 Hybrid Metaheuristics

Hybridization of metaheuristics involves a few major issues which may be classified as design and implementation. The former category concerns the hybrid algorithm itself, involving issues such as functionality and architecture of the algorithm. The implementation consideration includes the hardware platform, programming model and environment on which the algorithm is to be run. In this chapter, a difference is made between the design issues used to introduce hybridization and implementation issues that depend on the execution model of the algorithms.

1.2.1 Design Issues

The taxonomy will be kept as small as possible by proceeding in a hierarchical way as long as possible, but some choices of characteristics may be made independent of previous design choices, and thus will be specified as a set of descriptors from which a subset may be chosen.

1.2.1.1 Hierarchical Classification

the structure of the hierarchical portion of the taxonomy is shown in figure 1.1. A discussion about the hierarchical portion then follows. At the first level, one may distinguish between low-level and high-level hybridizations. The low-level hybridization addresses the functional composition of a single optimization method. In this hybrid class, a given function of a metaheuristic is replaced by another metaheuristic. In high-level hybrid algorithms, the different metaheuristics are self-contained. There is no direct relationship to the internal workings of a metaheuristic.

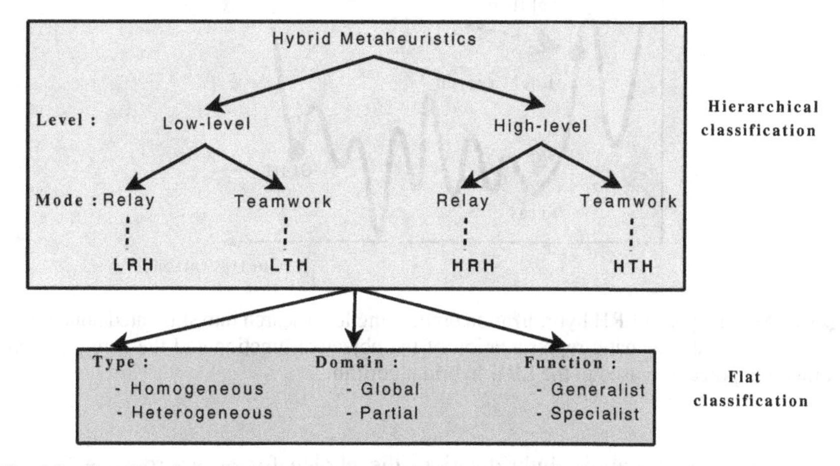

Fig. 1.1 Classification of hybrid metaheuristics in terms of design issues

In relay hybridization, a set of metaheuristics is applied one after another, each using the output of the previous as its input, acting in a pipeline fashion. Teamwork hybridization represents cooperative optimization models, in which many cooperating agents evolve in parallel; each agent carries out a search in a solution space.

Four classes are derived from this hierarchical taxonomy:

- **LRH (Low-level Relay Hybrid):** this class of hybrids represents algorithms in which a given metaheuristic is embedded into a S-metaheuristic (Single solution based metaheuristic) [?]. Few examples of hybrid metaheuristics belong to this class.

  Example 1.1. **Embedding local search into simulated annealing:** the main idea is to incorporate deterministic local search techniques into simulated annealing so that the Markov chain associated to simulated annealing explores only local optima [124]. The algorithm proceeds as follows: suppose the configuration is currently locally optimal. This is labeled `Start` in figure 1.2. A perturbation or a "kick" is applied to this configuration which significantly changes the current solution `Start`. After the kick, the configuration labeled `Intermediate` in the figure is reached. It is much better to first improve `Intermediate` by a local search and apply the accept/reject test of simulated annealing only afterwards. The local search takes us from `Intermediate` to the configuration labeled `Trial`, and then the accept/reject test is applied. If `Trial` is accepted, one has to find an interesting large change to `Start`. If `Trial` is rejected, return to `Start`. Many of the barriers (the "ridges") of the fitness landscape are jumped over in one step by the hybrid metaheuristic.

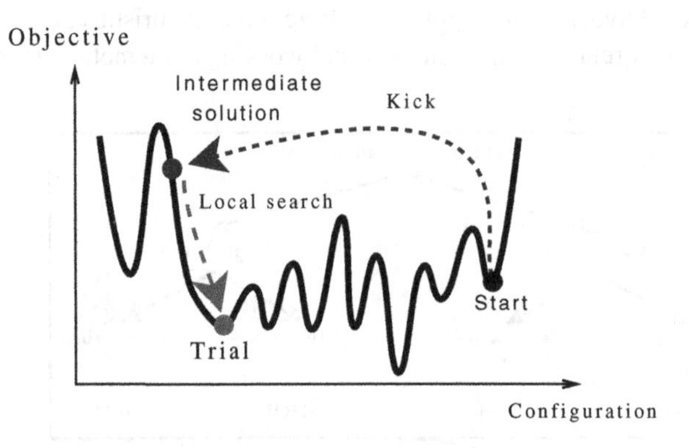

Fig. 1.2 An example of LRH hybridization embedding local search into simulated annealing. The figure gives a schematic representation of the objective function and the configuration modification procedure used in the LRH hybrid algorithm.

To implement the above hybridization, the choice for an appropriate "kick" should be adapted to both the optimization problem and the local search method used. For the traveling salesman problem, if the local search algorithm used is the 2-opt local search heuristic, the "kick" move must apply a k-change with $k > 2$ to prevent cycles. The "kick" operator must attain solutions which are always outside the neighborhood associated to the local search algorithm.

- **LTH (Low-level Teamwork Hybrid):** two competing goals govern the design of a metaheuristic: exploration and exploitation. Exploration is needed to ensure that every part of the space is searched enough to provide a reliable estimate of the global optimum. Exploitation is important since the refinement of the current solution will often produce a better solution. P-metaheuristics (Population based metaheuristics) [?] (e.g. evolutionary algorithms, scatter search, particle swarm, ant colonies) are powerful in the exploration of the search space, and weak in the exploitation of the solutions found.

  Therefore, most efficient P-metaheuristics have been coupled with S-metaheuristics such as local search, simulated annealing and tabu search, which are powerful optimization methods in terms of exploitation. The two classes of algorithms have complementary strengths and weaknesses. The S-metaheuristics will try to optimize locally, while the P-metaheuristics will try to optimize globally. In LTH hybrid, a metaheuristic is embedded into a

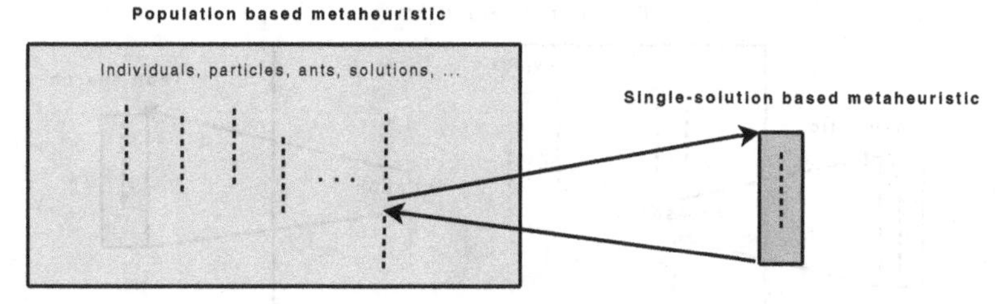

Fig. 1.3 Low-level Teamwork Hybrid (LTH). S-metaheuristics are embedded into P-metaheuristics.

P-metaheuristic[1] (Fig. 1.3). This class of hybrid algorithms is very popular and has been applied successfully to many optimization problems. Most of the state-of-the art of P-metaheuristics integrate S-metaheuristics.

Example 1.2. **Embedding S-metaheuristics into evolutionary algorithms:** when an evolutionary algorithm is used as a global optimizer, its standard operators may be augmented with the ability to perform local search. Instead of using a blind operator acting regardless of the fitness of the original individual and the operated one, an operator which is a heuristic that considers an individual as the origin of its search applies itself, and finally replaces the original individual by the enhanced one (see figure 1.4). The use of local search with evolutionary algorithms is also inspired by biological models of learning and evolution. EAs take many cues from mechanisms observed in natural evolution. Similarly, models of learning are often equated with techniques for local optimization [148]. Research on the interaction between evolution and learning had naturally led computer scientists to consider interactions between evolutionary algorithms and local optimization [20].

The genetic operators replaced or extended are generally mutation[2] and crossover.

- **mutation:** the local search algorithm may be a simple local search [157] [174] [100], tabu search [70] [111] [171], simulated annealing algorithm [36] [180] or any S-metaheuristic (e.g. threshold accepting, guided local search). This kind of operators is qualified *lamarckian*[3]. In the lamarckian model, an individual is replaced by the local optima found, contrary to the *baldwin* model where the local optima is just used to evaluate the individual. In several occasions, LTH has provided better results than other methods on difficult problems. For instance, good results have been obtained on the graph coloring problem combining a genetic algorithm with tabu search [71]. A local search algorithm which uses problem-specific knowledge may be incorporated into

[1] This class of hybrid metaheuristics includes *memetic algorithms*.

[2] Also known as evolutionary local search algorithms.

[3] The name is an allusion to Jean Batiste de Lamarck's contention that phenotype characteristics acquired during lifetime can become heritable traits.

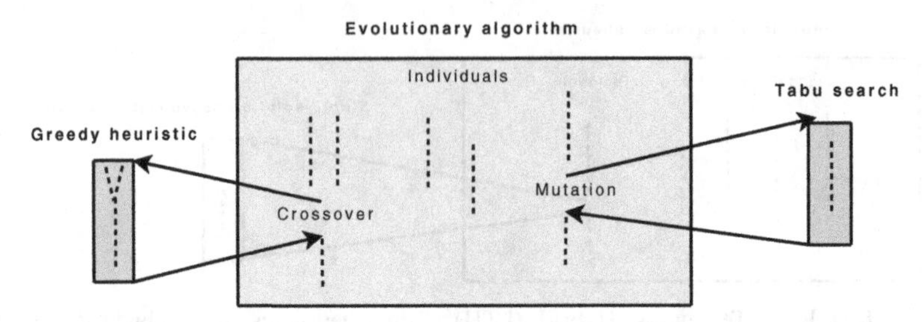

Fig. 1.4 Illustration of a LTH hybrid. For instance, a tabu search is used as a mutation operator and a greedy heuristic as a crossover operator into a genetic algorithm.

the genetic operators [37]. Questions concerning the best use of local search with a genetic algorithm have been addressed in [86].

- **Crossover:** classical crossover operators do not use any heuristic information about a specific application domain. They are blind operators. One can introduce heuristic crossover in order to account for problem-specific information [82]. For instance, greedy heuristics for the crossover operator have shown to improve EAs results when applied to job-shop scheduling, set covering, and traveling salesman problems [52].

  Many crossover operators including heuristic information have been proposed for continuous optimization:

  - Heuristic crossover where the offspring has the following form $x' = u(x_2 - x_1) + x_2$ where u is a uniform random value in $[0,1]$, x_1 and x_2 are the two parents with the condition that x_2 is better than x_1 [181]. This heuristic crossover uses the objective function in determining the direction of the search.
  - Simplex crossover where more than two parents are selected, the worst (resp. the best) individuals x_2 (resp. x_1) are determined. The centroid of the group c is then computed without taking into account the solution x_2. The offspring has the following form $x' = c + (c - x_2)$ [143].

This hybrid model can be used to improve any P-metaheuristic: ant colonies [161] [156], genetic programming [135], particle swarm optimization, and so forth. The S-metaheuristic has been introduced to intensify the search. Let us notice that the scatter search metaheuristic already includes an improvement procedure which is based on S-metaheuristics [49].

The main problem encountered in this class of hybrids is *premature convergence*. Indeed, if the hybridization is applied at each iteration, very competitive solutions will be generated in the beginning of the search which will cause an eventual premature convergence. Conditional hybridization is carried out to prevent this phenomenon by applying the combination:

- **Static manner:** for instance the combination is performed at a given frequency. The hybridization is applied once for a given number of iterations.

  - **Adaptive manner:** when a given event occurs during the search the hybridization is performed. For instance, if there is no improvement of the search for a given number of iterations.

- **HRH (High-level Relay Hybrid):** in HRH hybrids, self-contained metaheuristics are executed in a sequence. For example, the initial solution of a given S-metaheuristic may be generated by another optimization algorithm. Indeed, the initial solution in S-metaheuristics has a great impact on their performances. A well known combination scheme is to generate the initial solution by greedy heuristics which are in general of less computing complexity than iterative heuristics (Fig. 1.5).

  This scheme may be also applied to P-metaheuristics, but a randomized greedy heuristic must be applied to generate a diverse population (Fig. 1.5). Greedy heuristics are in general deterministic algorithms and then they generate always the same solution. On the other hand, the diversity of the initial population has a great impact on the performance of P-metaheuristics. This hybrid scheme is carried out explicitly in the scatter search metaheuristic.

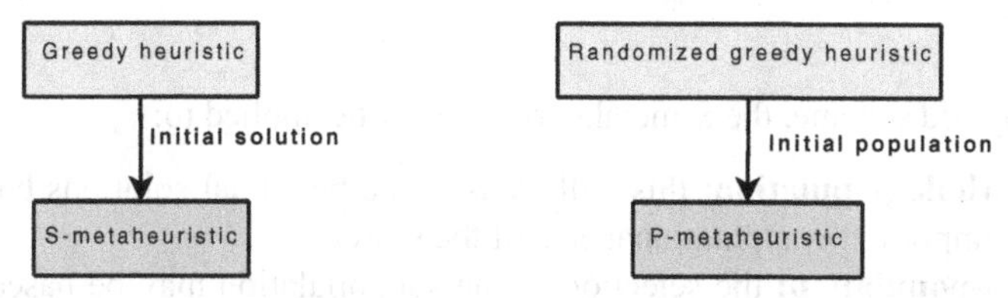

Fig. 1.5 High-level Relay Hybridization. (Left) Generation of the initial solution of a S-metaheuristic by a greedy algorithm. (Right) Generation of the initial population of a P-metaheuristic by a randomized greedy heuristic.

Combining in the HRH scheme P-metaheuristics with S-metaheuristic is also largely applied. It is well known that P-metaheuristics are not well suited for fine-tuning structures which are very close to optimal solutions. Indeed, the strength of P-metaheuristics is in quickly locating the high performance regions of vast and complex search spaces. Once those regions are located, it may be useful to apply S-metaheuristics to the high performance structures evolved by the P-metaheuristic.

A fundamental practical remark is that after a certain amount of time, the population is quite uniform and the fitness of the population is no longer decreasing. The odds to produce fitter individuals are very low. That is, the process has fallen into a basin of attraction from which it has a low probability to escape.

The exploitation of the already found basin of attraction to find as efficiently as possible the optimal point in the basin is recommended. It is experimentally clear that the exploitation of the basin of attraction that has been found may be more efficiently performed by another algorithm than by a P-metaheuristic. Hence, it is much more efficient to use a S-metaheuristic such as a hill-climbing or tabu search (see figure 1.6). The HRH hybridization may use a greedy heuristic to generate a good initial population for the P-metaheuristic (see figure 1.6). At the end of a simulated annealing search, it makes sense to apply local search on the best found solution to ensure that it is a local optima.

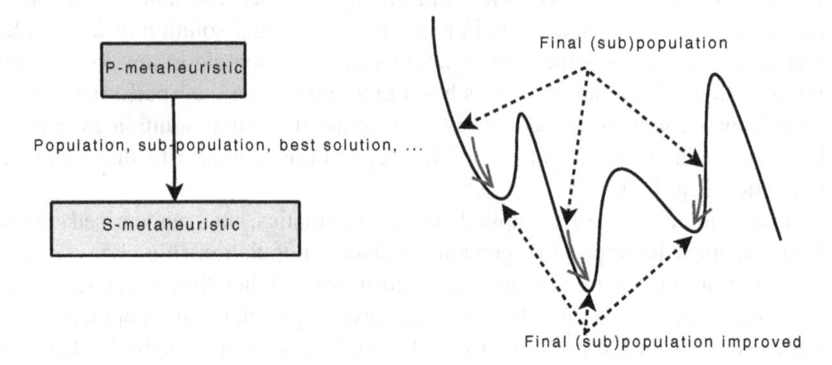

Fig. 1.6 High-level Relay Hybridization. There may be more than two algorithms to be pipelined.

In this hybrid scheme, the S-metaheuristics may be applied to:

- **The whole population:** this will leads to the best final solutions but with a more important computational cost of the search.
- **A sub-population:** the selection of the subpopulation may be based on the diversity of the solutions. This is a good compromise between the complexity of the search and the quality of the final results.
- **The best solution of the population:** the S-metaheuristic is applied once on the best solution of the obtained population. This procedure will reduce the search time but does not ensure to find the best solution.

A path relinking strategy may be applied to a population or a set of elite solutions found by a metaheuristic [6]. Path relinking may be seen as an intensification task over a given population of solutions.

Example 1.3. **HRH hybrid evolutionary algorithms:** many research works of the literature have used the idea of HRH hybridization for EAs. In [121] [166], the considered hybrid scheme introduces respectively simulated annealing and tabu search to improve the population obtained by a GA. In [132], the author introduces hill-climbing to improve the results obtained by an ES. In [118], the algorithm proposed starts from simulated annealing and uses GAs to enrich the solutions found. Experiments performed on the graph partitioning problem using the tabu search algorithm exploiting the result found by a GA give better results than a search performed either by the GA, or the tabu search alone [166].

- **HTH (High-level Teamwork Hybrid):** the HTH[4] scheme involves several self-contained algorithms performing a search in parallel, and cooperating to find an optimum. Intuitively, HTH will ultimately perform at least as well as one algorithm alone, more often perform better, each algorithm providing information to the others to help them.

Example 1.4. **Island model for genetic algorithms:** the first HTH hybrid model has been proposed for genetic algorithms (GAs). This is the well known island model[5]. The population in this model is partitioned into small subpopulations by geographic isolation. A GA evolves each subpopulation and individuals can migrate between subpopulations (Fig. 1.7). This model is controlled by several parameters: the topology that defines the connections between subpopulations, the migration rate that controls the number of migrant individuals, the replacement strategy used, and a migration interval that affects how often migration occurs. In some island models, the individuals really migrate and therefore leaves empty space in the original population. In general, the migrated individuals remain in the original population (i.e. pollination model [155]).

Let us present some pioneering island models for GAs. Tanese proposed a GA based HTH scheme that used a 4-D hypercube topology to communicate individuals from one subpopulation to another [170]. Migration is performed at uniform periods of time between neighbor subpopulations along one dimension of the hypercube. The migrants are chosen probabilistically from the best individuals of the subpopulation and they replace the worst individuals in the receiving subpopulation.

Cohoon, Hedge, Martin and Richards proposed a HTH based on the theory of "punctuated equilibria" [41]. A linear placement problem was used as a benchmark and experimented using a mesh topology. They found that the algorithm with migration outperforms the algorithm without migration and the standard GA. This work was later extended using a VLSI design problem (graph partitioning) on a 4-D hypercube topology [42] [43].

Belding in [19] attempted to extend the Tanese's work using the Royal Road continuous functions. Migrants individuals are sent to a random selected subpopulation, rather than using a hypercube topology. The global optimum was found more often when migration (i.e. cooperating GAs) occurred than in completely isolated cases (i.e. non-cooperating GAs).

Afterwards, the HTH hybrid model has been generalized to other P-metaheuristics and S-metaheuristics. Indeed, the HTH hybrid model has also been applied to simulated annealing [61], genetic programming [114], evolution strategies [178], ant colonies [123], scatter search [50], tabu search [62], and so on.

---

[4] HTH hybrids is referred as *multiple interacting walks* [176] *multi-agent algorithms* [24], and *cooperative search algorithms* [40] [39] [90] [92] [172].

[5] Also known as migration model, diffusion model, and coarse grain model.

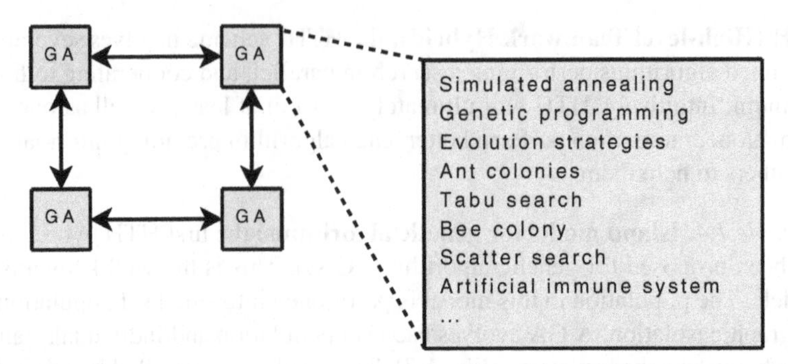

Fig. 1.7 The island model of genetic algorithms as an example of High-level Teamwork Hybrid (HTH). The same model has been used with different topologies for simulated annealing, genetic programming, evolution strategy, ant colony, tabu search, bee colony, artificial immune system, etc.

1.2.1.2 Flat Classification

Homogeneous/Heterogeneous: in homogeneous hybrids , all the combined algorithms use the same metaheuristic. Hybrid algorithms such as the island model for GAs belong to this class of hybrids. The homogeneous metaheuristics may differ in the initialization of their (Fig. 1.8):

- **Parameters:** in general, different parameters are used for the algorithms. For instance, in the HTH hybrid scheme which is based on tabu search, the algorithms may be initialized with different tabu list sizes [179]; different crossover and mutation probabilities may be used in evolutionary algorithms, etc.
- **Search components:** given a metaheuristic, one can use different strategies for any search component of the metaheuristic, such as the representation of solutions, objective function approximations [59] [151], initial solutions, search operators (neighborhood, mutation, crossover, ...), termination criteria, etc.

Using different parameters or search components into a given metaheuristic will increase the robustness of the hybrid algorithm.

Example 1.5. **Heterogeneous hybrids:** in heterogeneous algorithms , different metaheuristics are used (Fig. 1.9). A heterogeneous HTH algorithm based on genetic algorithms and tabu search has been proposed in [46] to solve a network design problem. The population of the GA is asynchronously updated by multiple tabu search algorithms. The best solutions found by tabu search algorithms build an elite population for the GA.

The GRASP method (Greedy Randomized Adaptive Search Procedure) may be seen as an iterated heterogeneous HRH hybrid, in which local search is repeated from a number of initial solutions generated by a randomized greedy heuristic [66] [65]. The method is called adaptive because the greedy heuristic takes into account the decisions of the precedent iterations [64].

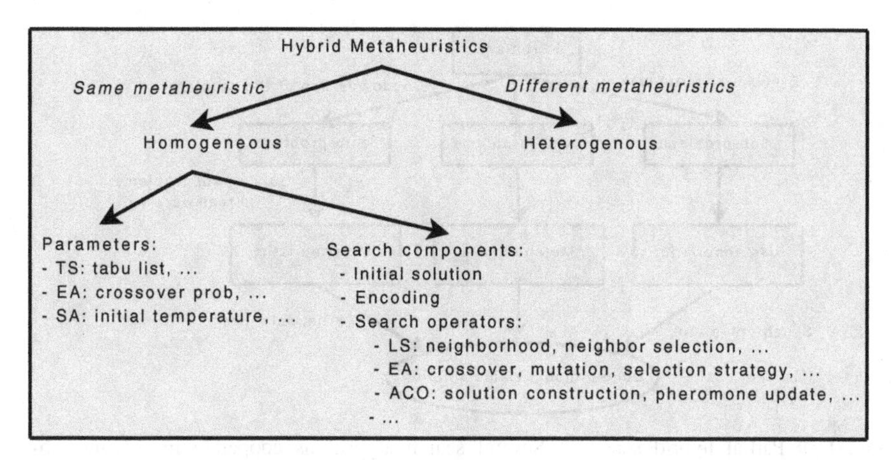

Fig. 1.8 Homogeneous versus heterogeneous hybrid metaheuristics. Some illustrative examples of parameters and search components are illustrated.

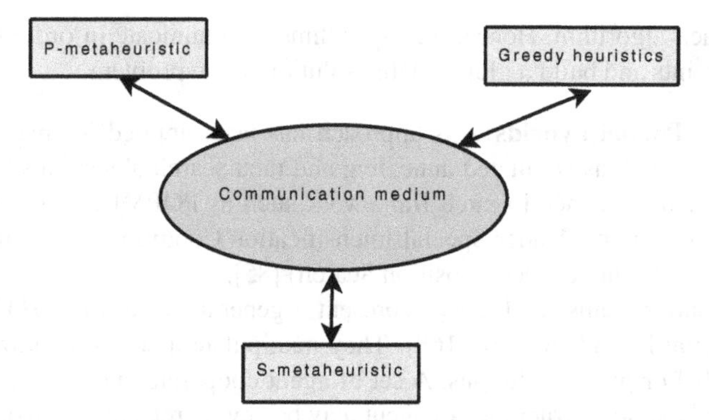

Fig. 1.9 High-level Teamwork Hybridization HTH (heterogeneous, global, general). Several search algorithms cooperate, co-adapt, and co-evolve a solution.

Global/Partial: from another point of view, one can also distinguish two kinds of cooperation: global and partial. In global hybrids , all the algorithms explore the same whole search space. The goal is here to explore the space more thoroughly. All the above mentioned hybrids are *global* hybrids, in the sense that all the algorithms solve the whole optimization problem. A global HTH algorithm based on tabu search has been proposed in [47], where each tabu search task performs a given number of iterations, then broadcasts the best solution. The best of all solutions becomes the initial solution for the next phase.

In partial hybrids, the problem to be solved is decomposed a priori into subproblems, each one having its own search space (Fig. 1.10). Then, each algorithm is dedicated to the search in one of these sub-spaces. Generally speaking, the subproblems are all linked with each others, thus involving constraints between optima

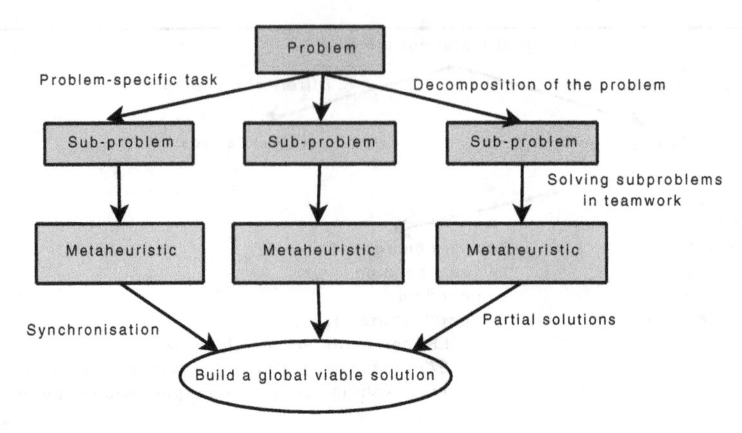

Fig. 1.10 Partial hybrid schemes. Several search algorithms cooperate in solving subproblems. A synchronization is performed to build a global solution from the partial solutions found.

found by each algorithm. Hence, the algorithms communicate in order to respect these constraints and build a global viable solution to the problem.

Example 1.6. **Partial hybrids:** this approach has been applied for many specific metaheuristics such as simulated annealing and tabu search algorithms [158]. It is also a part of more general search framework such as POPMUSIC (Partial OPtimization Metaheuristic Under Special Intensification Conditions [160] and VNDS (Variable Neighborhood Decomposition Search) [85].

Asynchronous teams (A-Teams) represent a general model for a HTH heterogeneous partial hybrid strategy [168]. They manipulate a set of solutions, which may be global or partial solutions. A set of agent cooperate via a blackboard system, a shared memory structure. An agent may be any search algorithm or operator, which consists in picking a (partial) solution from the blackboard, transforming it and sending back the result.

An example of application of partial homogeneous HTH has been done for the job-shop scheduling problem [93]. The search algorithm is a GA. Each GA evolves individuals of a specie which represent the process plan for one job. Hence, there are as many cooperating GAs as there are jobs. The communication medium collects fitted individuals from each GA, and evaluates the resulting schedule as a whole, rewarding the best process plans.

Decomposition techniques based on partitioning time have been used to solve many problems such as the production lot-sizing (partitioning of time) [63]. Decomposition techniques based on partitioning a geographical region have been largely applied to optimization problems associated with Euclidean distances such as the TSP [110], the VRP, and the P-median problem [159].

Example 1.7. **Partitioning a continuous objective function :** a function f is separable if:

$$(argmin_{x_1} f(x_1,...),...,argmin_{x_n} f(...,x_n)) = argmin(f(x_1,x_2,...,x_n))$$

It follows that the function f can be optimized in a parallel way using n independent algorithms. Each algorithm will solve a 1-D optimization problem.

Generalist/Specialist: all the above mentioned hybrids are *general hybrids* , in the sense that all the algorithms solve the same target optimization problem. *Specialist hybrids* combine algorithms which solve different problems. The COSEARCH generic model belongs to this class of hybrids (Fig. 1.11). COSEARCH manages the cooperation of a search agent (a local search), a diversifying agent and an intensifying agent. The three agents exchange information via a passive coordinator called the adaptive memory[6]. A main key point of the COSEARCH approach is the design of this memory which focus on high quality regions of the search and avoid attractive but deceptive areas. The adaptive memory contains a history of the search; it stores information about the already visited areas of the search space and about the intrinsic nature of the good solutions already found. When diversifying, the local search agent receives starting solutions in unexplored regions; when intensifying, the search agent receives an initial solution in a promising region. The diversifying agent refers to the adaptive memory (information about the explored areas) to yield a solution from an unexplored region. The intensifying agent refers to the adaptive memory (information about promising regions) to produce a promising starting solution.

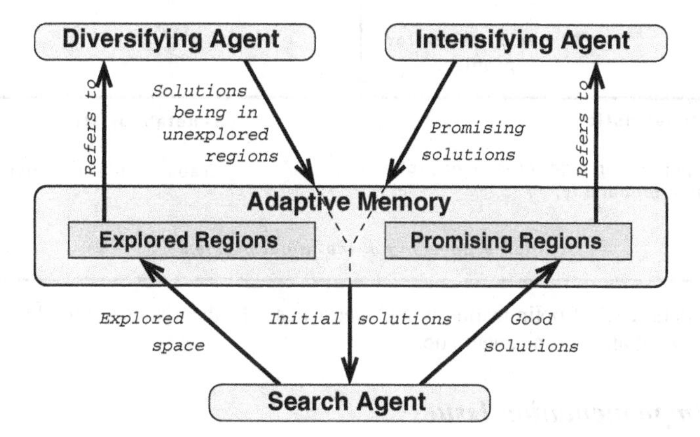

Fig. 1.11 The COSEARCH HTH specialist hybrid model for metaheuristics. Several search algorithms solve different problems.

Example 1.8. **COSEARCH for the quadratic assignment problem:** an example of the application of the COSEARCH approach has been developed in [164] to solve the quadratic assignment problem (QAP). A parallel tabu search is used to solve the QAP, while a genetic algorithm makes a diversification task, which is formulated as

[6] The concept of adaptive memory has been proposed in the domain of combinatorial optimization [162]. It is similar to the concept of blackboard in the field of Artificial Intelligence [60].

another optimization problem. The frequency memory stores information relative to all the solutions visited during the tabu search. The genetic algorithm refers to the frequency memory to generate solutions being in unexplored regions.

Another approach of specialist hybrid HRH heuristics is to use a heuristic to optimize another heuristic, i.e. find the optimal values of the parameters of the heuristic (Fig. 1.12). This approach is known as *meta-optimization*. For instance, it has been used to optimize simulated annealing and noisy methods (NM) by GA [115], ant colonies (AC) by GA [1], simulated annealing based algorithms by a GA [79], and a GA by a GA [153][7]. In [153], the three parameters optimized are the crossover rate, inversion rate, and mutation rate. The individuals of the population associated to the optimizer consist of three integers representing the mutation rate, inversion rate, and crossover rate. The fitness of an individual is taken to be the fitness of the best solution that the GA can find in the entire run, using these parameters.

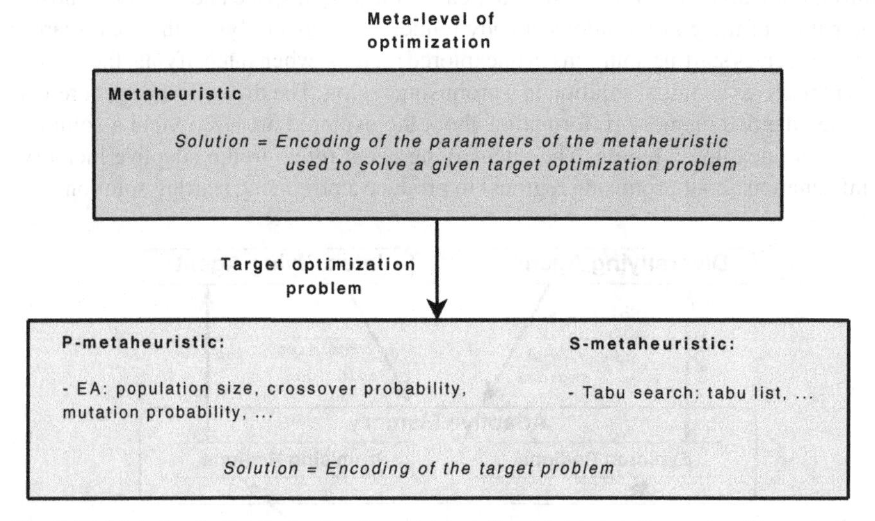

Fig. 1.12 Meta-level of optimization in metaheuristics. Metaheuristics are used to optimize the parameters of another metaheuristic.

1.2.2 Implementation Issues

The structure of the taxonomy concerning the implementation issues is shown in figure 1.13. A discussion about this taxonomy then follows.

1.2.2.1 Dedicated versus General-Purpose Computers

Application specific computers differ from general purpose ones in that they usually only solve a small range of problems, but often at much higher rates and lower cost.

[7] This procedure is also called meta-evolution.

Their internal structure is tailored for a particular problem, and thus can achieve much higher efficiency and hardware utilization than a processor which must handle a wide range of tasks.

In the last years, the advent of programmable logic devices has made easier to build specific computers for metaheuristics such as simulated annealing [3] and genetic algorithms [149]. A general architecture acting as a template for designing a number of specific machines for different metaheuristics (SA, TS, etc) may be constructed [2]. The processor is built with XILINX FPGAs and APTIX interconnection chips. Experiments evaluating a simulated annealing algorithm to solve the traveling salesman problem achieved a speedup of about 37 times over an IBM RS6000 workstation. To our knowledge, this approach has not been yet proposed for hybrid metaheuristics.

Nowadays, the use of GPU (Graphical Processing Unit) devices is more and more popular in many application domains. Indeed, those devices are integrated into many workstations to deal with visualization tasks. The idea is to exploit those available resources to improve the effectiveness of hybrid metaheuristics.

1.2.2.2 Sequential versus Parallel

Most of the proposed hybrid metaheuristics are sequential programs. According to the size of problems, parallel implementations of hybrid algorithms have been considered. The easiness to use a parallel and distributed architecture has been acknowledged for the HTH hybrid model.

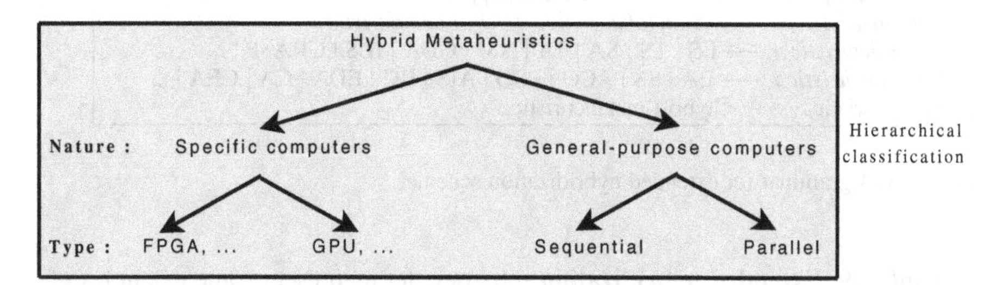

Fig. 1.13 Classification of hybrid metaheuristics (implementation issues)

1.2.3 A Grammar for Extended Hybridization Schemes

Given a set of metaheuristics A_i, a classification of basic hybridizations has been presented, in which the following notations can be described:

- $LRH(A_1(A_2))$ (homogeneous, heterogeneous) (partial, global) (specialist, general): the metaheuristic A_2 is embedded into the single-solution metaheuristic A_1.
- $HRH(A_1+A_2)$ (homogeneous, heterogeneous) (partial, global) (specialist, general): the self-contained metaheuristics A_1 and A_2 are executed in sequence.

- $LTH(A_1(A_2))$ (homogeneous, heterogeneous) (partial, global) (specialist, general): the metaheuristic A_2 is embedded into the population-based metaheuristic A_1.
- $HTH(A_1,A_2)$ (homogeneous, heterogeneous) (partial, global) (specialist, general): the self-contained metaheuristics A_1 and A_2 are executed in parallel and cooperate.

These hybridizations should be regarded as primitives that can be combined in different ways. The grammar given in figure 1.14 generalizes the basic hybridization schemes. One of the practical importance of the grammar is to specify the hybrid heuristic to use, if a metaheuristic problem solving tool is used.

<hybrid-metaheuristic> ⟶ <design-issues><implementation-issue>
<design-issues> ⟶ <hierarchical><flat>
<hierarchical> ⟶ <LRH> | <LTH> | <HRH> | <HTH>
<LRH> ⟶ LRH(<S-metaheuristic>(<metaheuristic>))
<LTH> ⟶ LTH(<P-metaheuristic>(<metaheuristic>))
<HRH> ⟶ HRH(<metaheuristic> + <metaheuristic>)
<HTH> ⟶ HTH(<metaheuristic>)
<HTH> ⟶ HTH(<metaheuristic>, <metaheuristic>)
<flat> ⟶ (<type> , <domain> , <function>)
<type> ⟶ homogeneous | heterogeneous
<domain> ⟶ global | partial
<function> ⟶ general | specialist
<implementation-issue> ⟶ *<specific computers>* | *<general-purpose computers>*
<specific computers> ⟶ FPGA | GPU | ...
<general-purpose computers> ⟶ sequential | parallel
< metaheuristic > ⟶ *<S-metaheuristic>* | *<P-metaheuristic>*
<S-metaheuristic> ⟶ LS | TS | SA | TA | NM | GDA | ILS | GRASP | ...
<P-metaheuristic> ⟶ EA | SS | ACO | PSO | AIS | BC | EDA | CA | CEA | ...
<metaheuristic> ⟶ <hybrid-metaheuristic>

Fig. 1.14 A grammar for extended hybridization schemes

Example 1.9. **Extended hybridization schemes:** let us present some examples of extended hybridization schemes (Fig. 1.15). Boese et al. [25] suggested an adaptive multi-start (AMS) approach, which may be seen as a HRH(LS + LTH(GA(LS))) scheme. First, AMS generates a set of random starting solutions and runs an LS algorithm for each solution to find corresponding local optima. Then, AMS generates new starting solutions by combining features of the T best local optima seen so far, with T being a parameter of the approach. This mechanism bears some resemblance to GAs, but differs in that many solutions (instead of just two) are used to generate the new starting solutions. New local optima are obtained by running the LS algorithm from these new starting solutions, and the process iterates until some stop criterion is met.

D. Levine has used a HTH(HRH(GH+LTH(GA(LS)))) hierarchical scheme in his PhD to solve set partitioning problems. Efficient results have been obtained with a parallel static implementation in solving big sized problems in real world applications (airline crew scheduling) [117]. At the first level, a HTH hybrid based on the island model of parallel genetic algorithms is used. The initial population of each GA was generated by a greedy heuristic (the Chvatal heuristic [38]), and a local search algorithm was used to improve the solutions at each generation of the GA. The same hybrid scheme with a sequential implementation has been used in [26] to solve the traveling salesman problem. The local search algorithms used are the well known 2-opt and or-opt heuristics. The author reported some interesting results on the 442 and 666-city problems. He found the optimum of the 442-city problem, and a solution within 0.04% of the optimum for the 666-city problem.

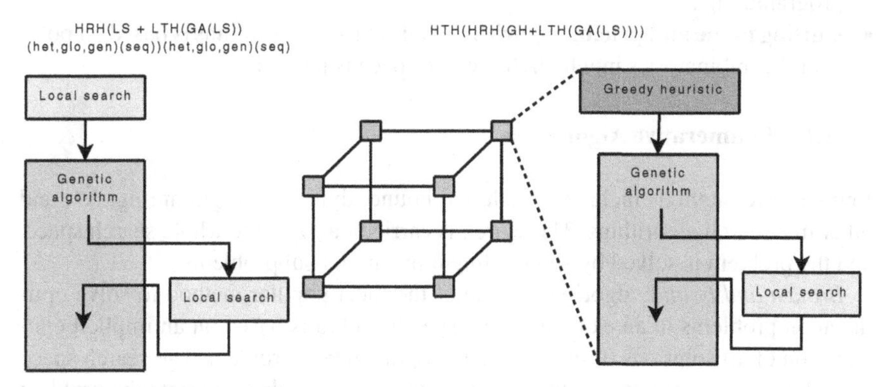

Fig. 1.15 Extended hybridization schemes

The objective of this chapter is far from providing an exhaustive list of research works using hybrid metaheuristics. Following this grammar, more than 125 annotated hybrid metaheuristics may be found in [163]. This shows the usefulness of the taxonomy.

1.3 Combining Metaheuristics with Mathematical Programming

Metaheuristics and exact algorithms are complementary optimization strategies in terms of the quality of solutions and the search time used to find them. In the last few years, solving exactly important optimization problems using for example integer programming techniques has improved dramatically. Moreover, the availability of efficient optimization software, libraries and frameworks for mathematical programming and high-level modeling languages will lead to more hybrid approaches combining metaheuristics and exact optimization algorithms. In the next section, the main mathematical programming exact approaches that can be used to solve optimization problems are presented. Then, an instantiation and extension of our classification to hybrid schemes combining mathematical programming approaches and metaheuristics is presented.

1.3.1 Mathematical Programming Approaches

The main mathematical programming approaches may be classified as follows:

- **Enumerative algorithms:** this class of algorithms contains tree search algorithms such as branch and bound and dynamic programming. They are based on a divide and conquer strategy to partition the solution space into subproblems and then optimizing individually each subproblem.
- **Relaxation and decomposition methods:** this class of methods are based on relaxation techniques such as the Lagrangian relaxation [69], and decomposition methods such as the Bender's decomposition and the continuous semi-definite programming .
- **Cutting plane and pricing algorithms:** this class of algorithms is based on polyhedral combinatorics in which the search space is pruned.

1.3.1.1 Enumerative Algorithms

Enumerative methods include branch and bound, dynamic programming, A*, and other tree search algorithms. The search is carried out over the whole search space, and the problem is solved by subdividing it in simpler subproblems.

Branch and bound algorithm is one of the most popular method to solve optimization problems in an exact manner. The algorithm is based on an implicit enumeration of all solutions of the considered optimization problem. The search space is explored by dynamically building a tree whose root node represents the problem being solved and its whole associated search space. The leaf nodes are the potential solutions and the internal nodes are subproblems of the total solution space. The size of the subproblems is increasingly reduced as one approaches the leaves.

The construction of such a tree and its exploration are performed using two main operators: *branching* and *pruning* (Fig. 1.16). The algorithm proceeds in several iterations during which the best found solution is progressively improved. The

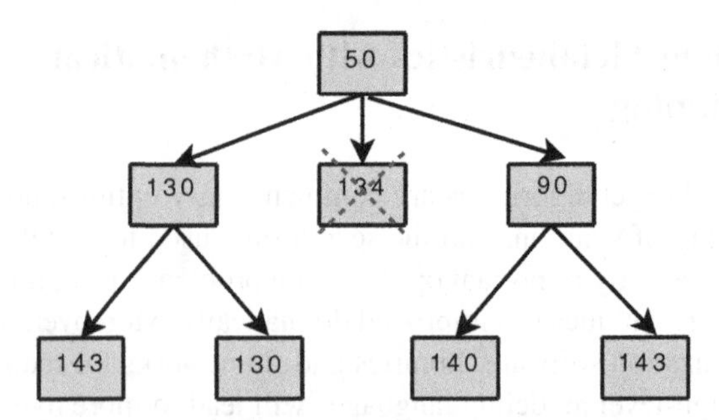

Fig. 1.16 The branch and bound algorithm. This figure shows the nodes actually explored in the example problem, assuming a depth-first and left-to-right search strategy. The subtree rooted at the second node on level 2 is pruned because the cost of this node (134) is greater than that of the cheapest solution already found (130).

generated nodes and not yet treated are kept in a list whose initial content is limited to only the root node. The two operators intervene at each iteration of the algorithm. The branching strategy determines the order in which the branches are explored. Many branching strategies may be applied such as the *depth-first*, the *breadth-first*, and the *best-first* strategies. The *pruning* strategy eliminates the partial solutions that do not lead to optimal solutions. This is done by computing the lower bound associated to a partial solution. If the lower bound of a node (partial solution) is greater than the best solution found so far or a known upper bound of the problem, the exploration of the node is not needed. The algorithm terminates if there are no more nodes to branch or all nodes are eliminated. Hence, the most important concepts in designing an efficient branch and bound algorithm are the quality of the bounds and the branching strategy.

Example 1.10. **Branch and bound algorithm on the TSP:** let us consider the TSP problem. A straightforward method for computing a lower bound on the cost of any solution may be the following:

$$\frac{1}{2}\sum_{v \in V} \text{sum of the costs of the two least cost edges adjacent to } v$$

For the example shown in figure 1.17, the lower bound is associated to the edges $(A,D),(A,B),(B,A),(B,E),(C,B),(C,A),(D,A),(D,C),(E,B),(E,D)$ and then is equal to 17.5. In the search tree (Fig. 1.17), each node represents a partial solution. Each partial solution is represented by the set of associated edges (i.e. edges that must be in the tour) and the non associated edges (i.e. set of edges that must not be on the tour). The branching consists in generating two children nodes. A set of additional excluding and including edges is associated to each child. Two rules may be applied. An edge (a,b) must be included if its exclusion makes it impossible for a or b to have two adjacent edges in the tour. An edge (a,b) must be excluded if its inclusion causes for a or b to have more than two adjacent edges in the tour or would complete a non-tour with edges already included. The pruning consists first in computing the lower bounds for each child. For instance, if the edge (A,E) is included and the edge (B,C) is excluded, the lower bound will be associated to the following selected edges $(A,D),(A,E),(B,A),(B,E),(C,D),(C,A),(D,A),(D,C),(E,B),(E,A)$ and is equal to 20.5. If the lower bound associated to a node is larger than the known upper bound, the node is proved to be unable to generate an optimal solution and then is not explored. A best-first search heuristic is considered in which the child with the smaller lower bound is explored first. The upper bound is updated each time a new complete solution is found with a better cost.

The *dynamic programming* (DM) approach is based on the recursive division of a problem into simpler subproblems. This procedure is based on the *Bellman's principle* which says that "the sub-policy of an optimal policy is itself optimal with regard to the start and end states" [21].

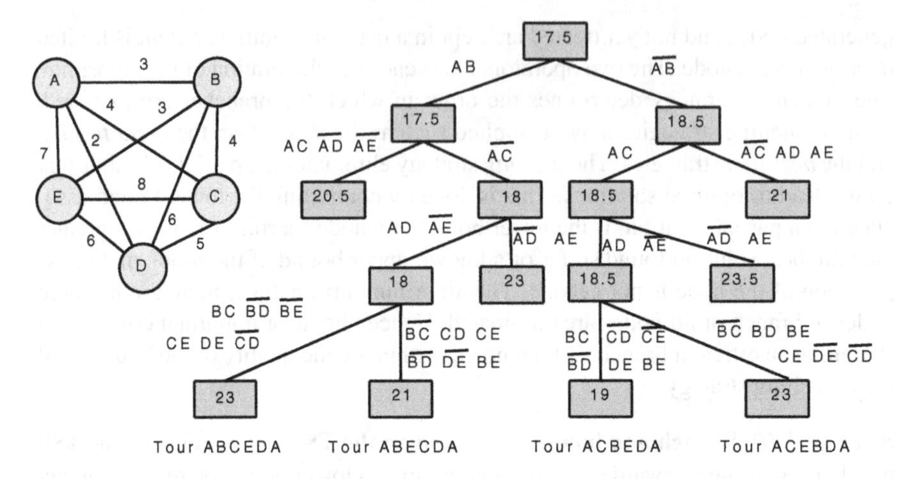

Fig. 1.17 Illustration of the branch and bound algorithm on the traveling salesman problem

Designing a dynamic programming procedure for a given problem needs the definition of the following components [23]:

- Define the *stages* and the *states*. A problem can be divided into a number of stages N. A number of states are associated to each stage.
- Define the cost of the initial stage and states. There is an initial state of the system x_0.
- Define the recursive relation for a state at stage k in terms of states of previous stages. The system takes the state x_k at the stage k. At the k stage, the state of the system change from x_k to x_{k+1} using the following equation

$$x_{k+1} = f_k(x_k, u_k)$$

where u_k is a control that takes values from a given finite set, which may depends on the stage k. The transition from the state k to $k+1$ involves a cost $g_k(x_k, u_k)$. The final transition from $N-1$ to N involves the terminal cost $G(x_N)$. The functions f_k, g_k and G must be determined.

Given a control sequence $(u_1, u_2, ..., u_{N-1})$, the corresponding state sequence will be $(x_0, ..., x_N)$ which is determined from the initial state x_0 using the equation below. In dynamic programming, the objective is to find the optimal control sequence minimizing the total cost:

$$G(x_N) + \sum_{k=0}^{N-1} g_k(x_k, u_k)$$

DP have been successfully applied to knapsack, planning and routing-type problems, in which it is easy to define efficient recursive relationships between stages.

Example 1.11. **Dynamic programming for the $\{0,1\}$−knapsack problem:** let us consider the following instance for the knapsack problem with a total capacity equal to 5 (see table 1.1).

Table 1.1 An instance for the knapsack problem with a capacity of 5

Item (i)	Weight (w_i)	Utility (u_i)
1	2	65
2	3	80
3	1	30

The stages are represented by the items. The number of stages are then equal to the number of items (3). The state y_i at stage i represents the total weight of items i and all following items in the knapsack. The decision at stage i corresponds to how many items i to place in the knapsack. Let us call this value k_j. This leads to the following recursive formulas: let $f_j(y_j)$ be the value of using y_j units of capacity for items j and following. Let $\lfloor a \rfloor$ represents the largest integer less than or equal to a.

$$f_3(y_i) = 30.y_i$$

$$f_j(y_i) = max_{k_i \leq \lfloor \frac{y_i}{w_i} \rfloor} \{u_i k_i + f_{i+1}(y_i - w_i k_i)\}$$

1.3.1.2 Relaxation and Decomposition Methods

Relaxation methods consist in relaxing a strict requirement in the target optimization problem. In general, a given strict requirement is simply dropped completely or substituted by another one which is more easily satisfied. The most used relaxation techniques are the LP-relaxation and the Lagrangian relaxation. In addition to their use on solving optimization problems, relaxation methods are also used to generate bounds.

Linear programming relaxation: linear programming relaxation (LP-relaxation) is a straightforward approach which consists in ignoring the integrity constraints of an integer program (IP). Once the integrity constraints are dropped, the problem can be solved using LP solvers. This gives a lower bound for the problem. If the solution found satisfies the integer constraints (generally not true), it will be considered as the optimal solution for the IP program. If the relaxed problem is infeasible, then so is the IP program. LP-relaxation is widely used in branch and bound algorithms to solve IP problems in which the branching is performed over the fractional variables.

Lagrangian relaxation: Lagrangian relaxations are widely used to generate tight lower bounds for optimization problems. The main idea is to remove some constraints and incorporate them in the objective function. For each constraint, a penalty function is associated. The choice of which constraints are handled in the objective function is important. More complicated constraints to satisfy are preferable as they generate an easiest problem to solve. Given the following LP problem:

Max $c^T x$
s.t. $Ax \leq b$
with $x \in \mathbb{R}^n$ and $A \in \mathbb{R}^{m,n}$

The set of constraints A is split into two sets: $A_1 \in R^{m_1,n}$ and $A_2 \in R^{m_2,n}$, where $m_1 + m_2 = m$. Then, the subset of constraints A_2 is integrated into the objective function which gives the following Lagrangian relaxation of the original problem:

Max $c^T x + \lambda^T (b_2 - A_2 x)$
s.t. $A_1 x \leq b_1$
with $x \in \mathbb{R}^n$, $A_1 \in \mathbb{R}^{m_1,n}$ and $A_2 \in \mathbb{R}^{m_2,n}$

where $\lambda = (\lambda_1, ..., \lambda_{m_2})$ are non negative weights which penalize the violated constraints A_2. The efficiency of Lagrangian relaxation depends on the structure of the problem; there is no general theory applicable to all problems. Lagrangian relaxation may find bounds which are tighter than the LP-relaxation. The problem is solved iteratively until optimal values for the multipliers are found. One of the main issues in the Lagrangian relaxation is the generation of the optimal multipliers. This difficult problem can be solved by metaheuristics.

In practice, decomposition methods are used to solve large IP problems. Among the numerous decomposition approaches one can refer to Bender's decomposition and Dantzig-Wolfe decomposition.

Bender's decomposition: the Bender's decomposition algorithm is based on the notion of complicated variables. It consists in fixing the values of complicated variables and solves the resulting reduced problem iteratively [22]. Given a MIP problem :

Max $c^T x + h^T y$
s.t. $Ax + Gy \leq b$
with $x \in \mathbb{Z}^n_+$ and $y \in \mathbb{R}^p_+$

If the set of variables x is fixed, the following linear program is obtained

$$z_{LP}(x) = max\{hy / Gy \leq b - Ax\}$$

and its dual

$$min\{u(b - Ax) / uG \geq h, u \in R^m_+\}$$

If the dual polyhedron is assumed to be not empty and bounded, the MIP model can be formulated as follows:

$$z = max_{x \in Z^n_+}(cx + min_{i \in 1,...,T}(u^i(b - Ax)))$$

This model can be reformulated as:

$$z = max\{\eta / \eta \leq u^i(b - Ax), i \in 1, ...T, x \in Z^n_+\}$$

Then, the algorithm finds cutting planes based on the dual problem. The cutting planes are added to the problem and the problem is re-solved.

1.3.1.3 Branch and Cut and Price Algorithms

The objective of the following popular techniques is to generate tighter IP relaxations.

Cutting plane: cutting plane approaches have been proposed in 1958 by Gomory [81]. The use of cuts can improve greatly branch and bound algorithms. In general, cutting plane algorithms consist in iteratively adding some specific constraints to the LP-relaxation of the problem. Those constraints represent restrictions to the problem so that the linear programming polytope closely approximates the polyhedron represented by the convex hull of all feasible solutions of the original IP problem. A good survey of branch and cut algorithms and their use for different families of optimization problems may be found in [108] [131].

Column generation: column generation has been first applied by Gilmore and Gomory [77]. Column generation (i.e. Dantzig-Wolfe decomposition) generates a decomposition of the problem into a master and subproblems (Fig. 1.18). A good survey may be found in [12].

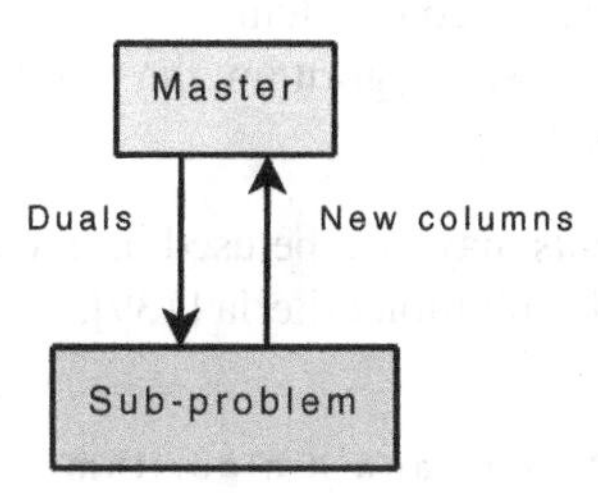

Fig. 1.18 Branch and price approach

1.3.2 *Classical Hybrid Approaches*

Exact MP algorithms are known to be time and/or memory consuming. In general they cannot be applied to large instances of difficult optimization problems. On one hand their combination with metaheuristics may improve the effectiveness of heuristic search methods (i.e. getting better solutions). On the other hand, this type of combination allows the design of more efficient exact methods (i.e. finding optimal solutions in shorter time). The following sections illustrate, for each class of hybrids belonging to the presented taxonomy, some hybridization schemes combining exact MP algorithms and metaheuristics.

1.3.2.1 Low-Level Relay Hybrids (LRH)

This class of algorithms represents hybrid schemes in which a metaheuristic approach (resp. exact approach) is embedded into an exact approach (resp. S-metaheuristic approach) to improve the search strategy. In this usual combination, a given metaheuristic or exact algorithm solves a problem of a different nature of the considered optimization problem.

Embedding S-metaheuristics into exact algorithms: indeed, metaheuristics may solve many search problems involved in the design of an exact method such as the node selection strategy, upper bound generation, column generation (Fig. 1.19):

- **Bounding:** providing an upper bound associated to a node of the branch and bound algorithm can be designed using a metaheuristic. Indeed, the partial solution is completed by a given metaheuristic and then a local upper bound is provided.
- **Cutting:** in the branch and cut algorithm, the cutting plane generation problem is a crucial part of the algorithm: the part that looks for valid inequalities that cut off the current non-feasible linear program (LP) solution. Metaheuristics may be used in this *separation procedure*. For instance, this approach has been proposed for the CVRP (Capacitated Vehicle Routing Problem) [10]. Some metaheuristics (e.g. tabu search, greedy heuristics) have been designed to extract a set of violated capacity constraints of the relaxed problem.
- **Pricing:** in the branch and price algorithm, the pricing of columns may be carried out by a metaheuristic [67].

Some metaheuristic ingredients may also be used in tree search algorithms such as the concepts of tabu lists and aspiration criteria [139].

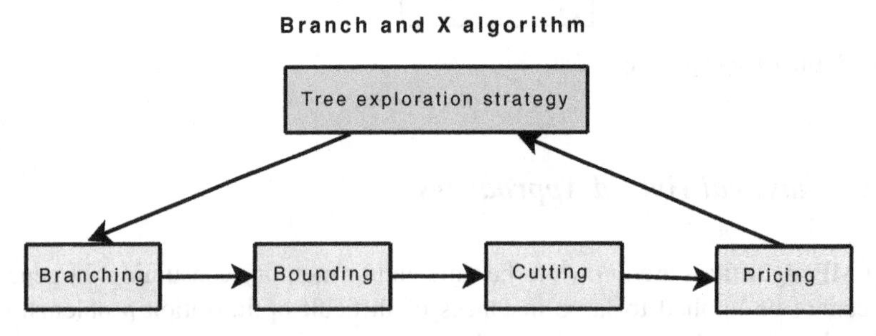

Fig. 1.19 LRH cooperation in which a metaheuristic can be used in the design of some search components of branch and X family of algorithms (e.g. branch and bound, branch and cut, branch and price): selection of the node to explore, generation of an upper bound, cutting plane generation, column generation selection, etc.).

Embedding exact algorithms into S-metaheuristics: many combinations may be designed in which exact algorithms are embedded into search components of S-metaheuristics.

Very large neighborhoods: S-metaheuristics may be improved using very large neighborhoods. The concept of large neighborhoods is also used in the ILS (perturbation operator [120]) and the VNS metaheuristics. Mathematical programming approaches may be used to search efficiently those large neighborhoods to find the best or an improving solution in the neighborhood (Fig. 1.20). Algorithms such as branch and bound, dynamic programming [138], network flow algorithms [56], and matching algorithms [83] have been used to explore large neighborhoods defined for different important optimization problems. Some hybrid schemes explore the *whole* neighborhood while other neighborhood search algorithms explore a *subset* of the neighborhood. If no polynomial time algorithm exists to search the whole neighborhood, a partial search is generally performed.

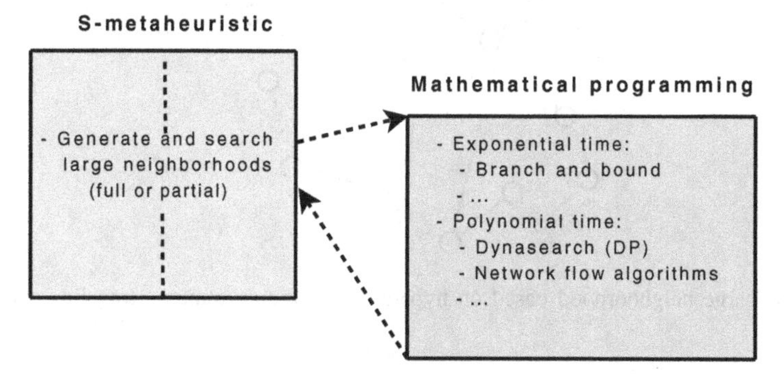

Fig. 1.20 LRH cooperation where a mathematical programming approach can be used for the efficient search of a very large neighborhood into S-metaheuristics.

Example 1.12. **Hyperopt:** the hyperopt S-metaheuristic explores only a subset of a very large neighborhood [27]. The hybrid algorithm has been used to solve the asymmetric traveling salesman problem. The move operator is based on hyperedges which represent subpaths of the tour. A hyperedge $H(i,j)$ is represented by its start node i, end node j and length k. A k-hyperopt move consists in deleting two hyperedges $H(i_1,i_{k+1})$ and $H(j_1,j_{k+1})$ of length k. It is supposed that $H(i_1,i_{k+1})\bigcap H(j_1,j_{k+1}) = \phi$, i.e. the hyperedges share no common edges. Then, the move operator adds edges to the hyperedges $H(i_{k+1},j_1)$ and $H(j_{k+1},i_1)$ to construct a feasible tour (Fig. 1.21). The size of the hyperedge neighborhood grows exponentially with k. The neighborhood search algorithm is reduced to a smaller TSP problem. The algorithms used are based on enumeration for small k and a dynamic programming algorithm for medium values of k.

Some search concepts of exact algorithms may be used in S-metaheuristics. Efficient mathematical programming approaches that generate "good" lower bounds exist

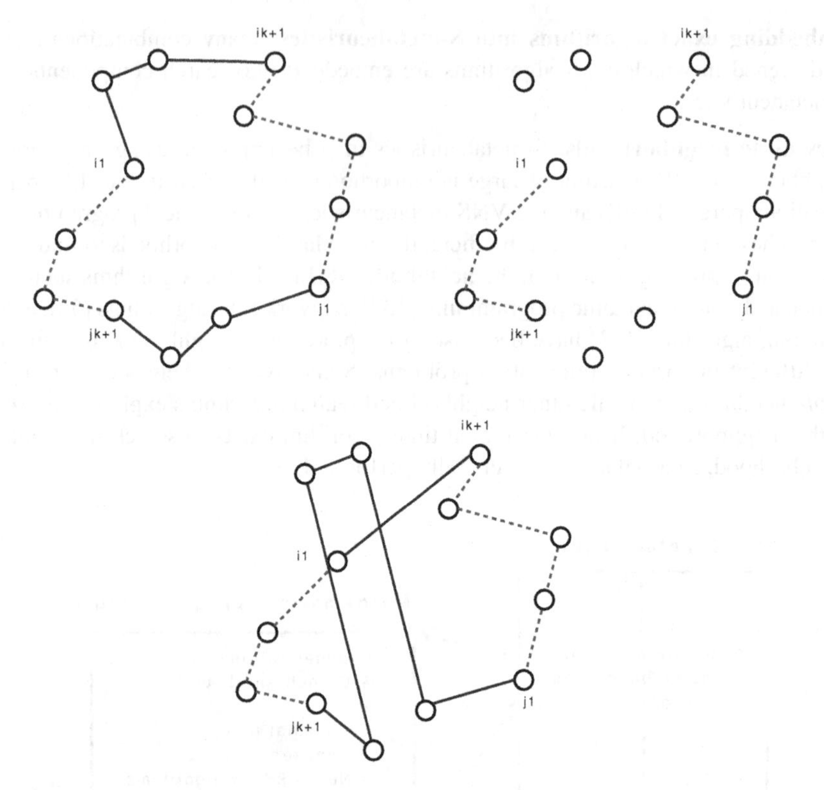

Fig. 1.21 Large neighborhood based on hyperedge for the asymmetric traveling salesmen problem

for many optimization problems. Information related to lower bounds, obtained for example by Lagrangian relaxation, can be exploited into a metaheuristic to intensify the search in promising regions of the search space. For instance, information based on Lagrangian multipliers are exploited to guide the metaheuristic in solving the set covering problem [17]. Lower bounds have been used in S-metaheuristics such as tabu search to improve the search [88] [56].

1.3.2.2 Low-Level Teamwork Hybrids (LTH)

Recall that in this class of hybrid algorithms, a search component of a P-metaheuristic is replaced by another optimization algorithm. Concerning the combination of P-metaheuristics and MP algorithms, two main hybrid approaches may be considered: exact search hybrid algorithms in which a P-metaheuristic is embedded into an exact algorithm, and heuristic search algorithms in which an exact algorithm is embedded into a P-metaheuristic.

Embedding a P-metaheuristic into an Exact Algorithm: as mentioned previously in this chapter, the questions arising in designing a branch and bound algorithm are:

- **Branch ordering:** how the problem to solve (node of the tree) is decomposed into subproblems? On which variable the next branching is applied? Indeed, near-optimal solutions obtained by metaheuristics may guide the branch and bound to apply an efficient branch ordering by giving preference to branches which share common values with near-optimal solutions.
  The node selection problem in tree search based on branch and bound may be solved by metaheuristics. For instance, genetic programming approaches have been used to deal with the node selection problem [113].
- **Variable selection:** in which subproblem (child node) the search will be performed in the next step? What value should first be assigned to the branching variable? Information obtained from branch and bound tree can be used by heuristic algorithms to determine a better strategy for variable selection [113].

An exact algorithm constructs partial solutions which are used to define a search space for a metaheuristic. Then, the obtained results are exploited in order to refine the bounds or generate the columns into a branch and cut algorithm.

Example 1.13. **Local branching :** the *local branching* exact approach has been proposed in [68]. It uses the principle of local search heuristics. The search space is partitioned by introducing branching conditions expressed through (invalid) linear inequalities called local branching cuts. Let us consider a MIP problem with $\{0,1\}$ variables. The $k-opt$ neighborhood is considered. The main principle of the local branching method is to iteratively solve a subproblem corresponding to the neighborhood $k-opt$ of a partial solution s. Two partitions are then considered: $p_1 = \{x \in \{0,1\}^n / \Delta(x,s) \leq k\}$ and $p_2 = \{x \in \{0,1\}^n / \Delta(x,s) \geq k+1\}$, where Δ represents the Hamming distance, and n the size of the problem. The problem associated to p_1 is solved. A new subproblem is generated if an improved solution is found. Otherwise, the other problem is solved using the standard procedure.

Embedding an exact algorithm into P-metaheuristic: in this hybrid scheme, some search components of a P-metaheuristic induce optimization problems which are solved by exact algorithms (Fig. 1.23).

Example 1.14. **Exact algorithms into recombination operators:** exact algorithms may be integrated into recombination operators of P-metaheuristics such as evolutionary algorithms to find the best offspring from a large set of possibilities. The induced problem `Recombination`(S_1,S_2) is defined to generate the best offsprings from the parents S_1 and S_2. A common idea is to keep the common elements of the parents and explore all the other possibilities to generate better offsprings (Fig. 1.22). For instance, a branch and bound algorithm (resp. dynamic programming) has been used into the crossover operator of a genetic algorithm in solving permutation problems [48] (resp. [182]). For some specific problems, polynomial exact algorithms may also be used such as minimum spanning tree algorithms [95], matching algorithms in a bipartite graph [4] [11] for optimized crossover operators.

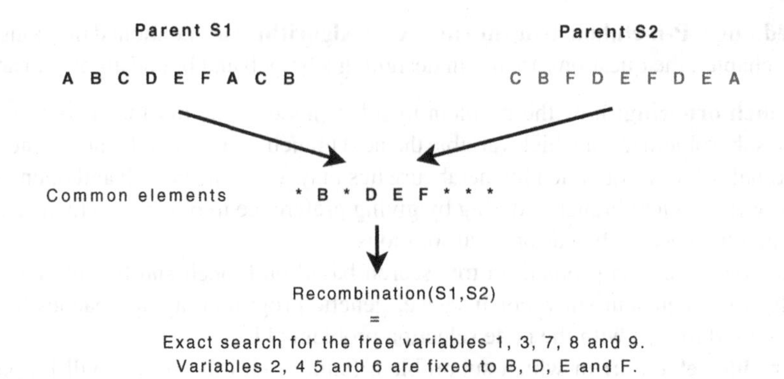

Fig. 1.22 Using exact algorithms into recombination operators (e.g. crossover) of P-metaheuristics

Large neighborhood search algorithms integrated in P-metaheuristics belong typically to the LTH class of hybrids. For instance, the mutation operator in EAs can also be substituted by MP algorithms which explore large neighborhoods (Fig. 1.23).

Exact decoding: exact algorithms can also be used as decoders of incomplete solutions carried out by metaheuristics. This hybrid strategy is applied in the case where metaheuristics use an incomplete encoding for the problem. Once a good incomplete solution is found, exact algorithms complete optimally the missing part of the encoding.

Exact search ingredients: some search ingredients of exact algorithms can also be used in P-metaheuristics:

- **Lower bounds:** the use of lower bounds into a P-metaheuristic can improve the search. Lower bounds have been used in the construction phase of the ant colonies P-metaheuristic to solve the quadratic assignment problem (QAP) [122]. The well known Gilmore-Lawler lower bound and the values of the dual variables are used to order the locations during the construction phase. The impact of the location in a given QAP instance depends on the value of its associated dual variable. The concept of bounds has been used into evolutionary algorithms for the mutation and the crossover operators [169] [57]. Indeed, partial solutions that exceed a given bound are deleted. The bounds are computed using the linear and Lagrangian relaxation, and tree search methods.
- **Partial solutions :** the evaluated partial solutions (subproblems) maintained by the branch and bound family of algorithms may provide interesting initial solutions to improve. The evaluation of those partial solutions will guide the metaheuristics to more promising regions of the search space [122]. The partial solution with the least cost lower bound suggests a promising region by giving additional information.

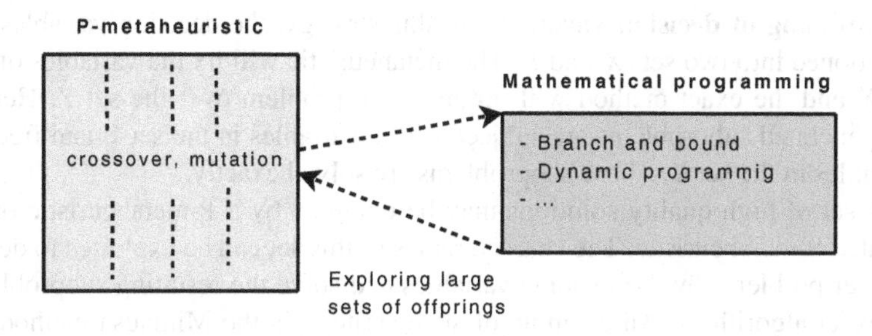

Fig. 1.23 LTH heuristic cooperation: exact algorithms are used as search components of a P-metaheuristic (e.g. recombination, mutation).

1.3.2.3 High-Level Relay Hybrids (HRH)

This class of cooperation, where self-contained algorithms are used in sequence, is the most popular in practice. This may be seen as a pre-processing or a post-processing step. Some information is provided in sequential between the two families of algorithms (metaheuristics and MP algorithms) (Fig. 1.25).

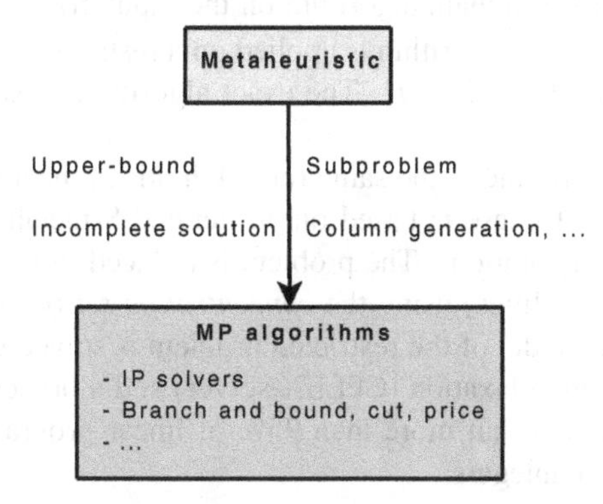

Fig. 1.24 HRH cooperation: information provided by metaheuristics to MP algorithms

Information provided by metaheuristics: in the case where the information is provided by the metaheuristics, the most natural and trivial hybrid approach is to start with a metaheuristic to find a "good" *upper bound* which will be used by a MP algorithm in the bounding phase (Fig. 1.25). Indeed, the efficiency of the search (pruning phase) is largely dependent on the quality of the upper bound.

Using the characteristics of generated high quality solutions, metaheuristics can be used to reduce the size of the original problem. Then, the exact method can be applied to solve the reduced problem. This approach is interesting for optimization problems where "good solutions" share many components [8]. This allow to reduce the problem into a much smaller problem which can be solved exactly by state-of-the-art mathematical programming algorithms. The reduction phase may concern:

- **Partitioning of decision variables:** in this strategy, the decision variables are partitioned into two sets X and Y. The metaheuristic will fix the variables of the set X and the exact method will optimize the problem over the set Y. Hence, the generated subproblems are subject to free variables in the set Y and freezed variables in the set X. Those subproblems are solved exactly.

  A set of high quality solutions may be obtained by a P-metaheuristic or an iterated S-metaheuristic. The characteristics of this set can be exploited to define smaller problems by fixing some variables and solve the resulting subproblems by exact algorithms. An example of such strategy is the Mimausa method for the quadratic assignment problem [125]. The method builds at each iteration a subproblem by fixing k decision variables and solves it by a branch and bound algorithm.

  Example 1.15. **Reducing problems by metaheuristics to be solved by MP algorithms:** analyzing the landscape for the TSP problem, one can observe that local optimum solutions share many edges with the global optimum and they are concentrated in the same region of the search space (big valley structure) [165]. This characteristic has been exploited in [44] to design one of the most efficient heuristic for the TSP: the tour merging heuristic. The tour merging heuristic consists of two phases: the first phase generates a set T of "good" tours using the Lin-Kernigham algorithm on the input graph $G = (V,E)$. Then, a dynamic programming algorithm is applied on a restricted graph $G' = (V,E')$, where $E' = \{e \in E / \exists t \in T, e \in t\}$. The exact algorithm solves instances up to 5000 cities.

  For the p-median problem, the same remark holds in the analysis of its landscape [146]. The first phase is based on an iterated S-metaheuristic using different random initial solutions. The problem is reduced in terms of the number of nodes (location facilities) using the *concentration set* (CS) concept. The integer programming model of the restricted problem is solved using respectively a linear programming relaxation (CPLEX solver) and a branch and bound. The authors exploit the fact that more than 95% of linear programming relaxation optimal solutions are integers.

  For continuous optimization, HRH hybrid schemes are very popular. For instance, a hybrid method combining tabu search and the simplex algorithm provides interesting results in solving complex continuous functions [35].

- **Domain reduction:** this strategy consists in reducing the domain of values that the decision variables can take. The metaheuristic will perform a domain reduction for the decision variables and then an exact method is used over the reduced domains. For instance, a GA may be used to find promising ranges for decision variables and then tree search algorithms are considered to find the optimal solution within those ranges [129].

Information provided by exact algorithms: in the case where the information is provided by an exact algorithm, many hybrid approaches may be designed:

- **Partial solutions:** partial solutions are first provided by an exact algorithm which are then completed by a metaheuristic.

- **Problem reduction:** in this strategy, a problem reduction is carried out by an exact algorithm. For instance, a tree search algorithm has been used to reduce the size of a nurse scheduling problem [58]. Then, a tabu search strategy is applied to solve the problem within a simplified objective function [58].
- **Relaxed optimal solutions and their duals:** the optimal solutions for relaxed formulation (e.g. LP-relaxation, Lagrangian relaxation) of the problem and its duals may be exploited by metaheuristics.

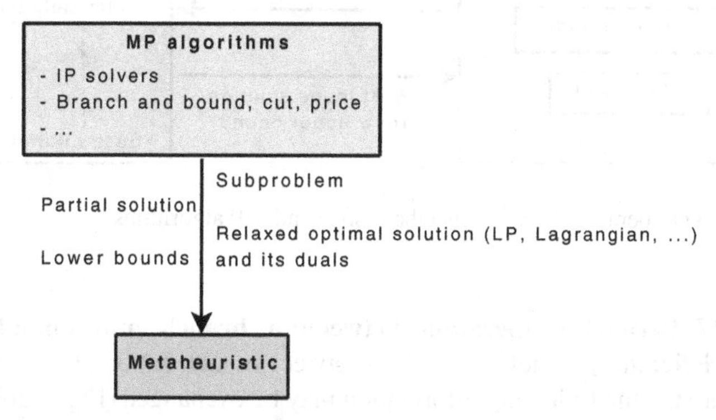

Fig. 1.25 HRH cooperation: information provided by MP algorithms to metaheuristics

Example 1.16. **LP-relaxations as an input for metaheuristics:** information gathered from solutions obtained by LP-relaxations of MIP problems may be used as an input for a metaheuristic. A straightforward approach is the "dive and fix" strategy, where the value of a subset of the integer variables are fixed and the resulting LP problem is solved. This strategy iterates until the LP finds an integer solution. This will restrict the search space of metaheuristics in promising regions. This idea has been used to design an efficient hybrid approach for the 0-1 multidimensional knapsack problem [175]. Many linear relaxation of the MIP formulation of the problem including different constraints on the number of elements of the knapsack are solved exactly. The obtained solutions are exploited to generate initial solutions for multiple tabu search metaheuristics.

1.3.2.4 High-Level Teamwork Hybrids (HTH)

Few strategies belong to this class of hybrids which combines metaheuristics and MP algorithms in a parallel cooperative way. However this is a promising class of hybrids. A set of agents representing metaheuristics and MP algorithms are solving global, partial or specialist optimization problems and exchanging useful informations. The majority of proposed approaches fall in the class of partial and specialist hybrids. Indeed, the search space is generally too large to be solved by an exact approach. One of the main issues in the HTH hybrid is the information exchanged between metaheuristics and MP algorithms. The different complementary algorithms solving different problems may exchange any information gathered during the search to improve the efficiency and the effectiveness of the hybrid approach: solution(s), subproblems, relaxed optimal solutions and its duals, upper bounds, lower bounds, optimal solutions for subproblems, partial solutions, etc.

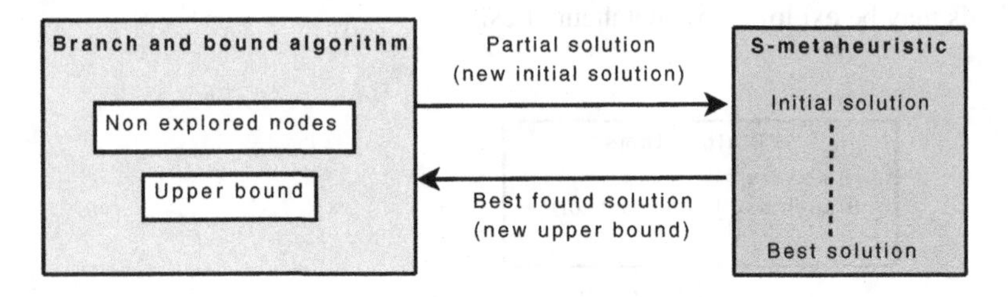

Fig. 1.26 HTH cooperation between metaheuristics and MP algorithms

Example 1.17. **Parallel cooperation between a branch and bound and a S-metaheuristic:** in a parallel cooperation between branch & bound algorithms and a S-metaheuristic, the following information may be exchanged (Fig. 1.26):

- **From a branch & bound algorithm to a S-metaheuristic:** a subproblem of the branch and bound (node of the tree, partial solution) with least-cost lower bound may be used by a S-metaheuristic to generate an initial solution. The lower bound is used to predict potential interesting search regions. This process may be initiated as a diversification search, when the classical "intensification" process is terminated. Indeed, this partial solution provides a promising area for a S-metaheuristic to explore. The non explored node list maintained by a branch and bound provides a metaheuristic with new initial solutions.
- **From a S-metaheuristic to a branch & bound algorithm:** the best solution identified so far by a metaheuristic may be used in branch and bound algorithms for a better pruning of the search tree. Indeed, better is the upper bound, more efficient is the pruning of the search tree. This information is exchanged each time the best solution found is improved.

In generalist and global hybrids, where all the algorithms are solving the same target problem, the space of design is reduced. For instance, a parallel HTH hybrid which consists in combining a branch and bound algorithm with simulated annealing has been designed [134]. The SA algorithm sends improved upper bounds to the exact algorithm. Any integer bound obtained by the B&B execution is passed to SA and used as an alternative reheated solution.

In specialist hybrids, where the algorithms are solving different problems, many strategies may be proposed (Fig. 1.27). For instance, a parallel cooperation between a local search metaheuristic and a column generation (branch and price) algorithm

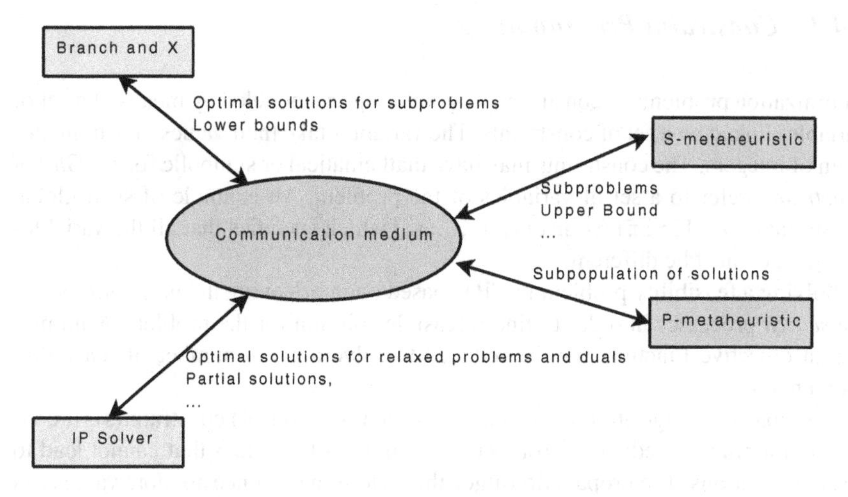

Fig. 1.27 Specialist HTH cooperation between S-metaheuristics and MP algorithms

to solve the VRP problem has been proposed [33]. The local search algorithm is used to generate new columns for a branch and cut algorithm.

Extending the grammar, presented in section 1.14, with hybrid schemes combining metaheuristics with exact optimization algorithms has been presented in [102]. More than 60 annotated hybrid approaches are detailed in the paper. Other examples of combining metaheuristics with exact algorithms may be found in the survey papers [140] [55] [102].

1.4 Combining Metaheuristics with Constraint Programming

Constraint programming (CP) is a modeling and an exact[8] search paradigm based on constraint satisfaction techniques which are largely used in the artificial intelligence community [9]. CP has been applied successfully to many combinatorial optimization problems with tightly-constrained search problems, while metaheuristics perform well for under-constrained optimization problems.

Nowadays, more and more hybrid approaches combining metaheuristics and constraint programming are used to solve optimization problems. Indeed, metaheuristics and constraint programming are complementary search and modeling approaches, which may be combined naturally to solve optimization problems in a more efficient manner [72]. One of the main advantages of using constraint programming is its flexibility. Models are based on a declarative programming paradigm. Hence, the addition/deletion of new constraints in the model is straightforward.

[8] The term complete is always used in the CP community.

1.4.1 Constraint Programming

Optimization problems in constraint programming are modeled by means of a set of variables linked by a set of constraints. The variables take their values on a finite domain of integers. The constraints may have mathematical or symbolic forms. *Global constraints* refer to a set of variables of the problem. An example of such global constraints is `all_different`$(x_1,x_2,...,x_n)$ which specifies that all the variables $x_1,x_2,...,x_n$ must be different.

Solving a feasibility problem in CP is based on interleaving the *propagation* and the *search* processes in order to find a feasible solution for the problem. Minimizing an objective function may be reduced to solve a given number of feasibility problems.

A propagation algorithm is associated to each (or a set of) constraint(s). It consists in filtering (or reducing) from variable domains the values that cannot lead to feasible solutions. The propagation algorithm is terminated once no more values can be eliminated from the variable domains.

Once the propagation phase is finished, there may remain some inconsistent values in the variable domains. Therefore, a *search* algorithm is launched. The search algorithm is based on a tree search procedure where a branching step is applied by partitioning the current problem into subproblems. Branching may be done by instantiating a given variable to a feasible value of its domain or adding a new constraint.

The questions arising in designing a search algorithm in CP are more or less similar to those of branch and bound algorithms:

- **Branch ordering:** how the problem to solve (node of the tree) is splitted into subproblems when the propagation algorithm is inconclusive? On which variable the branching is applied next?
- **Variable selection:** in which subproblem (child node) the search continue next? What value should be first assigned to the branching variable?

Example 1.18. **A CP model for Sudoku:** nowadays, the Sudoku logic game is very popular. The principle of the game is to fill a 9×9 grid so that each row and each column contains the numbers from 1 to 9. Moreover, each of the nine 3×3 boxes contains the numbers from 1 to 9. A partially completed grid is provided as an input for each game setting. A CP model using the Gecode solver may be the following:

The `must_be_distinct` constraint has been used to model the three constraints of the problem (row, column, 3*3 boxes). The rest of the model represents the input setting of the game. It assigns the predefined values to squares of the grid (otherwise by default it is 0). Figure 1.28 illustrates a solution for a given game input.

Algorithm 1. CP model for Sudoku

```
class Sudoku < Gecode::Model
def initialize(predefined_values)
# Create the squares representing the integer variables
@squares = int_var_matrix(9, 9, 1..9)
# Distinctness constraint
9.times do |i|
# All rows must contain distinct numbers
@squares.row(i).must_be.distinct
# All columns must contain distinct numbers
@squares.column(i).must_be.distinct
# All 3x3 boxes must contain distinct numbers
@squares.minor((i % 3) * 3, 3, (i / 3) * 3, 3).must_be.distinct
end
# Place the constraints from the predefined squares on them
predefined_values.row_size.times do |i|
predefined_values.column_size.times do |j|
unless predefined_values[i,j].zero?
@squares[i,j].must == predefined_values[i,j]
end
end
```

Input game

	2	6				8	1	
3			7		8			6
4				5				7
	5		1		7		9	
		3	9		5	1		
	4		3		2		5	
1				3				2
5			2		4			9
	3	8				4	6	

Solution of the game

7	2	6	4	9	3	8	1	5
3	1	5	7	2	8	9	4	6
4	8	9	6	5	1	2	3	7
8	5	2	1	4	7	6	9	3
6	7	3	9	8	5	1	2	4
9	4	1	3	6	2	7	5	8
1	9	4	8	3	6	5	7	2
5	6	7	2	1	4	3	8	9
2	3	8	5	7	9	4	6	1

Fig. 1.28 Illustration of the Sudoku game

1.4.2 Classical Hybrid Approaches

Many combination schemes show that the hybridization of metaheuristics and CP is fruitful for some optimization problems. The following sections illustrate, for each class of hybrids belonging to the presented taxonomy, some hybridization schemes between constraint programming algorithms and metaheuristics. Some illustrative examples may also be found in [72].

1.4.2.1 Low-Level Relay Hybrids (LRH)

As within mathematical programming approaches, constraint programming may be used to explore large neighborhoods in S-metaheuristics (full or partial). Indeed, when the propagation tends to reduce the search space, CP is an efficient approach in modeling the expression of neighborhoods and exploring very large neighborhoods with side constraints [154]. Two different types of exploration may be applied:

- **Neighborhoods with expensive testing of feasibility:** neighborhoods around the current solution are defined by adding side constraints to the original problem. Checking the feasibility for all side constraints by CP may be efficient. Indeed, the feasibility test of solutions may be an expensive task. The propagation algorithms of CP reduce the size of neighborhoods.
- **Large neighborhoods:** optimizing the exploration of the neighborhood with in-lined constraint checks. For instance, the problem of searching very large neighborhoods is tackled with a constraint programming solver in the resolution of vehicle routing problems [137]. A CP model has been proposed for the neighborhood, where every feasible solution represents a neighbor. A given subset of decision variables may also be fixed [8]. A CP search has been carried out over the uninstantiated variables to solve a scheduling problem. A similar approach has been proposed in [154] for a vehicle routing problem, and in [30] for a job-shop scheduling problem.

1.4.2.2 Low-Level Teamwork Hybrids (LTH)

In this class of LTH hybrids between metaheuristics and CP, two main categories may be distinguished: exact search hybrid algorithms in which a metaheuristic is embedded into constraint programming, and heuristic search algorithms in which constraint programming is embedded into a P-metaheuristic.

Embedding metaheuristics into constraint programming: metaheuristics may be used to improve the search algorithm in CP. The following hybrid approaches may be applied to converge more quickly to the optimal solution or approximating "good" solutions:

- **Node improvement:** metaheuristics may be applied to partial solutions of the tree to improve or repair the nodes of the search tree. A greedy approach may also explore a set of paths from a node of the search tree. Then, CP search continue from the improved solutions [31].
- **Discrepancy-based search algorithms:** this approach generates near-greedy paths in a search tree. This approach has been used in limited discrepancy search [87] and dynamic backtracking [78]. Lookahead evaluation of greedy algorithms may also be used over the nodes of the search tree [31].
- **Branch ordering:** metaheuristics may be applied to answer the following question: which child node to investigate first when diving deeper into the node. Metaheuristics may be used to give a preference to a branch that is consistent

with the near-optimal solution. Indeed, the use of metaheuristics produces a better variable ordering and then will speedup the tree search. This approach has been used for solving satisfiability problems [152].

Metaheuristics may also be considered to solve relaxed problem. At each node, the subproblem is relaxed by removing some constraints. The violated constraints in the obtained solution will form the basis for branching. for instance, this approach has been used to solve a scheduling problems [109] [130].

- **Variable selection:** metaheuristics are used for variable selection at each node of the tree. This strategy consists in reducing the list of candidates. A straightforward strategy has been used in [64]. Let $v_1, v_2, ..., v_n$ be the possible branches in a decreasing order of preference (lower bound $h(v_i)$). The strategy consists in selecting the v_i branches such as $h(v_i) \leq h(v_1) + \alpha(h(v_n) - h(v_1))$ where $\alpha \in [0,1]$ is a parameter. Those branches constitute the RCL list (Restricted Candidate Lists), whereas the other branches are not explored.
- **Branching restriction:** metaheuristics may be used to filter the branches of the tree-search node. This hybrid scheme has been proposed to solve a scheduling problem [32].

CP can construct partial solutions which are used to define a search space for a metaheuristic. Then, the results obtained are used in order to refine the bounds or columns to generate in a branch and cut algorithm.

Embedding constraint programming into P-metaheuristics: some search components of a P-metaheuristic induce optimization problems which are solved by CP. For instance, some recombination operators such as crossover in EAs may be optimized using CP. In addition to the recombination operators, large neighborhood search algorithms based on CP can be integrated into unary operators of P-metaheuristics such as the mutation in EAs.

Some search ingredients of constraint programming algorithms can also be used into P-metaheuristics. For instance, the use of lower bounds into a P-metaheuristic can improve the search. The partial solutions (subproblems) maintained by CP may provide to metaheuristics interesting initial solutions to metaheuristics. The evaluation of those partial solutions will guide the metaheuristics to more promising regions of the search space. The partial solution with the least cost lower bound suggests a promising region.

CP algorithms can also be used as decoders of indirect representations carried out by metaheuristics. This strategy may be applied once the metaheuristics use indirect encoding which represent incomplete solutions of the problem. This strategy is efficient when the decoding involves complex constraints to satisfy.

1.4.2.3 High-Level Relay Hybrids (HRH)

In this class of hybrids, self-contained metaheuristics are used in conjunction with CP in a pipeline manner. Metaheuristics are considered as a pre-processing or a post-processing step for CP.

Information provided by metaheuristics: in the case where the informations are provided by metaheuristics, similar information exchanges as with mathematical programming algorithms may be used: upper bounds, incomplete solutions, subproblems, etc.

Information provided by constraint programming: in the case where the information is provided by CP, the same information as with mathematical programming[9] may be considered: partial solutions (i.e. subproblems), optimal solutions for relaxed problems, etc.

For instance, heuristic search-based hybrid scheme may be applied in solving some generated subproblems by CP. A subset of variables are assigned values using a complete search approach. This approach has been used for scheduling problems [133] and routing problems [154]. This hybrid scheme has been also proposed to solve satisfiability (SAT) problems, where a depth-bounded tree search is carried out and a local search procedure is applied at nodes reaching the depth-limit [84]. CP can be also applied to an incomplete formulation of the problem. For instance, all (or a set of) feasible solutions are generated by a CP strategy. Then, a metaheuristic will be applied to improve feasible solutions represented by the leaves of the CP tree.

1.4.2.4 High-Level Teamwork Hybrids (HTH)

Few hybrid HTH strategies combining CP and metaheuristics have been investigated. This class constitutes a promising way to develop efficient solvers and optimization algorithms. The architecture of this class of hybrids may be viewed as a set of agents implementing different strategies (CP, metaheuristics, MP) in solving the target problem, and different subproblems and relaxed problems. Those agents

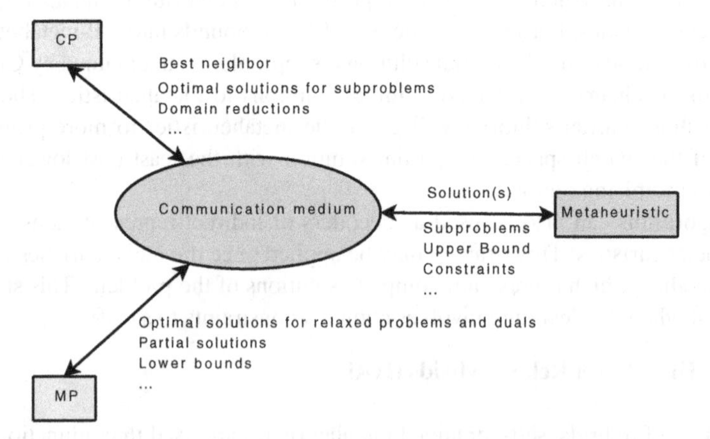

Fig. 1.29 HTH cooperation between metaheuristics, MP and CP strategies

[9] However, the duals cannot be considered.

will exchange information on the search. For an exact approach, the objective is to speedup the search in obtaining an optimal solution (efficiency). For a heuristic strategy, the objective is also to improve the quality of the obtained solutions (effectiveness). The information exchanged may include: solution(s), subproblems, relaxed optimal solutions, upper bounds , lower bounds , optimal solutions for subproblems, partial solutions, etc (Fig. 1.29).

1.5 Hybrid Metaheuristics with Machine Learning and Data Mining

Combining metaheuristics with data mining and machine learning techniques represents another way to improve the efficiency and effectiveness of the optimization algorithms based on metaheuristics.

1.5.1 Data Mining Techniques

Data mining (DM), also known as knowledge discovery in databases (KDD), is the process of automatically exploring large volumes of data (e.g. instances described according to several attributes), to extract interesting knowledge (patterns). In order to achieve this goal, data mining uses computational techniques from statistics, machine learning and pattern recognition .

Various data mining tasks can be used depending on the desired outcome of the model. Usually a distinction is made between supervised and unsupervised learning. Classical tasks of supervised learning are (Fig. 1.30):

- **Classification:** examining the attributes of a given instance to assign it to a predefined category or class.
- **Classification rule learners:** discovering a set of rules from the data which forms an accurate classifier.

The most common tasks of unsupervised learning are:

- **Clustering:** partitioning the input data set into subsets (clusters), so that data in each subset share common aspects. The partitioning is often indicated by a similarity measure implemented by a distance.
- **Association rule learners:** discovering elements that occur in common within a given data set.

The feature selection task objective consists in reducing the number of attributes (i.e. dimensionality of the data set). Feature selection is often considered as a necessary preprocessing step to analyze data characterized by a large number of attributes. It allows to improve the accuracy of the extracted models. Two models of feature selection exist depending on whether the selection is coupled with a learning scheme or not. The first one, the *filter model* , which carries out the feature subset selection and the learning (e.g. classification, clustering) in two separate phases, uses a

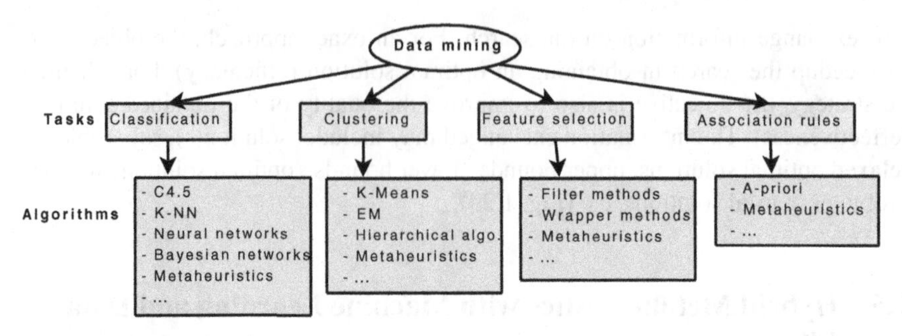

Fig. 1.30 Some data mining tasks and associated algorithms

measure that is simple and fast to compute. The second one, the *wrapper method*, which carries out the feature subset selection and learning in the same process, engages a learning algorithm to measure the accuracy of the extracted model. From the effectiveness point of view, wrapper methods are clearly advantageous, since the features are selected by optimizing the discriminate power of the finally used learning algorithm. However, their drawback is a more important computational cost.

Metaheuristics have been largely used to solve data mining tasks with a great success. However, using data mining techniques to improve the efficiency and effectiveness of metaheuristics, which is our concern in this chapter, is less studied. This hybridization scheme can be viewed as knowledge extraction and integration into metaheuristics. This knowledge may take different forms. Figure 1.31 describes some ways to integrate knowledge into metaheuristics.

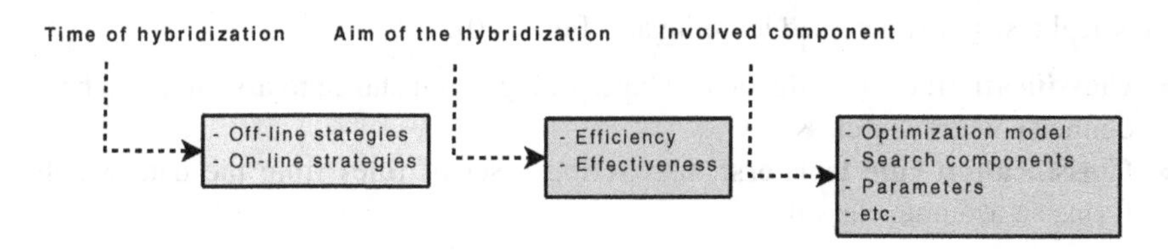

Fig. 1.31 Some ways integrating knowledge into metaheuristics

Three criteria will be used to refine our classification [104]:

- **Time of extracting the knowledge:** two kinds of hybridizations can be distinguished depending on the time of extracting the used knowledge. Hybridizations which extract the knowledge before the search starts are called *off-line knowledge* strategies and combinations where the knowledge is extracted dynamically during the search are described as *on-line knowledge* strategies.
- **Aim of the hybridization:** either the combination allows to improve the efficiency of the algorithm by reducing the search time, or the combination is used to improve the effectiveness of the algorithm leading to better quality of solutions. The efficiency may be carried out by approximating the objective function

or reducing the size of the search space. The effectiveness may be improved by incorporating some knowledge into the search components or by updating the parameters of the metaheuristics in an adaptive way. Of course, a given hybridization may improve both criteria: efficiency and effectiveness.
- **Involved component:** a metaheuristic is composed of different search components. Hybridization can occur in any search component such as encoding of solutions, initialization of solutions, search variation operators (e.g. mutation, crossover, neighborhood), etc. It may also be used to fix the parameters of the algorithm or defining the optimization problem to solve (e.g. objective function).

1.5.2 *Main Schemes of Hybridization*

In the following sections, some hybridization schemes between metaheuristics and data mining techniques are presented according to each class of the general taxonomy.

1.5.2.1 Low-Level Relay Hybrid (LRH)

Traditional S-metaheuristics, greedy or multi-start strategies (e.g. GRASP algorithm) do not use any information on the search of previous iterations to initialize the next search even if the tabu search algorithm uses the concept of memory to guide the search. Hence, some knowledge may be introduced in those families of metaheuristics.

Optimization model: the extracted knowledge may be used to transform the target optimization problem. For instance, in the ART (Adaptive Reasoning Technique) on-line approach, the search memory is used to learn the behavior of a greedy algorithm [136]. Some constraints are added to the problem. Those constraints are generated from the non interesting visited solutions according to the values associated to their decision variables. Similar to the tabu list strategy, those constraints are dropped after a given number of iterations.
Parameters setting: another LRH hybrid approach provides a dynamic and adaptive setting of the parameters of a S-metaheuristic. Indeed, knowledge extracted during the search may serve to change dynamically at run time the values of some parameters such as the size of the tabu list in tabu search, the temperature in simulated annealing.

This dynamic setting may also concern any search component of a S-metaheuristic such as the neighborhood and the stopping criteria.

1.5.2.2 Low-Level Teamwork Hybrids (LTH)

This hybrid scheme is very popular in P-metaheuristics.

Search components: a straightforward LTH hybrid approach consists in using data mining techniques in recombination operators of P-metaheuristics. In this class of hybrids, the knowledge extracted during the search is incorporated into the recombination operators for the generation of new solutions (Fig. 1.32). From a set of solutions (e.g. current population, elite solutions), some models are extracted which may be represented by classification rules, association rules, decision trees, etc. Those models (patterns) will participate in the generation of new solutions to intensify or diversify the search.

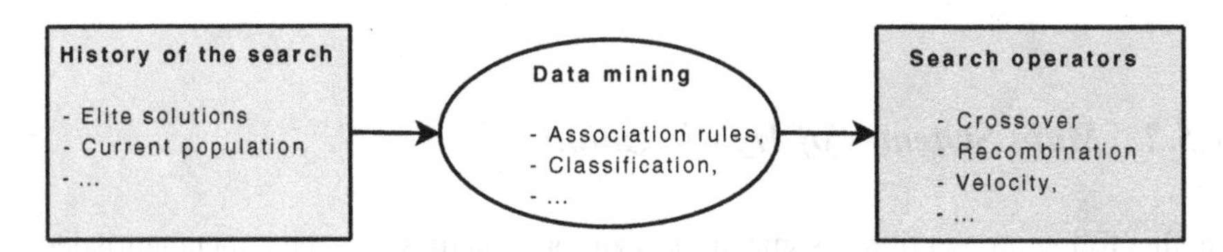

Fig. 1.32 Extracting knowledge from the history of the search and its use into search operators.

Example 1.19. **Integrating knowledge into recombination operators:** in this hybrid scheme, a set of decision rules describing the generated solutions are extracted. For instance, classification rules describing the best and worst individuals of the current population are extracted [127]. Those rules are generated using the *AQ learning algorithm*, a general decision rules learning algorithm (Fig. 1.33). The extracted rules are incorporated into the crossover operator of an evolutionary algorithm to reduce the search space for the offsprings (Fig. 1.34). The obtained results indicate that those learnable evolution models allow to speedup the search and improve the quality of solutions [127].

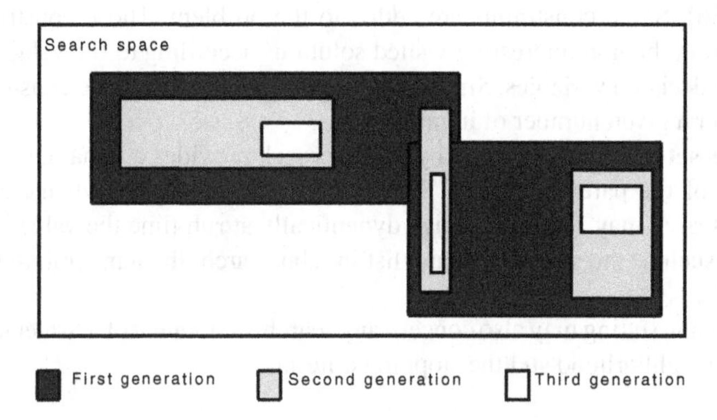

Fig. 1.33 Reduction of the search space for the offsprings using a learnable evolution model

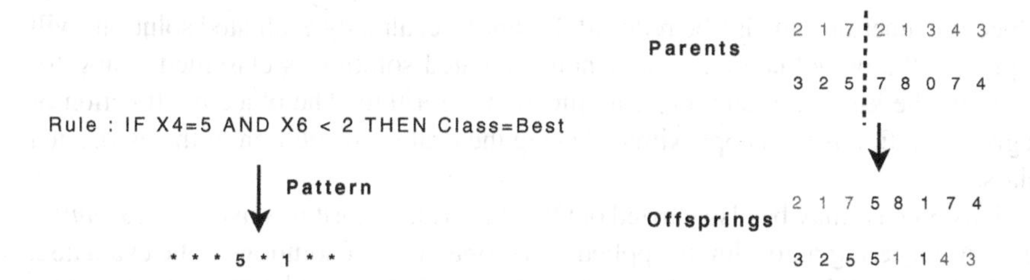

Fig. 1.34 Crossover operator using the induced rule as a pattern. For instance, the extracted pattern (...5.1..) is included into the offsprings.

EDA (Estimation of Distribution Algorithms) can also be considered as LTH hybrids using estimated probability distributions to generate new solutions. Similarly, *cultural algorithms* use high quality individuals to develop beliefs constraining the way in which individuals are transformed by genetic operators [144]. In cultural algorithms, beliefs are formed based on each entity's individual experiences. The reasoning behind this is that cultural evolution allows populations to learn and adapt at a rate faster than pure biological evolution. Importantly, the learning which takes place individually by each entity is passed on the remainder of the group, allowing learning to take place at a much faster rate.

Civilized genetic algorithms constitute another LTH hybrid approach integrating concepts from machine learning [150]. They differ from Darwinian evolution as they keep information of the population in order to avoid doing the same errors. The knowledge is dynamically updated during the successive generations. They have been applied to binary encodings in which a preprocessing step using a genetic algorithm is carried out to obtain a diverse population.

Parameter setting: a dynamic setting of the parameters of a P-metaheuristic can be carried out by a data mining task. Any parameter of a P-metaheuristic, such as the mutation and crossover probabilities in evolutionary algorithms, the pheromone update in ant colonies, and the velocity update in particle swarm optimization, can be modified dynamically during the search. Indeed, knowledge extracted during the search may serve to change dynamically at run time the values of those parameters. For instance, the initialization of the mutation rate may be adjusted adaptively by computing the progress of the last applications of the mutation operator [173] [91]. Hence, it becomes possible to determine the probabilities of application of a given operator in an adaptive manner where more efficient an operator is, more important its probability of application will be. Another approach could be to analyze in details the new individuals generated by operators (in terms of quality and diversity) using clustering algorithms. This would give valuable information that can help to set the new application probabilities.

Optimization model: many optimization problems such as engineering design problems are concerned by expensive objective functions. In this hybrid scheme, supervised classification algorithms can be used to approximate the objective function during the search. The number of solutions to evaluate according to the real objective function can also be reduced. In this case, already evaluated solutions will represent the predefined classes. A non evaluated solution is classified, using for example the k-nearest neighbor classification algorithm. The objective function of a given solution is then approximated using the evaluated solution of the associated class.

This process may be also carried out by clustering algorithms using *fitness imitation*. A clustering algorithm is applied on a population of solutions to be evaluated. Each cluster will have a representative solution. Only the solution that represents the cluster is evaluated [142] [112] [99]. Then, the objective function of other solutions of the cluster is estimated in respect to its associated representative[10] (Fig. 1.35). Different clustering techniques may be used such as K-means and fuzzy c-means.

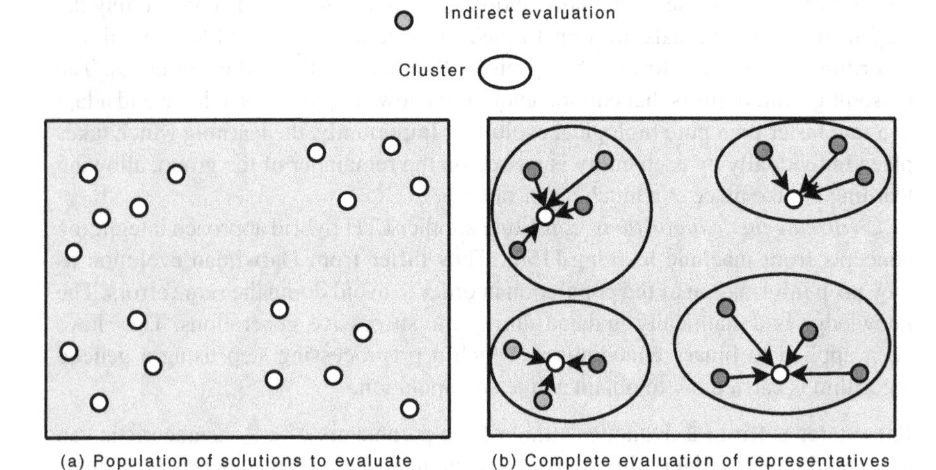

Fig. 1.35 Evaluating a solution by using the representative of its cluster (fitness imitation)

1.5.2.3 High-Level Relay Hybrid (HRH)

In this HRH hybrid approach, a priori knowledge is first extracted from the target optimization problem. Then, this knowledge is used into the metaheuristic for a more efficient search. The previously acquired knowledge may be obtained from previous experimentations, an expert, landscape analysis, etc. Many schemes may be introduced into this traditional hybrid class.

Search components: for instance, data mining algorithms may be applied for the initialization of solutions. Instead of generating the initial solutions randomly, problem knowledge can be used to generate solutions which integrate "good" patterns.

[10] This scheme is called fitness imitation or fitness inheritance.

Example 1.20. **Any-time learning algorithm in dynamic optimization:** a genetic algorithm has been initialized with a case-based reasoning in a tracker / target simulation with a periodically changing environment [141]. Case-based initialization (learning agent) allows the system to automatically bias the search of the GA toward relevant areas of the search space in a changing environment (dynamic optimization problem). This scheme may be seen as a general approach to continuous learning in a changing environment. The learning agent continuously tests search strategies using different initial solutions. This process allows the update of the knowledge base on the basis of the obtained results. This knowledge base generated by a simulation model will be used by any search agent.

Parameter setting: the same hybrid scheme may be used within the initialization of the parameters of any metaheuristic. A difficult part in designing metaheuristics deals with the setting of their parameters. Indeed, many parameters compose metaheuristics such as the probability of application of a given operator, the tabu list, the size of the population or the number of iterations? An empirical approach consists in both running several times the metaheuristic with different parameters values and trying to select the best values. If the number of trials or the number of parameters is important, determining the best set of parameters may require some statistical analysis. This may be seen as a data mining help (Fig. 1.36).

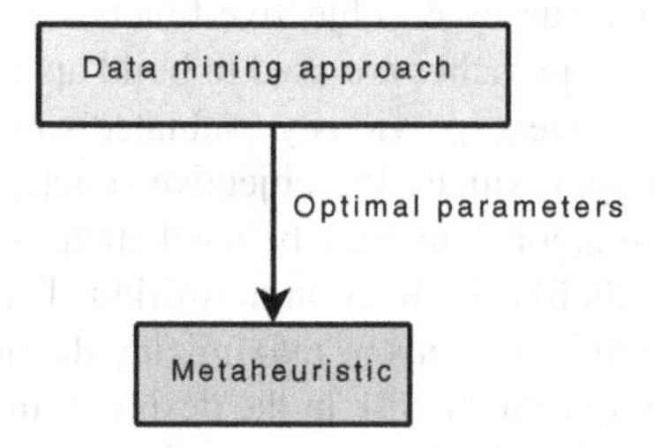

Fig. 1.36 Setting the parameters of a metaheuristic using a data mining approach

Optimization model: data mining techniques can also be used in decomposing the optimization problem handled by a metaheuristic. For instance, in optimization problems dealing with Euclidean distances, such as vehicle routing and the P-median optimization problems, clustering algorithms may be used to decompose the input space into subspaces. Metaheuristics are then used to solve those subproblems associated to the subspaces. Finally, a global solution is built using partial final solutions.

Example 1.21. **Clustering routing problems:** some efficient techniques in solving routing problems (e.g. TSP, VRP) decompose the operational space into subspaces using clustering algorithms such as the K-means or the EM (Expectation Maximization) algorithm (Fig. 1.37). Indeed, a metaheuristic is then used to solve the different subproblems. This approach is interesting for very large problems instances.

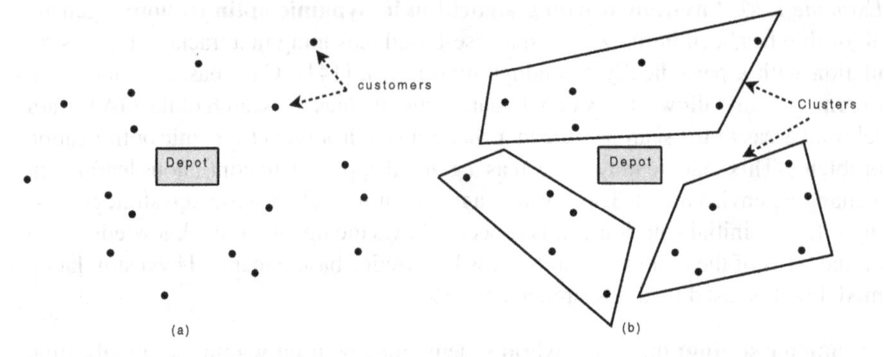

Fig. 1.37 Decomposing an optimization problem using clustering algorithms. (a) Instance of the VRP problem. (b) Clustering the customers and then applying a TSP metaheuristic to the subproblems.

A popular off-line hybrid scheme for expensive objective function consists in approximating the objective function of the problem. Indeed, in many complex real-life applications, the objective function is very expensive to compute. The main objective of this hybrid approach is to improve the efficiency of the search. The approximation can be used either for expensive objective functions or multi-modal functions. A comprehensive survey on objective function approximations may be found in [98]. Data mining approaches are used to build approximate models of the objective function. In this context, previously evaluated solutions are learned by a data mining algorithm to approximate the objective function of other individuals (Fig. 1.38). Many learning algorithms may be used such as neural networks (e.g. multi-layer perceptrons, radial-basis-function networks). The main issue here is to obtain a "good" approximation in terms of maximizing the quality and minimizing the computing time. Many questions arise in the design of this hybrid scheme such as: which proportion of the visited solutions are evaluated using the approximation, and at what time or in which component of the search algorithm the approximation is used.

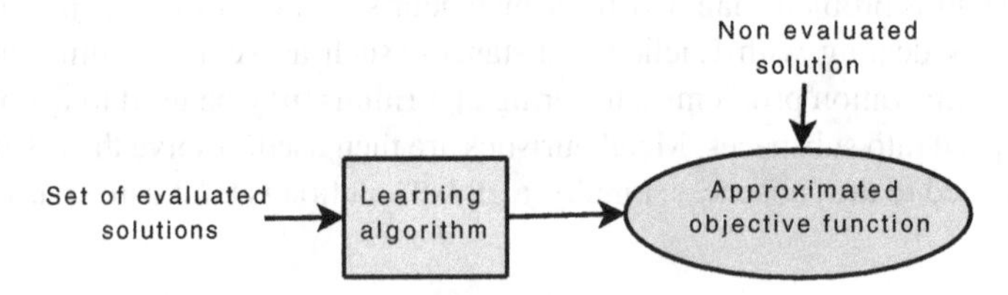

Fig. 1.38 Data mining approach to approximate the objective function

1.5.2.4 High-Level Teamwork Hybrid (HTH)

A HTH approach is a hybrid scheme in which a dynamically acquired knowledge is extracted in parallel during the search in cooperation with a metaheuristic (Fig. 1.39). Any on-line learning algorithms can be used to extract knowledge from informations provided by metaheuristics such as elite solutions, diversified set of good solutions, frequency memory, recency memory, etc. From this input, the data mining agent extracts useful information to be used by metaheuristics to improve the search. Any statistical indicator for landscape analysis of a problem may also be used.

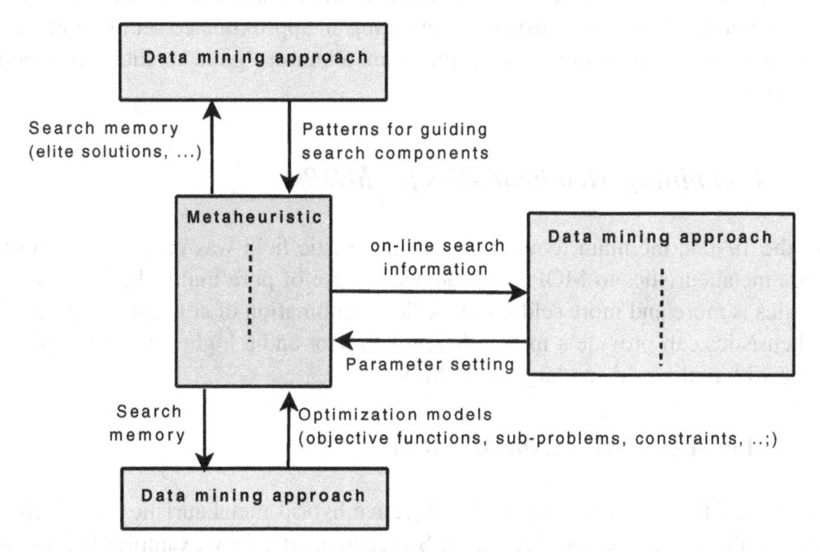

Fig. 1.39 On-line knowledge extraction and its use by a metaheuristic

Example 1.22. **Data mining in population management of a P-metaheuristic:** using the same scheme of cooperation, data mining approaches can be used to manage the population of a P-metaheuristic. Managing a population deals with the intensification and the diversification tasks of a metaheuristic. Diversification may be carried out by injecting new individuals into the population during the search. In order to lead the search to promising search spaces it could be also interesting to regularly introduce individuals that are built based on information of the past encountered high-quality solutions.

Such an approach has been proposed in the CIGAR (Case Injected Genetic AlgoRithm) algorithm [119]. The aim of CIGAR is to provide periodically to the genetic algorithm solutions that suit to similar instances / problems. Hence, a classification task is carried out to find similar instances in a case base. CIGAR has been successfully applied to several problems such as the job-shop scheduling and circuit modeling. For instance, a combination of a GA with the A-priori algorithm has been used to discover interesting subroutines for the oil collecting vehicle routing problem [51]. The obtained sub-routes are inserted into the new individuals of the population. Another illustrative example is the combination of the GRASP heuristic with A-priori like algorithms to extract promising patterns from elite solutions [145].

1.6 Hybrid Metaheuristics for Multi-objective Optimization

The taxonomy for hybrid metaheuristics presented in this chapter holds in solving multi-objective optimization problems (MOPs). However, the design of hybrid metaheuristics for MOP needs an adaptation for the reason that in multi-objective optimization the main goal consists in generating an approximated set of Pareto solutions whereas in mono-objective optimization a unique "good" solution is aimed to be generated.

1.6.1 Combining Metaheuristics for MOPs

Until the 1990's, the main focus in the metaheuristic field was on the application of pure metaheuristics to MOPs. Nowadays, the use of pure multi-objective metaheuristics is more and more seldom. A skilled combination of concepts of different metaheuristics can provide a more efficient behavior and a higher flexibility when dealing with real-world and large-scale MOPs.

1.6.1.1 Low-Level Relay Hybrids (LRH)

This class of hybrids represents multi-objective hybrid metaheuristics in which a given metaheuristic is embedded into a S-metaheuristic. Few examples belong to this class since S-metaheuristics are not well adapted to approximate the whole Pareto set of a MOP into a single run.

Example 1.23. **An adaptive hybrid metaheuristic:** a multi-objective tabu search hyper-heuristic may be used to optimize the use of different S-metaheuristics [29]. This hybrid approach, tested on timetabling and space allocation, uses a tabu list of S-metaheuristics which is updated by adding the last used S-metaheuristic and/or the worst one, in terms of performance. Hence, this hybrid approach will adapt dynamically the search according to the performance of various S-metaheuristics. More efficient multi-objective S-metaheuristics will be more frequently used during the search.

1.6.1.2 Low-Level Teamwork Hybrids (LTH)

P-metaheuristics (e.g. evolutionary algorithms, scatter search, particle swarm, ant colonies) are powerful in the approximation of the whole Pareto set while

S-metaheuristics are efficient in the intensification of the search around the obtained approximations. Indeed, S-metaheuristics need to be guided to solve MOPs.

Therefore, most efficient multi-objective P-metaheuristics have been coupled with S-metaheuristics such as local search, simulated annealing and tabu search, which are powerful optimization methods in terms of exploitation of the Pareto sets approximations. The two classes of metaheuristics have complementary strengths and weaknesses. Hence, LTH hybrids in which S-metaheuristics are embedded into P-metaheuristics have been applied successfully to many MOPs. Indeed, many state-of-the art hybrid schemes are P-metaheuristics integrating S-metaheuristics.

Example 1.24. **Multi-objective evolutionary local search algorithm:** many multi-objective hybrid metaheuristics proposed in the literature deal with hybridization between P-metaheuristics (e.g. evolutionary algorithms) and S-metaheuristics (e.g. local search). For instance, the well-known genetic local search[11] algorithms are popular in the multi-objective optimization community [167] [96] [94] [75]. The basic principle consists of incorporating a local search algorithm during an evolutionary algorithm search. The local search part could be included by replacing the mutation operator, but it can also be added after each complete generation of the evolutionary algorithm [15]. The classical structure of a multi-objective genetic local search (MOGLS) algorithm is shown in figure 1.40, which depicts the relationships between the evolutionary multi-objective (EMO) component and the local search one.

The local search algorithm can be applied in a given direction (i.e. weighted aggregation of the objectives) [94]. In order to adapt the basic local search algorithm to the multi-objective case, one may take into account the Pareto dominance relation [15]. The algorithm works with a population of non-dominated solutions *PO*. The hybridization process consists in generating the neighborhood of each solution of the Pareto set approximation *PO*. The new generated non dominated neighbors are inserted into the approximation Pareto set *PO*. Solutions belonging to the Pareto set *PO* and dominated by a new introduced solution are deleted. This process is reiterated until no neighbor of any Pareto solution is inserted into the Pareto set *PO*. The Pareto local search algorithm is described below:

Algorithm 2. Template of the Pareto guided local search (PLS) algorithm.

Input: an approximated Pareto set PO;
repeat
  $S' = PO$;
  Generate the neighborhood PN_x for each solution x of S' ;
  Let PO be the set of non-dominated solutions of $S' \cup PN_x$;
until $PO{=}S'$ (the population has reached the local optima)
Output: Pareto set PO

[11] Called also *memetic*.

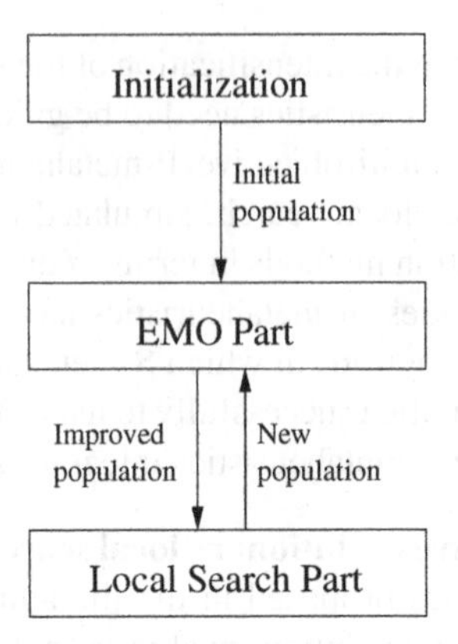

Fig. 1.40 Generic form of multi-objective genetic local search algorithms (MOGLS)

1.6.1.3 High-Level Relay Hybrids (HRH)

In HRH hybrids, self-contained multi-objective metaheuristics are executed in a sequence. A classical HRH for MOP is the application of an intensification strategy (e.g. path relinking, S-metaheuristic) on the approximation of the Pareto set obtained by a P-metaheuristic [53] [106].

Example 1.25. **Target aiming Pareto search - the TAPAS algorithm:** S-metaheuristics can be combined with any multi-objective metaheuristic to improve the quality of a Pareto approximation. First, a multi-objective metaheuristic (e.g. any P-metaheuristic) is used to generate a good approximation P of the Pareto set in terms of diversity. The design of the TAPAS algorithm was motivated by the need to improve the approximation P in terms of convergence towards the optimal Pareto set. Indeed, any S-metaheuristic algorithm can be applied to improve the quality of this approximation [105].

In the TAPAS algorithm, a S-metaheuristic l_i (e.g. tabu search[12]) is applied to each solution s_i of the initial Pareto set P. A specific mono-objective function o_i is defined for each search l_i. The defined objective function o_i must take into account the multiplicity of the S-metaheuristics invoked. Indeed, two S-metaheuristics should not examine the same region of the objective space, and the entire area that dominates the Pareto approximation P should be explored in order to converge towards the optimal Pareto front. The definition of the objective o_i is based on the partition of the objective space O according to the approximation P (see figure 1.41):

$$A_D = \{s \in O / \exists s' \in P, s' \prec s\}$$
$$A_{ND} = \{s \in O / \forall s' \in P, (s' \nprec s) \text{ and } (s \nprec s')\}$$
$$A_S = \{s \in O / \nexists s' \in P, s \prec s'\}$$
$$A_P = \{s \in O / \exists s_1, s_2 \in P, (s \prec s_1) \text{ and } (s \prec s_2)\}$$

[12] An efficient S-metaheuristic for the target problem should be selected.

Each solution $s_i \in P$ is associated with a part A^i_S of A_S. If l_i is able to generate a feasible solution in A^i_S, then the approximation is improved according to the convergence, without decreasing the diversity.

To guide the search, a goal g_i is given to each S-metaheuristic l_i, with g_i being the point that dominates all points of A^i_S. In cases where certain coordinates of g_i cannot be defined (e.g. the extremities of P), a lower bound for the missing coordinates should be used. For an objective f_m, the goal g_p is computed as follows:

$$f_m(g_p) = arg\ min_{\{f_m(s')/(s' \in P) \text{ and } (f_m(s') < f_m(s))\}} (f_m(s') - f_m(s))$$

Then, the objective o_i is stated as follows:

$$min(\sum_{j=1}^{M} |f_j(s) - f_j(g_i)|^r)^{1/r}$$

When a S-metaheuristic l_i reaches the goal g_i or when it finds a solution that dominates g_i, it stops and produces an archive a_i which contains all the current solutions that are non-dominated. When all the S-metaheuristics l_i are terminated, a new Pareto approximation set P' is formed by the Pareto union of all a_i. Because P' might be improved by another application of S-metaheuristics, the complete process is iterated until P' does not differ from P.

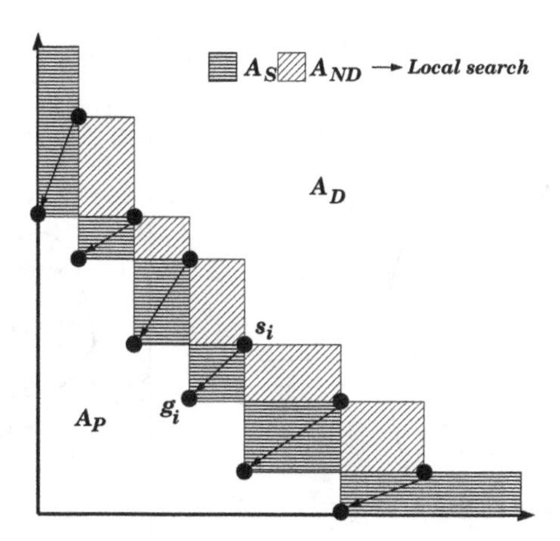

Fig. 1.41 The hybrid TAPAS algorithm for multi-objective optimization: the goal g_i of a solution s_i is defined in function of s_i neighbors in the objective space.

Example 1.26. **Filling the gap of a Pareto approximation with path-relinking:** path relinking can be combined with any multi-objective metaheuristic to intensify the search around a Pareto approximation. First, a multi-objective metaheuristic (e.g. any P-metaheuristic) is used to generate a good approximation of the Pareto set. Then, path relinking concept can be applied to connect the non-dominated solutions of the approximated Pareto set [16] [18] [97]. The design questions which must be considered are:

- **Selection of the initial and the guiding solutions:** this design question concerns the choice of the pair of solutions to connect. For instance, a random selection from the approximated Pareto set may be applied [16]. Otherwise, some criteria must be used to choose the initial and the guiding solutions: distance between solutions (e.g. distance in the decision or the objective space), quality of the solutions (e.g. best solution according to a reference point or weighted aggregation), etc.
- **Path generation:** many paths may be generated between two solutions. One has to establish which path(s) has to be explored and selected. Among other concepts, a neighborhood operator and a distance measure in the decision space have to be defined. For instance, the shortest paths may be generated according to the selected neighborhood operator [16]. Let us consider x as the current solution and y as the guiding solution. The neighborhood N of x is generated with the following constraint: $\forall z \in N, d(z,x) < d(y,x)$. From this neighborhood, only the non-dominated solutions may be selected to be potential solutions of the future paths (see figure 1.42). The process is iterated, until a complete path from x to y is generated. Many paths may also be considered. However, generating all possible paths may be computationally expensive. Moreover, the non-dominated solutions may also be selected to participate to a Pareto local search algorithm as shown in figure 1.43 [16].

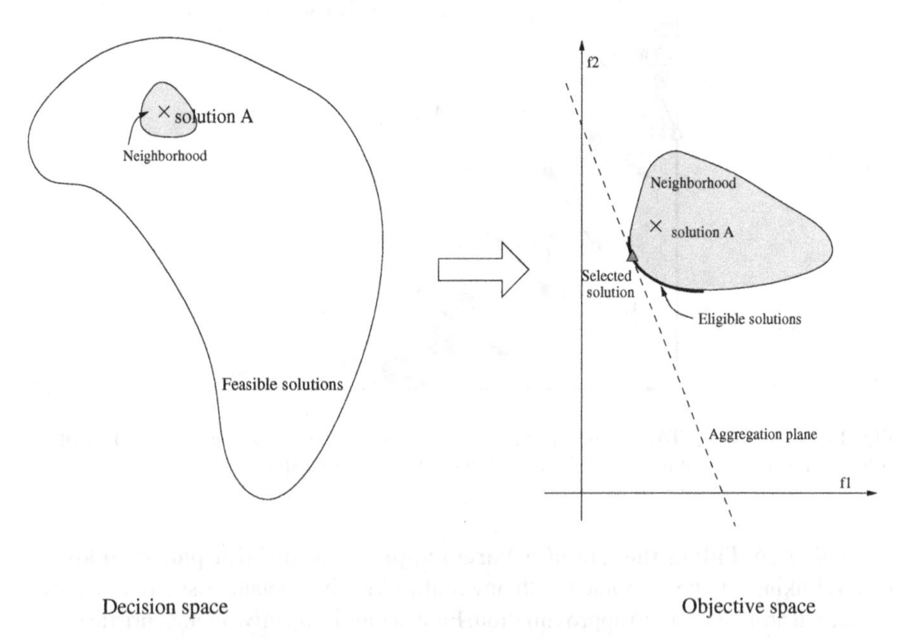

Fig. 1.42 Path Relinking algorithm filling the gap between two non-dominated solutions of an approximation Pareto set: neighborhood exploration

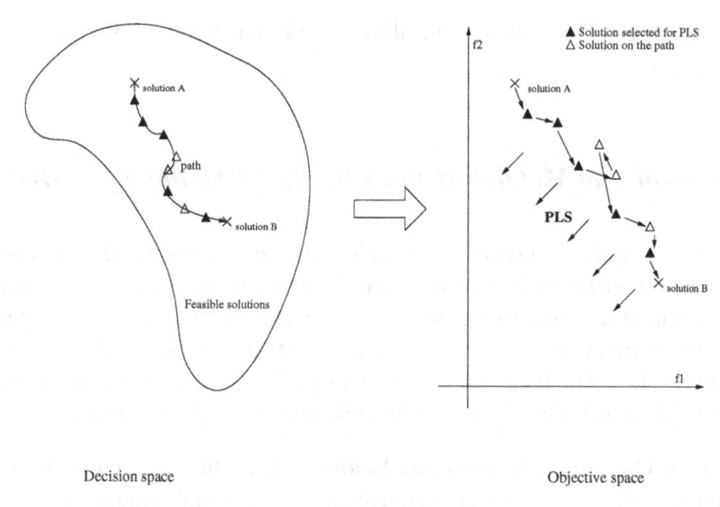

Fig. 1.43 Path relinking algorithm combined with a Pareto local search (PLS) algorithm

1.6.1.4 High-Level Teamwork Hybrid (HTH)

As previously shown in this chapter, HTH hybrids scheme involves several self-contained multi-objective metaheuristics performing a search in parallel and cooperating to find a Pareto set approximation.

Example 1.27. **Cooperative multi-objective evolutionary algorithms:** a growing interest is dedicated to design and implement parallel cooperative metaheuristics to solve multi-objective problems. The majority of designed parallel models in the literature are evolutionary algorithms [80] [101] [126] [147]. In multi-objective evolutionary algorithms, the individuals are selected from either the population, the Pareto archive or both of them. In the multi-objective island model, different strategies are possible. For instance, the newcomers replace individuals selected randomly from the local population that do not belong to the local Pareto archive. Another strategy consists in ranking and grouping the individuals of the local population into Pareto fronts using the non-dominance relation. The solutions of the worst Pareto front are thus replaced by the new arrivals. One can also make use of the technique that consists in merging the immigrant Pareto front with the local one, and the result constitutes the new local Pareto archive. The number of emigrants can be expressed as a fixed or variable number of individuals, or as a percentage of individuals from the population or the Pareto archive. The choice of the value of such parameter is crucial. Indeed, if it is low the migration process will be less efficient as the islands will have the tendency to evolve in an independent way. Conversely, if the number of emigrants is high, the EAs will likely to converge to the same solutions (premature convergence).

Although most of works on parallel multi-objective metaheuristics are related to evolutionary algorithms, there are also proposals related to alternative methods, such us tabu search [7], simulated annealing [5] [34], ant colonies [54] and memetic algorithms [15].

1.6.2 Combining Metaheuristics with Exact Methods for MOP

Another recent popular issue is the cooperation between multi-objective metaheuristics and exact optimization algorithms. Some hybrid schemes mainly aim at providing Pareto optimal sets in shorter time, while others primarily focus on getting better Pareto set approximations. In a multi-objective context, only few studies tackle this type of approaches. The main interest is to adapt the classical mono-objective hybrids presented in sections 1.3 and 1.4 to multi-objective optimization.

Example 1.28. **Combining branch and bound with multi-objective metaheuristics:** an investigation of several cooperative approaches combining multi-objective branch and bound [177] and multi-objective metaheuristics can be considered for MOPs [13]. Let us consider the bi-objective flow-shop scheduling problem, a multi-objective metaheuristic which approximates the Pareto set of the problem [14], and a bi-objective branch and bound which has been designed to solve the bi-objective flow-shop scheduling problem [116].

Three hybrid schemes combining an exact algorithm with a multi-objective metaheuristic may be considered [13]:

- **Metaheuristic to generate an upper bound:** the first HRH hybrid exact scheme is a multi-objective exact algorithm (e.g. branch and bound) in which the Pareto set approximation is used to speedup the algorithm (Fig. 1.44). The Pareto set approximation is considered as a good upper bound approximation for the multi-objective exact algorithm. Hence, many nodes of the search tree can be pruned by the branch and bound algorithm. This is a multi-objective adaptation of a classical cooperation found in the mono-objective context (see section 1.3). The time required to solve a given problem instance is smaller if the distance between the Pareto front approximation and the Pareto optimal front is small. If the distance is null, the exact algorithm will serve to prove the optimality of the Pareto set approximation. Even if this hybrid approach reduces the search time needed to find the Pareto optimal set, it does not allow to increase considerably the size of the solved instances.
- **Exact algorithm to explore very large neighborhoods:** in this hybrid heuristic approach, the exact multi-objective algorithm is used to explore large neighborhoods of a Pareto solution. The main idea is to reduce the search space explored by the exact algorithm by pruning nodes when the solution in construction is too far from the initial Pareto solution.

  Let us consider a permutation based representation for the bi-objective flow-shop scheduling problem, and an insertion neighborhood operator. The exact algorithm is allowed to explore the neighborhood of the initial Pareto solution in which the solutions are within a distance less or equal to δ_{max} (Fig. 1.45). The

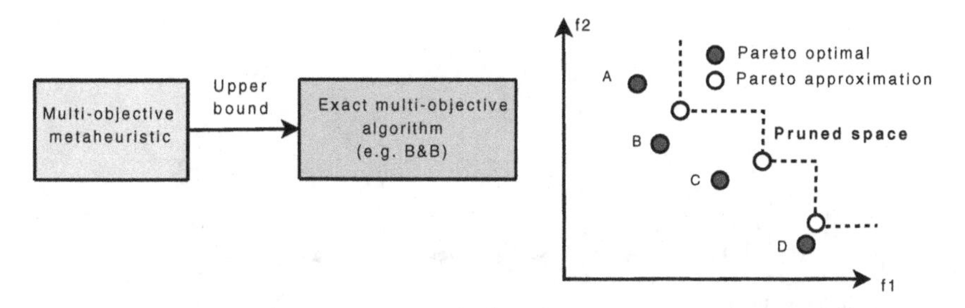

Fig. 1.44 A HRH exact hybrid scheme in which a multi-objective metaheuristic generates an upper bound Pareto set to an exact multi-objective algorithm (e.g. branch and bound).

size of the insertion-based neighborhood is $\Theta(n^2)$, where n represents the number of jobs. Hence, the size of the search space explored by the exact algorithm is exponential and may be approximated by $\Theta(n^{2\delta_{max}})$. Then, the distance δ_{max} must be limited, especially for instances with large number of jobs.

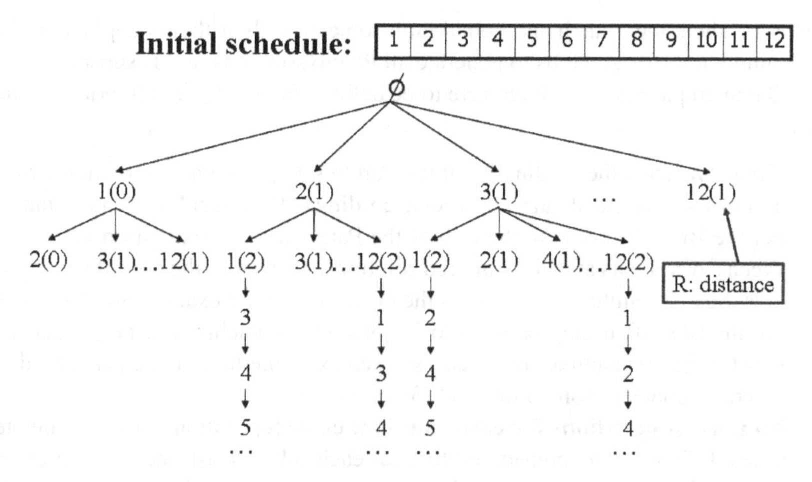

Fig. 1.45 A hybrid heuristic scheme in which an exact algorithm explores very large neighborhoods of the multi-objective metaheuristic

- **Exact algorithm to solve subproblems:** in this hybrid heuristic approach, the exact multi-objective algorithm solve subproblems which are generated by the multi-objective metaheuristic. A given region of the decision space is explored by the exact algorithm. Figure 1.46 shows an example of such hybridization. Let us consider an initial Pareto solution composed of 10 jobs $(a,b,...,i,j)$ which is obtained by the multi-objective metaheuristic. Subproblems of a given size (e.g. 4) are explored by the exact algorithm (e.g. the subproblem defined by the non-freezed jobs d,e,f,g). The first phase consists in placing the three first jobs at the beginning of the schedule. Moreover, the branch and bound algorithm places the three last jobs at the end of the schedule (a job j placed in queue is symbolized

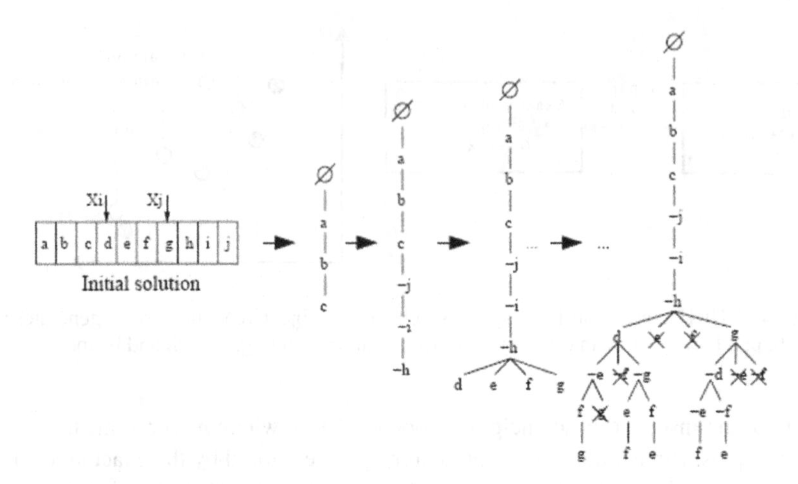

Fig. 1.46 A hybrid heurisitic scheme in which an exact algorithm solves subproblems generated by a multi-objective metaheuristic

by $-j$). Then, the branch and bound multi-objective algorithm is applied on the remaining non-freezed jobs to generate all Pareto solutions in this subspace.

The main parameters which have to be defined for an efficient hybrid scheme are:

- **Partition sizes:** the cardinality of the Pareto set approximation obtained by a multi-objective metaheuristic varies according to the target MOP and instance. For the BOFSP problem, the size of the Pareto set approximation varies between several tens and two hundred solutions. Moreover, the size of partitions must be also limited according to the efficiency of the exact method at hand. For the BOFSP, it may be fixed to 25 jobs for 10 machines instances and 12 jobs for the 20 machines instances, so each exact method can be performed in several seconds or some minutes [13].
- **Number of partitions for each solution:** enough partitions of the complete schedule have to be considered to treat each job at least once by the exact method. Moreover, it is interesting to superpose consecutive partitions to allow several moves of a same job during optimization. Then, a job which is early scheduled could be translated at the end of the schedule by successive moves. On the other side, more partitions are considered, more important the computational time is. For instance, for the BOFSP, 8 partitions for the 50−jobs instances, 16 partitions for the 100−jobs and 32 partitions for the 200−jobs instances may be considered [13].

Example 1.29. **Combining branch and cut with multi-objective metaheuristics:** this example investigates the solution of a multi-objective routing problem, namely the bi-objective covering tour problem (BOCTP), by means of a hybrid HRH strategy involving a multi-objective metaheuristic and a single-objective branch-and-cut algorithm. The BOCTP aims to determine a minimal length tour for a subset of nodes while also minimizing the greatest distance between the nodes of another set and the nearest visited node. The BOCTP can be formally described as follows (Fig. 1.48): let $G = (V \cup W, E)$ be an undirected graph, where $V \cup W$ is the vertex set, and $E = \{(v_i, v_j)/v_i, v_j, V \cup W, i < j\}$ is the edge set. Vertex v_1 is a depot, V is the set of vertices that can be visited, $T \subseteq V$ is the set of vertices that must be visited ($v_1 \in T$), and W is the set of vertices that must be covered. A distance matrix $C = (c_{ij})$, satisfying triangle inequality, is defined for E. The BOCTP consists of defining a tour for a subset of V, which contains all the vertices from T, while at the same time optimizing the following two objectives: (i) the minimization of the tour length and (ii) the minimization of the cover. The cover of a solution is defined as the greatest distance between a node $w \in W$, and the nearest visited node $v \in V$.

The BOCTP problem has been extended from the mono-objective covering tour problem (CTP). The CTP problems consists in determining a minimum length tour for a subset of V that contains all the vertices from T, and which covers every vertex w from W that is covered by the tour (i.e. w lies within a distance c from a vertex of the tour, where c is a user defined parameter). A feasible solution for a small instance is provided in figure 1.48. One generic application of the CTP involves designing a tour in a network whose vertices represent points that can be visited, and from which the places that are not on the tour can be easily reached. In the bi-objective covering tour problem BOCTP, the constraint on the cover has been replaced by an objective in which the covering distance is minimized [107].

Let us consider a multi-objective metaheuristic to solve the BOCTP problem which approximates the Pareto set [107], and a branch and cut algorithm to solve the mono-objective CTP problem [76]. The branch and cut algorithm may be considered as a black box optimization tool whose inputs are a subset of V, the set W, and a cover, and whose output is the optimal tour for the CTP. The branch and cut algorithm first relaxes the integrality conditions on the variables and the connectivity constraints of the integer linear programming model. Integrality is then gradually restored by means of a branch and bound mechanism. Before initiating branching at any given node of the search tree, a search is conducted for violated constraints, including the initially relaxed connectivity constraints and several other families of valid constraints. Several classes of valid inequalities have been considered such as dominance constraints, covering constraints, sub-tour elimination constraints, and 2-matching inequalities [76].

In the hybrid approach, the multi-objective metaheuristic generates a Pareto set approximation, which is used to build subproblems; these subproblems are then solved using the branch and cut algorithm (Fig. 1.48). Subproblem construction is a key point of the cooperative design, given that prohibitive computational times result if the subsets of V are too large. By limiting their size and giving the branch and cut algorithm access to the information extracted from the Pareto set approximation, the method makes solving the subproblems relatively easy for the branch-and-cut algorithm. Two procedures for building the subproblems can be considered [107]:

- **One objective improvement by an exact algorithm:** the main purpose of the first construction procedure is to improve the solutions found by the multi-

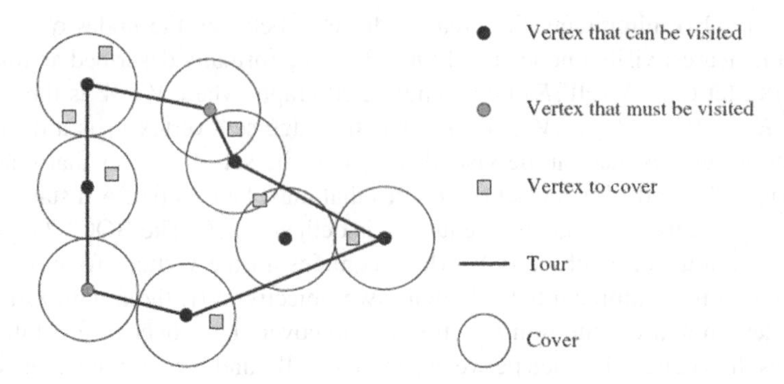

Fig. 1.47 The covering tour problem: an example of a solution

objective metaheuristic in terms of the tour length objective without modifying the cover value. It accomplishes this goal by investigating the possibility that some elements of the set of visited vertices $\tilde{V}$ can be replaced by sets of vertices $R \subseteq V \backslash \tilde{V}$ so that the cover value $\tilde{c}$ provided by the couple (v_t, v_c) remains unchanged (Fig. 1.48a). A vertex $v_k \in \tilde{V}$ can be replaced by a set R if and only if: (i) No subset of R can replace v_k; (ii) No vertex from R can provide a better cover: $\forall v_i \in R, c_{tc} \leq c_{ic}$; (iii) There must be a vertex from $\tilde{V}$ or from R that can replace v_k for every vertex of W that can be covered by v_k. Therefore, $\forall v_l \in W \backslash \{v_c\}$, such that $c_{kl} \leq \tilde{c}$, where the following condition must be true: $\exists v_n \in R \cup (\tilde{V} \backslash \{v_k\}), c_{nl} \leq \tilde{c}$.

Replacing a node of $\tilde{V}$ by a subset R tends to become easier as the cardinality of R increases. However, in practice, condition (i) limits the candidate subsets. The larger the R set, the higher the cost of the test. Certainly, if the size of the set used for the branch and cut algorithm is very large, the algorithm will require too much computational time. Therefore, in practice, the cardinality of R is limited to one or two elements.

For each solution s of the Pareto set approximation, a problem is built as follows. The set V_I of vertices that can be visited is created by the union of $\tilde{V}$ and all subsets of V with a cardinality of 1 or 2 that can replace a vertex of $\tilde{V}$. The set W of vertices that must be covered remains unchanged. Here, the parameter c is equal to the cover of s.

- **Region exploration by an exact algorithm:** in the first construction procedure it is unlikely that all the feasible covers corresponding to Pareto optimal solutions will be identified. These unidentified solutions must always be situated between two solutions of the approximation, although not always between the same two solutions. Thus, it is reasonable to assume that new Pareto optimal solutions may be discovered by focusing searches in the area of the objective space between two neighboring solutions. The second procedure aims to build sets of vertices in order to identify potentially Pareto optimal solutions whose cover values were not found by the multi-objective metaheuristic. Let A and B be two neighboring solutions in the approximation sets found by the evolutionary algorithm (i.e. there are no other solutions between A and B). A (resp. B) is a solution with a cover c_A (resp. c_B) which visits the vertices of the set V_A (resp. V_B). Assuming that $c_A < c_B$, the branch and cut algorithm can be executed on a set V_{II}, built according to both V_A and V_B, with the first cover $\tilde{c}$ which is strictly smaller than c_B as a parameter (Fig. 1.48b). If $\tilde{c}$ is equal to c_A, there is no need to execute the branch and cut algorithm.

It appears that neighboring solutions in the Pareto set have a large number of vertices in common. Thus, V_{II} contains V_A and V_B. This inclusion insures that the branch and cut algorithm will at least be able to find the solution A, or a solution with the same cover but a better tour length in cases for which the tour on V_A is not optimal. The following process is used to complete V_{II}: for every feasible cover c, so that $c_A < c < c_B$, vertices are added to V_{II} in order to obtain a subset of V_{II} with c as a cover. The algorithm below provides the procedure for constructing the set V_{II}.

Algorithm 3. Construction of the set V_{II}.

$V_{II} = V_A \cup V_B$;
for all c so that $c_A < c < c_B$ **do**
  for $v_l \in W$ **do**
    $V_{II} = \cup V_{II}\{v_k \in V \backslash V_{II} / c_{kl} \leq c\}$
  end for
end for

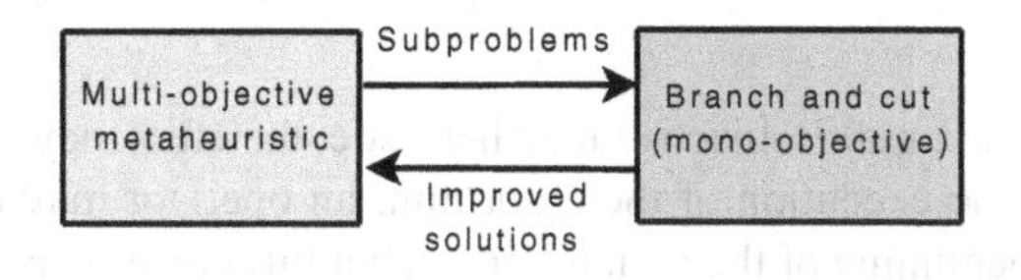

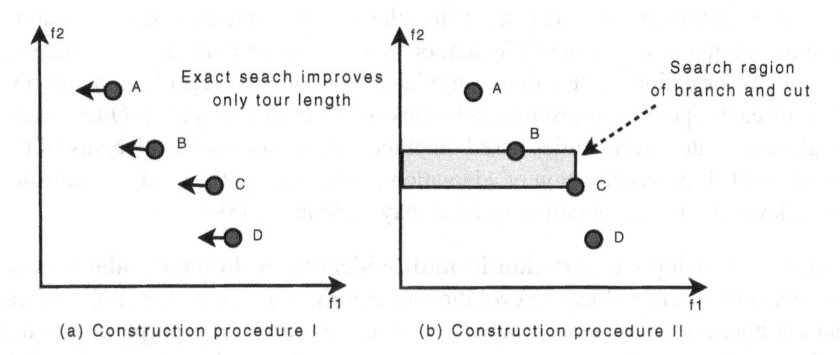

Fig. 1.48 Combining a mono-objective branch and cut algorithm and a multi-objective metaheuristic to solve the bi-objective covering tour problem

1.6.3 Combining Metaheuristics with Data Mining for MOP

Most of the classical combinations of metaheuristics with machine learning and data mining techniques (e.g. feature selection, classification, clustering, association rules) which have been applied to mono-objective optimization (see section 1.5) can be generalized to multi-objective optimization:

- Search operators (e.g. recombination operators in P-metaheuristics, neighborhoods in S-metaheuristics).
- Optimization models (e.g. approximation of the objectives functions, generation of sub-problems, new constraints).
- Parameter setting of the metaheuristics.

Example 1.30. **Search operators:** integrating knowledge into search operators is the most popular scheme in this class of hybrids. For instance, in a P-metaheuristic (e.g. evolutionary algorithm), a set of decision rules describing the best and worst individuals of the current population may be extracted. The extracted rules are incorporated into the crossover operator of an evolutionary algorithm to generate solutions sharing the characteristics of non-dominated solutions and avoiding those of dominated solutions.

This principle can be applied to multi-objective optimization in the following way [103]: a set of rules that describes why some individuals dominate others (positive rules) and why some individuals are dominated by others (negative rules[13]) are extracted using the *C4.5 classifier*. Offsprings that match the positive rules and do not match the negative rules are generated. The obtained results indicate that those learnable evolution models allow to speedup the search and improve the quality of solutions.

Parameter setting: in a multi-objective metaheuristic, the efficiency of an operator may change during the execution of the algorithm: an operator may offer a better convergence at the beginning of the metaheuristic, but this convergence may be improved later with another operator. The success of an operator may also depend on the instance of the problem. This motivates the use of adaptive operator probabilities to automate the selection of efficient operators. The adaptation can be done by exploiting information gained, either implicitly or explicitly, regarding the current ability of each operator to produce solutions of better quality [173]. Other methods adjust operator probabilities based on other criteria, such as the diversity of the population [45]. A classification of adaptation on the basis of the used mechanisms, and the level at which adaptation operates may be found in [89].

Example 1.31. **Adaptive mutation in multi-objective evolutionary algorithms:** let us consider a multi-objective evolutionary algorithm in which the choice of the mutation operators is done dynamically during the search. The purpose is to use simultaneously several mutation operators during the EA, and to change automatically the probability selection of each operator according to its effectiveness [15].

[13] Negative knowledge.

So the algorithm always uses more often the best operators than the others. Let us remark that a similar approach could be defined with other operators (e.g. crossover, neighborhoods, hybrid strategies).

Initially, the same probability is assigned to each mutation operator: $Mu_1, \ldots, Mu_k$. Those probabilities are equal to the same ratio $P_{Mu_i} = 1/(k * P_{Mu})$, where k is the number of mutation operators, and P_{Mu} is the global mutation probability. At each iteration, the probabilities associated to the mutation operators are updated according to their average progress. To compute the progress of the operators, each mutation Mu_i applied to the individual I is associated with a progress value:

$$\Pi(I_{Mu_i}) = \begin{cases} 1 \text{ if } I \text{ is dominated by } I_{Mu_i} \\ 0 \text{ if } I \text{ dominates } I_{Mu_i} \\ \frac{1}{2} \text{ otherwise (non comparable solutions)} \end{cases}$$

where I_{Mu_i} is the solution after mutation (Fig. 1.49).

At the end of each generation of the EA, an average progress $Progress(Mu_i)$ is assigned to each operator Mu_i. Its value is the average progress of $\Pi(I_{Mu_i})$ computed with each solution modified by the mutation Mu_i:

$$Progress(Mu_i) = \frac{\sum \Pi(I_{Mu_i})}{||Mu_i||}$$

where $||Mu_i||$ is the number of applications of the mutation Mu_i on the population. The new selection probabilities are computed proportionally to these values:

$$P_{Mu_i} = \frac{Progress(Mu_i)}{\sum_{j=1}^{k} Progress(Mu_j)} \times (1 - k \times \delta) + \delta$$

where δ is the minimal selection probability value of the operators.

This approach of progress computation compares two solutions with their dominance relation. However, a comparison only between I and I_{Mu_i} is not sufficient. Firstly, if the two individuals I and I_{Mu_i} are non comparable, the quality of the mutation cannot be evaluated. For instance, in figure 1.50, the progress Π of the two mutation operators applied on the solution $\triangle$ is the same $(1/2)$. However, the observation of the whole Pareto front shows that the second mutation operator performs better since the generated solution by the second mutation operator is Pareto optimal whereas the solution generated by the first is not.

Secondly, if the generated individual dominates the initial individual, the progress realized cannot be measured with precision. For instance, in figure 1.51, the progress Π of the two mutation operators applied on the solution $\triangle$ is the same (1). However, the observation of the whole population shows that the second mutation operator performs much better than the first one.

These problems can be tackled in the case of evolutionary algorithms using selection by ranking. The progress value can be replaced by:

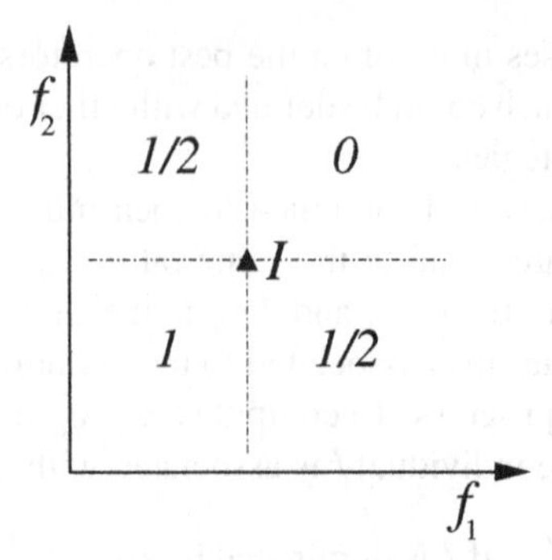

Fig. 1.49 Progress value of $\Pi(I_{Mu_i})$ for mutation operators

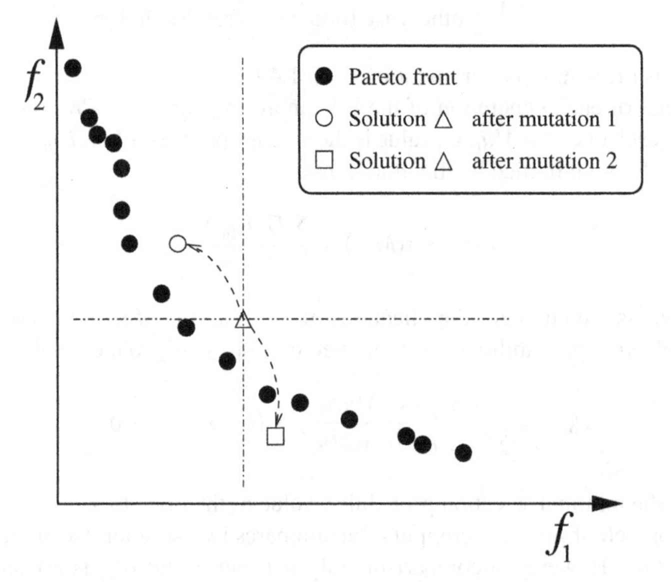

Fig. 1.50 Evaluation of the quality of the mutation operators

$$\Pi(I_{Mu_i}) = \left(\frac{R_k I}{R_k I_{Mu_i}}\right)^{\beta}$$

where $R_k I_{Mu_i}$ is the rank of the solution after mutation, $R_k I$ is the rank of the solution before mutation, and β is how much the progress made by mutation operators is encouraged (e.g. $\beta = 2$).

The evaluation of the progress of the mutation operators can be still improved by supporting the progresses realized on good solutions. In fact, these progresses are generally more interesting for the front progression than progresses made on bad solutions (Fig. 1.52). So an elitist factor $E_f I_{Mu_i}$ has been introduced into the last progress indicator to favor progresses made on good solutions:

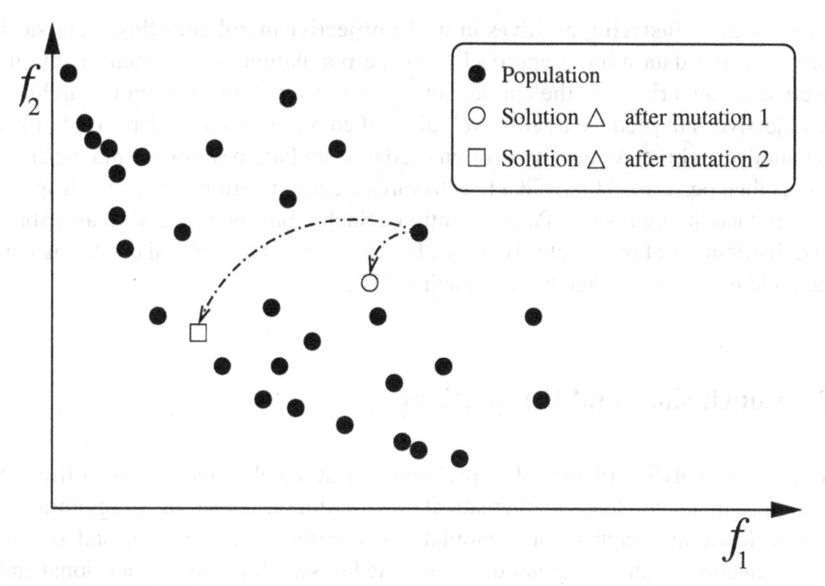

Fig. 1.51 Computing the progress realized by different mutation operators

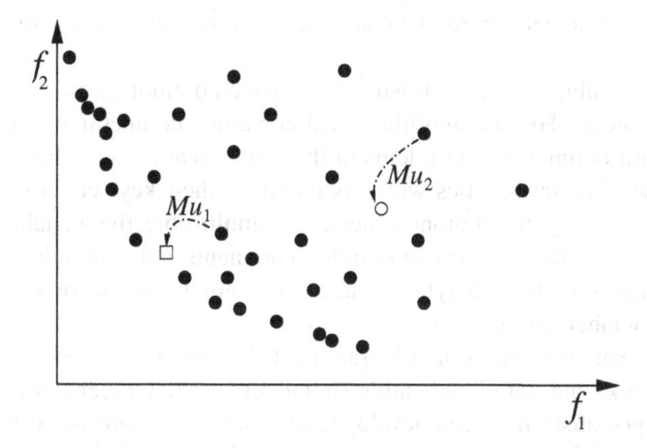

Fig. 1.52 Progress realized by mutation operators on good quality solutions

$$\Pi(I_{Mu_i}) = E_f I_{Mu_i} \times \left(\frac{R_k I_{Mu_i}}{R_k I}\right)^{\beta}$$

with $E_f I_{Mu_i} = (R_k I_{Mu_i})^{-1}$. Then, the average progress of a mutation Mu_i is defined as follows:

$$Progress(Mu_i) = \frac{\sum \Pi(I_{Mu_i})}{\sum E_f I_{Mu_i}}$$

Some hybrid schemes are specific to multi-objective metaheuristics such as introducing data mining tasks in the search component dealing with elitism.

Example 1.32. **Clustering archives in multi-objective metaheuristics:** a classical approach using data mining approaches in the population management of multi-objective metaheuristics is the application of clustering algorithms on the archive. The objective is to produce a set of well diversified representatives Pareto solutions in a bounded archive. An archive is often used to store Pareto solutions and the clustering is then performed to avoid a bias towards a certain region of the search space and to reduce the number of Pareto solutions. Such a bias would lead to an unbalanced distribution of the Pareto solutions. For instance, a hierarchical clustering can be applied using the average linkage method [183].

1.7 Conclusions and Perspectives

The efficient solving of complex problems must involve ideas from different paradigms: metaheuristics, mathematical programming, constraint programming, machine learning, graph theory, parallel and distributed computing, and so on. Pure metaheuristics are not generally well suited to search in high-dimensional and complex landscapes. Hybrid metaheuristics represent actually the most efficient algorithms for many classical and real-life difficult problems. This is proven by the huge number of efficient hybrid metaheuristics proposed to solve a large variety of problems.

Nowadays, combining metaheuristics becomes a common strategy to solve optimization problems. Hybrid algorithms will constitute competitive candidates for solving difficult optimization problems in the future years. As we have developed a unified view of metaheuristics which is based on their key search components, one can say that designing a mono-objective or multi-objective metaheuristic can be reduced to select the most suited search components and combining them. This design approach is naturally a hybrid one, and it is not under the control of a single paradigm of metaheuristics[14].

A unified taxonomy, based on a hierarchical (low level versus high level, relay versus teamwork) and flat classification (homogeneous/heterogeneous, global/partial, general/specialist), has been developed to describe in terms of design and implementation the different hybridization schemes of metaheuristics with:

- **Metaheuristics:** combining P-metaheuristics with S-metaheuristics has provided very powerful search algorithms. Pure P-metaheuristics such as evolutionary algorithms, scatter search, and ant colonies are generally not well suited to fine-tuned search in highly combinatorial spaces. P-metaheuristic are more efficient in terms of diversification (i.e. exploration) in the search space. Hence, they need to be combined with more intensification-based (i.e. exploitation-based) search algorithms which are generally based on S-metaheuristics (e.g. local search, tabu search).

[14] Using this design approach, it is worthwhile to speak about hybrid metaheuristics as any metaheuristic will be a hybrid one!.

- **Mathematical programming:** in the last decade, there has been an important advance in designing efficient exact methods in the operations research community (e.g. integer programming). There are many opportunities to design hybrid approaches combining metaheuristics and exact methods. Indeed, the two approaches have complementary advantages and disadvantages (e.g. efficiency and effectiveness).
- **Constraint programming:** over the last years, interest on combining metaheuristics and constraint programming has risen considerably. The availability of high-level modeling languages and software solvers for constraint programming will lead to more hybrid approaches which capture the most desirable features of each paradigm.
- **Data mining:** nowadays, using metaheuristics to solve data mining and machine learning problems becomes common. But the challenge is the incorporation of machine learning and data mining techniques into metaheuristics. The major interest in using machine learning and data mining techniques is to extract useful knowledge from the history of the search in order to improve the efficiency and the effectiveness of metaheuristics. Both positive and negative knowledge must be extracted. In fact, most of the actual works focus only on positive knowledge [128].

The main drawback of hybridization is the introduction of new parameters which define the hybrid scheme. The setting of those parameters is non trivial. A crucial question that has to be addressed in the future is an aid for the efficient design of hybrid metaheuristics, in which the automatic setting of parameters must be investigated [73] [25]. Indeed, it will be interesting to guide the user to define the suitable hybrid scheme to solve a given problem. It will be also interesting to define "adaptive" cooperation mechanisms which allows to select dynamically the optimization methods according to convergence or other criteria such as diversity. Some approaches such as the COSEARCH [164] or "hyper-heuristics" [28] have been proposed to deal with this problem. Those approaches are dedicated to choose the right heuristic for the right operation at the right time during the search. It must be noted that the those hybrid approaches operate in the heuristic space, as opposed to most implementations of meta-heuristics, which operate in the solution space. This principle is relatively new, although the concept of "optimizing heuristics" is not a recent one.

Using the software framework ParadisEO, it is natural to combine metaheuristics which have been developed under the framework to design S-metaheuristics (under ParadisEO-MO), P-metaheuristics (under ParadisEO-EO), and multi-objective metaheuristics (under ParadisEO-MOEO). Still a work to do for combining metaheuristics with exact optimization and machine learning algorithms. The coupling of software frameworks dealing with the three classes of algorithms (i.e. metaheuristics, exact and machine learning algorithms) is an important issue for the future. This enables to reduce the complexity of designing and implementing hybrid approaches and make them more and more popular.

It will be also interesting to deeply explore parallel models for hybrid methods. Parallel schemes ideally provide novel ways to design and implement hybrid algorithms by providing parallel models of the algorithms. Hence, instead of merely parallelizing and finely tuning a sequential hybrid algorithm which has limited capabilities to be parallelized, teamwork hybrid schemes are inherently suited to parallel environments.

References

1. Abbattista, F., Abbattista, N., Caponetti, L.: An evolutionary and cooperative agent model for optimization. In: IEEE Int. Conf. on Evolutionary Computation, ICEC 1995, Perth, Australia, pp. 668–671 (December 1995)
2. Abramson, D., Logothetis, P., Postula, A., Randall, M.: Application specific computers for combinatorial optimisation. In: Australien Computer Architecture Workshop, Sydney, Australia (February 1997)
3. Abramson, D.A.: A very high speed architecture to support simulated annealing. IEEE Computer 25, 27–34 (1992)
4. Aggarwal, C.C., Orlin, J.B., Tai, R.P.: An optimized crossover for the maximum independent set. Operations Research 45, 226–234 (1997)
5. Agrafiotis, D.K.: Multiobjective optimization of combinatorial libraries. Technical report, IBM J. Res. and Dev. (2001)
6. Aiex, R.M., Binato, S., Ramakrishna, R.S.: Parallel GRASP with path relinking for job shop scheduling. Parallel Computing 29, 393–430 (2003)
7. Al-Yamani, A., Sait, S., Youssef, H.: Parallelizing tabu search on a cluster of heterogeneous workstations. Journal of Heuristics 8(3), 277–304 (2002)
8. Applegate, D., Cook, W.: A computational study of the job-shop scheduling problem. ORSA Journal on Computing 3, 149–156 (1991)
9. Apt, K.: Principles of constraint programming. Cambridge University Press (2003)
10. Augerat, P., Belenguer, J.M., Benavent, E., Corberan, A., Naddef, D.: Separating capacity constraints in the CVRP using tabu search. European Journal of Operational Research 106(2), 546–557 (1998)
11. Balas, E., Niehaus, W.: Optimized crossover-based genetic algorithms for the maximum cardinality and maximum weight clique problems. Journal of Heuristics 4(2), 107–122 (1998)
12. Barnhart, C., Johnson, E.L., Nemhauser, G.L., Savelsbergh, M.W.P., Vance, P.H.: Branch-and-price: column generation for huge integer programs. Operations Research 46(316) (1998)
13. Basseur, M., Lemesre, J., Dhaenens, C., Talbi, E.-G.: Cooperation Between Branch and Bound and Evolutionary Approaches to Solve a Bi-Objective Flow Shop Problem. In: Ribeiro, C.C., Martins, S.L. (eds.) WEA 2004. LNCS, vol. 3059, pp. 72–86. Springer, Heidelberg (2004)
14. Basseur, M., Seynhaeve, F., Talbi, E.-G.: Design of multi-objective evolutionary algorithms: Application to the flow-shop scheduling problem. In: Congress on Evolutionary Computation, CEC 2002, Honolulu, Hawaii, USA, pp. 1151–1156 (May 2002)
15. Basseur, M., Seynhaeve, F., Talbi, E.-G.: Adaptive mechanisms for multi-objective evolutionary algorithms. In: Congress on Engineering in System Application, CESA 2003, Lille, France, pp. 72–86 (2003)
16. Basseur, M., Seynhaeve, F., Talbi, E.-G.: Path Relinking in Pareto Multi-Objective Genetic Algorithms. In: Coello Coello, C.A., Aguirre, A.H., Zitzler, E. (eds.) EMO 2005. LNCS, vol. 3410, pp. 120–134. Springer, Heidelberg (2005)
17. Beasley, J.E.: OR-Library: Distributing test problems by electronic mail. Journal of the Operational Research Society 41(11), 1069–1072 (1990)
18. Beausoleil, R.P.: Mutiple criteria scatter search. In: 4th Metaheuristics International Conference (MIC 2001), Porto, Portugal, pp. 539–544 (2001)

19. Belding, T.: The distributed genetic algorithm revisted. In: Eshelmann, D. (ed.) Sixth Int. Conf. on Genetic Algorithms. Morgan Kaufmann, San Mateo (1995)
20. Belew, R.K., McInerny, J., Schraudolph, N.N.: Evolving networks: Using genetic algorithms with connectionist learning. In: Langton, C.G., Taylor, C., Doyne Farmer, J.D., Rasmussen, S. (eds.) Second Conf. on Artificial Life, pp. 511–548. Addison-Wesley, USA (1991)
21. Bellman, R.: Dynamic programming. Princeton University Press, NJ (1957)
22. Benders, J.F.: Partitioning procedures for solving mixed-variables programming problems. Numerische Mathematik 4, 238–252 (1962)
23. Bertsekas, D.P.: Network optimization: Continuous and discrete models. Athena Scientific, MA (1998)
24. Boese, K.D.: Models for iterative global optimization. PhD thesis. University of California, Los Angeles (1996)
25. Boese, K.D., Kahng, A.B., Muddu, S.: New adaptive multi-start techniques for combinatorial global optimizations. Operations Research Letters 16(2), 101–113 (1994)
26. Braun, H.: On Solving Traveling Salesman Problems by Genetic Algorithms. In: Schwefel, H.-P., Männer, R. (eds.) PPSN 1990. LNCS, vol. 496, pp. 129–133. Springer, Heidelberg (1991)
27. Burke, E.K., Cowling, P.I., Keuthen, R.: Effective Local and Guided Variable Neighbourhood Search Methods for the Asymmetric Travelling Salesman Problem. In: Boers, E.J.W., Gottlieb, J., Lanzi, P.L., Smith, R.E., Cagnoni, S., Hart, E., Raidl, G.R., Tijink, H. (eds.) EvoIASP 2001, EvoWorkshops 2001, EvoFlight 2001, EvoSTIM 2001, EvoCOP 2001, and EvoLearn 2001. LNCS, vol. 2037, pp. 203–312. Springer, Heidelberg (2001)
28. Burke, E.K., Kendall, G., Newall, J., Hart, E., Ross, P., Schulemburg, S.: Hyperheuristics: An emerging direction in modern search technology. In: Handbook of Metaheuristics. Kluwer Academic Publishers (2003)
29. Burke, E.K., Landa Silva, J.D., Soubeiga, E.: Hyperheuristic approaches for multiobjective optimisation. In: 5th Metaheuristics International Conference (MIC 2003), Kyoto, Japan (August 2003)
30. Caseau, Y., Laburthe, F.: Disjunctive scheduling with task intervals. Technical Report LIENS-95-25, Ecole Normale Supérieure de Paris, France (1995)
31. Caseau, Y., Laburthe, F.: Heuristics for large constrained routing problems. Journal of Heuristics 5, 281–303 (1999)
32. Cesta, A., Cortellessa, G., Oddi, A., Policella, N., Susi, A.: A Constraint-Based Architecture for Flexible Support to Activity Scheduling. In: Esposito, F. (ed.) AI*IA 2001. LNCS (LNAI), vol. 2175, pp. 369–390. Springer, Heidelberg (2001)
33. Chabrier, A., Danna, E., Le Pape, C.: Coopération entre génération de colonnes sans cycle et recherche locale appliquée au routage de véhicules. In: Huitièmes Journées Nationales sur la Résolution de Problèmes NP-Complets, JNPC 2002, Nice, France (May 2002)
34. Chang, C.S., Huang, J.S.: Optimal multiobjective SVC planning for voltage stability enhancement. IEE Proceedings on Generation, Transmission and Distribution 145(2), 203–209 (1998)
35. Chelouah, R., Siarry, P.: A hybrid method combining continuous tabu search and Nelder-Mead simplex algorithms for the global optimization of multiminima functions. European Journal of Operational Research 161(3), 636–654 (2004)
36. Chen, H., Flann, N.S.: Parallel simulated annealing and genetic algorithms: A space of hybrid methods. In: Davidor, Y., Schwefel, H.-P., Manner, R. (eds.) Third Conf. on Parallel Problem Solving from Nature, PPSN 1994, Jerusalem, Israel, pp. 428–436. Springer (October 1994)
37. Chu, P.C.: A genetic algorithm approach for combinatorial optimization problems. PhD thesis. University of London, London, UK (1997)
38. Chvatal, V.: A greedy heuristic for the set covering problem. Mathematics of Operations Research 4(3), 233–235 (1979)

39. Clearwater, S.H., Hogg, T., Huberman, B.A.: Cooperative problem solving. In: Huberman, B.A. (ed.) Computation: The Micro and the Macro View, pp. 33–70. World Scientific (1992)
40. Clearwater, S.H., Huberman, B.A., Hogg, T.: Cooperative solution of constraint satisfaction problems. Science 254, 1181–1183 (1991)
41. Cohoon, J., Hedge, S., Martin, W., Richards, D.: Punctuated equilibria: A parallel genetic algorithm. In: Grefenstette, J.J. (ed.) Second Int. Conf. on Genetic Algorithms, pp. 148–154. MIT, Cambridge (1987)
42. Cohoon, J.P., Martin, W.N., Richards, D.S.: Genetic Algorithms and Punctuated Equilibria. In: Schwefel, H.-P., Männer, R. (eds.) PPSN 1990. LNCS, vol. 496, pp. 134–141. Springer, Heidelberg (1991)
43. Cohoon, J.P., Martin, W.N., Richards, D.S.: A multi-population genetic algorithm for solving the k-partition problem on hypercubes. In: Belew, R.K., Booker, L.B. (eds.) Fourth Int. Conf. on Genetic Algorithms, pp. 244–248. Morgan Kaufmann, San Mateo (1991)
44. Cook, W., Seymour, P.: Tour merging via branch-decomposition. INFORMS Journal on Computing 15(3), 233–248 (2003)
45. Coyne, J., Paton, R.: Genetic Algorithms and Directed Adaptation. In: Fogarty, T.C. (ed.) AISB-WS 1994. LNCS, vol. 865, pp. 103–114. Springer, Heidelberg (1994)
46. Crainic, T.G., Nguyen, A.T., Gendreau, M.: Cooperative multi-thread parallel tabu search with evolutionary adaptive memory. In: 2nd Int. Conf. on Metaheuristics, Sophia Antipolis, France (July 1997)
47. Crainic, T.G., Toulouse, M., Gendreau, M.: Synchronous tabu search parallelization strategies for multi-commodity location-allocation with balancing requirements. OR Spektrum 17, 113–123 (1995)
48. Crainic, T.G., Toulouse, M.: Parallel strategies for metaheuristics. In: Glover, F.W., Kochenberger, G.A. (eds.) Handbook of Metaheuristics, pp. 475–513. Springer (2003)
49. Cung, V.-D., Mautor, T., Michelon, P., Tavares, A.: A scatter search based approach for the quadratic assignment problem. In: IEEE Int. Conf. on Evolutionary Computation, ICEC 1997, Indianapolis, USA (April 1997)
50. Cung, V.-D., Mautor, T., Michelon, P., Tavares, A.: Recherche dispersée parallèle. In: Deuxième Congrés de la Société Francaise de Recherche Opérationnelle et d'Aide à la Décision, ROADEF 1999, Autrans, France (January 1999)
51. Dalboni, F.L., Ochi, L.S., Drummond, L.M.D.: On improving evolutionary algorithms by using data mining for the oil collector vehicle routing problem. In: Int. Network Optimization Conf., INOC 2003, Paris, France (October 2003)
52. Davis, L.: Job-shop scheduling with genetic algorithms. In: Grefenstette, J.J. (ed.) Int. Conf. on Genetic Algorithms and their Applications, Pittsburgh, pp. 136–140 (1985)
53. Deb, K., Goel, T.: A hybrid Multi-Objective Evolutionary Approach to Engineering Shape Design. In: Zitzler, E., Deb, K., Thiele, L., Coello Coello, C.A., Corne, D. (eds.) EMO 2001. LNCS, vol. 1993, pp. 385–399. Springer, Heidelberg (2001)
54. Delisle, P., Krajecki, M., Gravel, M., Gagné, C.: Parallel implementation of an ant colony optimization metaheuristic with OpenMP. In: 3rd European Workshop on OpenMP (EWOMP 2001), pp. 8–12 (2001)
55. Dimitrescu, I., Stutzle, T.: Combinations of local search and exact algorithms. In: Evo Workshops, pp. 211–223 (2003)
56. Dowsland, K.A.: Nurse scheduling with tabu search and strategic oscillation. European Journal of Operational Research 106, 393–407 (1998)
57. Dowsland, K.A., Herbert, E.A., Kendall, G.: Using tree search bounds to enhance a genetic algorithm approach to two rectangle packing problems. European Journal of Operational Research 168(2), 390–402 (2006)
58. Dowsland, K.A., Thomson, J.M.: Solving a nurse scheduling problem with knapsacks, networks and tabu search. Journal of Operational Research Society 51, 825–833 (2000)

59. Eby, D., Averill, R., Punch, W., Goodman, E.: Evaluation of injection island model GA performance on flywheel design optimization. In: Int. Conf on Adaptive Computing in Design and Manufacturing, Devon, UK, pp. 121–136. Springer (1998)
60. Engelmore, R.S., Morgan, A.: Blackboard systems. Addison-Wesley (1988)
61. De Falco, I., Del Balio, R., Tarantino, E.: An analysis of parallel heuristics for task allocation in multicomputers. Computing 59(3), 259–275 (1997)
62. De Falco, I., Del Balio, R., Tarantino, E., Vaccaro, R.: Improving search by incorporating evolution principles in parallel tabu search. In: IEEE Conference on Evolutionary Computation, pp. 823–828 (1994)
63. Federgruen, A., Tzur, M.: Time-partitioning heuristics: Application to one warehouse, multi-item, multi-retailer lot-sizing problems. Naval Research Logistics 46, 463–486 (1999)
64. Feo, T.A., Resende, M.G.C.: Greedy randomized adaptive search procedures. Journal of Global Optimization 6, 109–133 (1995)
65. Feo, T.A., Resende, M.G.C., Smith, S.H.: A greedy randomized adaptive search procedure for maximum independent set. Operations Research 42, 860–878 (1994)
66. Feo, T.A., Venkatraman, K., Bard, J.F.: A GRASP for a difficult single machine scheduling problem. Computers and Operations Research 18, 635–643 (1991)
67. Filho, G.R., Lorena, L.A.N.: Constructive genetic algorithm and column generation: An application to graph coloring. In: APORS 2000 Conf. of the Association of the Asian-Pacific Operations Research Societies within IFORS (2000)
68. Fischetti, M., Lodi, A.: Local branching. Mathematical Programming B 98(23-47) (2003)
69. Fisher, M.L.: An application oriented guide to lagrangian relaxation. Interfaces 15, 399–404 (1985)
70. Fleurent, C., Ferland, J.A.: Genetic hybrids for the quadratic assignment problem. DIMACS Series in Discrete Mathematics and Theoretical Computer Science 16, 173–188 (1994)
71. Fleurent, C., Ferland, J.A.: Genetic and hybrid algorithms for graph coloring. Annals of Operations Research 63(3), 437–461 (1996)
72. Focacci, F., Laburthe, F., Lodi, A.: Local search and constraint programming. In: Handbook of Metaheuristics. International Series in Operations Research and Management Science. Kluwer Academic Publishers, Norwell (2002)
73. Fonlupt, C., Robillard, D., Preux, P., Talbi, E.-G.: Fitness landscape and performance of metaheuristics. In: Meta-Heuristics - Advances and Trends in Local Search Paradigms for Optimization, pp. 255–266. Kluwer Academic Press (1999)
74. Gao, B., Liu, T.-Y., Feng, G., Qin, T., Cheng, Q.-S., Ma, W.-Y.: Hierarchical taxonomy preparation for text categorization using consistent bipartite spectral graph copartitioning. IEEE Transactions on Knowledge and Data Engineering 17(9), 1263–1273 (2005)
75. Gen, M., Lin, L.: Multiobjective hybrid genetic algorithm for bicriteria network design problem. In: The 8th Asia Pacific Symposium on Intelligent and Evolutionary Systems, Cairns, Australia, pp. 73–82 (December 2004)
76. Gendreau, M., Laporte, G., Semet, F.: The covering tour problem. Operations Research 45, 568–576 (1997)
77. Gilmore, P.C., Gomory, R.E.: A linear programming approach to the cutting stock problem. Operations Research 9, 849–859 (1961)
78. Ginsberg, M.L.: Dynamic backtracking. Journal of Artificial Intelligence Research 1, 25–46 (1993)
79. Golden, B., Pepper, J., Vossen, T.: Using genetic algorithms for setting parameter values in heuristic search. Intelligent Engineering Systems Through Artificial Neural Networks 1, 9–32 (1998)
80. Golovkin, I.E., Louis, S.J., Mancini, R.C.: Parallel implementation of niched Pareto genetic algorithm code for x-ray plasma spectroscopy. In: Congress on Evolutionary Computation, CEC 2002, pp. 1820–1824 (2002)

81. Gomory, R.E.: Outline of an algorithm for integer solutions to linear programs. Bulletin AMS 64, 275–278 (1958)
82. Grefenstette, J.J.: Incorporating problem specific knowledge into genetic algorithms. In: Davis, L. (ed.) Genetic Algorithms and Simulated Annealing, Research Notes in Artificial Intelligence, pp. 42–60. Morgan Kaufmann, San Mateo (1987)
83. Gutin, G.M.: Exponential neighborhood local search for the traveling salesman problem. Computers and Operations Research 26(4), 313–320 (1999)
84. Habet, D., Li, C.-M., Devendeville, L., Vasquez, M.: A Hybrid Approach for SAT. In: Van Hentenryck, P. (ed.) CP 2002. LNCS, vol. 2470, pp. 172–184. Springer, Heidelberg (2002)
85. Hansen, P., Mladenovic, M., Perez-Britos, D.: Variable neighborhood decomposition search. Journal of Heuristics 7(4), 330–350 (2001)
86. Hart, W.E.: Adaptive global optimization with local search. PhD thesis. University of California, San Diego (1994)
87. Harvey, W.D., Ginsberg, M.L.: Limited discrepancy search. In: IJCAI Int. Joint Conference on Artificial Intelligence, pp. 607–613. Morgan Kaufmann (1997)
88. Hindi, K.S., Fleszar, K., Charalambous, C.: An effective heuristic for the CLSP with setup times. Journal of the Operations Research Society 54, 490–498 (2003)
89. Hinterding, R., Michalewicz, Z., Eiben, A.-E.: Adaptation in evolutionary computation: A survey. In: Proceedings of the IEEE Conference on Evolutionary Computation, Indianapolis, USA, pp. 65–69 (April 1997)
90. Hogg, T., Williams, C.: Solving the really hard problems with cooperative search. In: 11th Conf. on Artificial Intelligemce AAAI 1993, pp. 231–236. AAAI Press (1993)
91. Hong, T.-P., Wang, H.-S., Chen, W.-C.: Simultaneous applying multiple mutation operators in genetic algorithm. Journal of Heuristics 6(4), 439–455 (2000)
92. Huberman, B.A.: The performance of cooperative processes. Physica D 42, 38–47 (1990)
93. Husbands, P., Mill, F., Warrington, S.: Genetic Algorithms, Production Plan Optimisation and Scheduling. In: Schwefel, H.-P., Männer, R. (eds.) PPSN 1990. LNCS, vol. 496, pp. 80–84. Springer, Heidelberg (1991)
94. Ishibuchi, H., Murata, T.: A multi-objective genetic local search algorithm and its application to flowshop scheduling. IEEE Transactions on Systems, Man, and Cybernetics - Part C: Applications and Reviews 28(3), 392–403 (1998)
95. Jahuira, C.A.R., Cuadros-Vargas, E.: Solving the TSP by mixing GAs with minimal spanning trees. In: First Int. Conf. of the Peruvian Computer Society, Lima, Peru, pp. 123–132 (2003)
96. Jaszkiewicz, A.: Genetic local search for multiple objective combinatorial optimization. Technical Report RA-014/98. Institute of Computing Science, Poznan University of Technology (1998)
97. Jaszkiewicz, J.: Path relinking for multiple objective combinatorial optimization: TSP case study. In: The 16th Mini-EURO Conference and 10th Meeting of EWGT (Euro Working Group Transportation) (2005)
98. Jin, Y.: A comprehensive survey of fitness approximation in evolutionary computation. Soft Computing 9(1), 3–12 (2005)
99. Jin, Y., Sendhoff, B.: Reducing Fitness Evaluations Using Clustering Techniques and Neural Network Ensembles. In: Deb, K., et al. (eds.) GECCO 2004. LNCS, vol. 3102, pp. 688–699. Springer, Heidelberg (2004)
100. Jog, P., Suh, J.Y., Van Gucht, D.: The effects of population size, heuristic crossover and local improvement on a genetic algorithm for the traveling salesman problem. In: 3rd Int. Conf. Genetic Algorithms. Morgan Kaufmann, USA (1989)
101. Jones, B.R., Crossley, W.A., Lyrintzis, A.S.: Aerodynamic and aeroacoustic optimization of airfoils via parallel genetic algorithm. Journal of Aircraft 37(6), 1088–1098 (2000)
102. Jourdan, L., Basseur, M., Talbi, E.-G.: Hybridizing exact methods and metaheuristics: A taxonomy. European Journal of Operational Research (2008) (to appear)

103. Jourdan, L., Corne, D.W., Savic, D.A., Walters, G.A.: Preliminary Investigation of the 'Learnable Evolution Model' for Faster/Better Multiobjective Water Systems Design. In: Coello Coello, C.A., Hernández Aguirre, A., Zitzler, E. (eds.) EMO 2005. LNCS, vol. 3410, pp. 841–855. Springer, Heidelberg (2005)
104. Jourdan, L., Dhaenens, C., Talbi, E.-G.: Using Datamining Techniques to Help Metaheuristics: A Short Survey. In: Almeida, F., Blesa Aguilera, M.J., Blum, C., Moreno Vega, J.M., Pérez Pérez, M., Roli, A., Sampels, M. (eds.) HM 2006. LNCS, vol. 4030, pp. 57–69. Springer, Heidelberg (2006)
105. Jozefowiez, N.: Modélisation et résolution approchée de problèmes de tournées multi-objectif. PhD thesis. University of Lille, Lille, France (2004)
106. Jozefowiez, N., Semet, F., Talbi, E.-G.: Parallel and Hybrid Models for Multi-Objective Optimization: Application to the Vehicle Routing Problem. In: Guervós, J.J.M., Adamidis, P.A., Beyer, H.-G., Fernández-Villacañas, J.-L., Schwefel, H.-P. (eds.) PPSN 2002. LNCS, vol. 2439, pp. 271–280. Springer, Heidelberg (2002)
107. Jozefowiez, N., Semet, F., Talbi, E.-G.: The bi-objective covering tour problem. Computers and Operations Research 34, 1929–1943 (2007)
108. Juenger, M., Reinelt, G., Thienel, S.: Practical problem solving with cutting plane algorithms in combinatorial optimization. DIMACS Series in Discrete Mathematics and Theoretical Computer Science 20, 111–152 (1995)
109. Kamarainen, O., El Sakkout, H.: Local Probing Applied to Scheduling. In: Van Hentenryck, P. (ed.) CP 2002. LNCS, vol. 2470, pp. 155–171. Springer, Heidelberg (2002)
110. Karp, R.M.: Probabilistic analysis of partitioning algorithms for the traveling salesman problem in the plane. Mathematics of Operations Research 2, 209–224 (1977)
111. Kim, H., Hayashi, Y., Nara, K.: The performance of hybridized algorithm of genetic algorithm simulated annealing and tabu search for thermal unit maintenance scheduling. In: 2nd IEEE Conf. on Evolutionary Computation, ICEC 1995, Perth, Australia, pp. 114–119 (December 1995)
112. Kim, H.-S., Cho, S.-B.: An efficient genetic algorithm with less fitness evaluation by clustering. In: Congress on Evolutionary Computation, CEC 2001, pp. 887–894. IEEE Press (2001)
113. Kostikas, K., Fragakis, C.: Genetic Programming Applied to Mixed Integer Programming. In: Keijzer, M., O'Reilly, U.-M., Lucas, S., Costa, E., Soule, T. (eds.) EuroGP 2004. LNCS, vol. 3003, pp. 113–124. Springer, Heidelberg (2004)
114. Koza, J., Andre, D.: Parallel genetic programming on a network of transputers. Technical Report CS-TR-95-1542. Stanford University (1995)
115. Krueger, M.: Méthodes d'analyse d'algorithmes d'optimisation stochastiques à l'aide d'algorithmes génétiques. PhD thesis, Ecole Nationale Supèrieure des Télécommunications, Paris, France (December 1993)
116. Lemesre, J., Dhaenens, C., Talbi, E.-G.: An exact parallel method for a bi-objective permutation flowshop problem. European Journal of Operational Research (EJOR) 177(3), 1641–1655 (2007)
117. Levine, D.: A parallel genetic algorithm for the set partitioning problem. PhD thesis. Argonne National Laboratory, Illinois Institute of Technology, Argonne, USA (May 1994)
118. Lin, F.T., Kao, C.Y., Hsu, C.C.: Incorporating genetic algorithms into simulated annealing. In: Proc. of the Fourth Int. Symp. on AI, pp. 290–297 (1991)
119. Louis, S.J.: Genetic learning from experiences. In: Congress on Evolutionary Computations, CEC 2003, Australia, pp. 2118–2125 (2003)
120. Lourenco, H.R.: Job-shop scheduling: Computational study of local search and large-step optimization methods. European Journal of Operational Research 83, 347–367 (1995)
121. Mahfoud, S.W., Goldberg, D.E.: Parallel recombinative simulated annealing: A genetic algorithm. Parallel Computing 21, 1–28 (1995)

122. Maniezzo, V.: Exact and approximate nondeterministic tree-search procedures for the quadratic assignment problem. INFORMS Journal on Computing 11(4), 358–369 (1999)
123. Mariano, C.E., Morales, E.: A multiple objective ant-q algorithm for the design of water distribution irrigation networks. In: First Int. Workshop on Ant Colony Optimization, ANTS 1998, Brussels, Belgium (1998)
124. Martin, O.C., Otto, S.W., Felten, E.W.: Large-step markov chains for the TSP: Incorporating local search heuristics. Operation Research Letters 11, 219–224 (1992)
125. Mautor, T., Michelon, P.: Mimausa: A new hybrid method combining exact solution and local search. In: Second Int. Conf. on Metaheuristics, Sophia-Antipolis, France (1997)
126. Meunier, H., Talbi, E.-G., Reininger, P.: A multiobjective genetic algorithm for radio network optimization. In: Proceedings of the 2000 Congress on Evolutionary Computation (CEC 2000), La Jolla, CA, USA, pp. 317–324. IEEE Press (2000)
127. Michalski, R.S.: Learnable evolution model: Evolutionary processes guided by machine learning. Machine Learning 38(1), 9–40 (2000)
128. Minsky, M.: Negative expertise. International Journal of Expert Systems 7(1), 13–19 (1994)
129. Nagar, A., Heragu, S.S., Haddock, J.: A metaheuristic algorithm for a bi-criteria scheduling problem. Annals of Operations Research 63, 397–414 (1995)
130. Narayek, A., Smith, S., Ohler, C.: Integrating local search advice into a refinment search solver (or not). In: CP 2003 Workshop on Cooperative Constraint Problem Solvers, pp. 29–43 (2003)
131. Nemhauser, G., Wolsey, L.: Integer and combinatorial optimization. Wiley (1999)
132. Nissen, V.: Solving the quadratic assignment problem with clues from nature. IEEE Transactions on Neural Networks 5(1), 66–72 (1994)
133. Nuijten, W., Le Pape, C.: Constraint based job scheduling with ILOG scheduler. Journal of Heuristics 3, 271–286 (1998)
134. Nwana, V., Darby-Dowman, K., Mitra, G.: A cooperative parallel heuristic for mixed zero-one linear programming. European Journal of Operational Research 164, 12–23 (2005)
135. O'Reilly, U.-M., Oppacher, F.: Hybridized crossover-based techniques for program discovery. In: IEEE Int. Conf. on Evolutionary Computation, ICEC 1995, Perth, Australia, pp. 573–578 (December 1995)
136. Patterson, R., Rolland, E., Pirkul, H.: A memory adaptive reasoning technique for solving the capacitated minimum spanning tree problem. Journal of Heuristics 5, 159–180 (1999)
137. Pesant, G., Gendreau, M.: A view of local search in constraint programming. Journal of Heuristics 5, 255–279 (1999)
138. Potts, C.N., Velde, S.L.: Dynasearch-iterative local improvement by dynamic programming. Technical Report TR. University of Twente, Netherlands (1995)
139. Prestwich, S.: Combining the scalability of local search with the pruning techniques of systematic search. Annals of Operations Research 115, 51–72 (2002)
140. Puchinger, J., Raidl, G.R.: Combining Metaheuristics and Exact Algorithms in Combinatorial Optimization: A Survey and Classification. In: Mira, J., Álvarez, J.R. (eds.) IWINAC 2005. LNCS, vol. 3562, pp. 41–53. Springer, Heidelberg (2005)
141. Ramsey, C.L., Grefenstette, J.J.: Case-based initialization of genetic algorithms. In: Fifth Int. Conf. on Genetic Algorithms, pp. 84–91 (1993)
142. Rasheed, K., Vattam, S., Ni, X.: Comparison of methods for developing dynamic reduced models for design optimization. In: CEC 2002 Congress on Evolutionary Computation, pp. 390–395 (2002)
143. Renders, J.-M., Bersini, H.: Hybridizing genetic algorithms with hill-climbing methods for global optimization: two possible ways. In: First IEEE International Conference on Evolutionary Computation, pp. 312–317 (1994)

144. Reynolds, R.G., Michalewicz, Z., Peng, B.: Cultural algorithms: Computational modeling of how cultures learn to solve problems-an engineering example. Cybernetics and Systems 36(8), 753–771 (2005)
145. Ribeiro, M., Plastino, A., Martins, S.: Hybridization of GRASP metaheuristic with data mining techniques. Journal of Mathematical Modelling and Algorithms 5(1), 23–41 (2006)
146. Rosing, K.E., ReVelle, C.S.: Heuristic concentration: Two stage solution construction. European Journal of Operational Research 97(1), 75–86 (1997)
147. Rowe, J., Vinsen, K., Marvin, N.: Parallel GAs for multiobjective functions. In: Proc. of the 2nd Nordic Workshop on Genetic Algorithms and Their Applications (2NWGA), pp. 61–70 (1996)
148. Rumelhart, D.E., Hinton, G.E., Williams, R.J.: Learning representations by backpropagating errors. Nature 323, 533–536 (1986)
149. Salami, M., Cain, G.: Genetic algorithm processor on reprogrammable architectures. In: Fifth Annual Conference on Evolutionary Programming, EP 1996. MIT Press, San Diego (1996)
150. Sebag, M., Schoenauer, M., Ravise, C.: Toward civilized evolution: Developing inhibitions. In: Bäck, T. (ed.) Seventh Int. Conf. on Genetic Algorithms, pp. 291–298 (1997)
151. Sefraoui, M., Periaux, J.: A Hierarchical Genetic Algorithm Using Multiple Models for Optimization. In: Deb, K., Rudolph, G., Lutton, E., Merelo, J.J., Schoenauer, M., Schwefel, H.-P., Yao, X. (eds.) PPSN 2000. LNCS, vol. 1917, pp. 879–888. Springer, Heidelberg (2000)
152. Sellmann, M., Ansótegui, C.: Disco - novo - gogo: Integrating local search and complete search with restarts. In: The Twenty-First National Conference on Artificial Intelligence and the Eighteenth Innovative Applications of Artificial Intelligence Conference, Boston, USA (2006)
153. Shahookar, K., Mazumder, P.: A genetic approach to standard cell placement using meta-genetic parameter optimization. IEEE Trans. on Computer-Aided Design 9(5), 500–511 (1990)
154. Shaw, P.: Using Constraint Programming and Local Search Methods to Solve Vehicle Routing Problems. In: Maher, M.J., Puget, J.-F. (eds.) CP 1998. LNCS, vol. 1520, pp. 417–431. Springer, Heidelberg (1998)
155. Sprave, J.: A unified model of non-panmictic population structures in evolutionary algorithms. In: Proc. of the 1999 Congress on Evolutionary Computation, Piscataway, NJ, vol. 2, pp. 1384–1391. IEEE Press (1999)
156. Stutzle, T., Hoos, H.H.: The MAX-MIN ant system and local search for combinatorial optimization problems: Towards adaptive tools for global optimization. In: 2nd Int. Conf. on Metaheuristics, Sophia Antipolis, France, pp. 191–193. INRIA (July 1997)
157. Suh, J.Y., Van Gucht, D.: Incorporating heuristic information into genetic search. In: 2nd Int. Conf. Genetic Algorithms, pp. 100–107. Lawrence Erlbaum Associates, USA (1987)
158. Taillard, E.: Parallel iterative search methods for vehicle routing problem. Networks 23, 661–673 (1993)
159. Taillard, E.: Heuristic methods for large centroid clustering problems. Journal of Heuristics 9(1), 51–74 (2003)
160. Taillard, E., Voss, S.: POPMUSIC: Partial optimization metaheuristic under special intensification conditions. In: Essays and Surveys in Metaheuristics, pp. 613–629. Kluwer Academic Publishers (2002)
161. Taillard, E.D., Gambardella, L.: Adaptive memories for the quadratic assignment problem. Technical Report 87-97. IDSIA, Lugano, Switzerland (1997)
162. Taillard, E.D., Gambardella, L.M., Gendreau, M., Potvin, J.-Y.: Adaptive memory programming: a unified view of metaheuristics. European Journal of Operational Research 135(1), 1–16 (2001)
163. Talbi, E.-G.: A taxonomy of hybrid metaheuristics. Journal of Heuristics 8, 541–564 (2002)

164. Talbi, E.-G., Bachelet, V.: COSEARCH: A parallel cooperative metaheuristic. Journal of Mathematical Modelling and Algorithms (JMMA) 5(2), 5–22 (2006)
165. Talbi, E.-G., Fonlupt, C., Preux, P., Robillard, D.: Paysages de problèmes d'optimisation et performances des méta-heuristiques. In: Premier Congrés de la Société Francaise de Recherche Opérationnelle et Aide à la Décision ROAD, Paris, France (January 1998)
166. Talbi, E.G., Muntean, T., Samarandache, I.: Hybridation des algorithmes génétiques avec la recherche tabou. In: Evolution Artificielle, EA 1994, Toulouse, France (September 1994)
167. Talbi, E.-G., Rahoual, M., Mabed, M.H., Dhaenens, C.: A Hybrid Evolutionary Approach for Multicriteria Optimization Problems: Application to the Flow Shop. In: Zitzler, E., Deb, K., Thiele, L., Coello Coello, C.A., Corne, D.W. (eds.) EMO 2001. LNCS, vol. 1993, pp. 416–428. Springer, Heidelberg (2001)
168. Talukdar, S., Baerentzen, L., Gove, A., De Souza, P.: Asynchronous teams: cooperation schemes for autonomous agents. Journal of Heuristics 4(4), 295–321 (1998)
169. Tamura, H., Hirahara, A., Hatono, I., Umano, M.: An approximate solution method for combinatorial optimization-hybrid approach of genetic algorithm and lagrangean relaxation method. Trans. Soc. Instrum. Control Engineering 130, 329–336 (1994)
170. Tanese, R.: Parallel genetic algorithms for a hypercube. In: Proc. of the Second Int. Conf. on Genetic Algorithms, pp. 177–183. MIT, Cambridge (1987)
171. Thiel, J., Voss, S.: Some experiences on solving multiconstraint zero-one knapsack problems with genetic algorithms. INFOR 32(4), 226–242 (1994)
172. Toulouse, M., Crainic, T., Gendreau, M.: Communication issues in designing cooperative multi-thread parallel searches. In: Osman, I.H., Kelly, J.P. (eds.) Meta-Heuristics: Theory and Applications, pp. 501–522. Kluwer Academic Publishers (1996)
173. Tuson, A., Ross, P.: Adapting operator settings in genetic algorithms. Evolutionary Computation 6(2), 161–184 (1998)
174. Ulder, N.L.J., Aarts, E.H.L., Bandelt, H.-J., Van Laarhoven, P.J.M., Pesch, E.: Genetic Local Search Algorithms for the Traveling Salesman Problem. In: Schwefel, H.-P., Männer, R. (eds.) PPSN 1990. LNCS, vol. 496, pp. 109–116. Springer, Heidelberg (1991)
175. Vasquez, M., Hao, J.-K.: A hybrid approach for the 0-1 multidimensional knapsack problem. In: Proceedings of the International Joint Conference on Artificial Intelligence, IJCAI, pp. 328–333 (2001)
176. Verhoeven, M.G.A., Aarts, E.H.L.: Parallel local search. Journal of Heuristics 1(1), 43–65 (1995)
177. Visée, M., Teghem, J., Pirlot, M., Ulungu, E.L.: Two-phases method and branch and bound procedures to solve knapsack problem. Journal of Global Optimization 12, 139–155 (1998)
178. Voigt, H.-M., Born, J., Santibanez-Koref, I.: Modelling and Simulation of Distributed Evolutionary Search Processes for Function Optimization. In: Schwefel, H.-P., Männer, R. (eds.) PPSN 1990. LNCS, vol. 496, pp. 373–380. Springer, Heidelberg (1991)
179. Voss, S.: Tabu search: Applications and prospects. In: Network Optimization Problems, pp. 333–353. World Scientific, USA (1993)
180. Wang, L.-H., Kao, C.-Y., Ouh-young, M., Chen, W.-C.: Molecular binding: A case study of the population-based annealing genetic algorithms. In: IEEE Int. Conf. on Evolutionary Computation, ICEC 1995, Perth, Australia, pp. 50–55 (December 1995)
181. Wright, A.H.: Genetic algorithms for real parameter optimization. In: Foundation of Genetic Algorithms, pp. 205–218. Morgan Kaufmann (1991)
182. Yagiura, M., Ibaraki, T.: Metaheuristics as robust and simple optimization tools. In: IEEE Int. Conf. on Evolutionary Computation, ICEC 1996, pp. 541–546 (1996)
183. Zitzler, E., Thiele, L.: Multiobjective evolutionary algorithms: A comparative case study and the strength Pareto approach. IEEE Trans. on Evolutionary Computation 3(4), 257–271 (1999)

Chapter 2
Hybrid Metaheuristics for Dynamic and Stochastic Vehicle Routing

Ulrike Ritzinger and Jakob Puchinger

Abstract. Recent developments in telematics, such as the wide spread use of positioning services and mobile communication technologies, allow the exact monitoring of vehicles. These advances build the basis for automatic real-time fleet management systems. To be successful such systems have to rely on optimization algorithms for solving dynamic and stochastic vehicle routing problems based on ingredients such as historical data, stochastic modeling, machine learning, fast shortest-path calculation, fast construction heuristics, and exact and (meta)heuristic optimization methods. This book documents the growing interest in and success of hybrid metaheuristics. They are often used to solve complex and large real-world optimization problems, combining advantages from various fields of computer science and mathematical optimization. Within this chapter the application of such methods for the dynamic and stochastic vehicle routing problem is studied. After a general introduction in this field, the main commonalities of dynamic and stochastic vehicle routing problems are described and a short overview of classical algorithms for these problems is given. Then, in the third part hybrid metaheuristics for dynamic problems vehicle routing problems are be described. The third part focusses on stochastic problems. The fourth part examines the combination of dynamic and stochastic problems. The chapter is concluded with an outlook towards future developments in the field as well as promising open research areas.

2.1 Introduction

In todays globalized economy fast, reliable but also flexible supply chains are among the main factors for successful enterprises. In real-world applications it is most often the case that some information about the future is available (stochastic information) and that known information is revealed over time (dynamic information). For example, an estimation for the customer demand is given at the beginning, whereas

Ulrike Ritzinger · Jakob Puchinger
Mobility Department, Austrian Institute of Technology
e-mail: {ulrike.ritzinger_fl, jakob.puchinger}@ait.ac.at

E.-G. Talbi (Ed.): Hybrid Metaheuristics, SCI 434, pp. 77–95.
springerlink.com

the actual demand will be revealed at a later date. Additionally, recent developments in telematics, such as the wide spread use of positioning services and mobile communication technologies, allow the exact monitoring of vehicles [33]. These advances build the basis for automatic real-time fleet management systems and extensive and detailed data collection. To be successful such systems have to rely on optimization algorithms for solving dynamic and stochastic vehicle routing problems based on ingredients such as historical data, stochastic models, machine learning, fast shortest-path calculation, fast construction and insertion heuristics, and exact and (meta)heuristic optimization methods.

In dynamic real-world vehicle routing applications it is fundamental that short-term decisions are accurate and made quickly, while long-term decision need to ensure a certain quality standard. Moreover, stochastic models based on previously collected data can be used in order to provide some information about upcoming events. The combination of the dynamic and stochastic problem further extends the high complexity of their static and deterministic counterparts. Therefore, such problems are especially amenable to hybrid optimization approaches, combining the advantages of different techniques. The growing interest in hybrid metaheuristics and their success is well documented in [13]. They are often used to solve complex and large real-world optimization problems, combining advantages from various fields of computer science and mathematical optimization. In the recent vehicle routing literature there is an increasing number of successful application of hybrid metaheuristics [32].

Here, we examine the application of hybrid methods to dynamic and stochastic vehicle routing problems (VRP). In a first part, we give a short overview of vehicle routing variants and a classification depending on the nature of the available data. Additionally, it includes a short summary of various algorithms for solving dynamic and stochastic VRPs. In the second part, we examine hybrid metaheuristics for dynamic problem variants, where not all information is available in advance. In the third part of this chapter, we focus on problems that consider a priori stochastic information about possible future events and progresses. The fourth part of this chapter, examines dynamic problems where some stochastic information about future events is available. We conclude with an outlook towards future developments in the field as well as promising open research areas.

2.2 Dynamic and Stochastic Vehicle Routing Problems

The Vehicle Routing Problem (VRP) is a well-known and extensively studied combinatorial optimization problem [79], [35]. In the last years, the interest in solving real world applications of the VRP has grown tremendously as information technologies now allow the gathering of relevant information about available vehicles and scheduled requests in real time. This new problem class, where information is handled at the time it arrives, known as dynamic VRP (DVRP), has recently received increased attention from the research community [52], [27]. In the DVRP not all relevant information is known at the time of route construction and the information may change during the execution of the planned routes. Recent summaries about current developments in DVRP can be found in [75] and [66]. Additionally, in real world applications it is often the case, that some information about the future, for example about travel times or arising requests or demand, is available. This information can be incorporated into the route planning process by modeling stochastic VRPs [31], [81], [47].

2.2.1 Vehicle Routing Problem Variants and Available Information

The literature discerns various classical VRP variants that are further complemented by numerous real-world applications with additional requirements and constraints. A detailed description of numerous VRP variants can be found in [79] and [35].

In the classic VRP formulation, a set of vertices representing customers or cities, a set of arcs where each arc is associated with travel costs and a set of vehicles, stationed at a depot are considered. The aim is to construct vehicle routes where each vertex is visited exactly once, all vehicle routes start and end in the depot and the total travel costs are minimized. A more detailed definition of the classic VRP can be found in [49]. In the following, the most important VRP variants are presented.

The simplest and most studied variant is the Capacitated VRP (CVRP). It is known to be NP-hard and generalizes the well known Traveling Salesman Problem (TSP) [79]. In the CVRP, the aim is to serve all customer demands with a given fleet of vehicles located at a single depot where the capacity of the vehicles is restricted.

A related variant is the distance-constrained VRP where the route length is restricted by a maximum tour length and the VRP with Time Windows (VRPTW) is an extension of the CVRP in which the service at the customer has to start within a given time window.

Another variant is the VRP with Backhauls (VRPB) where the customer set is split into two subsets. The first subset contains of customers which require a product to be delivered (*linehaul*), whereas the second subset contains customers where a given quantity of a product has to be picked up (*backhaul*). Here, the precedence of the customers must be considered.

In the VRP with Pickup and Delivery (PDP) each customer is split into two different locations, where the goods have to be picked-up at one location and delivered to the other one. Another variant of the PDP, where people are transported instead of goods, is called the Dial-a-Ride Problem (DARP). In this problem class, additional constraints for user convenience are introduced [19].

2.2.1.1 Information Availability

In the literature, problem variants are discerned according to the availability and certainty of information [75]. A problem can be seen as *static* or *dynamic* depending on the availability of information before the start of the optimization process. Depending on the certainty of the available information problems can be considered as *deterministic* or *stochastic*. The combination of these characteristics yields four categories of routing problems.

The first category are *static and deterministic* problems. Here, all relevant information is completely known at the beginning of the route planning process and no changes take place during the execution of the routes. This leads to the classic static VRPs and solution methods ranging from exact methods to metaheuristics. These problems have been extensively studied in the literature [79], [20], [50], [63].

The next category encompasses *static and stochastic* problems, the relevant information is known a priori, but some parts of it are afflicted with a given uncertainty. Some information is given as random variables, and the aim is to generate solutions optimizing the expected value of the objective function. Commonly, stochastic programming methods are applied to such problems [69].

In *dynamic and deterministic* problems, the available information at the beginning of the planning process is incomplete and there is no information about future events. In the literature, these problems are often referred to as real-time or online optimization problems. Most commonly, some information is already known before the planning horizon starts, but other parts of the information are revealed or change during the execution phase. Solution methods for the dynamic VRP can range from reoptimization algorithms over fast insertion heuristics to queuing theory based algorithms, depending on the degree of dynamism [54], [55], [27].

In the forth category, *dynamic and stochastic* problems, relevant information is revealed throughout the planning horizon, but additionally stochastic information about the future, most commonly gathered from historical data, is available. To deal with stochastic information, solution methods are either based on sampling approaches where possible future scenarios are included [6], [7] or considering stochastic information explicitly [26], [76].

2.2.1.2 Stochastic Information

As described above, there are problem categories which have some information about future events available. This means, there exists an estimation about the occurrence of possible future events. This often happens in real world applications where stochastic information can be obtained from historical data. There are several types of stochastic information which can be incorporated in the optimization process of VRPs [31]. In the following section, a differentiation about the most common types of stochastic information is given.

Travel times are elementary data in VRPs, thus, it is important to provide authentic travel times for the considered network. *Stochastic Travel Times* are random variables and can be used to deal with uncertainties occurring in real world environment, like time dependent travel times, seasonal effects, car accidents, bad weather or working zones. In [49], the VRP with stochastic travel times is described in detail. Sometimes time dependent travel times defined in [27] and stochastic travel times are combined as proposed in [29].

Another extensively studied problem is the VRP with *Stochastic Demand* [10]. Here, the actual demand of customers is not known in advance, but it is known as a random variable which follows a known probability distribution. This problem arises in practical applications where unknown amount of goods have to be either delivered or collected. Commonly, this problem is solved by a two stage approach where first, all routes are constructed a priori,and later, when the demand becomes known, the returns to the depot for refilling are planned. The aim is to construct routes with minimum expected routing costs, comprising the costs of the route and the return trips to the depot.

The next problem class, VRP with *Stochastic Customers*, mainly arises in a dynamic environment where not all customer requests are known a priori, but reveal during the day of operation. In this case, stochastic information about the expected number of customer requests is considered and incorporated into the optimization algorithm, as for example shown in [7] or [42]. Another possibility of stochastic customer requests in the DARP is presented in [74] where the requests for the return trips of patients are stochastic.

The last category considers problems with multiple uncertainties where lots of information is assumed to be uncertain and modeled as stochastic variables. Thus, not only travel times or requests are stochastic, but also the location and time of requests, as well as, cancellation of requests, vehicle break downs or traffic jams. In [6] and [40] problems with rich stochastic information are described, and in [26] and [3] practical applications using stochastic information are presented.

2.2.2 Algorithms for Solving Dynamic Vehicle Routing Problems

In the dynamic VRP (DVRP), not all relevant information is known at the time the routes are planned, but it is revealed throughout the execution of the scheduled routes. An extensive overview on DVRP can be found in [75].

One common strategy for solving DVRPs is to apply a static algorithm to the already known data at the beginning for computing an initial solution, and whenever new information becomes known the current solution is updated. In order to solve the static problem, solution concepts based on exact procedures, like Linear Assignment [28] or Column Generation [18], can be applied. Another concept is rule based decision making where decision rules are defined and applied whenever dynamic events occur [70], [53]. Local Search (LS) approaches compute a feasible starting solution first, usually by repeated insertion operations, and then apply some improvement techniques. For example, *best insertion* is used to add a request to the current solution and after this *cross-exchange*, *or-opt* or *interroute exchange* moves are applied to improve the solution, as in [67], [17], [15].

The most widely applied methods for solving DVRPs are metaheuristics [61], [80]. They include Ant Colony Optimization Algorithms applied in [62], Evolutionary Algorithms used in [37] or Variable Neighborhood Search implemented in [14]. A very popular metaheuristic for the DVRP is Tabu Search as shown in [56] and [57]. In many cases, a parallel implementation is used to improve computational time of reoptimization, as presented in [4] and [30].

2.2.3 Algorithms for Solving Stochastic Vehicle Routing Problems

In stochastic optimization problems knowledge about the uncertainty of certain aspects is known a priori. This can be knowledge about future requests or travel times. In this section the focus is on different methods dealing with stochastic information.

Markov Decision Processes (MDP) are often used to model stochastic VRPs, for example in [23], a VRP with stochastic demand is formulated as a MDP. In [78], a MDP is used as well to solve a VRP with customer requests which arise with a certain probability. Other approaches use an approximate dynamic programming approach [68], [76] for solving VRP or fleet management problems.

In Stochastic Programming , mathematical programming and stochastic models are combined for solving optimization problems with uncertainties. The aim is to determine a feasible solution for all possible outcomes and the optimization of the expected value of the objective function [12]. A detailed introduction to stochastic programming in the context of transportation and logistics is given in [69].

Another method to deal with stochastic information is *Sampling*. This strategy considers already known and stochastic information and generates possible future scenarios by drawing them from a given probability distribution. A novel approach, using *Sampling* is described in [7], where a VRP with stochastic customer requests is solved. Further approaches using *Sampling* can be found in [6] and [26].

2.3 Hybrid Metaheuristics for Dynamic Problems

Dynamic problems have usually been solved using reoptimization or fast insertion techniques depending on the amount of time available for reacting to new events. One of the earliest works presenting a reoptimization based hybrid metaheuristic in the dynamic vehicle routing context is the algorithm by Jih and Hsu [44] combining a Dynamic Programming (DP) algorithm and a Genetic Algorithm (GA) for the single-vehicle PDP with time windows and capacity constraints. The dynamic programming component is executed for a certain amount of time. It will either return an optimal solution or multiple partially constructed routes. Those partial solutions are used as initial population of a genetic algorithm. The hybrid approach was able to improve the results of the non-hybrid methods.

More recent algorithms combine fast insertion heuristics with background optimization techniques, allowing an almost immediate response to dynamic events but utilizing possibly available time for improving solution quality. Another promising hybridization technique for dynamic problems are various parallelization variants in general. Depending on the specific problem characteristics such as the degree of dynamism [52], and response time requirements either of the variants makes sense.

2.3.1 Parallelization Approaches

Parallel optimization methods are often applied to complex large scale VRPs, a recent survey on parallel solution methods in the context of vehicle routing can be found in [21].

Several variants of a parallel Tabu Search (TS) heuristic for the dynamic multi-vehicle Dial-a-Ride Problem (DARP) are proposed in Attanasio et al. [4]. The authors motivate the use of a parallel approach with high running times of classical methods. The objective of the dynamic DARP to fulfill as many requests as possible with the available number of vehicles. Whenever a request can be added to the current solution without violating the problem constraints, it is accepted. Therefore a fast mechanism for checking the possible acceptance of a request is required. The parallel TS approach is applied to generate a starting solution based on already known requests and to perform a background optimization after a new request has been inserted. The fast insertion procedure is performed randomly inserting the new request in the current solution for every thread. If a feasible solution is found, the insertion is possible. If this is not the case the parallel TS with independent thread is run with parameters set to focus on feasibility. The presented computational experiments show that parallelization significantly increases the amount of served requests in real-world instances.

Khouadjia et al. [48] present a multi-swarm based optimization algorithm for the VRP with dynamic requests. The method consists of a particle swarm optimization approach with interacting swarms, thereby maintaining population diversity. New customers are inserted into existing routes using a method resembling the ejection chain approach. In addition to parallelization a low-level hybridization using *2-opt* as local improvement heuristic is implemented. The results of the novel approach are significantly better than the ones obtained by the current state of the art for the dynamic vehicle routing problem.

The main advantage of applying parallel optimization methods in a dynamic context is speed. Especially in the case when new information becomes available a fast reaction is of great importance and the ability to guarantee fast response times will decide over the applicability of the optimization approaches.

2.3.2 Other Hybridization Approaches

Most other hybridization approaches are based on the principle of combining fast insertion techniques with longer running, usually metaheuristic background optimization methods. In some of these approaches further hybridizations are developed in order to achieve higher quality results in comparison to more classical algorithms.

Alvarenga et al. [1] extend a hybrid Column Generation Genetic Algorithm approach for solving the static VRP [2] towards the dynamic case. The algorithm is based on a set partitioning formulation of the VRP. In such an integer programming formulation, every vehicle route is explicitly modeled and an optimal solution is a set of those routes minimizing the objective function. Explicitly representing VRP instances of realistic sizes as set packing problems would require enormous computational resources, therefore techniques such as Column Generation and Branch and Price are often used [65]. Alvarenga et al. are generating and iteratively refining a subset of the routes using a Genetic Algorithm (GA). The resulting restricted set partitioning problem is finally solved using an integer programming solver. This hybrid approach is extended to the dynamic case by applying a fast insertion heuristic to integrate new requests in all the individuals of the GA before restarting the integer programming solver. Computational experiments show the effectiveness of this approach and the advantages over reoptimization using the static version of the algorithm.

Fabri and Recht [24] solve a capacitated DARP where all customers are occurring dynamically using a combination of an A^*-algorithm and Tabu Search (TS). Their approach extends a hybrid heuristic proposed by Caramia et al. [16]. New requests are inserted by applying a fast procedure. First, single vehicle routes are created by representing the single vehicle DARP as Shortest Path Problem (SPP) and solving it using the A^*-algorithm. In a second step, the routes are then heuristically assigned to the vehicles. Several TS variants are then presented for optimizing the routes between two-occurring requests. The presented computational results show that the additional optimization significantly increases the solution quality.

Creput et al. [22] propose a novel approach combining a Self Organizing Map (SOM) with an Evolutionary Algorithm (EA) for solving the VRP with dynamic requests. Creput et al. describe the SOM as a center-based clustering algorithm preserving the density and the topology of the data distribution. The approach is based on applying SOM to the Travelling Salesman Problem (TSP). Cities of the TSP are mapped to the SOM network, local moves increasingly approach the vertices of the SOM network to the cities. By mapping the SOM vertices to the closest city a solution to the TSP is generated. This approach is extended to the VRP by embedding it into an solution pool based iterative improvement algorithm. New customers are added to the existing routes by simple insertion satisfying the relative route duration constraints. Extensive computational results show the advantages of the presented approach.

Berbeglia et al. [8] consider a dynamic dial-a-ride problem (DARP) in which some requests are static and the others arrive in real time. In this work, a hybrid algorithm is introduced which combines an exact constraint programming (CP) algorithm and a tabu search (TS) heuristic. A crucial point in dynamic DARP is to determine whether a new incoming request can be accepted and satisfied or not. Generally, the TS algorithm manages the insertion of new requests well when the problem is not too tightly constrained and CP is fairly effective in proving infeasibility in tight settings. Thus, the idea is to combine the advantages of these two methods. In their approach, the CP algorithm returns either a feasible solution for a given instance or proves that none exists. To tackle the dynamic aspects, the CP algorithm for the static DARP in [9] is extended by additional constraints to state that the solution must consider the partial routes followed up to now. The TS algorithm constructs a feasible starting solution, continually optimizes the current solution and also tries to insert new requests. Thus, when a new request arrives the CP and TS algorithm run in parallel for insertion. The TS algorithm uses three scheduling schemes determining the arrival times, begin of service times and departure times for each request vertex, which has a considerable impact on the algorithm performance. Concluding, it is shown that the hybrid algorithm clearly outperforms each of the two algorithms when executed separately.

2.4 Hybrid Metaheuristics with Stochastic Information

There are several examples of hybrid metaheuristics incorporating stochastic information. Similar to methods for other problem classes, many of the proposed methods combination different search algorithms. Another possibility is to combine approximated and exact stochastic models for computing the expected value of the objective function.

2.4.1 Hybridization of Search Techniques

Hvattum and Løkketangen [38] consider the stochastic Inventory Routing Problem (IRP). This problem is a combination of inventory management, vehicle routing, and stochastic demands. The problem is solved by applying the progressive hedging algorithm to a scenario tree representation of the problem. The problem is originally modeled using a Markov Decision Process (MDP). Based on the observation that most probably it is sufficient to consider a finite horizon, the authors propose to approximate the MDP by using a scenario tree based integer programming formulation (STP). The authors adapt the Progressive Hedging Algorithm (PHA) [73] to the STP. The PHA decomposes the scenario tree and solves the scenarios separately as subproblems and iteratively joins them using penalty terms in the adapting the respective objective functions. The subproblems are solved using a Greedy Randomized Adaptive Search Procedures (GRASP) based approach [41]. Although calibration of the PHA is reportedly difficult, results of the combined PHA and GRASP approach are more robust than the ones by any of the presented methods examined separately.

In Laporte et al. [51], a capacitated Arc-Routing Problem with stochastic demand (CARPSD) is considered. This problem arises for example in practical applications like garbage collection. Here, the customer locations are known in advance but the demand is random and not known until the location is reached. Thus, a capacity constraint violation can happen at some point in the planned solution. In this case, the vehicle has to interrupt its route and empty the vehicle by going to the dump site early and returns to the point of failure to restart its route. The objective is to minimize the costs of the planned route and the expected costs of the recourse action. The CARPSD is conventionally formulated as a stochastic program, where a first-stage solution is computed, realization of random variables are revealed and then a recourse action is applied. In contrast to that, an alternative approach is proposed where a first-stage solution is constructed which will take the expected cost of the recourse action into account. This is realized by an Adaptive Large Neighborhood Search (ALNS) heuristic. The garbage quantities are assumed to be independent random variables with known probability distribution and expected costs of recourse are defined for the discrete and continuous case. A construction heuristic, *stochastic path scanning*, is proposed. Several removal and insertion heuristics, destroying and repairing the current solution are also used. At each iteration one of these heuristics is selected based on the *roulette-wheel selection principle*. The acceptance criterion for a new solution and the stopping criterion are obtained by using an annealing-based search framework. The approach was tested on self generated instances, based on instances for CARP from Golden et al. [34]. The comparison between the deterministic and stochastic case shows clearly that improved results could be computed with the presented approach.

Another significant stochastic routing problem is the Probabilistic Traveling Salesman Problem (PTSP) [43], and since nature inspired intelligence became increasingly popular, Marinakis and Marinaki present a hybrid algorithm based on nature inspired approaches in [58]. Here, a hybrid scheme incorporating Particle Swarm Optimization (PSO) [46] and two further metaheuristics, Greedy Randomized Adaptive Search Procedure (GRASP) [72] and Expanding Neighborhood Search (ENS) [59] is introduced. The combination of PSO and GRASP is used to produce as good as possible initial populations, and the ENS strategy, speeds up the optimization process. The performed computational experiments demonstrate that the proposed approach leads to an effective handling of the PTSP, resulting in fast computational run-times and good results for very large problem instances.

Rei et al. [71] solve the single VRP with stochastic demands by combining Monte-Carlo sampling and local branching [25]. The authors consider an a priori optimization setting based on a two-stage stochastic programming model of the problem. In the first stage of the stochastic program a route visiting all customers once is constructed. In the second stage the route is executed with the actual demands and possibly necessary predetermined recourse actions. Starting from an optimal solution to the original first stage problem, the approach partitions the search space using the local branching principle in an iterative multi-descent search. The subproblems are solved to optimality or until a certain time limit is reached using the L-shaped method. The expected value of the recourse action is approximated using Monte-Carlo sampling. The computational results show the competitiveness of the presented approach, yielding comparable solution quality in significantly less run-time than an exact solution approach.

Mendoza et al. [60] solve the multi-compartment vehicle routing problem with stochastic demands (MC-VRPSD) using a Memetic Algorithm (MA) combining a genetic algorithm with local search including novel evaluation and repair procedures taking into account the stochastic nature of the problem. In a first step the problem is modeled as two-stage stochastic program, where the recourse actions consist of trips back to depot in order to reload the vehicle. The authors then propose an MA for solving the problem using an approximation of the expected cost as objective function. The initial solutions are created using a stochastic best insertion heuristic. A combination of relocate and *2-opt* moves are used as local improvement operators. Finally, the repair and evaluation of individuals is done by applying a stochastic extension of the split algorithm [5]. The presented algorithm is evaluated on random test instances as well as benchmark instances from the literature, showing an improved performance compared to the state of the art.

2.4.2 Objective Function Hybridization

Bianchi et al. [11] focus on the most commonly studied problem in this class, the VRP with stochastic demand (VRPSD). The idea is, to analyze the hybridization of different approximations of the objective function (minimizing the expected costs of the tour) with well known metaheuristics for this problem. The aim is to test the impact of interleaving the exact VRPSD objective function with the a priori tour length as an approximation. Therefore, five well known metaheuristics, Simulated Annealing, Iterated Local Search (ILS), Tabu Search (TS), Ant Colony Optimization (ACO) and Evolutionary Algorithms (EA) are presented. All considered metaheuristics use the common OrOpt Local search (LS) as proposed in [81], in order to obtain meaningful comparisons. The basic operator in the OrOpt LS considers a starting tour and moves sets of consecutive customers from one position in the tour to another one. For the computation of the moving costs two types of approximation schemes are described. In the *VRPSD approximation scheme* the costs are composed of the savings from extracting the customers from the tour and the costs of inserting them back, whereas the *TSP approximation scheme* only computes the length difference of the tour. The starting solution for all metaheuristics are generated by the *Farthest Insertion Construction Heuristic* [45]. The first hybridization shows the impact of using approximate move costs in local search, by running each proposed metaheuristic with the *VRPSD approximation scheme* and the *TSP approximation scheme*. The tests show that metaheuristics which use the local search as a black box (EA, ACO, ILS) perform better with the *TSP approximation*, while the other perform better with the *VRPSD approximation*. The second hybridization further explores the TSP objective function. Therefore, they expand the best algorithms determined in the first hybridization (ILS, EA) with the *3-opt LS for TSP* [77] and show significant improvements of the performance. The proposed approach is evaluated on self generated instances, where four factors are considered: the customer position, the capacity over demand ratio, variance of the stochastic demand and the number of customer, where the size of the instances is between 50 and 200 customers, with the customers uniformly distributed or grouped in clusters. As a conclusion it is shown, that the new hybrid approach clearly outperforms the state of the art.

2.5 Hybrid Metaheuristics for Dynamic and Stochastic Problems

In most dynamic real-world application, data is gathered and stored allowing to develop stochastic models for predicting future events. The hybrid approaches developed here can be discerned in two groups. On the one hand those working with single solutions and incorporating the stochastic knowledge directly in the optimization procedure. On the other hand approaches relying on solution pools, where multiple solutions are generated based on sampling, which are then further reconciled into a single solution. Such approaches can also be used to provide multiple solutions as suggestions to human dispatchers responsible for taking final decisions.

2.5.1 Single Solution Approaches

Hvattum et al. [39] consider a dynamic and stochastic VRP, based on a case from a large distribution company. Stochastic information about customer requests, like the location and its demand and the frequency of appearance, is gathered from historical data. It is shown how this problem can be formulated as a multistage stochastic programming problem with recourse. Dynamic events are captured by dividing the time horizon into a specified number of intervals and construct a plan for each interval using the currently known requests, in contrast to [7], who proposed an event-driven model. They developed a dynamic stochastic hedging heuristic which uses sample scenarios. In each time interval the solution from the previous interval plus the requests which became known during the past interval are considered and a plan for the current time interval is constructed. Sample scenarios are solved as static VRPs and the customers which are visited most frequently are determined iteratively. Then, the solution is built by assigning the request to the vehicles according to a ranking which states which customer is serviced first most often. The algorithm was tested against a myopic dynamic heuristic and was able to reduce travel distances significantly. In [40], they present a branch-and-regret heuristic which is based on the approach described above to additionally tackle the stochastic demand of customer, thus, the customer location is already known, but not its demand. Improvements with this new approach are shown as well as the capability to cope with different stochastic information.

Attanasio et al. [3] present a real-time fleet management system for a same-day courier service, describing forecast and optimization methodologies. A *forecast* and an *allocation* module are introduced, where the forecast module generates reliable near future predictions of travel times and demand and hands this information to the allocation module which is responsible for the assignment of customer requests to the couriers and the relocation of idle couriers. In order to make reliable predictions they divided the service area into geographical zones and time periods, which show typical traffic and demand patterns. Both, demand and travel time forecasting are based on a classical decomposition approach followed by an artificial neural networks which incorporates real time information. The demand forecasting specifies the number of requests from one region to another one at a given time period and the travel time forecasting provides expected travel times from one address to another address during a time period. The methodology for the travel time forecasting also takes traffic and real time information into account and is trained through artificial neural networks. When a new request passes the feasibility check the algorithm post optimization procedure tries to improve the current solution in the background. Therefore a parallel implementation of the Tabu Search (TS) algorithm is proposed. As a result of applying this system the efficiency of the couriers raised and less dispatchers for the fleet management are needed.

Schilde et al. [74] investigate whether using stochastic information about future requests can improve the solution quality of a dynamic stochastic Dial-a-Ride Problem (DSDARP). Here, a special type of customer request is considered as with a certain probability, a request for patient transportation to the hospital creates a corresponding transportation request to return the patient back home on the same day. In order to investigate the benefit of using stochastic information about return trips two approaches for the dynamic DARP are implemented and extended to deal with stochastic information. The Variable Neighborhood Search (VNS) approach introduced by [64] is adapted for the dynamic case (DVNS) and then extended to a dynamic S-VNS due to the S-VNS concept proposed in [36], and the Multiple Plan Approach (MPA) and Multiple Scenario Approach (MSA) described by [7] are applied. Computational results show that incorporating stochastic information about return trips yields better solution quality than the myopic methods, and that it is most beneficial to consider possible return transports in the very near future (up to 20 minutes).

2.5.2 *Algorithms Based on Solution Pools*

Bent et al. [7] consider a dynamic VRP with time windows (VRPTW) with stochastic customers. The aim is, to investigate how to exploit stochastic information about customers to accept and serve as many requests as possible, thus to miss fewer requests. This is achieved with a Multiple Plan Approach (MPA) dealing with dynamic customer request. Then, a related approach, called Multiple Scenario Approach (MSA) is introduced. It significantly outperforms the MPA by exploiting stochastic information about customers. The MPA is an event driven approach where a pool of plans is maintained. Because only one specific plan can be executed, just one plan is chosen from the pool and routing plans that correspond to the current situation are generated. The selection is done by a consensus function that selects the plan most similar to the other plans in the pool. To take advantage of the stochastic information, the MSA generates new routing plans for scenarios that include a priori known requests and possible future events, which are obtained by sampling their probability distributions. Experimental results show that using stochastic information yields significantly better results and that selecting plans by consensus function brings benefits for problems with many stochastic customers.

In a subsequent paper, Bent et al. [6] examine the MSA, described in [7], for a dynamic VRP with stochastic information about customer requests. In contrast to [7], the behavior of the MSA on a less constrained but more stochastic problem is studied, in order to capture long-distance mail services. The difference is, that now customer, customer locations and service times are random variables and that the focus is on the objective function, i.e., the aim is to minimize the total travel distance. Additionally, the MSA is improved by using stochastic information to delay the departure of vehicles in order to place new stochastic requests. To optimize the plans a large neighborhood search is used basically. In the case Large Neighborhood Search (LNS) does not find any improvements also a *nearest neighbor* heuristic is applied. Experimental results on a variety of models show that with this approach travel distance could be reduced scientifically.

2.6 Conclusion

Dynamic and stochastic VRPs are currently the most challenging class of problems in the vehicle routing area, especially if they are tackled in the real-world context where instances are often much larger than those usually considered in the literature. In most cases it is not possible to solve such problems in an exact way due to the limitation of computational resources and time. Furthermore, only partial, incomplete, and uncertain information is available, thereby making the quest for optimality impossible. The surveyed literature shows that in recent years hybrid metaheuristics have been increasingly successfully applied to such complex problems. Most often the diverse aspects of the problems can be solved by applying a combination of methods sometimes even requiring an interdisciplinary approach.

In our survey, we have discerned between hybrid metaheuristics for solving vehicle routing problems with three different types of available information: dynamic, stochastic, dynamic and stochastic. These three variants have then been further divided according to predominant hybridization principles. In the case of dynamic problems, where the major challenge is the realization of quick response times, parallelization approaches and fast insertion combined with background optimization were the most commons patterns of hybridization. When only a priori stochastic information is considered, we found that most commonly different complementing search algorithms are combined. Another form of hybridization was the combination of exact and approximate computation of the objective function. The combination of dynamic and stochastic information is most often addressed using sampling based approaches, the proposed algorithms are most often working with a single solution, but important results have also been obtained with methods based on solution pools where the diversity comes from solving multiple scenarios. Usually a single solution is then recommended, but such approaches can become necessary when multiple solutions have to be recommended to a human decision maker. Depending on application specific requirements and the nature of the available information an appropriate approach will have to be chosen.

The area of dynamic and stochastic optimization in general and vehicle routing in particular is recently getting increasing attention. This is mainly due to technological advances in fields such as telematics and computing but also in algorithmic advances in the vehicle routing area. In our opinion, multi-disciplinary approaches combining strong stochastic modeling and combinatorial optimization skills will gain importance able to solve complex academic and real-world problems.

Acknowledgements. This work, partially funded by the Austrian Federal Ministry for Transport, Innovation and Technology (BMVIT) within the strategic program FIT-IT ModSim under grant 822739 (project HealthLog).

References

1. Alvarenga, G., de Abreu Silva, R., Mateus, G.: A hybrid approach for the dynamic vehicle routing problem with time windows. In: Fifth International Conference on Hybrid Intelligent Systems (HIS 2005), 7 pages (November 2005)
2. Alvarenga, G., Mateus, G., de Tomi, G.: A genetic and set partitioning two-phase approach for the vehicle routing problem with time windows. Computers & Operations Research 34(6), 1561–1584 (2007); Part Special Issue: Odysseus 2003 Second International Workshop on Freight Transportation Logistics
3. Attanasio, A., Bregman, J., Ghiani, G., Manni, E.: Real-time fleet management at ecourier ltd. In: Sharda, R., Voss, S., Zeimpekis, V., Tarantilis, C.D., Giaglis, G.M., Minis, I. (eds.) Dynamic Fleet Management. Operations Research/Computer Science Interfaces Series, vol. 38, pp. 219–238. Springer, US (2007)
4. Attanasio, A., Cordeau, J.-F., Ghiani, G., Laporte, G.: Parallel tabu search heuristics for the dynamic multi-vehicle dial-a-ride problem. Parallel Comput. 30, 377–387 (2004)
5. Beasley, J.: Route first–cluster second methods for vehicle routing. Omega 11(4), 403–408 (1983)
6. Bent, R.W., Van Hentenryck, P.: Dynamic vehicle routing with stochastic requests. In: International Joint Conference On Artificial Intelligence, vol. 18, pp. 1362–1363 (2003)
7. Bent, R.W., Van Hentenryck, P.: Scenario-based planning for partially dynamic vehicle routing with stochastic customers. Oper. Res. 52, 977–987 (2004)
8. Berbeglia, G., Cordeau, J.-F., Laporte, G.: A hybrid tabu search and constraint programming algorithm for the dynamic dial-a-ride problem. INFORMS Journal on Computing (2011)
9. Berbeglia, G., Pesant, G., Rousseau, L.-M.: Checking the feasibility of dial-a-ride instances using constraint programming. Transportation Science 45, 399–412 (2011)
10. Bertsimas, D.J.: A vehicle routing problem with stochastic demand. Oper. Res. 40, 574–585 (1992)

11. Bianchi, L., Birattari, M., Chiarandini, M., Manfrin, M., Mastrolilli, M., Paquete, L., Rossi-Doria, O., Schiavinotto, T.: Hybrid metaheuristics for the vehicle routing problem with stochastic demands. Journal of Mathematical Modelling and Algorithms 5, 91–110 (2006), doi:10.1007/s10852-005-9033-y
12. Birge, J.R., Louveaux, F.: Introduction to Stochastic Programming. Springer (1997)
13. Blum, C., Puchinger, J., Raidl, G.R., Roli, A.: Hybrid metaheuristics in combinatorial optimization: A survey. Applied Soft Computing 11(6), 4135–4151 (2011)
14. Bock, S.: Real-time control of freight forwarder transportation networks by integrating multimodal transport chains. European Journal of Operational Research 200(3), 733–746 (2010)
15. Branchini, R.M., Armentano, V.A., Løkketangen, A.: Adaptive granular local search heuristic for a dynamic vehicle routing problem. Computers and Operations Research 36(11), 2955–2968 (2009)
16. Caramia, M., Italiano, G., Oriolo, G., Pacifici, A., Perugia, A.: Routing a fleet of vehicles for dynamic, combined pickup and delivery services. In: Proceedings of the Symposium on Operation Research 2001, pp. 3–8. Springer (2001)
17. Chen, H.-K., Hsueh, C.-F., Chang, M.-S.: The real-time time-dependent vehicle routing problem. Transportation Research Part E: Logistics and Transportation Review 42(5), 383–408 (2006)
18. Chen, Z.-L., Xu, H.: Dynamic column generation for dynamic vehicle routing with time windows. Transportation Science 40, 74–88 (2006)
19. Cordeau, J.-F., Laporte, G.: The dial-a-ride problem: models and algorithms. Annals of Operations Research 153, 29–46 (2007)
20. Cordeau, J.-F., Laporte, G., Savelsbergh, M.W., Vigo, D.: Vehicle routing. In: Barnhart, C., Laporte, G. (eds.) Transportation. Handbooks in Operations Research and Management Science, vol. 14, ch. 6, pp. 367–428. Elsevier (2007)
21. Crainic, T.G.: Parallel solution methods for vehicle routing problems. In: Golden, et al. (eds.), [35], p. 589 (2008)
22. Créput, J.-C., Hajjam, A., Koukam, A., Kuhn, O.: Self-organizing maps in population based metaheuristic to the dynamic vehicle routing problem. Journal of Combinatorial Optimization, 1–22 (2011)
23. Dror, M., Laporte, G., Trudeau, P.: Vehicle routing with stochastic demands: Properties and solution frameworks. Transportation Science 23(3), 166–176 (1989)
24. Fabri, A., Recht, P.: On dynamic pickup and delivery vehicle routing with several time windows and waiting times. Transportation Research Part B: Methodological 40(4), 335–350 (2006)
25. Fischetti, M., Lodi, A.: Local branching. Mathematical Programming 98, 23–47 (2003)
26. Flatberg, T., Hasle, G., Kloster, O., Nilssen, E.J., Riise, A.: Dynamic and stochastic vehicle routing in practice. In: Sharda, R., Voss, S., Zeimpekis, V., Tarantilis, C.D., Giaglis, G.M., Minis, I. (eds.) Dynamic Fleet Management. Operations Research/Computer Science Interfaces Series, vol. 38, pp. 41–63. Springer, US (2007), doi:10.1007/978-0-387-71722-7_3
27. Fleischmann, B., Gnutzmann, S., Sandvoss, E.: Dynamic vehicle routing based on online traffic information. Transportation Science 38, 420–433 (2004)
28. Fleischmann, B., Gnutzmann, S., Sandvoss, E.: Dynamic vehicle routing based on online traffic information. Transportation Science 38, 420–433 (2004)
29. Fu, L., Rilett, L.R.: Expected shortest paths in dynamic and stochastic traffic networks. Transportation Research Part B: Methodological 32(7), 499–516 (1998)
30. Gendreau, M., Guertin, F., Potvin, J.-Y., Taillard, E.: Parallel tabu search for real-time vehicle routing and dispatching. Transportation Science 33, 381–390 (1999)
31. Gendreau, M., Laporte, G., Séguin, R.: Stochastic vehicle routing. European Journal of Operational Research 88(1), 3–12 (1996)
32. Gendreau, M., Potvin, J.-Y., Bräumlaysy, O., Hasle, G., Løkketangen, A.: Metaheuristics for the vehicle routing problem and its extensions: A categorized bibliography. In: Golden, et al. (eds.) [35], pp. 143–169 (2008)

33. Goel, A.: Fleet Telematics. Operations Research/Computer Science Interfaces Series, vol. 40. Springer, US (2008)
34. Golden, B., Dearmon, J., Baker, E.: Computational experiments with algorithms for a class of routing problems. Computers & Operations Research 10(1), 47–59 (1983)
35. Golden, B., Raghavan, S., Wasil, E. (eds.): The Vehicle Routing Problem: Latest Advances and New Challenges. Operations Research/Computer Science Interfaces Series, vol. 43. Springer (2008)
36. Gutjahr, W.J., Katzensteiner, S., Reiter, P.: A VNS Algorithm for Noisy Problems and its Application to Project Portfolio Analysis. In: Hromkovič, J., Královič, R., Nunkesser, M., Widmayer, P. (eds.) SAGA 2007. LNCS, vol. 4665, pp. 93–104. Springer, Heidelberg (2007), doi:10.1007/978-3-540-74871-7_9
37. Haghani, A., Jung, S.: A dynamic vehicle routing problem with time-dependent travel times. Comput. Oper. Res. 32, 2959–2986 (2005)
38. Hvattum, L., Løkketangen, A.: Using scenario trees and progressive hedging for stochastic inventory routing problems. Journal of Heuristics 15, 527–557 (2009)
39. Hvattum, L.M., Løkketangen, A., Laporte, G.: Solving a dynamic and stochastic vehicle routing problem with a sample scenario hedging heuristic. Transportation Science 40(4), 421–438 (2006)
40. Hvattum, L.M., Løkketangen, A., Laporte, G.: A branch-and-regret heuristic for stochastic and dynamic vehicle routing problems. Networks 49(4), 330–340 (2007)
41. Hvattum, L.M., Løkketangen, A., Laporte, G.: Scenario tree-based heuristics for stochastic inventory-routing problems. INFORMS Journal on Computing 21(2), 268–285 (2009)
42. Ichoua, S., Gendreau, M., Potvin, J.-Y.: Exploiting knowledge about future demands for real-time vehicle dispatching. Transportation Science 40(2), 211–225 (2006)
43. Jaillet, P.: Probabilistic Traveling Salesman Problems. PhD thesis. Operations Research Center, MIT (February 1985)
44. Jih, W.-R., Yung-Jen Hsu, J.: Dynamic vehicle routing using hybrid genetic algorithms. In: Proceedings of 1999 IEEE International Conference on Robotics and Automation, vol. 1, pp. 453–458 (1999)
45. Johnson, D., McGeoch, L.: Experimental analysis of heuristics for the stsp. In: Gutin, G., Punnen, A., Du, D.-Z., Pardalos, P.M. (eds.) The Traveling Salesman Problem and Its Variations. Combinatorial Optimization, vol. 12, pp. 369–443. Springer, US (2004), doi:10.1007/0-306-48213-4_9
46. Kennedy, J., Eberhart, R.: Particle swarm optimization. In: Proceedings of IEEE International Conference on Neural Networks, vol. 4, pp. 1942–1948 (November 1995)
47. Kenyon, A.S., Morton, D.P.: Stochastic vehicle routing with random travel times. Transportation Science 37(1), 69–82 (2003)
48. Khouadjia, M.R., Alba, E., Jourdan, L., Talbi, E.-G.: Multi-Swarm Optimization for Dynamic Combinatorial Problems: A Case Study on Dynamic Vehicle Routing Problem. In: Dorigo, M., Birattari, M., Di Caro, G.A., Doursat, R., Engelbrecht, A.P., Floreano, D., Gambardella, L.M., Groß, R., Şahin, E., Sayama, H., Stützle, T. (eds.) ANTS 2010. LNCS, vol. 6234, pp. 227–238. Springer, Heidelberg (2010)
49. Laporte, G.: The vehicle routing problem: An overview of exact and approximate algorithms. European Journal of Operational Research 59(3), 345–358 (1992)
50. Laporte, G.: Fifty years of vehicle routing. Transportation Science 43, 408–416 (2009)
51. Laporte, G., Musmanno, R., Vocaturo, F.: An adaptive large neighbourhood search heuristic for the capacitated arc-routing problem with stochastic demands. Transportation Science 44(1), 125–135 (2010)
52. Larsen, A.: The Dynamic Vehicle Routing Problem. PhD thesis. Technical University of Denmark, Kongens, Lyngby, Denmark (2000)
53. Larsen, A., Madsen, O., Solomon, M.: Partially dynamic vehicle routing - models and algorithms. Journal of the Operational Research Society 53, 637–646 (2002)
54. Larsen, A., Madsen, O.B., Solomon, M.M.: Classification of dynamic vehicle routing systems. In: Sharda, R., Voss, S., Zeimpekis, V., Tarantilis, C.D., Giaglis, G.M., Minis, I. (eds.) Dynamic Fleet Management. Operations Research/Computer Science Interfaces Series, vol. 38, pp. 19–40. Springer, US (2007)

55. Larsen, A., Madsen, O.B., Solomon, M.M.: Recent developments in dynamic vehicle routing systems. In: Sharda, R., Voss, S., Golden, B., Raghavan, S., Wasil, E. (eds.) The Vehicle Routing Problem: Latest Advances and New Challenges. Operations Research/Computer Science Interfaces Series, vol. 43, pp. 199–218. Springer, US (2008) doi:10.1007/978-0-387-77778-8_9
56. Liao, T.-Y.: Tabu search algorithm for dynamic vehicle routing problems under real-time information. Transportation Research Record: Journal of the Transportation Research Board 1882(1), 140–149 (2004)
57. Liao, T.-Y., Hu, T.-Y.: An object-oriented evaluation framework for dynamic vehicle routing problems under real-time information. Expert Systems with Applications 38(10), 12548–12558 (2011)
58. Marinakis, Y., Marinaki, M.: A hybrid multi-swarm particle swarm optimization algorithm for the probabilistic traveling salesman problem. Comput. Oper. Res. 37, 432–442 (2010)
59. Marinakis, Y., Migdalas, A., Pardalos, P.: Expanding neighborhood grasp for the traveling salesman problem. Computational Optimization and Applications 32, 231–257 (2005), doi:10.1007/s10589-005-4798-5
60. Mendoza, J.E., Castanier, B., Guéret, C., Medaglia, A.L., Velasco, N.: A memetic algorithm for the multi-compartment vehicle routing problem with stochastic demands. Computers & Operations Research 37(11), 1886–1898 (2010)
61. Mladenovic, N., Hansen, P.: Variable neighborhood search. Computers & Operations Research 24(11), 1097–1100 (1997)
62. Montemanni, R., Gambardella, L., Rizzoli, A., Donati, A.: Ant colony system for a dynamic vehicle routing problem. Journal of Combinatorial Optimization 10, 327–343 (2005), doi:10.1007/s10878-005-4922-6
63. Parragh, S.N.: Ambulance Routing Problems with Rich Constraints and Multiple Objectives. PhD thesis. University of Vienna, Department of Business Administration (2009)
64. Parragh, S.N., Doerner, K.F., Hartl, R.F.: Variable neighborhood search for the dial-a-ride problem. Comput. Oper. Res. 37, 1129–1138 (2010)
65. Pessoa, A., de Arago, M.P., Uchoa, E.: Robust branch-cut-and-price algorithms for vehicle routing problems. In: Golden, et al. (eds.) [35], pp. 297–325 (2008)
66. Pillac, V., Gendreau, M., Guéret, C., Medaglia, A.: A review of dynamic vehicle routing problems. Technical Report CIRRELT-2011-62. Centre interuniversitaire de recherche sur les reseaux denterprise, la logistique et le transport (CIRRELT), Montreal, Canada (2011)
67. Potvin, J.-Y., Xu, Y., Benyahia, I.: Vehicle routing and scheduling with dynamic travel times. Computers and Operations Research 33(4), 1129–1137 (2006), Part Special Issue: Optimization Days 2003
68. Powell, W.B.: Approximate Dynamic Programming: Solving the Curses of Dimensionality. Wiley (2007)
69. Powell, W.B., Topaloglu, H.: Stochastic programming in transportation and logistics. In: Ruszczynski, A., Shapiro, A. (eds.) Stochastic Programming. Handbooks in Operations Research and Management Science, vol. 10, pp. 555–635. Elsevier (2003)
70. Regan, A.C., Mahmassani, H.S., Jaillet, P.: Evaluation of dynamic fleet management systems: Simulation framework. Transportation Research Record: Journal of the Transportation Research Board 1645(1), 176–184 (1998)
71. Rei, W., Gendreau, M., Soriano, P.: A hybrid monte carlo local branching algorithm for the single vehicle routing problem with stochastic demands. Transportation Science 44(1), 136–146 (2010)
72. Resende, M., Ribeiro, C.: Greedy randomized adaptive search procedures. In: Glover, F., Kochenberger, G. (eds.) Handbook of Metaheuristics. International Series in Operations Research & Management Science, vol. 57, pp. 219–249. Springer, New York (2003), doi:10.1007/0-306-48056-5_8
73. Rockafellar, R.T., Wets, R.J.-B.: Scenarios and policy aggregation in optimization under uncertainty. Mathematics of Operations Research 16(1), 119–147 (1991)

74. Schilde, M., Doerner, K., Hartl, R.: Metaheuristics for the dynamic stochastic dial-a-ride problem with expected return transports. Computers & Operations Research 38(12), 1719–1730 (2011)
75. Schorpp, S.: Dynamic Fleet Management for International Truck Transportation focusing on Occasional Transportation Tasks. PhD thesis. University of Augsburg, Faculty of Economics and Business Administration (2010)
76. Simão, H.P., Day, J., George, A.P., Gifford, T., Nienow, J., Powell, W.B.: An approximate dynamic programming algorithm for large-scale fleet management: A case application. Transportation Science 43(2), 178–197 (2009)
77. Stützle, T., Hoos, H.H.: Analyzing the run-time behaviour of iterated local search for the tsp. In: III Metaheuristics International Conference. Kluwer Academic Publishers (1999)
78. Thomas, B.W., White, I.: Chelsea C. Anticipatory route selection. Transportation Science 38(4), 473–487 (2004)
79. Toth, P., Vigo, D. (eds.): The vehicle routing problem. Society for Industrial and Applied Mathematics (2001)
80. Voss, S., Osman, I.H., Roucairol, C. (eds.): Meta-Heuristics: Advances and Trends in Local Search Paradigms for Optimization. Kluwer Academic Publishers, Norwell (1999)
81. Yang, W.-H., Mathur, K., Ballou, R.H.: Stochastic vehicle routing problem with restocking. Transportation Science 34, 99–112 (2000)

Chapter 3
Combining Two Search Paradigms for Multi-objective Optimization: Two-Phase and Pareto Local Search

Jérémie Dubois-Lacoste, Manuel López-Ibáñez, and Thomas Stützle

Abstract. In this chapter, we review metaheuristics for solving multi-objective combinatorial optimization problems, when no information about the decision maker's preferences is available, that is, when problems are tackled in the sense of Pareto optimization. Most of these metaheuristics follow one of the two main paradigms to tackle such problems in a heuristic way. The first paradigm is to rely on Pareto dominance when exploring the search space. The second paradigm is to tackle several single-objective problems to find several solutions that are non-dominated for the original problem; in this case, one may exploit existing, efficient single-objective algorithms, but the performance depends on the definition of the set of scalarized problems. There are also a number of approaches in the literature that combine both paradigms. However, this is usually done in a relatively ad-hoc way. In this chapter, we review two conceptually simple methods representative of each paradigm: Pareto local search and Two-phase local search. The hybridization of these two strategies provides a general framework for engineering stochastic local search algorithms that can be used to improve over the state-of-the-art for several, widely studied problems.

3.1 Introduction

Optimization problems appear in many different real-world situations of high social, environmental or economic relevance. Often, such problems are evaluated according to various conflicting objectives. Not surprisingly, multi-objective optimization is attracting significant efforts from researchers, such that nowadays it has become a mature field of research. Many early studies on the development of methods and algorithms for multi-objective problems use an *a priori* approach where several

Jérémie Dubois-Lacoste · Manuel López-Ibáñez · Thomas Stützle
IRIDIA, CoDE, Université Libre de Bruxelles (ULB), Brussels, Belgium
e-mail: {jeremie.dubois-lacoste,manuel.lopez-ibanez}@ulb.ac.be
stuetzle@ulb.ac.be

E.-G. Talbi (Ed.): Hybrid Metaheuristics, SCI 434, pp. 97–117.
springerlink.com

objectives are aggregated into a single objective function. However, when no information is known about the decision maker's preferences one must rely on an *a posteriori* approach . In this case, the possible solutions to the problem are evaluated in the *Pareto sense*, using what is called Pareto dominance (see Section 3.2). When optimizing a multi-objective problem in the Pareto sense, the goal is to return a set of mutually non-dominated solutions, among which the decision maker can then choose the final solution to implement.

In this chapter, we focus on problems that are *combinatorial*. Solutions for such problems are made of discrete components and analytical methods cannot be used to solve them. For many relevant problems, no algorithm is known to find the optimum in polynomial time w.r.t. the size of the problem. Formally speaking, these problems are NP-hard [29]. Multi-objective combinatorial optimization problems (MCOPs) are always at least as hard as their single-objective counterparts. Often, multi-objective versions of a problem are NP-hard even if the single-objective version of the problem is not. An example is the shortest path problem, where the single-objective variant can be solved in polynomial time but the multi-objective version is NP-hard [28]. When dealing with such NP-hard problems, the typical sizes of the instances prevent the use of exact algorithms, and one must rely on *heuristic* algorithms. A heuristic algorithm is designed to return solutions in polynomial time (better said, quickly enough to be practical), without ensuring that they are optimal.

We discuss heuristic methods for tackling MCOPs with an a posteriori approach. From a very abstract perspective, most algorithms follow one of two main search paradigms: algorithms can be *dominance*-based or *scalarization*-based. Dominance-based methods tackle multi-objective problems by using some form of Pareto dominance relationship among solutions. Scalarization-based methods transform the multi-objective problem into a set of single-objective problems. By solving these single-objective problems, scalarization-based algorithms can provide solutions to the original multi-objective problem. In addition, several algorithms also combine some elements from these two search paradigms, that is, they are hybrid search methods.

For each of the two search paradigms, we present a representative local search framework. As a representative of dominance-based methods we discuss Pareto local search (PLS) [51], and as a representative of scalarization-based methods we present two-phase local search (TPLS) [53]. We describe these two methods together with their latest associated developments as well as similar methods from the literature; we show that their hybridization provides a general framework for multi-objective combinatorial optimization that leads to high performing algorithms.

This chapter is structured as follows. We introduce in Section 3.2 the basics of multi-objective optimization in the Pareto sense. Next, we present the two different search paradigms and the PLS and TPLS frameworks, in Sections 3.3 and 3.4, respectively. In Section 3.5, we explain how TPLS and PLS can be hybridized into a general framework and we review some hybrid algorithms in the literature. Section 3.6 reviews results that have been obtained by this framework. We conclude and highlight some directions for future research in Section 17.6.

3.2 Preliminaries

When tackling a multi-objective problem in the Pareto sense, the notion of optimal solution from single-objective optimization does not apply anymore. A solution s in the multi-objective case is better than another solution r if s is better than r for at least one objective and not worse for any of the remaining ones. If none of the two solutions is better than the other, they represent two different trade-offs of the objectives values that, without knowledge of the decision maker's preferences, are considered to be indifferent. The goal of an algorithm tackling a multi-objective problem in the Pareto sense is then to return all solutions representing different trade-offs among which the decision maker can choose the preferred one.

More formally, let us consider an MCOP with p objectives, all to be minimized. We call $\mathbf{f}(s)$ the vector of the objective function values of solution s; $f_k(s)$ denotes the specific value of objective k, that is, the k-th component of the objective function vector $\mathbf{f}(s)$. We call $N(s)$ the set of all neighbors of solution s.

Definition 3.1 (Dominance). A solution s_1 is said to dominate a solution s_2 ($s_1 \prec s_2$) if and only if $f_k(s_1) \leq f_k(s_2)\, \forall k = 1,\ldots,p$ and $\exists\, j \in \{1,\ldots,p\}$ such that $f_j(s_1) < f_j(s_2)$.

Definition 3.2 (Weak dominance). A solution s_1 is said to weakly dominate a solution s_2 ($s_1 \preceq s_2$) if and only if $f_k(s_1) \leq f_k(s_2)\, \forall k = 1,\ldots,p$.

Definition 3.3 (Incomparable solutions). Solutions s_1 and s_2 are said to be incomparable ($s_1 \parallel s_2$) if and only if neither $s_1 \npreceq s_2$ nor $s_2 \npreceq s_1$, and $s_1 \neq s_2$.

The fact that solutions can be incomparable is a fundamental difference to single-objective optimization, where a total ordering of the solutions exists.

Definition 3.4 (Pareto global optimum solution). Let S denote the set of all feasible solutions. A solution $s_1 \in S$ is a Pareto global optimum if and only if $\nexists\, s_2 \in S$ such that $s_2 \prec s_1$. Such solutions are also called *efficient*.

Definition 3.5 (Pareto local optimum set). A set S' of solutions is a Pareto local optimum if $\forall s \in S', \forall s_1 \in N(s), \exists s_2 \in S'$ verifying $s_2 \preceq s_1$.

Definition 3.6 (Pareto front). Let S denote the set of all feasible solutions. A set S' is a Pareto global optimum set if and only if it contains all the Pareto global optimum solutions of S and only these solutions. The set of objective vectors of the Pareto global optimum is called the Pareto front.

Several Pareto global optimum solutions can have the same objective vector in the Pareto front. In practice, it is common to return only one solution for each objective vector. Such a set is also called strict Pareto global optimum set [52] or strictly Pareto optimal set [21].

The dominance relations can be extended to sets of solutions. We present here the weak dominance relation on sets, which we use in this chapter. In the following, A and B denote two sets of solutions.

Definition 3.7 (Weak dominance on sets). A set A is said to weakly dominate a set B ($A \lhd B$) if and only if $\forall s_i \in B, \exists s_j \in A$, such that $s_j \preceq s_i$, and $A \neq B$.

Definition 3.8 (Incomparable sets). A and B are said to be incomparable ($A \parallel B$) if and only if neither $A \lhd B$ nor $B \lhd A$, and $A \neq B$.

In the next two sections we present the two search paradigms more in detail and we review representative methods in the literature, focusing on PLS and TPLS.

3.3 Dominance-Based Multi-objective Optimization

We say that algorithms are dominance-based if they use some form of Pareto dominance for acceptance decisions on solutions. When comparing solutions using Pareto dominance, solutions may be mutually non-dominated; thus, there is only a partial order defined over solutions, which is a fundamental difference to the single-objective case. Also for this reason, dominance-based algorithms keep an archive of solutions instead of only a single solution as the best one found so far.

There are many algorithms for MCOPs that are purely dominance-based. We restrict our discussion to methods that are based on the iterative improvement of the set of non-dominated solutions by performing local search (or mutation) of solutions one at a time. We do not consider here population-based algorithms such as multi-objective evolutionary algorithms [10, 8] or multi-objective ant colony optimization algorithms [5, 27]. However, it should be noted that these algorithms also often make direct or indirect use of Pareto dominance for directing the search, in particular, in acceptance or selection decisions on solutions.

Pareto local search (PLS) [51] is a paradigmatic representative of dominance-based multi-objective algorithms that improve solutions one at a time by use of neighborhood search [52]. While the original motivation for proposing PLS was to study the connectedness of solutions [51], PLS turned out to be also an effective local search method for multi-objective problems. Independently, a very similar algorithm was proposed by Angel et al. [4].

PLS extends iterative improvement procedures from the single-objective case to the multi-objective case by changing the acceptance criterion. While in the single-objective case an iterative improvement algorithm accepts a new solution if it is better than the current one, in the multi-objective case PLS accepts a new solution to enter the archive only if it is not dominated by any solution in the archive. PLS takes care that the archive contains only non-dominated solutions by filtering out dominated ones.

Algorithm 4 illustrates the general framework of PLS. It is initialized by an initial set A of mutually non-dominated solutions, called *archive*. These solutions are initially marked as unexplored (line 2). PLS then iteratively applies the following steps. First, a solution s is selected among all unexplored ones (*selection step*, line 5). Then, some (or all) of the neighbors of s, are explored (*neighborhood exploration*) and all the neighbors that are accepted (*acceptance criterion*) w.r.t. the

Algorithm 4. Pareto Local Search

```
1: Input: An initial set of non-dominated solutions A
2: explored(s) := FALSE   ∀s ∈ A
3: A0 := A
4: repeat
5:    s := SelectSolution(A0)
6:    repeat
7:       s' := NeighborhoodExploration(s)
8:       if AcceptSolution(A, s') then
9:          explored(s') := FALSE
10:         A := Update(A, s')
11:      end if
12:   until (termination criterion)
13:   explored(s) := TRUE
14:   A0 := {s ∈ A | explored(s) = FALSE}
15: until A0 = ∅
16: Output: A
```

archive A are added to A (lines 8 to 11). Solutions in A that are dominated by the newly added solutions are removed (procedure Update in line 10). Once the termination criterion for the exploration of the neighborhood of s is met, s is marked as explored (line 13). When all solutions have been explored, and no more new non-dominated solutions can be discovered, the algorithm stops in a Pareto local optimum. Algorithm 4 is a generic outline and different variants of PLS can be obtained by different instantiations of the components SelectSolution, AcceptSolution and NeighborhoodExploration. In the original PLS algorithm, as proposed in [51], the three main components are implemented as follows.

Selection step. The next solution to be explored is selected uniformly at random from the unexplored ones in the archive.

Neighborhood exploration. The neighborhood of a solution is always explored entirely. This corresponds to the "best-improvement" neighborhood exploration in single-objective local search, i.e., all neighborhood solutions are explored before exploring the neighborhood of a new solution.

Acceptance criterion. The original PLS accepts any non-dominated solution for inclusion in the archive.

The method proposed by Angel et al. [4] is very similar to PLS. The difference lies in the selection step: in the method of Angel et al., contrary to PLS, the neighborhoods of all unexplored solutions are explored before updating the archive. It has a stronger exploration capability than PLS because the archive is not updated immediately after the neighborhood of a single solution has been explored. Due to this, neighbors of solutions that otherwise would have become dominated can be examined in addition to those of non-dominated solutions.

Liefooghe et al. [43, 42] study the performance of some variants of the PLS algorithm. They test variants of the selection step that are obtained by restricting the overall exploration with a limit on the number of solutions to be selected, and variants of the neighborhood exploration itself, combining different ways of scanning the neighborhood and different acceptance criteria. The several variants are compared experimentally by the authors. In their experimental setup, they choose the same, predefined computation time limit for all variants. Variants that would finish before this computation time limit are then restarted from scratch and a variant is judged by the final aggregate non-dominated set found across the multiple restarts. As a result, they highlight variants that are the most effective if the computation time is known a priori and the PLS algorithms is launched several times.

Another recent study of PLS is carried out by Dubois-Lacoste et al. [20]. The aim is to examine variants for the algorithmic components of PLS with a main focus on their impact on the *anytime behavior* [64] of PLS. Several variants are tested for the three main components of PLS. It is shown that some of the resulting combinations can significantly speed-up the convergence of PLS towards good Pareto front approximations.

The fact that PLS stops upon finding a Pareto local optimum set can be a disadvantage if the algorithm finishes while there is still computation time available. A possibility is to keep the non-dominated solutions found in an external archive and to restart PLS "from scratch" (as done in [42]). There were other extensions that aim at obtaining a more efficient search. Alsheddy and Tsang [1] proposed an extension of PLS that continues the search when a Pareto local optimum set is found without restarting from different solutions. The idea is based on the *guided local search* [63] strategy in the single-objective case: a penalty is applied to worsen the components of the objective vectors of solutions in the current archive, allowing the algorithm to escape from a Pareto local optimum set. Other strategies to continue the search focus on generating good solution(s) to restart the search. In particular, solutions mutated from the ones in the Pareto local optimum set can be used, resulting, in some sense in an extension of *iterated local search* [47] for single-objective problems. A study of such strategies was done by Drugan and Thierens [12]. In that paper, the authors showed that the best results on the bi-objective quadratic assignment problem are attained when restarting PLS from new solutions on a "path" between two solutions in the Pareto local optimum set (this path is constructed in a manner similar to *path-relinking* [32]). Geiger [30] proposed to apply a different neighborhood operator (w.r.t. the one used during the search) when PLS converges to a Pareto local optimum, allowing PLS to find possibly new non-dominated solutions. This idea can be seen as an extension of *variable neighborhood search* [36].

Some evolutionary algorithms are similar to PLS. PAES has been proposed by Knowles and Corne [40] as an algorithm whose simplicity should make it a baseline for comparison to more complex evolutionary algorithms. In PAES, a solution is selected in the current archive of non-dominated solutions and a mutation operator is applied to obtain a new candidate solution. This new candidate is potentially inserted in the archive, which is then updated to keep only non-dominated solutions. The archive size is kept limited by using an archive bounding strategy. Contrary to PLS, this algorithm does not have a natural stopping criterion since solutions are never marked as explored. Laumanns et al. proposed the SEMO algorithm [41], which is similar to PAES but solutions are selected from an archive whose size is not limited. Differently from PAES, SEMO marks solutions as explored analogously to

PLS. The authors also test variants of SEMO that differ in the selection step. These variants tend to balance the number of times solutions are selected for mutation, or they try to focus on the most recently found solutions.

An advantage of dominance-based algorithms is that they deal with an archive of solutions rather than a single solution. Therefore, they can return quickly numerous non-dominated solutions to the problem. However, this can also be a drawback since dealing with possibly many solutions can make the exploration of the search space slower in terms of the closeness to the Pareto front. In the next section, we present the scalarization-based search paradigm for multi-objective optimization, whose aim is to provide quickly few high-quality solutions whose objective vectors are close to the Pareto front.

3.4 Scalarization-Based Multi-objective Optimization

Scalarization-based algorithms rely on solving several single-objective problems in order to find various non-dominated solutions for a multi-objective problem. To do so, the multiple objective functions are *aggregated* into a single scalar function. In this way, the solutions can be compared by scalar values, resulting in a total ordering of solutions. In other words, an aggregation transforms the multi-objective problem into a (*scalarized*) single-objective problem, often called simply *scalarization*.

There are many ways of scalarizing multi-objective problems. However, most commonly few standard methods are used for their simplicity and desirable properties. We present here the most common methods in use in the context of heuristic algorithms; for other possibilities, we refer the interested reader to [21, 8].

- **Linear aggregation.** The weighted sum method defines a linear aggregation of the objectives that is commonly used to define the preferences of the decision maker with an a priori approach. It is also used to define scalarizations for tackling problems with an a posteriori approach and we also use it in this chapter in the experimental part. A weight vector is used to give a relative importance to each objective. Let us consider a solution s whose objective function vector is $\mathbf{f}(s) = (f_1(s), f_2(s), \ldots, f_p(s))$, and a weight vector $\lambda = (\lambda_1, \lambda_2, \ldots, \lambda_p)$. We assume, without loss of generality, that the components of $\mathbf{f}(s)$ are non-negative. The scalar value for this solution and this weight is then:

$$f_\lambda(s) = \sum_{1 \leq i \leq p} \lambda_i \cdot f_i(s).$$

Since the different components of the weight vector have an effect that is relative to the value of each other, there exist infinitely many different weight vectors that define the same scalarization. Therefore, it is common to use normalized weight vectors, whose components sum up to one. The advantage of considering a weighted sum is that an optimal solution for the scalarized problem is a *supported* non-dominated solution, that is, its objective vector is located on the convex hull of the Pareto front.

- **Tchebycheff aggregation.** The Tchebycheff method requires to define a *reference point*, r, that dominates any feasible solution. Some weights, additionally, can be assigned to each objective with a weight vector $\lambda = (\lambda_1, \lambda_2, \ldots, \lambda_p)$. The scalar value for a solution s is then:

$$f_\lambda(s) = \max\{\lambda_i \cdot |f_i(s) - f_i(r)|\}, \quad i = 1 \ldots p.$$

  Hence, the goal becomes to find a solution as close as possible to the reference point r, using the Tchebycheff distance as a measure of "closeness".
- **Lexicographic ordering.** This method requires to define an order of the objectives, in decreasing importance, and, thus, it defines a total order of the solutions: If some solutions have the same value for objectives 1 to k, objective $k+1$ is used to break ties. The number of different orders of the objectives is limited, and therefore is also the number of solutions that can be obtained with this method. For instance, for bi-objective problems there are only two possible orderings of the objectives and thus only two different scalarized problems can be defined, preventing to return more than two solutions. Thus, when tackling a problem in the Pareto sense, a lexicographic ordering can be used to provide some initial solutions only, and must be used in combination with another technique.

Solving scalarized problems is often done when the multi-objective problem is tackled a priori, that is, the decision maker is able to define the components of the weight vector before the optimization. If the problem is considered using an a posteriori approach, multi-objective algorithms can make use of scalarizations, but the weight vector must be varied *during* the optimization process by the algorithm. An advantage of scalarization-based algorithms is that they can make use of any algorithm known to solve the scalarized problems effectively, and make use of its effectiveness in the multi-objective context. The drawback is that solutions are found one at a time, and the total number of non-dominated solutions returned is at most the number of scalarizations considered, which may be small compared to what is desirable.

Two-phase local search (TPLS) is a representative example of a scalarization-based algorithm [53]. TPLS is a general algorithmic framework that, as the name suggests, is composed of two phases. In the first phase, a single-objective algorithm generates a high-quality solution for one or all objectives, and then one of these high-quality solutions serves as the starting point of the second phase, where a sequence of scalarizations is tackled. Each scalarization uses the best solution found by the previous scalarization as the solution to start from. TPLS will be successful if the underlying single-objective algorithms are high-performing, and if solutions that are close to each other in the solution space have also objective function vectors that are close to each other in the objective space.

Algorithm 5 presents the general framework of TPLS for a bi-objective problem. First, high-quality solutions are generated for each objective (lines 1 and 2) using dedicated single-objective algorithms SLS_1 and SLS_2, and added to the archive

Algorithm 5. General Framework for Two-Phase Local Search

1: $s_1 :=$ SLS$_1$()
2: $s_2 :=$ SLS$_2$()
3: $A :=$ Update(A, s_1)
4: $A :=$ Update(A, s_2)
5: **repeat**
6: $\lambda :=$ ChooseWeight(A)
7: $s' :=$ ChooseSeed(λ, A)
8: $s' :=$ SLS$_\Sigma$(s', λ)
9: $A :=$ Update(A, s')
10: **until termination criterion**
11: Filter(A)
12: **Output:** A

(lines 3 and 4). Then, a sequence of scalarizations is solved (lines 5 to 10), based on strategies to generate a weight vector (procedure ChooseWeight on line 6) and to define how the previous solutions can be used as seed for further scalarizations (procedure ChooseSeed on line 7). Solutions are generated using a single-objective algorithm that tackles scalarized problems, SLS$_\Sigma$. The archive is updated with the solutions obtained for each scalarization (line 9). Note that the archive could include more information than just the solutions themselves, e.g., it could also include the fact that solutions have already been used as seeds, for which scalarization they have been obtained, etc.

Algorithm 5 is a generic outline that covers both the original TPLS and recent developments, depending on how the procedures ChooseWeight and ChooseSeed are implemented. TPLS in its original form considers a regular sequence of weights. Two main weight-setting strategies have been originally proposed to define the order in which these weights are selected. The simplest way to define a sequence of scalarizations is to use a regular sequence of weight vectors from the first objective to the second (for instance $\lambda_1 = (1, 0.8, 0.6, 0.2, 0)$ with $\lambda_2 = 1 - \lambda_1$) or from the second objective to the first one. However, this introduces a bias towards the region of the objective space where the first scalarizations are performed, and against the region where the last ones are performed [53]. To avoid this effect, a *double* weight setting strategy has been proposed: first a sequence of scalarizations is performed from one objective to the other and then a second sequence of scalarizations is performed in the opposite direction.

Recently, an *adaptive anytime* strategy (AA-TPLS) has been proposed [16, 19] to further improve the TPLS method. Inspired by the dichotomic scheme of Aneja and Nair [3], AA-TPLS has been shown to lead to better results by adapting to the shape of the Pareto front. Moreover, AA-TPLS shows a very good anytime behavior, providing a high-quality approximation to the Pareto front at any time without requiring a predefined computation time limit, as the original TPLS does. To do so, the selection of the next scalarization to be performed is based on the *optimistic hypervolume improvement*, which measures the potential improvement that can be expected in terms of hypervolume [66].

TPLS is a general framework that relies on a simple idea: run several times a single-objective algorithm on weighted sum scalarizations and obtain several non-dominated solutions, one for each weight. Another general-purpose method based on scalarizations has been proposed by Borges [7]. This method uses the Tchebycheff distance, but not in the standard way explained earlier in this section. Instead, the goal for each single-objective problem is to find a new solution that maximizes the Tchebycheff distance between this new solution and the closest one in the archive. The idea is, as in TPLS, to obtain a set of well-spread solutions in the objective space.

Other methods have been proposed in the literature, more specific than the general idea behind TPLS. They are, however, general-purpose and can be applied to many different problems. These are extensions of single-objective metaheuristics to the multi-objective case. An example is Multi-Objective Tabu Search (MOTS), which is an extension of tabu search [31] to multi-objective problems proposed by Hansen [35]. MOTS keeps a set of non-dominated solutions and tries to improve each solution in a direction that moves its objective vector away from other non-dominated solutions. To do so, it updates the weights for a given solution based on all other non-dominated solutions (the closer solutions are, the higher is their mutual influence). The purpose of this behavior is to obtain a set of solutions as spread as possible in the objective space along the Pareto front. The optimization of solutions toward different directions is performed using tabu search principles, each solution dealing with its own tabu list.

There have been several adaptations of the Simulated Annealing (SA) principle to the multi-objective case. They usually use several runs of single-objective SA algorithms, and mainly differ by the acceptance rules of new solutions. The first SA for multi-objective algorithm, proposed by Serafini [58], uses the following acceptance criterion. If the new solution dominates the current one, this new solution is accepted to replace the current one. Otherwise, the acceptance probability is computed on a weighted sum of the objectives. Several runs are performed using different weight vectors, and some small random variations are applied to them each time a solution is considered. A similar method is MOSA, proposed by Ulungu et al. [61]. MOSA uses the same type of rule for the acceptance criterion and a similar set of predefined weight vectors to define the single-objective problems. However, MOSA is not only returning one solution per weight vector: every time a solution is accepted as the new current one, it is potentially inserted in a set of non-dominated solutions. Each run of the single-objective SA maintains its own set of non-dominated solutions, and the sets are merged and filtered in a last step. Suppapitnarm et al. [59] proposed another adaptation of the SA principle to the multi-objective case. The acceptance criterion is different from other SA adaptations. Their proposal uses a multiplicative function of the objectives, instead of a weighted sum, and a different *temperature* for each objective. The setting of the temperature does not follow a pre-scheduled decrease, but is automatically updated based on the variance of each objective among already accepted solutions. The algorithm is then restarted several times to provide several solutions.

Finally, various population-based methods such as evolutionary algorithms, have used scalarizations to direct the search towards the Pareto front. An early example is VEGA [57]; other examples include the algorithms proposed by Ishibuchi and Murata [37] and MOGLS of Jaszkiewicz [38]. Also ACO algorithms frequently use some form of scalarized aggregation, for example, for combining pheromone (or heuristic) information specific to each objective [5, 27, 46]. However, an overview of such population-based methods is beyond the scope of this chapter.

3.5 Hybridization of Search Paradigms

We have presented in Sections 3.3 and 3.4 two search paradigms for multi-objective optimization. Each of them has its particular advantages and drawbacks. Dominance-based algorithms can return quickly a large number of non-dominated solutions; however, they progress rather slowly towards the Pareto front and they may require a long computation time before reaching high-quality approximations to the Pareto front. Scalarization-based algorithms can exploit effective single-objective algorithms and they find quickly high-quality approximations to the Pareto front. However, they return only relatively few solutions and they may not be able to approximate well certain types of solutions. For example, heuristic algorithms based on weighted-sum scalarizations are not designed to identify non-supported solutions and, thus, they may leave "gaps" in the Pareto front approximation. Thus, combining both search paradigms can be profitable, in order to exploit their respective advantages and to avoid as much as possible their respective disadvantages.

Hybrid algorithms combining TPLS and PLS elements have been considered in the literature and have shown high performance. (Examples on the performance of such hybrids are given in Section 3.6.) The natural way of combining TPLS and PLS is to first use TPLS to generate a set S' of (few) non-dominated solutions that are a high-quality approximations to the Pareto front and then use the solutions in S' to seed PLS. In fact, such a hybrid TP+PLS algorithm is straightforwardly obtained from existing TPLS and PLS algorithms. A rudimentary form of a TP+PLS algorithm has been studied by Paquete and Stützle [53]; they use a restricted form of PLS, the component-wise step. Later, Lust and Teghem apply a TP+PLS algorithm that runs the PLS phase to completion [49]. More recently, Dubois-Lacoste et al. [15, 18] have presented applications of TP+PLS algorithms to bi-objective flow-shop problems and also considered the automatic configuration of TP+PLS [17]. In their applications, PLS is terminated based on a bound on the available computation time.

TP+PLS is an example of a sequential hybridization of algorithms from the dominance-based and the scalarization-based search paradigms. A second class of hybridizations considers iterative hybridizations where elements of the two search paradigms are alternately applied. In the following, we give a concise overview of some representative examples of sequential and iterative hybrids. For a more complete review, we refer the interested reader to [23].

3.5.1 Sequential Hybridization

Combining a scalarization-based and a dominance-based component by switching from one to the other is the most straightforward way of hybridizing the two search paradigms. This switching forms the basis of a *sequential* hybridization. A common usage of sequential hybrids is to first use an exact algorithm to solve scalarized problems to optimality and, thus, to provide some (or all) of the supported solutions. Then, in a second phase, a dominance-based component aims at finding some non-supported solutions. In the heuristic case, a scalarization-based components can provide a small set of high-quality solutions (not necessarily supported ones), and in a second step, a dominance-based component improves this set of solutions further. We describe next some representative examples of such sequential hybrids.

Hamacher and Ruhe [34] combined the two search paradigms to tackle the bi-objective minimum spanning tree problem. A sequence of scalarizations is solved to optimality in the first phase; this is well feasible given that the minimum spanning tree problem is polynomially solvable. In a second phase, the neighborhood of all solutions obtained from the scalarizations is explored to search for additional non-dominated solutions. Andersen et al. [2] proposed a similar approach. They tested restrictions that consider only solutions that are neighbors of *two* different solutions in the set, and show that it may be useful for large scale problems since the number of solutions to consider is small.

Ulungu and Tehgem [60] proposed the *two-phases method*. This is a scheme for exactly solving MCOPs that works as follows. In a first phase, the whole set of supported solutions is determined using weighted sum scalarizations defined by the dichotomic scheme of Aneja and Nair [3]. In a second phase, this set of supported solutions is used to provide bounds to algorithms such as branch & bound, to find all non-dominated solutions. Despite being developed for exact solving, this approach has also inspired developments for heuristic solvers [49, 19].

Gandibleux et al. [26] proposed an algorithm for the bi-objective assignment problem that combined the two search paradigms as follows. First, an exact algorithm finds several supported solutions (a polynomial-time algorithm is known for the scalarized problems), and then the set of solutions obtained is improved further by seeding with this set a dominance-based evolutionary algorithm, which is run for few iterations.

Parragh et al. [55] designed an hybrid algorithm to solve the multi-objective dial-a-ride problem. A variable-neighborhood search algorithm is used to tackle weighted sum scalarizations defined by a regular sequence of weight vectors. In a second, dominance-based phase, a path-relinking step is used to further improve the set of solutions.

Delorme et al. [11] combined a greedy randomized adaptive search procedure (GRASP) [24] with the strength Pareto evolutionary algorithm (SPEA) [67] to tackle the bi-objective set packing problem. GRASP is used to tackle a sequence of weighted sum scalarized problems, and then SPEA is used to improve further the set of solutions returned by the GRASP.

3.5.2 *Iterative Hybridization*

The second possibility is to use both search paradigms in an *iterative* way. In that case, typically a scalarization-based component is used for a specific step within a dominance-based algorithm. Such iterative algorithms are often implicit hybrids of the two search paradigms: researchers seek the best possible performance and include a scalarization-based component within a dominance-based algorithm (or vice-versa), without making explicit the general concept behind this combination.

Gandibleux et al. [25] proposed an algorithm called MOTS (not to be confused with the MOTS algorithm mentioned in the previous section) that uses the tabu search principle to push solutions towards local *ideal* points. Once a solution becomes the new current one in the tabu search process, its neighborhood is explored using Pareto dominance to add possible non-dominated solutions to the archive.

Czyzzak and Jaszkiewicz [9] proposed the Pareto Simulated Annealing (PSA). Several runs of a single-objective SA are performed *in parallel*. Each run of the single-objective SA is tackling a problem defined by a weight vector and when a new solution is accepted to be the current one, its neighborhood is explored to search for non-dominated solutions. The weight vectors that define the scalarized problems are updated to "escape" from the current solutions of the other runs (taking only the closest ones into account).

López-Ibáñez et al. [44] tested the combination of a tabu search algorithm with a multi-objective ACO, and another combination with an evolutionary algorithm (SPEA2 [65]). The ACO and the SPEA2 algorithms use the tabu search algorithm at each iteration to improve individual solutions by tackling scalarized problems defined from a regular sequence of weight vectors.

3.6 Experimental Results of TP+PLS

In this section, we present some exemplary computational results with a TP+PLS algorithm for bi-objective flowshop scheduling problems and we give an overview of other recent experimental results. The flow-shop scheduling problem [39] is of high relevance since it models a common type of machine scheduling environment in industry. In the flow-shop scheduling problem, a set of n jobs is processed on m machines. The most common objective is to minimize the *makespan* (denoted by $C_{\max}$), that is, the completion time of the last job on the last machine. Other common objectives are the minimization of the *total completion time* (denoted by *SFT*) and the minimization of the *total tardiness*, denoted by *TT*, or the *total weighted tardiness*, denoted by *WT*. The bPFSP is NP-hard and it is one of the most widely studied scheduling problems. (To be more exact, minimizing makespan is NP-hard for three or more machines, minimizing the total completion time is NP-hard for two or more machines, and minimizing the total tardiness is NP-hard already for a single machine [13, 29].) For more details on this problem we refer to [18, 50].

In our research, we have developed a TP+PLS algorithm for five bi-objective permutation flowshop scheduling problems (bPFSP) [14, 15, 18] that correspond to all combinations of the above mentioned objectives with the exception of for the combination of *TT* and *WT*. We have carefully engineered the various algorithmic components of TPLS and PLS. The TPLS part of the algorithm exploits an underlying *iterated greedy algorithm* that is state-of-the-art for the PFSP with makespan minimization [56] and that has been adapted to tackle efficiently the other objectives and weighted sum problems. The TPLS version that we used is the most recent one: AA-TPLS [16, 19]. We showed that its adaptive behavior yields better results than the classical TPLS since it can adapt the search to the shape of the Pareto front, rather irregular for bPFSPs. The PLS version that we used is the original one [51], it uses two different types of neighborhood operators; we have shown in earlier studies [14] that it is profitable for bPFSPs. For more details on the TP+PLS framework for bPFSP, the interested reader can refer to [18].

The TP+PLS framework has been extensively compared to MOSA, a multi-objective SA proposed by Varadharajan and Rajendran [62]. MOSA was shown to outperform other algorithms on bPFSPs for several combinations of objectives [50]. The experimental comparison between MOSA and TP+PLS has been done based on a re-implementation of MOSA under equal computation times. Exemplary results are given in Table 3.1, which shows the percentage of runs where the output set of the TP+PLS algorithm weakly dominates in the Pareto sense (see Def. 3.7; hereafter we say that a set that weakly dominates another is "better") the output set obtained by a run of MOSA, and, conversely, the average percentage of runs that the output set of MOSA is better than TP+PLS. These percentages are computed for each instance over 625 pairwise comparisons obtained from 25 runs for each algorithm, and averaged over the 10 instances of each size. The results given are clearly in favor of TP+PLS. While TP+PLS is better than MOSA for a large percentage of the comparisons, the opposite is very rarely the case. Note that a value of 0 means that MOSA is not able to produce in any run a non-dominated set better than the worst non-dominated set produced by TP+PLS in any of the 25 runs of the 10 instances of a given size. The percentages given in Table 3.1 show that for small instances of 20 jobs, MOSA and the TP+PLS algorithm are difficult to compare. The low percentages are explained by the fact that both algorithms often find the same non-dominated set, which is probably the optimal Pareto front. For these small instances, differences are not consistent across instances and combinations of objectives, and it cannot be said that any algorithm is clearly better than the other. Nevertheless, for all the remaining instances, Table 3.1 shows that TP+PLS clearly outperforms MOSA.

Despite the fact that the TP+PLS algorithm often dominates MOSA, it is interesting to study how different the Pareto front approximations provided by the two algorithms are. To do so we use a graphical tool, the empirical attainment function (EAF). The EAF gives the, empirically estimated, probability that an algorithm dominates an arbitrary area of the objective space [33]. By plotting the differences of the EAFs of two algorithms, one can graphically show where in the objective space, and how frequently, an algorithm performs better relative to the other. A more

Table 3.1 For each bi-objective problem (denoted by the two objectives in parenthesis), the left column shows the percentage of runs (computed over 25 runs per instance and averaged over 10 instances of the same size) in which an output set obtained by TP+PLS is better in the Pareto sense than an output set obtained by MOSA. The right column shows the percentage of runs for which an output set of MOSA is better than an output set of TP+PLS .

	PFSP-($C_{\max}$, *SFT*)		*PFSP-*($C_{\max}$, *TT*)		*PFSP-*($C_{\max}$, *WT*)		*PFSP-(SFT, TT)*		*PFSP-(SFT, WT)*	
nxm	TP+PLS	MOSA	TP+PLS	MOSA	TP+PLS	MOSA	TP+PLS	MOSA	TP+PLS	MOSA
20x5	4.66	5.83	6.1	1.34	14.95	0.18	10.19	26.31	0.02	20.15
20x10	1.87	9.2	0.07	0.26	0.02	0.06	0.19	0.63	0.03	0.07
20x20	0.13	1.23	1.27	1.57	1.99	2.32	3.63	5.55	4.2	10.09
50x5	89.49	0	84.33	0	79.22	0	98.13	0.08	33.67	0
50x10	72.92	0	63.17	0	63.24	0	94.07	0	20.53	0
50x20	75.94	0	61.11	0	63.01	0	5.79	0	14.72	0
100x5	84.97	0	70.5	0	67.12	0	93.66	2.54	9.72	0
100x10	76.94	0.05	69.86	0	37.49	0	95.38	0.58	16.84	0
100x20	73.17	0	63.29	0	23.81	0	97.35	0	15.31	0
200x10	18.04	0.16	24.5	0	4.15	0	91.77	3.72	0.02	0
200x20	15.16	0	37.83	0	0.25	0	78.23	6.28	1.04	0.02

detailed explanation of this graphical tool can be found in [45]. Fig. 3.1 shows the differences between the EAFs obtained by the hybrid TP+PLS algorithm and the MOSA algorithm, for one instance with makespan and total completion time minimization. A large gap is observed in favor of the TP+PLS algorithm over the one obtained by MOSA, indicating a clearly superior performance. Other instances and combinations of objectives show similar trends.

Table 3.1, Fig. 3.1 and additional results available in [18] show that the hybrid TP+PLS algorithm clearly outperforms the previously best-known algorithm, by a large gap, and for all combinations of makespan minimization, total completion time minimization, total and weighted tardiness minimization.

TP+PLS algorithms were also applied to the bi-objective traveling salesman problem (bTSP). The TSP and the bi-objective version of it are well-known NP-hard combinatorial problems widely used to assess the performance of optimization algorithms and metaheuristics [22, 54]. The goal in the bTSP is to find a Hamiltonian tour that minimizes the sum of the edge costs in the tour. In the bi-objective variant of the TSP, two cost values are assigned to each edge of a graph, and each of the two objective functions is computed with respect to the corresponding cost value.

Paquete and Stützle [53] proposed the Pareto double two-phase local search (PD-TPLS method) and applied it to the bTSP. In this sequential hybrid algorithm, the first phase uses the TPLS method (in its original form, that is, with a regular sequence of weights) to return a set of non-dominated solutions, and in the second phase the neighborhood of all these solutions is explored to find additional non-dominated solutions. In a sense, this so-called component-wise step implements a restricted version of PLS. The single-objective algorithm used to tackle the scalarized problems is an iterated local search algorithm, and the neighborhood operator

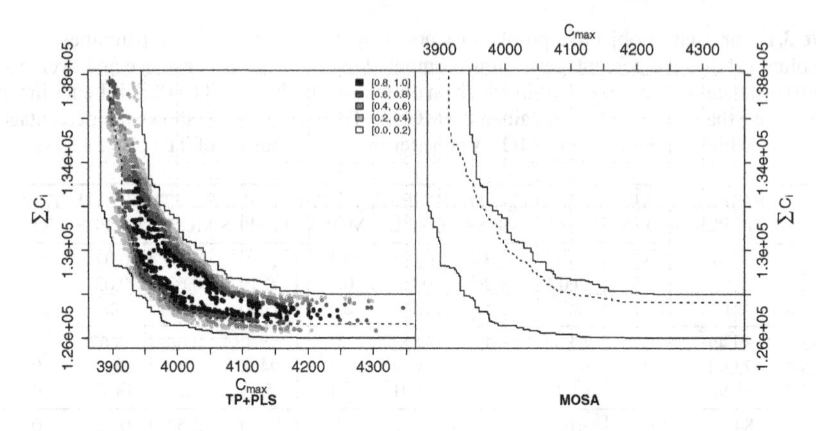

Fig. 3.1 Differences of the empirical attainment functions obtained over 25 runs in favor of the hybrid TP+PLS algorithm (left) and in favor of the MOSA algorithm (right), for the bPFSP with minimization of the makespan and of the total completion time, on an instance with 50 jobs and 20 machines.

is based on 2-opt moves. This hybrid algorithm has been compared favorably to the best algorithm known at that time, the MOGLS algorithm from Jaszkiewicz [38]. A more in-depth experimental study of PD-TPLS and other TPLS variants has been presented by Paquete and Stützle [54].

More recently, another TP+PLS algorithm has been applied to the bTSP by Lust and Teghem [49]. This algorithm is reported to outperform the PD-TPLS of Paquete and Stützle. The main differences are that (i) the single-objective algorithm used to tackle the scalarized problems is an effective implementation of the chained Lin-Kernighan heuristic [6], which is presumably more effective than the iterated local search algorithm used by the PD-TPLS, and (ii) a full version of PLS is used (more precisely, the version of Angel et al. [4]) instead of the restricted one in the PD-TPLS method. This algorithm is nowadays, to the best of our knowledge, the state of the art for the bTSP.

Finally, Lust and Teghem [48] tackled multi-objective multi-dimensional knapsack problems. In these problems, two or more types of profits are associated to each item, each type of profit representing a different objective, and the goal is to determine a subset of items to place in a knapsack that maximizes, in the Pareto sense, the sum of profits for each objective. In the multi-dimensional version, the knapsack has more than one capacity constraint, and a feasible solution needs to satisfy all capacity constraints. The TPLS phase of the algorithm finds solutions for each weighted sum scalarization using a greedy constructive heuristic. In the second phase, PLS uses a very-large neighborhood search that starts from the set of solutions obtained by TPLS. They consider the bi-objective variant using regularly distributed weights for the scalarizations, and a three-objective variant using random weights for the scalarizations. For the bi-objective multi-dimensional knapsack problem the proposed TP+PLS algorithm is a new state-of-the-art approach, clearly outperforming its competitors on widely tested benchmark instances.

3.7 Conclusion

In this chapter, we have reviewed heuristic methods for solving multi-objective combinatorial optimization problems in the Pareto sense. In particular, we reviewed methods rooted in the scalarization-based and the dominance-based search paradigms. For each of the two search paradigms, we presented a representative method, namely Pareto local search (PLS) as a dominance-based method and two-phase local search (TPLS) as a scalarization-based method. We have detailed the main algorithmic components that are required to define PLS and TPLS algorithms, and discussed how PLS and TPLS can be combined into hybrid TP+PLS algorithms. Some exemplary computational results have illustrated the high potential of TP+PLS algorithms. In fact, for several well-known multi-objective problems, including various bi-objective flowshop problems, the traveling salesman problem with two and three objectives and the bi-objective knapsack problem, TP+PLS algorithms are currently state of the art.

An important direction for future research is certainly to extend the current generation of TPLS, PLS and TP+PLS algorithms to three and more objectives. We also believe that an iterative hybridization instead of a sequential one could be beneficial in terms of flexibility and performance. The TP+PLS framework is also a clear candidate for the automatic configuration of high-performing multi-objective algorithms. The feasibility of such an endeavor has already been proven [17]. Regarding applications, we believe the hybrid TP+PLS framework could be applied to other well-known and widely studied problems, providing high-quality results without requiring a large implementation effort.

Acknowledgements. This work was supported by the META-X project, an *Action de Recherche Concertée* funded by the Scientific Research Directorate of the French Community of Belgium, and by the MIBISOC network, an Initial Training Network funded by the European Commission, grant PITN–GA–2009–238819. Manuel López-Ibáñez and Thomas Stützle acknowledge support from the Belgian F.R.S.-FNRS, of which they are a postdoctoral researcher and a Research Associate, respectively. The authors also acknowledge support from the FRFC project *"Méthodes de recherche hybrides pour la résolution de problèmes complexes"*.

References

1. Alsheddy, A., Tsang, E.: Guided Pareto local search and its application to the 0/1 multi-objective knapsack problems. In: Caserta, M., Voß, S. (eds.) Proceedings of MIC 2009 The 8th Metaheuristics International Conference. University of Hamburg, Hamburg (2010)
2. Andersen, K., Jörnsten, K., Lind, M.: On bicriterion minimal spanning trees: An approximation. Computers & Operations Research 23(12), 1171–1182 (1996)
3. Aneja, Y.P., Nair, K.P.K.: Bicriteria transportation problem. Management Science 25(1), 73–78 (1979)

4. Angel, E.: Approximating the Pareto curve with local search for the bicriteria TSP(1,2) problem. Theoretical Computer Science 310(1-3), 135–146 (2004)
5. Angus, D., Woodward, C.: Multiple objective ant colony optimization. Swarm Intelligence 3(1), 69–85 (2009)
6. Applegate, D., Cook, W., Rohe, A.: Chained Lin-Kernighan for large traveling salesman problems. INFORMS Journal on Computing 15(1), 82–92 (2003)
7. Borges, P.C.: CHESS - changing horizon efficient set search: A simple principle for multiobjective optimization. Journal of Heuristics 6(3), 405–418 (2000)
8. Coello Coello, C.A., Lamont, G.B., Van Veldhuizen, D.A.: Evolutionary Algorithms for Solving Multi-Objective Problems. Springer, New York (2007)
9. Czyżak, P., Jaszkiewicz, A.: Pareto simulated annealing - a metaheuristic technique for multiple objective combinatorial optimization. Journal of Multi-Criteria Decision Analysis 7(1), 34–47 (1998)
10. Deb, K.: Multi-Objective Optimization Using Evolutionary Algorithms. Wiley, Chichester (2001)
11. Delorme, X., Gandibleux, X., Degoutin, F.: Evolutionary, constructive and hybrid procedures for the bi-objective set packing problem. European Journal of Operational Research 204(2), 206–217 (2010)
12. Drugan, M.M., Thierens, D.: Path-Guided Mutation for Stochastic Pareto Local Search Algorithms. In: Schaefer, R., Cotta, C., Kołodziej, J., Rudolph, G. (eds.) PPSN XI. LNCS, vol. 6238, pp. 485–495. Springer, Heidelberg (2010)
13. Du, J., Leung, J.Y.T.: Minimizing total tardiness on one machine is NP-hard. Mathematics of Operations Research 15(3), 483–495 (1990)
14. Dubois-Lacoste, J.: A study of Pareto and Two-Phase Local Search Algorithms for Biobjective Permutation Flowshop Scheduling. Master's thesis. IRIDIA, Université Libre de Bruxelles, Belgium (2009)
15. Dubois-Lacoste, J., López-Ibáñez, M., Stützle, T.: Effective Hybrid Stochastic Local Search Algorithms for Biobjective Permutation Flowshop Scheduling. In: Blesa, M.J., Blum, C., Di Gaspero, L., Roli, A., Sampels, M., Schaerf, A. (eds.) HM 2009. LNCS, vol. 5818, pp. 100–114. Springer, Heidelberg (2009)
16. Dubois-Lacoste, J., López-Ibáñez, M., Stützle, T.: Adaptive "Anytime" Two-Phase Local Search. In: Blum, C., Battiti, R. (eds.) LION 4. LNCS, vol. 6073, pp. 52–67. Springer, Heidelberg (2010)
17. Dubois-Lacoste, J., López-Ibáñez, M., Stützle, T.: Automatic configuration of state-of-the-art multi-objective optimizers using the TP+PLS framework. In: Krasnogor, N., et al. (eds.) Proceedings of the Genetic and Evolutionary Computation Conference, GECCO 2011, pp. 2019–2026. ACM press, New York (2011)
18. Dubois-Lacoste, J., López-Ibáñez, M., Stützle, T.: A hybrid TP+PLS algorithm for bi-objective flow-shop scheduling problems. Computers & Operations Research 38(8), 1219–1236 (2011)
19. Dubois-Lacoste, J., López-Ibáñez, M., Stützle, T.: Improving the anytime behavior of two-phase local search. Annals of Mathematics and Artificial Intelligence 61(2), 125–154 (2011)
20. Dubois-Lacoste, J., López-Ibáñez, M., Stützle, T.: Pareto Local Search Algorithms for Anytime Bi-Objective Optimization. In: Hao, J.-K., Middendorf, M. (eds.) EvoCOP 2012. LNCS, vol. 7245, pp. 206–217. Springer, Heidelberg (2012)
21. Ehrgott, M.: Multicriteria optimization. Lecture Notes in Economics and Mathematical Systems, vol. 491. Springer, Berlin (2000)
22. Ehrgott, M., Gandibleux, X.: Approximative solution methods for combinatorial multicriteria optimization. TOP 12(1), 1–88 (2004)
23. Ehrgott, M., Gandibleux, X.: Hybrid metaheuristics for multi-objective combinatorial optimization. In: Blum, C., Blesa, M.J., Roli, A., Sampels, M. (eds.) Hybrid Metaheuristics: An Emergent Approach for Optimization, pp. 221–259. Springer, Berlin (2008)
24. Feo, T.A., Resende, M.G.C.: Greedy randomized adaptive search procedures. Journal of Global Optimization 6, 109–113 (1995)

25. Gandibleux, X., Mezdaoui, N., Fréville, A.: A Tabu Search Procedure to Solve Multiobjective Combinatorial Optimization Problem. In: Caballero, R., Ruiz, F., Steuer, R. (eds.) Advances in Multiple Objective and Goal Programming. Lecture Notes in Economics and Mathematical Systems, vol. 455, pp. 291–300. Springer, Heidelberg (1997)
26. Gandibleux, X., Morita, H., Katoh, N.: Use of a Genetic Heritage for Solving the Assignment Problem with Two Objectives. In: Fonseca, C.M., Fleming, P.J., Zitzler, E., Deb, K., Thiele, L. (eds.) EMO 2003. LNCS, vol. 2632, pp. 43–57. Springer, Heidelberg (2003)
27. García-Martínez, C., Cordón, O., Herrera, F.: A taxonomy and an empirical analysis of multiple objective ant colony optimization algorithms for the bi-criteria TSP. European Journal of Operational Research 180(1), 116–148 (2007)
28. Garey, M.R., Johnson, D.S.: Computers and Intractability: A Guide to the Theory of NP-Completeness. Freeman & Co., San Francisco (1979)
29. Garey, M.R., Johnson, D.S., Sethi, R.: The complexity of flowshop and jobshop scheduling. Mathematics of Operations Research 1, 117–129 (1976)
30. Geiger, M.J.: Decision support for multi-objective flow shop scheduling by the Pareto iterated local search methodology. Computers and Industrial Engineering 61(3), 805–812 (2011)
31. Glover, F.: Tabu search – Part I. INFORMS Journal on Computing 1(3), 190–206 (1989)
32. Glover, F.: A Template for Scatter Search and Path Relinking. In: Hao, J.-K., Lutton, E., Ronald, E., Schoenauer, M., Snyers, D. (eds.) AE 1997. LNCS, vol. 1363, pp. 13–51. Springer, Heidelberg (1998)
33. Grunert da Fonseca, V., Fonseca, C.M., Hall, A.O.: Inferential performance assessment of stochastic optimisers and the attainment function. In: Zitzler, E., Deb, K., Thiele, L., Coello Coello, C.A., Corne, D.W. (eds.) EMO 2001. LNCS, vol. 1993, pp. 213–225. Springer, Heidelberg (2001)
34. Hamacher, H.W., Ruhe, G.: On spanning tree problems with multiple objectives. Annals of Operations Research 52(4), 209–230 (1994)
35. Hansen, M.P.: Tabu search for multiobjective optimization: MOTS. In: Climaco, J. (ed.) Proceedings of the 13th International Conference on Multiple Criteria Decision Making (MCDM 1997), pp. 574–586. Springer (1997)
36. Hansen, P., Mladenovic, N.: Variable neighborhood search: Principles and applications. European Journal of Operational Research 130(3), 449–467 (2001)
37. Ishibuchi, H., Murata, T.: A multi-objective genetic local search algorithm and its application to flowshop scheduling. IEEE Transactions on Systems, Man, and Cybernetics – Part C 28(3), 392–403 (1998)
38. Jaszkiewicz, A.: Genetic local search for multi-objective combinatorial optimization. European Journal of Operational Research 137(1), 50–71 (2002)
39. Johnson, D.S.: Optimal two- and three-stage production scheduling with setup times included. Naval Research Logistics Quarterly 1, 61–68 (1954)
40. Knowles, J.D., Corne, D.: The Pareto archived evolution strategy: A new baseline algorithm for multiobjective optimisation. In: Proceedings of the 1999 Congress on Evolutionary Computation (CEC 1999), pp. 98–105. IEEE Press, Piscataway (1999)
41. Laumanns, M., Thiele, L., Zitzler, E.: Running time analysis of multiobjective evolutionary algorithms on pseudo-boolean functions. IEEE Transactions on Evolutionary Computation 8(2), 170–182 (2004)
42. Liefooghe, A., Humeau, J., Mesmoudi, S., Jourdan, L., Talbi, E.G.: On dominance-based multiobjective local search: design, implementation and experimental analysis on scheduling and traveling salesman problems. Journal of Heuristics 18(2), 317–352 (2011)
43. Liefooghe, A., Mesmoudi, S., Humeau, J., Jourdan, L., Talbi, E.G.: A Study on Dominance-Based Local Search Approaches for Multiobjective Combinatorial Optimization. In: Stützle, T., Birattari, M., Hoos, H.H. (eds.) SLS 2009. LNCS, vol. 5752, pp. 120–124. Springer, Heidelberg (2009)

44. López-Ibáñez, M., Paquete, L., Stützle, T.: Hybrid population-based algorithms for the bi-objective quadratic assignment problem. Journal of Mathematical Modelling and Algorithms 5(1), 111–137 (2006)
45. López-Ibáñez, M., Paquete, L., Stützle, T.: Exploratory analysis of stochastic local search algorithms in biobjective optimization. In: Bartz-Beielstein, T., Chiarandini, M., Paquete, L., Preuss, M. (eds.) Experimental Methods for the Analysis of Optimization Algorithms, pp. 209–222. Springer, Berlin (2010)
46. López-Ibáñez, M., Stützle, T.: The automatic design of multi-objective ant colony optimization algorithms. IEEE Transactions on Evolutionary Computation (2012) (accepted)
47. Lourenço, H.R., Martin, O., Stützle, T.: Iterated local search: Framework and applications. In: Gendreau, M., Potvin, J.Y. (eds.) Handbook of Metaheuristics, 2nd edn. International Series in Operations Research & Management Science, vol. 146, ch. 9, pp. 363–397. Springer, New York (2010)
48. Lust, T., Teghem, J.: The multiobjective multidimensional knapsack problem: a survey and a new approach. Arxiv preprint arXiv:1007.4063 (2010)
49. Lust, T., Teghem, J.: Two-phase Pareto local search for the biobjective traveling salesman problem. Journal of Heuristics 16(3), 475–510 (2010)
50. Minella, G., Ruiz, R., Ciavotta, M.: A review and evaluation of multiobjective algorithms for the flowshop scheduling problem. INFORMS Journal on Computing 20(3), 451–471 (2008)
51. Paquete, L., Chiarandini, M., Stützle, T.: Pareto local optimum sets in the biobjective traveling salesman problem: An experimental study. In: Gandibleux, X., et al. (eds.) Metaheuristics for Multiobjective Optimisation. Lecture Notes in Economics and Mathematical Systems, vol. 535, pp. 177–200. Springer (2004)
52. Paquete, L., Schiavinotto, T., Stützle, T.: On local optima in multiobjective combinatorial optimization problems. Annals of Operations Research 156, 83–98 (2007)
53. Paquete, L., Stützle, T.: A Two-Phase Local Search for the Biobjective Traveling Salesman Problem. In: Fonseca, C.M., Fleming, P.J., Zitzler, E., Deb, K., Thiele, L. (eds.) EMO 2003. LNCS, vol. 2632, pp. 479–493. Springer, Heidelberg (2003)
54. Paquete, L., Stützle, T.: Design and analysis of stochastic local search for the multiobjective traveling salesman problem. Computers & Operations Research 36(9), 2619–2631 (2009)
55. Parragh, S., Doerner, K.F., Hartl, R.F., Gandibleux, X.: A heuristic two-phase solution approach for the multi-objective dial-a-ride problem. Networks 54(4), 227–242 (2009)
56. Ruiz, R., Stützle, T.: A simple and effective iterated greedy algorithm for the permutation flowshop scheduling problem. European Journal of Operational Research 177(3), 2033–2049 (2007)
57. Schaffer, J.D.: Multiple objective optimization with vector evaluated genetic algorithms. In: Grefenstette, J.J. (ed.) ICGA 1985, pp. 93–100. Lawrence Erlbaum Associates (1985)
58. Serafini, P.: Simulated annealing for multiple objective optimization problems. In: Tzeng, G.H., Yu, P.L. (eds.) Proceedings of the 10th International Conference on Multiple Criteria Decision Making (MCDM 1991), vol. 1, pp. 87–96. Springer (1992)
59. Suppapitnarm, A., Seffen, K., Parks, G., Clarkson, P.: A simulated annealing algorithm for multiobjective optimization. Engineering Optimization 33(1), 59–85 (2000)
60. Ulungu, E., Teghem, J.: The two phases method: An efficient procedure to solve biobjective combinatorial optimization problems. Foundations of Computing and Decision Sciences 20(2), 149–165 (1995)
61. Ulungu, E., Teghem, J., Fortemps, P., Tuyttens, D.: MOSA method: a tool for solving multiobjective combinatorial optimization problems. Journal of Multi-Criteria Decision Analysis 8(4), 221–236 (1999)
62. Varadharajan, T.K., Rajendran, C.: A multi-objective simulated-annealing algorithm for scheduling in flowshops to minimize the makespan and total flowtime of jobs. European Journal of Operational Research 167(3), 772–795 (2005)
63. Voudouris, C., Tsang, E.: Guided local search and its application to the travelling salesman problem. European Journal of Operational Research 113(2), 469–499 (1999)

64. Zilberstein, S.: Using anytime algorithms in intelligent systems. AI Magazine 17(3), 73–83 (1996)
65. Zitzler, E., Laumanns, M., Thiele, L.: SPEA2: Improving the strength Pareto evolutionary algorithm for multiobjective optimization. In: Giannakoglou, K., Tsahalis, D., Periaux, J., Papaliliou, K., Fogarty, T. (eds.) Evolutionary Methods for Design, Optimisation and Control, pp. 95–100. CIMNE, Barcelona (2002)
66. Zitzler, E., Thiele, L.: Multiobjective Optimization Using Evolutionary Algorithms - A Comparative Case Study. In: Eiben, A.E., Bäck, T., Schoenauer, M., Schwefel, H.-P. (eds.) PPSN 1998. LNCS, vol. 1498, pp. 292–301. Springer, Heidelberg (1998)
67. Zitzler, E., Thiele, L.: Multiobjective evolutionary algorithms: A comparative case study and the strength Pareto evolutionary algorithm. IEEE Transactions on Evolutionary Computation 3(4), 257–271 (1999)

Part II
Combining Metaheuristics with (Complementary) Metaheuristics

Chapter 4
Hybridizing Cellular GAs with Active Components of Bio-inspired Algorithms

E. Alba and A. Villagra

Abstract. Cellular Genetic Algorithm (cGA) and Particle Swam Optimization (PSO) are two powerful metaheuristics being used successfully since their creation for the resolution of optimization problems. In this work we present two hybrid algorithms based on a cGA with the insertion of components from PSO. We aim to achieve significant numerical improvements in the results obtained by a cGA in combinatorial optimization problems. We here analyze the performance of our hybrids using a set of different problems. The results obtained are quite satisfactory in efficacy and efficiency.

4.1 Introduction

Research in exact algorithms, heuristics, and metaheuristics for solving combinatorial optimization problems is nowadays highly on the rise. Evolutionary Algorithms (EAs) are very popular optimization techniques [2], [4], [5]. They work by evolving a population of individuals (potential solutions), emulating the biological processes of selection, mutation, and recombination found in Nature, so that individuals (i.e., solutions) are improved. This family of techniques apply an iterative and stochastic process on a set of individuals (population), and are well-known good algorithms for exploring complex and large search spaces.

E. Alba
Department of Computer Science
University of Málaga, Spain
e-mail: eat@lcc.uma.es

A. Villagra
Emerging Technologies Laboratory
Universidad Nacional de la Patagonia Austral, Argentine
e-mail: avillagra@uaco.unpa.edu.ar

E.-G. Talbi (Ed.): Hybrid Metaheuristics, SCI 434, pp. 121–133.
springerlink.com

Most EAs in the classical literature are panmictic, although restricting the mating among individuals has appeared as an important research line. To this end, some kind of structure is added to panmictic (unrestricted) mating by defining neighborhoods among them. Among the many types of structured EAs (where the population is somehow decentralized), distributed and cellular algorithms are the most popular optimization tools [2], [6], [15], [21].

A cGA, is a class of a decentralized population in which the tentative solutions evolve in overlapped neighborhoods [19], [21]. In a cGA individuals are conceptually set in a toroidal mesh, and are allowed to recombine with nearby individuals. The overlaping of neighborhoods provides to cEA an implicit slow diffusion mechanism. The slow dispersion of the best solutions over the population is the cause of a good balance between the exploitation and the exploration efficacy, something really sought when an efficient and accurate algorithm is designed. Nevertheless, this characteristic produces a slow convergence to the optimum and thus decreases the efficiency of the algorithm. An open research line then consists in creating new algorithmic models which try to improve the efficiency of cEA by incorporating active components of other algorithms. This is not the only means to leverage the efficiency of a cEA, but it is a structured and novel way of approaching it. In this work we are introducing "active components" of PSO in a CGA, as an extension of a previous preliminary study [3].

PSO was introduced by Kennedy and Eberhart [11] in 1995 as a population-based stochastic search and optimization process. It originated from computer simulation of individuals (particles or living organisms) in a bird flock or fish school [22]. Over the last years, interest in hybrid metaheuristics has risen considerably in the field of optimization [9]. Combinations of algorithms such as several metaheuristics in a single technique have provided very powerful search procedures in the past. In this work we intend to generate new functional and efficient hybrid algorithms in a methodological and structured way. Thus, we will take out of PSO its very special particle movement equation and plug it into a canonical cGA, to get exploitation and exploration at the same time in one single algorithm. We propose two algorithms based on cGA by adding to it a mutation based on PSO. We test our algorithms with a set of combinatorial problems and compare our results with the canonical cGA. Results are quite satisfactory in efficacy and efficiency.

This chapter is organized as follows. In Section 4.2 we show basic concepts of cGA. In Section 4.3 we present the classic PSO algorithm. In Section 4.4 we describe our hybrid algorithms. In Section 4.5 we show the experiments and results, and finally in Section 4.6 we describe some conclusions and suggest future research lines.

4.2 Characterizing Cellular Genetic Algorithms

Genetic Algorithms (GAs) are a particular class of EAs. In turn, cGAs are a subclass of Genetic Algorithms (GAs) with a spatially structured population, i.e. the individuals can only mate with their neighboring individuals [1]. These overlapped small neighborhoods help in exploring the search space because the induced slow diffusion of solutions through the population provides a kind of exploration, while the exploitation takes place inside each neighborhood by genetic operators. In cGAs the population is usually structured in a 2D toroidal grid. The most commonly used neighborhood is called L5 [15], [21]. This neighborhood always contains five individuals: the considered one (*position*(*x*,*y*)) plus the North, East, West, and South individuals shown in Figure 4.1.

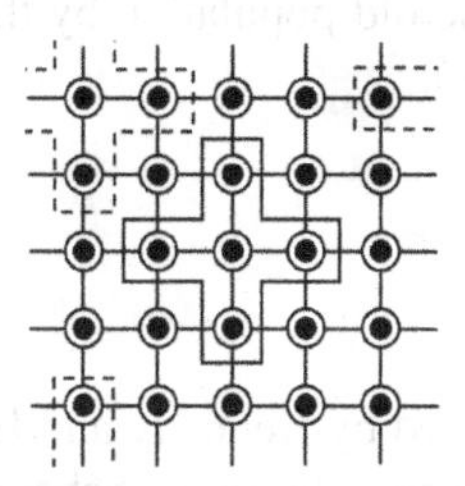

Fig. 4.1 Population arrangement (cGA)

In Algorithm 6 we present the pseudo-code of a canonical cGA. It starts by generating and evaluating an initial population. After that, genetic operators (selection, recombination, mutation, and replacement) are iteratively applied to each individual until the termination condition is met.

The population is structured in a two-dimensional (2-D) toroidal grid, and the neighborhood defined on it (line 6) contains five individuals. The considered

Algorithm 6. Pseudocode of a cGA

```
1: /* Algorithm parameters in 'cga' */
2: Steps-Up(cga)
3: for s ←— 1 to MAX_STEPS do
4:    for x ←— 1 to WIDTH do
5:       for y ←— 1 to HEIGHT do
6:          nList ←— ComputeNeigh (cga,position(x,y));
7:          parent1 ←— IndividualAt(cga,position(x,y));
8:          parent2 ←— LocalSelect(nList);
9:          /* Recombination */
10:         DPX1(cga.Pc,nList[parent1],nList[parent2],auxInd.chrom);
11:         /* Mutation */
12:         BitFlip(cga.Pm,auxInd.chrom);
13:         auxInd.fit ←— cga.Fit(Decode(auxInd.chrom));
14:         InsertNewInd(position(x,y),auxInd,[ if_not_worse ], cga, auxPop);
15:      end for
16:   end for
17:   cga.pop ←— auxPop;
18:   UpdateStatistics(cga)
19: end for
```

individual itself is always selected for being one of the two parents (line 7). The second parent is selected by Tournament Selection (line 8). Genetic operators are applied to individuals in lines 10 and 12. We use in this chapter a two point crossover operator (DPX1) and traditional binary mutation operator - *bit-flip*.

After applying these operators, the algorithm calculates the fitness value of the new individual (line 13) and inserts it on its equivalent place in the new population (line 14) only if its value is better or equal than the old one (always adding the new individual to the next population). After applying the above mentioned operators to the individuals we replace the old population by the new one (line 17), and we calculate some statistics (line 18).

4.3 Classic PSO

The PSO algorithm was developed by Kennedy and Eberhart in 1995 [11]. This is a population-based technique inspired by social behavior of the movement of flocks of birds or schools of fish. In PSO the potential solutions, called particles, "fly" through the problem space. All of the particles have fitness values based on their position and have velocities which direct the flight of the particles. PSO is initialized with a group of random particles (solutions) and then searches for the optima by updating generations. In every iteration, each particle updates its information (velocity and position) based on personal and social knowledge. The personal knowledge is obtained from the best solution (fitness) the particle has achieved so far. This value is called *pbest*. The social knowledge comes from the best value obtained so far for any particle in the population. This best value is a global best and called *gbest*.

After finding these values, the particle updates its velocity and position according to the following equations:

$$v_{n+1} = \omega_i v_n + \varphi_1 * rand * (pbest_n - x_n) + \varphi_2 * rand * (gbest_n - x_n) \tag{4.1}$$

$$x_{n+1} = x_n + v_{n+1} \tag{4.2}$$

ω_i is the inertia coefficient which slows velocity over time; v_n is the particle velocity; x_n is the current particle position in the search space; $pbest_n$ and $gbest_n$ are defined as the "personal" best and "global" best; *rand* is a random number between (0,1); φ_1, φ_2 are learning factors.

PSO was originally developed to solve real-value optimization problems. To extend the real-value version of PSO to a binary/discrete space, Kennedy and Eberhart [12] proposed a binary PSO (BPSO) method. In their model a particle will decide on "yes" or "no", "true" or "false", etc. also this binary values can be a representation of a real value in binary search space. In this binary version, the particle's personal best and global best is updated as in the continuous version. The velocities of the particles are defined in terms of probabilities that a bit will change to one. Using this definition, a velocity must be restricted within the range $[0,1]$. So a transformation is used to map all real valued of velocity to the range $[0,1]$ [12]. The normalization function used here is a sigmoid function s:

$$s(v_n) = \frac{1}{1+exp(-v_n)} \tag{4.3}$$

Also, the equation 7.1 is used to update the velocity vector of the particle. And the new position of the particle is obtained using the equation below:

$$x_{n+1} = \begin{cases} 1 & \text{if } r < s(v_n) \\ 0 & \text{otherwise} \end{cases} \tag{4.4}$$

where r is a uniform number in the range $[0,1]$.

4.4 Our Hybrid cGA Algorithms

We propose to insert concepts of PSO into the canonical cGA with the intention of improving its performance. The basic idea is to capture the positive main characteristics of a metaheuristic, PSO in this case, and inserted them into cGA.
To do this, we propose two hybrid algorithms called hyCP-local and hyCP-global. In both algorithms we will treat each individual as a particle. We maintain its velocity, position and information about its personal (pbest), and social (gbest) knowledge to update the information (velocity and position). Then a mutation based on PSO is used and the line 12 (mutation in the canonical cGA Algorithm 6) is replaced with the following lines:

```
1:  UpdateVelocity;
2:  UpdateIndividual (cga.Pm, auxInd.chrom);
```

The first line updates the velocity of the particle using equation 7.1. The second line modifies the individual taking into account the mechanism with the sigmoid function using equation 4.4. The pseudo-code of the algorithms proposed is described in Algorithm 7.

Both algorithms will apply this mutation based on PSO, with the difference that hyCP-local uses the local neighborhood (L5), and then selects one neighbor from there as *gbest*. For hyCP-global the global optimum of the all population is used as *gbest*.

4.5 Experiments and Analysis of Results

In this section we present the set of problems chosen for this study. We have chosen a representative set of problems to better study our proposal. The benchmarks

Algorithm 7. Pseudocode of hyCP-local and hyCP-global

```
1: /* Algorithm parameters */
2: Steps-Up(cga)
3: for s ←— 1 to MAX_STEPS do
4:    for x ←— 1 to WIDTH do
5:       for y ←— 1 to HEIGHT do
6:          nList ←— ComputeNeigh (cga,position(x,y));
7:          parent1 ←— IndividualAt(cga,position(x,y));
8:          parent2 ←— LocalSelect(nList);
            /* Recombination */
9:          DPX1(cga.Pc,nList[parent1],nList[parent2],auxInd.chrom);
            /* Mutation based on PSO */
10:         UpdateVelocity;
11:         UpdateIndividual (cga.Pm, auxInd.chrom);
12:         auxInd.fit ←— cga.Fit(Decode(auxInd.chrom));
13:         InsertNewInd(position(x,y),auxInd,[ if_not_worse ], cga, auxPop);
14:      end for
15:   end for
16:   cga.pop ←— auxPop;
17:   UpdateStatistics(cga)
18: end for
```

contains many different interesting features in optimization, such as epistasis, multimodality, and deceptiveness. The problems used are Massively Multimodal Deceptive Problem (MMDP) [8], Frequency Modulation Sounds (FMS) [20] , Multimodal Problem Generator (P-PEAKS) [10], COUNTSAT [7] (an instance of MAXSAT [16]), Error Correcting Code Design (ECC) [14], and Maximum Cut of a Graph (MAXCUT) [13]; The minimum tardy task problem (MTTP) [18]. Finally the OneMax Problem [17](or BitCounting). The problems selected for this bechmark are explained bellow:

Massively Multimodal Deceptive Problem (MMDP): made up of k deceptive subproblems (s_i) of 6 bits each one, whose value depends on the number of ones (*unitation*) a binary string has. The global optimum has a value of k and it is attained when every subproblem is composed of zero or six ones. We use here a instance of $k = 40$ subproblems, and its maximum value is 40.

Frequency Modulation Sounds problem (FMS): is defined as determining six real-parameters of frequency modulated sound model. The parameters are defined in the range $[-6.4, +6.35]$, and we encode each parameter into a 32 bit substring in the individual. The optimum value for this problem is 0.0.

Multimodal Problem Generator (P-PEAKS): the idea is to generate P random $N-$bit string that represent the location of P peaks in the search space. In this chapter, we have used an instance of $P = 100$ peaks of length $N = 100$ bits each. The maximum fitness value for this problem is 1.0.

COUNTSAT problem: is an instance of MAXSAT. In this problem the solution value is the number of clauses that are satisfied by $n-$bit input string. The optimum value is having all the variables set to 1. In this work, an instance of $n = 20$ variables has been used, with the optimum value of 6860.

Error Correcting Code Design Problem (ECC): we will consider a three-tuple (n,M,d), where n in the length of each codeword (number of bits), M is the number of codewords, and d is the minimum Hamming distance between any pair of codewords. We consider in the present chapter an instance where $M = 24$ and $n = 12$ which has a fitness value of 0.0674.

Maximum Cut of a Graph (MAXCUT): for coding the problem we use a binary string of length n. We have considered three different graph examples in this study. Two of them are randomly generated graphs of moderate sizes: a sparse one "cut20.01" and a dense one "cut20.09", both of them are made up of 20 vertices. The other instance is a scalable weighted graph of 100 vertices. The globally optimal solutions for this instances are 10.119812 for "cut20.01", 56.740 064 in the case of "cut20.09", and 1077 for "cut100".

Minimum tardy task problem (MTTP): we have used three different instances for analyzing the behavior of our algorithms with this function: "mttp20", "mttp100" and "mttp200" with sizes 20, 100, and 200, and known maximum fitness values of 0.02439, 0.005, and 0.0025, respectively.

OneMax problem (OneMax): is a simple problem that consists of maximizing the number of ones containing a string of bits. We consider an instance with a string length of n $(n = 100)$. The optimum solution is a string of n ones, i.e. all bits of the string are fixed into one.

Table 4.1 Parameterization used in our algorithms

Population Size	400 individuals
Selection of Parents	itself + Tournament Selection
Recombination	DPX1, $p_c = 1.0$
Bit Mutation	(Bit-flip for cGA), $p_m = 1/L$
Replacement	Replace If Not Worse
Inertia coefficient	w = 1
Leaning factors	$\varphi_1, \varphi_2 = 1$
Random value	$rand = UN(0,1)$

The common parameterization used for all algorithms is described in Table 4.1, where L is the length of the string representing the chromosome of the individuals. One parent is always the individual itself while the other one is obtained by using *Tournament Selection* (TS). The two parents are forced to be different in the same neighborhood.

In the recombination operator, we obtain just one offspring from two parents. The *DPX1* recombination is always applied (probability $p_c = 1.0$). The bit mutation probability is set to $p_m = 1/L$. The exceptions are COUNTSAT, where we use $p_m = (L-1)/L$ and the FMS problem, for which a value of $p_m = 1/(2*L)$ is used. These two values are needed because the algorithms had a negligible solution rate with the standard $p_m = 1/L$ probability in our preliminary set of experiments. We here measure hit rate as the number of experiments.

Table 4.2 Percentage of Success obtained by hyCP-local, hyCP-global, and cGA for a set of problems

Problem	hyCP-local	hyCP-global	cGA
ECC	**100%**	**100%**	**100%**
P-PEAKS	**100%**	**100%**	**100%**
MMDP	58%	**61%**	54%
FMS	**83%**	81%	25%
COUNTSAT	**80%**	36%	0%
"cut20.01"	**100%**	**100%**	**100%**
"cut20.09"	**100%**	**100%**	**100%**
"cut100"	38%	**48%**	45%
"mttp20"	**100%**	**100%**	**100%**
"mttp100"	**100%**	**100%**	**100%**
"mttp200"	**100%**	**100%**	**100%**
OneMax	**100%**	**100%**	**100%**

We will replace the considered individual on each generation by the newly created individual in the same neighborhood only if the offspring fitness is not worse than the selected individual. The cost of solving a problem is analyzed by measuring the number of evaluations of the objective function made during the search. The stop condition for all algorithms is to find a solution or to achieve a maximum of one millon function evaluations. The last three rows of Table 4.1 represent the values used only for the algorithms based on PSO. Throughout the paper all best values are **bolded**.

All algorithms are implemented in Java, and run on a 2.53 GHz Intel i5 processor under Windows 7.

In Table 4.2 we show the percentage of success in 150 independent runs for the three algorithms.

We can observe that the success rate for our hybrids is higher (or equal in some cases) than for cGA algorithm. Moreover, cGA obtained a very undesirable (0%) hit rate for the COUNTSAT problem.

In Table 4.3 the following information is shown. The first column (Problem) represents the name of the problem resolved, the second column (Best) the better found solution and then for each algorithm (hyCP-local, hyCP-global, and cGA) the number of evaluations (columns Evals) needed to solve each problem, and the time in ms consumed (columns Time). Finally, the last column (ANOVA|K-W) represents the p-values computed by performing ANOVA or Kruskal-Wallis tests as appropriate, on the time and evaluations results, in order to assess the statistical significance of them (columns Evals and Time). We will consider a 0.05 level of significance. Statistical significant differences among the algorithms are shown with symbols "(+)", while non-significance is shown with "(-)".

We can observe that our hybrid algorithms reduce the number of evaluations required to reach the optimum value and also in three problems (FMS, COUNTSAT, and "mttp20") these differences are statistical significant. Meanwhile, for the time required to obtain the optimum in general, cGA obtained the minimum values

Table 4.3 Results obtained by cGA and our hybrid algorithms for a set of problems

		hyCP-local		hyCP-global		cGA		ANOVA	K-W
Problem	Best	Evals	Time	Evals	Time	Evals	Time	Evals	Time
ECC	0.07	**141400**	3369	157600	4370	150000	**2512**	(-)	(+)
P-PEAKS	1.00	39600	**3126**	**38200**	3376	39200	3283	(-)	(+)
MMDP	40	182000	5051	211200	6457	**144000**	**2295**	(+)	(+)
FMS	0.00	462800	24269	**367400**	**22183**	646800	29326	(+)	(-)
COUNTSAT	6860	**348200**	**1848**	577200	3468	1000000	2342	(+)	(+)
"cut20.01"	10.12	**4800**	31	5200	33	5200	**26**	(-)	(-)
"cut20.09"	56.74	7600	**41**	**7000**	51	8000	49	(-)	(-)
"cut100"	1077	210800	5936	220800	6031	**180800**	**3742**	(-)	(+)
"mttp20"	0.0244	4800	31	**4600**	**27**	5600	28	(+)	(-)
"mttp100"	0.005	150200	1905	**141000**	1759	152800	**1084**	(-)	(-)
"mttp200"	0.0025	**440400**	10228	450800	10265	459600	**5487**	(-)	(+)
OneMax	500	199200	10236	228000	11683	**128200**	**3581**	(+)	(+)

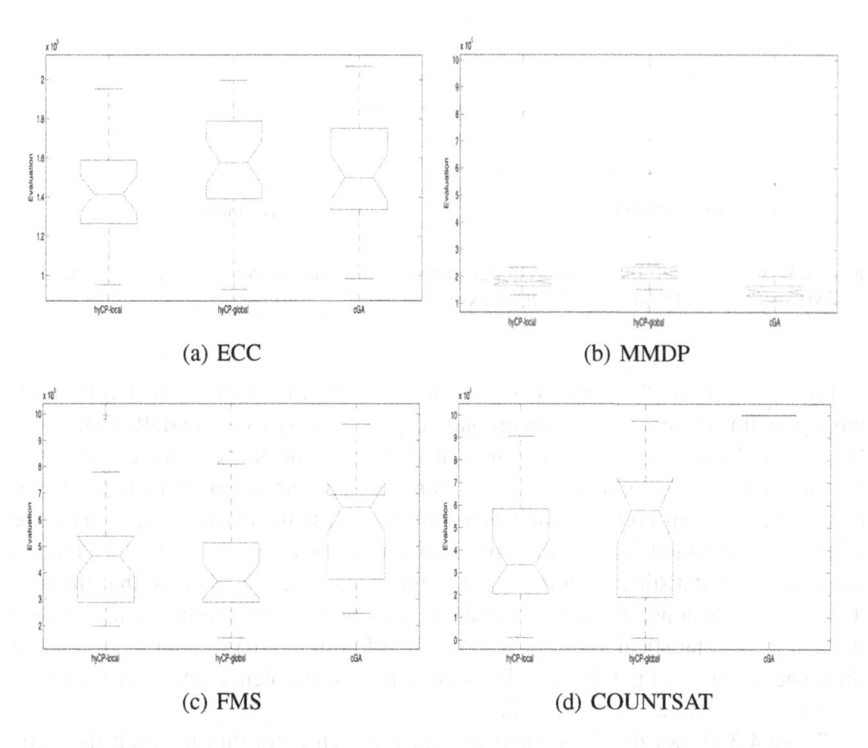

(a) ECC (b) MMDP (c) FMS (d) COUNTSAT

Fig. 4.2 Box-plots of the number of evaluations required for the algorithms considering: (a) ECC, (b) MMDP, (c) FMS, and (d) COUNTSAT problems

being slightly faster than the hybrids, who have a slight overhead compared to the canonical algorithm. This is an expected behavior since the mutation based on PSO requires some additional calculations to keep updated the particles, and to be able of using individual and social knowledge as appropriate.

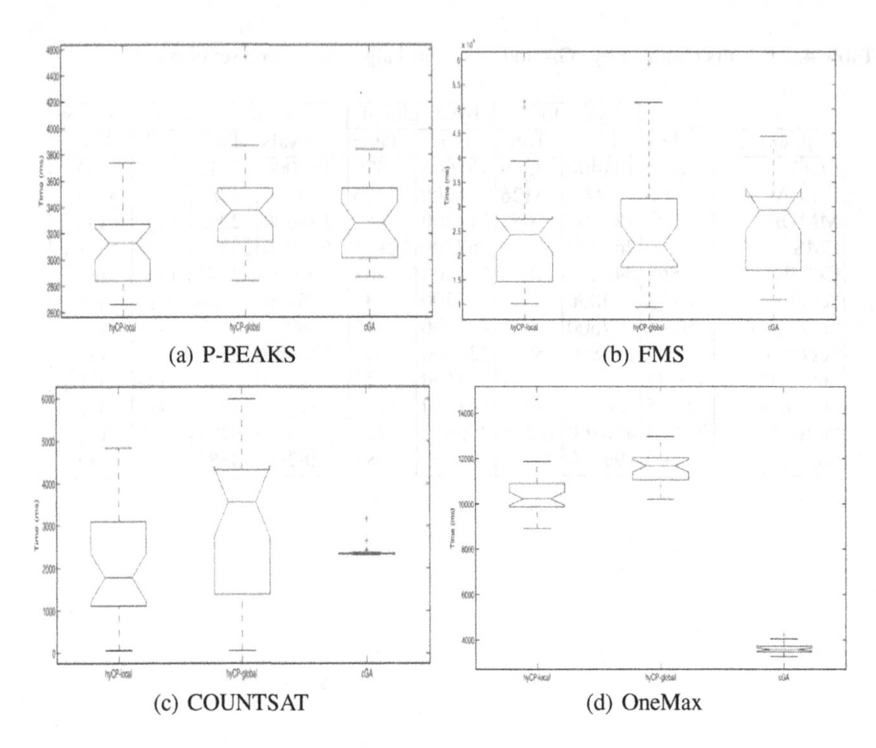

(a) P-PEAKS (b) FMS

(c) COUNTSAT (d) OneMax

Fig. 4.3 Box-plots of the Time required for the algorithms considering: (a)P-PEAKS, (b)FMS, (c) COUNTSAT, and (d) OneMax problems

Figure 4.2 shows the number of evaluations needed for each algorithm to reach the optimum value in four representative problems (ECC, MMDP, FMS, and COUNTSAT). We can observe the median values and how the results are distributed for four problems. In Figure 4.2(a) the minimum values are obtained by hyCP-local; nevertheless, the difference among the results are not statically significant. In Figure 4.2(b) cGA obtained the minimum median, and in this case the difference among the results are statistical significant. In Figure 4.2(c) we can observe that the median value is obtained by hyCP-global and also in this case the difference among the results are statistical significant. Finally, in Figure 4.2(d) we can see a marked difference in favor of hyCP-local. Recall that for this problem cGA never found the optimal value, while our hybrids did so.

Figure 4.3 shows the Time (ms) needed for each algorithm to reach the optimum value in four representative problems (P-PEAKS, FMS, COUNTSAT, and OneMax). In Figures 4.3(a) and (c) we can observe that our hybrid hyCP-local obtained the minimum median values in this two cases and also in both cases the difference among the results are statically significant. In Figure 4.3(b) hyCP-global obtained the minimum median values but the difference among the results are not statically significant. Finally, in Figure 4.3(d) cGA obtained the minimum values and the difference among the results are statically significant.

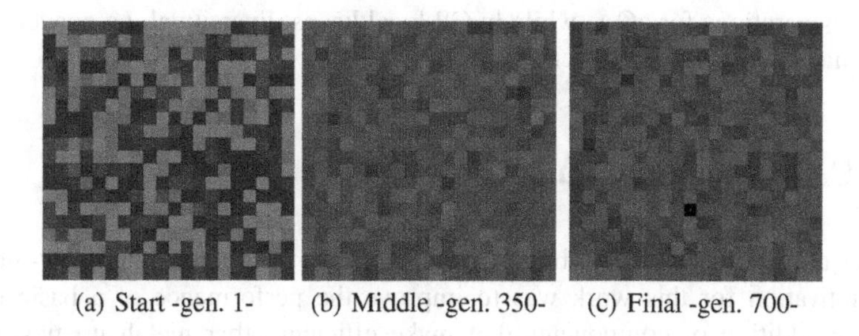

Fig. 4.4 Evolution of the search with hyCP-local on a 20x20 grid for COUNTSAT at different moments: (a)Start, (b)Middle, and (c) Final

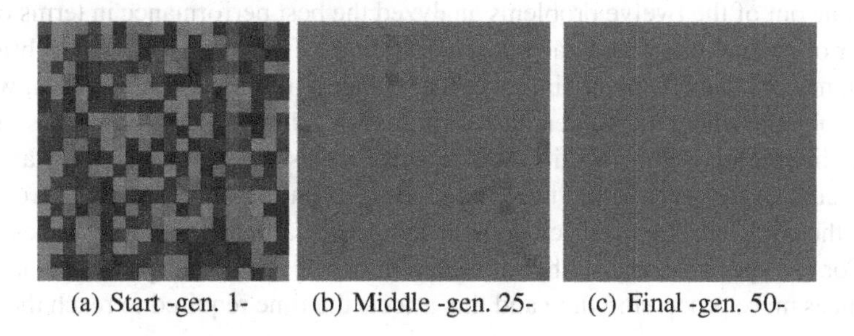

(a) Start -gen. 1-
(b) Middle -gen. 25-
(c) Final -gen. 50-

Fig. 4.5 Evolution of the search with cGA on a 20x20 grid for COUNTSAT at different moments: (a)Start, (b)Middle, and (c) Final

Now we analyze the evolution from the point of view of the phenotype diversity. The images in the Figures 4.4(a), (b), and (c) are taken from the beginning, the middle, and the final of a typical execution of our hybrid hyCP-local when solving COUNTSAT, and the images in the Figures 4.5(a), (b), and (c) are taken from the beginning, the middle, and the final of a typical execution of cGA when solving the same instance.

Figure 4.4, and Figure 4.5 show three snapshots of the fitness distribution in the population in different stages of the search. The different scales of grey represent different fitness values and, the darker the better. We can observe how the diversity (different scales of grey) decreases during the execution as expected, but in a very smooth manner, allowing reaching the global optimum.

In Figure 4.4(a) the generation number is 1 and the best found value is 6806, for Figure 4.4 (b) the generation number is 350 and the best found value is 6841. Finally, in Figure 4.4(c) the generation number is 700, and in this case, best found value is 6860.

In Figure 4.5(a) the generation number is 1 and the best found value is 6806, for Figure 4.4 (b) the generation number is 25 and the best found value is 6841. Finally, in Figure 4.4(c) the generation number is 50, and in this case, best found value is 6841. We can observe that the cGA algorithm loses genetic diversity much faster than hyCP-local. This was the reason for plotting just what happens in the first 50 generations for cGA, while hyCP-local keeps diversity along convergence to optimal solutions much better.

4.6 Conclusions and Further Work

In this work we present two hybrid algorithms, called hyCP-local and hyCP-global. The motivation for this work was to improve the performance of a basic cGA with the addition of components that make efficient other metaheuristics, with the idea of getting even better results than those already obtained by the core technique.

In nine out of the twelve problems analyzed the best performance in terms of the number of evaluations was obtained by our hybrids. This means that our hybridization framework can effectively improve the efficiency of the basic cGA. Meanwhile, in all problems where the percentage of success in achieving the optimum is different, our hybrid algorithms obtained the highest success percentage (so they are also very accurate). As regards the time required to reach the optimum values, only in five of the twelve problems discussed our hybrids obtain the minimum values. This behavior is expected because the introduction of the PSO concepts into cGA also introduces more processing time and this affects the time required to reach the optimum values.

In this chapter we used concepts of PSO, but other metaheuristics will be considered in further works like ACO (Ant Colony Optimization), SA (Simulated Annealing), and VNS (Variable Neighborhood Search), among others. Thus, what we have created is really a framework on how a family of techniques (cGA) can be hybridized in a structured manner. In fact, this framework could be used on other pairs of algorithms to ease the construction of new techniques of higher unseen efficiency and accuracy.

Acknowledgements. This research was partially funded by Spanish TIN2008-06491-C04-01 and P07-TIC-03044. The second author acknowledges the constant support provided by the Universidad Nacional de la Patagonia Austral.

References

1. Alba, E., Dorronsoro, B.: Cellular Genetic Algorithms. Springer (2008)
2. Alba, E., Tomassini, M.: Parallelism and evolutionary algorithms. IEEE Transactions on Evolutionary Computation 6(5), 443–462 (2002)
3. Alba, E., Villagra, A.: Inserting active components of particle swarm optimization in cellular genetic algorithms. In: EVOLVE A Bridge between Probability, Set Oriented Numerics and Evolutionary Computation (2011)
4. Bäck, T.: Evolutionary Algorithms in Theory and Practice: Evolution Strategies, Evolutionary Programming, Genetic Algorithms. Oxford University Press (1996)
5. Bäck, T., Fogel, D.B., Michalewicz, Z. (eds.): Handbook of Evolutionary Computation. Oxford University Press (1997)

6. Cantú-Paz, E.: Eficient and Accurate Parallel Genetic Algorithms, 2nd edn. Book Series on Genetic Algorithms and Evolutionary Computation, vol. 1. Kluwer Academic (2000)
7. Droste, S., Jansen, T., Wegener, I.: A natural and simple function which is hard for all evolutionary algorithms. In: 3rd SEAL, pp. 2704–2709 (2000)
8. Goldberg, D., Deb, K., Horn, J.: Massive multimodality, deception, and genetic algorithms. In: Männer, R., Manderick, B. (eds.) Int. Conf. Parallel Prob. Solving from Nature, PPSN II, pp. 37–46 (1992)
9. Hart, W., Krasnogor, N., Smith, J.: Recent Advances in Memetic Algorithms. Springer (2005)
10. De Jong, K., Potter, M., Spears, W.: Using problem generators to explore the effects of epistasis. In: 7th Int. Conf. Genetic Algorithms, pp. 338–345. Morgan Kaufmann (1997)
11. Kennedy, J., Eberhart, R.: Particle Swarm Optimization. In: IEEE Int. Conf. Neural Netw., vol. 4, pp. 1942–1948 (1995)
12. Kennedy, J., Eberhart, R.: A Discrete Binary Version of the Particle Swarm Algorithm. A discrete binary version of the particle swarm algorithm (1997)
13. Khuri, S., Bäck, T., Heitkötter, J.: An evolutionary approach to combinatorial optimization problems. In: 22nd Annual ACM C.S. Conf., pp. 66–73 (1994)
14. MacWilliams, F., Sloane, N.: The Theory of Error-Correcting Codes. North-Holland (1977)
15. Manderick, B., Spiessens, P.: Fine-grained parallel genetic algorithm. In: Schaffer, J.D. (ed.) 3rd ICGA, pp. 428–433. Morgan Kaufmann (1989)
16. Papadimitriou, C.: Computational Complexity. Adison-Wesley (1994)
17. Schaffer, J.D., Eshelman, L.J.: On Crossover as an Evolutionary Viable Strategy. In: Belew, R.K., Booker, L.B. (eds.) Proceedings of the 4th ICGA, pp. 61–68. Morgan Kaufmann (1991)
18. Stinson, D.: An Introduction to the Design and Analysis of Algorithms. The Charles Babbage Research Centre, St. Pierre (1985)
19. Tomassimi, M.: The parallel genetic cellular automata: Application to global function optimization. In: Albrecht, R.F., Reeves, C.R., Steele, N.C. (eds.) International Conference on Artificial Neural Networks and Genetic Algorithms, pp. 385–391. Springer, Heidelberg (1993)
20. Tsutsui, S., Fujimoto, Y.: Forking genetic algorithm with blocking and shrinking modes. In: Forrest, S. (ed.) 5th ICGA, pp. 206–213 (1993)
21. Whitley, D.: Cellular genetic algorithms. In: Forrest, S. (ed.) 5th ICGA, p. 658. Morgan Kaufmann (1993)
22. Wilson, E.O.: Sociobiology: The New Systhesis. Belknap Press (1975)

Chapter 5
Hybridizations of GRASP with Path-Relinking

Paola Festa and Mauricio G.C. Resende

Abstract. A greedy randomized adaptive search procedure (GRASP) is a metaheuristic for combinatorial optimization. GRASP heuristics are multistart procedures which apply local search to a set of starting solutions generated with a randomized greedy algorithm or semi-greedy method. The best local optimum found over the iterations is returned as the heuristic solution. Path-relinking is a search intensification procedure that explores paths in the neighborhood solution space connecting two good-quality solutions. A local search procedure is applied to the best solution found in the path and the local optimum found is returned as the solution of path-relinking. The hybridization of path-relinking and GRASP adds memory mechanisms to GRASP. This chapter describes basic concepts of GRASP, path-relinking, and the hybridization of GRASP with path-relinking.

5.1 Introduction

A combinatorial optimization problem can be defined by a finite ground set $E = (1,\ldots,n)$, a set of feasible solutions $F \subseteq 2^E$, and an objective function $f : 2^E \mapsto \mathbb{R}$. In this chapter, we consider optimization problems in their minimization form, where an optimal solution $S^* \in F$ is sought such that $f(S^*) \leq f(S)$, for all $S \in F$. The ground set E, the set of feasible solutions F, and the objective function f are defined for each specific problem. Many combinatorial optimization problems are computationally intractable, i.e. they fall into the category of NP-hard problems [32].

Paola Festa
Department of Mathematics and Applications, University of Napoli Federico II,
80126 Napoli, Italy
e-mail: paola.festa@unina.it

Mauricio G.C. Resende
Algorithms and Optimization Research Department, AT&T Labs Research,
Florham Park, NJ 07932 USA
e-mail: mgcr@research.att.com

E.-G. Talbi (Ed.): Hybrid Metaheuristics, SCI 434, pp. 135–155.
springerlink.com

Much progress has been made in the direction of exact methods for combinatorial optimization, such as branch and bound, branch and cut, and dynamic programming [72, 76]. These methods, however, suffer from the *curse of dimensionality*, i.e. they tend to break down as the size of the instance being solved increases. Likewise, approximation algorithms [74, 75], which provide a guaranteed suboptimal solution to hard combinatorial optimization problems, have also experienced significant progress. Although interesting in theory, approximation algorithms are often outperformed in practice by more straightforward heuristics with no particular performance guarantees.

Metaheuristics [33, 36] are general high-level procedures that coordinate simple heuristics and rules to find good (often optimal) approximate solutions to combinatorial optimization problems. They include genetic algorithms, simulated annealing, tabu search, scatter search, ant colonies, variable neighborhood search, GRASP, and path-relinking . There are many ways to classify metaheuristics. These include, trajectory-based versus population-based, nature-inspired versus non-nature inspired, memoryless versus memory-based, etc. Genetic algorithms, for example, are nature-inspired, population-based, with memory. Tabu search are trajectory-based with memory. GRASP is trajectory-based.

Hybrid metaheuristics combine one or more algorithmic ideas from different metaheuristics and sometimes even from outside the traditional field of metaheuristics. The main motivation to hybridize metaheuristics is to make up for the shortcomings of one metaheuristic with special characteristics of the other. In this chapter, we consider the hybridization of two metaheuristics: GRASP and path-relinking.

GRASP, or greedy randomized adaptive search procedures [25, 26, 30, 31, 59], is a metaheuristic for combinatorial optimization. GRASP heuristics are multistart procedures which apply local search to a set of starting solutions generated with a randomized greedy algorithm or semi-greedy method. The best local optimum found over the iterations is returned as the heuristic solution. Since GRASP iterations are independent of one another, GRASP heuristics do not make use of solutions produced throughout the search, i.e. they do not have any memory mechanism.

One way to add memory to GRASP is its hybridization with path-relinking. Path-relinking [35, 60, 63] is a search intensification procedure that explores paths in the neighborhood solution space connecting two good-quality solutions. A local search procedure is applied to the best solution found in the path and the local optimum found is returned as the solution of path-relinking.

This chapter describes basic concept of GRASP, path-relinking, and the hybridization of GRASP with path-relinking. In Section 5.2 we describe the main building blocks of GRASP. In Section 5.3 we consider path-relinking and, in Section 5.4, address issues related to the hybridization of GRASP with path-relinking and evolutionary path-relinking. A hybridization of GRASP with path-relinking and Lagrangean relaxation is discussed in Section 5.5. In Section 5.6 we consider parallel implementation of GRASP with path-relinking heuristics. Finally, concluding remarks are made in Section 5.7.

5.2 GRASP

Given a feasible solution $S \in F$ of a combinatorial optimization problem, a neighborhood $N(S)$ of S is a subset of F such that each element in $N(S)$ is "close" to S and can be obtained applying some elementary operation (or move) to S that changes one or more components of S. Consider the search space graph $G = (F,M)$, where the node set F is the set of feasible solutions and the edges in the set M correspond to moves in the neighborhood structure, i.e. $(S,S') \in M$ if and only if $S,S' \in F$, $S \in N(S')$, and $S' \in N(S)$.

Local search seeks a locally optimum solution in G, i.e. a solution $\hat{S} \in F$ such that $f(\hat{S}) \leq f(S)$, for all $S \in N(\hat{S})$. It starts from some solution $S^0 \in F$. At any iteration k, it seeks an improving solution $S^{k+1} \in N(S^k)$ such that $f(S^{k+1}) < f(S^k)$. On one hand, if a *first-improving* strategy is used, any improving solution S^{k+1} can be accepted. On the other hand, when a *best-improving* strategy is adopted, the improving solution S^{k+1} is the best-valued in the neighborhood, i.e. $f(S^{k+1}) = \min\{f(S) : S \in N(S^k)\}$. Local search terminates when a locally optimum solution is found. The effectiveness of local search depends strongly on the structure of the solution space graph $G = (F,M)$, the objective function f, and the starting solution $S^0 \in F$.

When designing a local search algorithm, one has the flexibility to design different neighborhoods and to select different starting solutions. Usually there is less flexibility in selecting an objective function. Some attention is needed in the design of neighborhoods since the complexity of each iteration k of local search is $O(|N(S^k)|)$. A neighborhood that is exponentially large will result in a local search with exponentially large computational complexity. Another cause of exponential computational complexity in local search is an exponentially small reduction in the objective function value when moving from a solution to a neighbor.

Since it is possible to select the starting solution S^0, a possible strategy is a *multi-start* algorithm, where local search is applied to a series of starting solutions $S_1^0, S_2^0, \ldots, S_q^0$ and the best local optimum found by the procedure is returned.

A straightforward way to implement such a multi-start algorithm is to generate each starting solution at random. A drawback to this approach is the fact that the quality of randomly-generated solutions is not very good and the number of moves needed to reach a global optimum is usually large. Not only does this result in long running times, it also increases the chance that local search will encounter a suboptimal local optimum along the way and get trapped there. The number of local optima with better cost than a randomly generated solution is usually larger than the number of local optima with better cost than a greedy solution.

A greedy algorithm builds a solution to a combinatorial optimization problem, one element of the ground set at a time. Given a partial solution, all possible candidate elements of the ground set (i.e. those elements that can be added to the partial solution without causing infeasibility) are ranked according to a myopic benefit associated with their inclusion in the solution and the next element to be added to the solution is one among the best-valued. Using a greedy algorithm to generate starting solutions for a multi-start algorithm is not recommended since the generated solutions would differ very little one from another. However, a good characteristic

Fig. 5.1 Pseudo-code of a generic GRASP

```
begin GRASP
1  f* ← ∞;
2  while stopping criterion not satisfied do
3      S ← RandomizedGreedy(·);
4      if S is not feasible then
5          S ← Repair(S);
6      end-if
7      S ← LocalSearch(S);
8      if f(S) < f* then
9          S* ← S;
10         f* ← f(S);
11     end-if
12 end-while
13 return S*;
end
```

of greedy solutions is their quality. Usually, fewer moves are needed to go from a greedy solution to a locally optimum than what is needed to go to a local optimum from a randomly generated solution.

A tradeoff between a greedy solution and a random solution is a *semi-greedy* or *randomized greedy* solution [38]. A semi-greedy heuristic is also a constructive procedure that builds a solution, one element of the ground set at a time. Like a greedy algorithm, in a semi-greedy algorithm, all possible candidate elements are ranked according to a myopic benefit associated with their inclusion in the solution. Instead of selecting one among the best-valued elements as the next one to be added to the solution, a restricted candidate list (RCL) is built with a set of good-valued candidates. One element from the RCL is selected at random and is added to the partial solution.

Hart and Shogan [38] proposed a multi-start procedure that uses a semi-greedy method but without local search. GRASP is a multi-start procedure which uses a semi-greedy method to generate starting solutions for local search. Since solutions produced by the algorithm of Hart and Shogan are not necessarily local optima, GRASP solutions are almost always better than semi-greedy solutions.

Figure 5.1 shows pseudo-code for a generic GRASP. GRASP iterations are carried out in lines 2 to 12. In line 3, the procedure attempts to build a feasible semi-greedy solution. Since this is not always possible because there is no backtracking in the greedy algorithm, a repair procedure may have to be applied in line 5 to achieve feasibility. An example of such a case can be seen in the GRASP for the generalized quadratic assignment problem of Mateus et al. [46]. A feasible solution S is used as the starting solution for the local search in line 7. If the local optimum S is better than the incumbent, then, in lines 9 and 10, it is saved as S^* and its objective function value as f^*. In line 13, the best solution found over all GRASP iterations is returned as the GRASP solution.

Fig. 5.2 Pseudo-code of the semi-greedy GRASP construction phase

```
begin GreedyRandomized
1  S ← ∅;
2  Initialize set of candidates C;
3  Evaluate the incremental cost of candidates;
4  while C ≠ ∅ do
5      Build the RCL;
6      Select s ∈ RCL at random;
7      S ← S ∪ {s};
8      Update C;
9      Reevaluate the incremental costs;
10 end-while
11 return S or indication that S is infeasible;
end
```

5.2.1 GRASP Construction

The GRASP construction phase in line 3 of the pseudo-code of Figure 5.1 combines greedy and randomized characteristics. The first implementations of GRASP made use of the semi-greedy algorithms of Hart and Shogan [38]. Figure 5.2 shows pseudo-code for a generic version of the semi-greedy algorithm of Hart and Shogan.

The semi-greedy construction builds a solution S, one element at a time. In line 1 of the pseudo-code, solution S is initialized empty. The elements of the ground set than can be feasibly added to the solution are called candidates. This set is initialized in line 2 and the costs of adding each candidate element to the solution is determined in line 3. The solution is built in the loop in lines 4 to 10. This loop is repeated while there remain candidate elements. When $C = \emptyset$, solution S can be either feasible or not. In the case that S is infeasible, a repair procedure will need to be called in the main GRASP procedure. Otherwise S is returned in line 11. In line 5, a restricted candidate list (RCL) is set up from which an element s is selected at random in line 6. This element is added to the partial solution in line 7. In line 8 the candidate set C is updated to reflect the inclusion of s in S. Finally, in line 9 the incremental costs are computed for each element of C.

Hart and Shogan [38] proposed two ways to construct the RCL. The first, called *cardinality based* , takes as input a parameter k and places the k elements with best incremental cost in the RCL. The second scheme is called *value based*. Let c^{min} and c^{max} denote, respectively, the minimum and maximum incremental cost of the candidate elements and let α be a real number in the interval $[0,1]$. A threshold $\tau = c^{min} + \alpha \cdot (c^{max} - c^{min})$ is computed and all candidate elements having incremental cost at most τ are placed in the RCL. Notice that the parameter α controls the amount of randomness and greediness in the construction process. If $\alpha = 0$, the construction is purely greedy. If $\alpha = 1$, the construction is random. By controlling the value of α, the algorithm designer can control how much greediness and/or randomness characterizes the construction which in turn controls the intensification and diversification of the search.

One way to mix intensification and diversification is to randomly generate a different α at each GRASP iteration. Prais and Ribeiro [54] proposed a scheme they call *Reactive GRASP* in which the parameter α is self-tuned to favor values which resulted in better quality solutions in previous GRASP iterations. They define $\Psi = \{\alpha_1, \ldots, \alpha_m\}$ to be the set of possible values for α. Initially, the probability of choosing a value α_i is $p_i = 1/m$, $i = 1, \ldots, m$. Furthermore, let f^* be the objective function value of the incumbent solution and let A_i be the average value of all solutions found using $\alpha = \alpha_i$, $i = 1, \ldots, m$. The selection probabilities are periodically recomputed by taking $p_i = q_i / \sum_{j=1}^{m} q_j$, with $q_i = f^*/A_i$ for $i = 1, \ldots, m$. The value of q_i will be larger for values of $\alpha = \alpha_i$ that lead, on average, to the best solutions. Larger values of q_i correspond to more suitable values for the parameter α. The probabilities associated with these more appropriate values will then increase when they are reevaluated. This reactive strategy is not limited to semi-greedy procedures where membership in the RCL depends on relative quality. It can be extended to the other greedy randomized construction schemes, all of which need to balance greediness with randomization.

In addition to the semi-greedy construction scheme, other alternative greedy randomized construction criteria have been proposed. Three such alternatives are the *random plus greedy*, the *sampled greedy* [61], and the construction by *cost perturbation* [15] schemes.

In random plus greedy, the first p components of the constructed solution are selected at random, one at a time. The remaining components are then added to the solution in a greedy fashion. In this scheme, parameter p controls the amount of randomness and/or greediness in the solution. Small values of p result in a greedy-like construction while large values of p correspond to a random-like construction.

Sampled greedy also makes use of a parameter p to control the amount of greediness and/or randomness in the construction process. At each step of sampled greedy construction process the procedure builds a RCL by sampling $\min\{p, |C|\}$ elements of the candidate set C. The incremental cost associated with adding each element of the RCL into the solution is evaluated. An element with the best-valued incremental cost is added to the partial solution. The balance between greediness and randomness is controlled by the value of parameter p. Small values of p lead to solutions constructed in a more random fashion while large values of p lead to solutions constructed in a more greedy fashion.

Construction by cost perturbation makes use of the problem data to balance the amount of randomness and greediness in the construction process. Some construction algorithm, such as, for example, an approximation algorithm, is applied to the problem where the data is randomly perturbed. The constructed solution is then evaluated using the original data. This way, by controlling the amount of perturbation, the construction will result in either a more random construction or a more greedy one.

5.2.2 Other Local Search Strategies

In addition to the first-improvement and best-improvement local search scheme described earlier in Section 5.2, other hybrid schemes have been proposed. These involve the replacement of the above mentioned local search schemes with more sophisticated local improvement methods, such as variable neighborhood descent [7, 23, 44, 67, 68], variable neighborhood search [15, 29], tabu search [16, 19, 40, 73], simulated annealing [18, 42], iterated local search [69], and very large scale neighborhood search [34].

5.2.3 Stopping Criteria

As any multi-start procedure, GRASP iterates until some stopping criterion is satisfied. Such criteria could be maximum number of iterations, maximum number of iterations without improvement of the incumbent solution, maximum running time, or solution quality at least as good as a given target value. With the exception of the last criterion, all other rules suffer from the same drawback, i.e. they cannot provide any information regarding the quality of the solution returned.

Stochastic-based stopping rules for GRASP and similar stochastic local search algorithms have been proposed, e.g. [9, 13, 21, 39, 49], but computational studies with these proposals are lacking.

Ribeiro et al. [66] study the distribution of solution values obtained by two GRASP procedures. For both procedures, the authors show that these solution values fit a normal distribution. With this observation they propose a probabilistic stopping rule for GRASP.

Let $f_1, f_2, \ldots, f_k$ be a sample formed by the first k solution values generated by GRASP. Furthermore, let μ^k and σ^k be, respectively, the estimated mean and the standard deviation of the sample. Define X to be the random variable representing the value of the local minimum found at each iteration. We assume that $X \sim N(\mu^k, \sigma^k)$, i.e. X is normally distributed with mean μ^k and standard deviation σ^k. Let $f_X^k(\cdot)$ and $F_X^k(\cdot)$ be, respectively, the probability density and the cumulative probability distribution function of X. If UB^k is the smallest solution value over the first k GRASP iterations, the probability of finding a solution at least as good UB^k in the next iteration can be estimated as $F_X^k(UB^k) = \int_{-\infty}^{UB^k} f_N^x(\tau)d\tau$. This probability is always reevaluated when the incumbent solution improves. It is reevaluated periodically even if no change in the value of the incumbent is observed. For a given threshold value β, the Ribeiro et al. probabilistic stopping rule is to stop the GRASP iterations whenever $F_X^k(UB^k) \leq \beta$. The pseudo-code in Figure 5.3 shows a GRASP with the probabilistic stopping rule.

Fig. 5.3 Pseudo-code of a generic GRASP with a probabilistic stopping rule

```
begin GRASP(β)
1  f* ← ∞; k ← 0;
2  repeat
3      S ← RandomizedGreedy(·);
4      if S is not feasible then
5          S ← Repair(S);
6      end-if
7      S ← LocalSearch(S);
8      if f(S) < f* then
9          S* ← S;
10         f* ← f(S);
11     end-if
12     k ← k+1;
13     f_k ← f(S);
14     UB^k ← f(S*);
15     Update μ^k and σ^k of f_1,...,f_k;
16     Compute F_X^k(UB^k) = ∫_{-∞}^{UB^k} f_N^x(τ)dτ;
17 until F_X^k(UB^k) < β
18 return S*;
end
```

5.3 Path-Relinking

From Section 5.2 recall the search space graph $G = (F, M)$, where the node set F is the set of feasible solutions and the edges in the set M correspond to moves in the neighborhood structure, i.e. $(S, S') \in M$ if and only if $S, S' \in F$, $S \in N(S')$, and $S' \in N(S)$. Given two solutions $S, T \in F$, the *path-relinking* operator [35] explores a path $\mathscr{P}(S,T)$ in G connecting S and T with the objective of finding solutions $S^* \in \mathscr{P}(S,T)$ for which $f(S^*) < \min\{f(S), f(T)\}$. If both S and T are good-quality solutions, then one can think of path-relinking as a search intensification procedure, which explores regions of the solution space spanned by both S and T.

Suppose path-relinking is to be done between two solutions $S \in F$ and $T \in F$. Let S be called the *initial* solution and T the *guiding* solution. One or more paths connecting these solutions in G can be explored. Local search can be applied to the best solution in each of these paths since there is no guarantee as to the local optimality of the best solution in the path.

Let $S' \in F$ be some solution in $\mathscr{P}(S,T)$. During path-relinking not all solutions in $N(S')$ are allowed to follow S' on the path $\mathscr{P}(S,T)$. Path-relinking restricts the choice to those solutions in $N(S')$ that share more attributes, or elements, with T than S' does. We denote by $N_T(S')$ this restricted neighborhood which consists of all neighbors of S' obtained by introducing into S' attributes of T not present in S'. To select the solution that follows S' on $\mathscr{P}(S,T)$, the most common choice is the greedy choice, i.e. the best-valued solution in $N_T(S')$.

Fig. 5.4 Pseudo-code of a greedy path-relinking operator

```
begin PathRelinking(S,T)
1  f* ← min{f(S), f(T)};
2  S* ← argmin{f(S), f(T)};
3  S' ← S;
4  while |Δ(S',T)| > 1 do
5      S_δ = argmin{f(Ŝ) | Ŝ ∈ N_T(S')};
6      if f(S_δ) < f* then
7          S* ← S_δ;
8          f* ← f(S_δ);
9      end-if
10     S' ← S_δ;
11 end-while
12 S* ← LocalSearch(S*);
13 return S*;
end
```

Let $\Delta(S',T)$ be the set of attributes present in T but not in S'. Introducing in S' any element $\delta \in \Delta(S',T)$ leads to a solution $S_\delta \in N_T(S')$ that can be reached by traversing edge $(S',S_\delta) \in M$. Figure 5.4 shows a pseudo-code for a basic greedy path-relinking operator. This operator scans a path from the initial solution S to the guiding solution T. In the first two lines, the best solution S^* and its value f^* are initialized and in line 3 the current solution S' is initialized to the initial solution S. The loop from line 4 to line 11 is repeated while there are attributes in the guiding solution that are not present in the current solution S'. Among all solutions in the restricted neighborhood $N_T(S')$ of S', a best-valued solution S_δ is selected in line 5. If this solution is the best seen so far, it and its value are recorded in lines 7 and 8. The current solution S' is updated in line 10 to S_δ. After examining the entire path from S to T, local search is applied to the best solution in line 12 and the resulting local optimum is returned as the solution of path-relinking in line 13.

5.3.1 *Flavors of Path-Relinking*

The scheme shown in the pseudo-code of Figure 5.4 can be implemented as different variants of path-relinking, including forward, backward, back and forward, mixed, and greedy randomized. In *forward* path-relinking, the starting solution S' is such that $S' = \mathrm{argmax}\{f(S), f(T)\}$. Conversely, in *backward* path-relinking, the starting solution S' is such that $S' = \mathrm{argmin}\{f(S), f(T)\}$. When carrying out path-relinking, the neighborhood of the initial solution is explored more thoroughly than that of the guiding solution. Since the quality of the initial solution in backward path-relinking is better than that of the initial solution in forward path-relinking, backward path-relinking usually performs better than forward path-relinking. Better yet is *back and forward* path-relinking, where a backward path-relinking is applied first and then a forward path-relinking follows. Back and forward path-relinking finds, by

Fig. 5.5 Pseudo-code of mixed path-relinking operator

```
begin MixedPathRelinking(S,T)
1  f* ← min{f(S), f(T)};
2  S* ← argmin{f(S), f(T)};
3  S' ← S;
4  while |Δ(S',T)| > 1 do
5      S_δ = argmin{f(Ŝ) | Ŝ ∈ N_T(S')};
6      if f(S_δ) < f* then
7          S* ← S_δ;
8          f* ← f(S_δ);
9      end-if
10     T' ← S_δ;
11     S' ← T;
12     T ← T';
13 end-while
14 S* ← LocalSearch(S*);
15 return S*;
end
```

definition, solutions that are at least as good as either backward or forward path-relinking, but at the expense of taking about twice as long as either.

In contrast to back and forward path-relinking, a less expensive way to explore the neighborhoods of the initial and guiding solutions is with *mixed path-relinking* [35, 65]. In mixed path-relinking, the roles of initial and guiding solutions are exchanged after each move. This way, two paths are generated, one emanating from the initial solution and the other from the guiding solution. The paths eventually meet at some solution about half way between the two input solutions. A pseudo-code for mixed path-relinking is shown in Figure 5.5.

If ties are broken deterministically in greedy path-relinking, the procedure will always generate the same path when applied to a given input pair $\{S,T\}$. Since the number of paths connecting the input pair grows exponentially with $|\Delta(S,T)|$, exploring a single path can be limiting. *Greedy randomized adaptive* path-relinking [12, 24] uses a semi-greedy move selection strategy that enables exploration of different paths when applied to the same input pair. Instead of making the greedy move choice as in line 5 of the pseudo-code in Figure 5.4, greedy randomized adaptive path-relinking builds a restricted candidate list of moves, one of which is selected at random to lead to the next solution along the path.

Good-quality solutions tend to be located near other good-quality solutions. Consequently good solutions found by path-relinking are usually found near S or T. Resende et al. [56] showed this was the case for the max-min diversity problem (see Figure 5.6). In *truncated* path-relinking, only a partial path is explored. The search is limited to solutions where only a small portion of the attributes of the guiding solution are introduced and consequently the running time to apply path-relinking is reduced.

Mateus et al. [46] observed that path-relinking can fail when $N_T(S') = \emptyset$ in line 5 of the pseudo-code of Figure 5.4. In such a case a repair procedure is applied to S' in an attempt to move from S' to some solution S'' such that $N_T(S'') \neq \emptyset$.

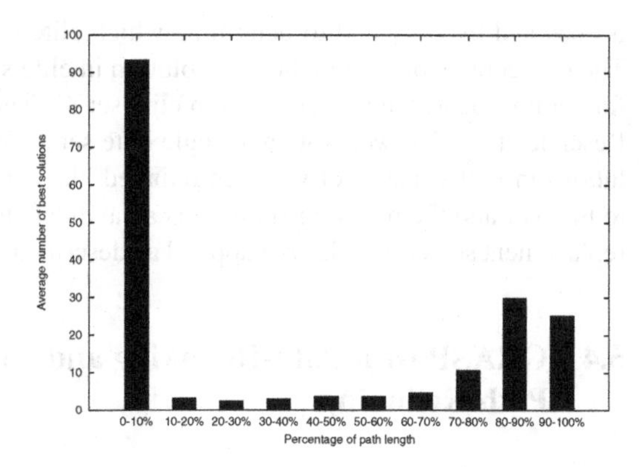

Fig. 5.6 Average number of best solution found at different depths of the path from the initial solution to the guiding solution on instances of the max-min diversity problem [56].

5.3.2 Path-Relinking and Elite Sets

An *elite set* or *pool* $\mathscr{E}$ of solutions is a fixed-size set of good-quality and diverse solutions. The quality of a solution S is with respect to its objective function value $f(S)$ while the diversity between two solutions S and T is with respect to $\Delta(S,T)$. When initially populating $\mathscr{E}$, a candidate solution S is inserted into $\mathscr{E}$ if it differs from all other solutions already in $\mathscr{E}$, i.e. if $|\Delta(S,T)| \neq 0$, for all $T \in \mathscr{E}$.

If a solution S is inserted into $\mathscr{E}$ when it is already full, it must replace some solution $T \in \mathscr{E}$. A candidate solution S is inserted into $\mathscr{E}$ if one of the following two conditions is satisfied:

1. $f(S) < f(T)$ for all $T \in \mathscr{E}$;
2. Condition (1) does not hold but $f(S) < f(T)$ for some $T \in \mathscr{E}$ and $|\Delta(S,T)| > \varepsilon$ for all $T \in \mathscr{E}$, where ε is an input parameter used to control the diversity of the elite solutions.

Once a solution S is accepted to enter the elite set, it must replace a solution $T \in \mathscr{E}$. T should be such that its replacement by S in $\mathscr{E}$ results in an elite set with smaller average objective function value and minimizes the impact on diversity of $\mathscr{E}$. A strategy [61] that achieves this goal is to select, among all solutions $T \in \mathscr{E}$ that have worse objective function value than S, the one that is most similar to S, i.e. select

$$T = \underset{T' \in \mathscr{E}}{\operatorname{argmin}} \{ |\Delta(S,T')| \text{ such that } f(T') > f(S) \}.$$

One way to combine path-relinking and elite sets is through *evolutionary* path-relinking [61]. Given an initial elite set, evolutionary path-relinking evolves the elite set applying the path-relinking operator among pair of elite set solutions. Two variants of evolutionary path-relinking have been proposed. The first, proposed in Resende and Werneck [61], works with a series of elite sets. At step k, pairs of solutions in $\mathscr{E}_k$ are relinked one pair at a time. The resulting solution of each path-relinking operation is a candidate for inclusion in elite set $\mathscr{E}_{k+1}$. The acceptance and replacement selection rules described above are used to determine if a candidate is accepted by $\mathscr{E}_{k+1}$ and to determine which elite solution in $\mathscr{E}_{k+1}$ it will replace. The procedure stops when the best solution in elite set $\mathscr{E}_{k+1}$ has the same objective function value as the best solution in elite set $\mathscr{E}_k$. The second scheme, proposed by Resende et al. [56] works with a single elite set $\mathscr{E}$. While there remain pairs of solutions in $\mathscr{E}$ that have not yet been relinked, the path-relinking operator is applied to the pair and the resulting solution is a candidate to enter $\mathscr{E}$. The acceptance and replacement selection rules are applied as described above.

5.4 GRASP with Path-Relinking and Evolutionary Path-Relinking

Laguna and Martí [41] proposed the first hybridization of GRASP with path-relinking. In their implementation, the elite set is made up of only three solutions. Each GRASP solution (local minimum obtained by the local search procedure) is relinked with a randomly chosen elite set solution. If the solution resulting from the path-relinking operator is better than the best elite solution, it replaces the worst elite solution.

Since 1999, much work has been done to improve the hybridization of GRASP with path-relinking [58, 59]. The pseudo-code in Figure 5.7 is a template for implementation of GRASP with path-relinking heuristics. The iterations of GRASP with path-relinking are carried out in lines 2 to 17. Lines 3 to 7 comprise the two phases of GRASP, producing a locally optimal solution S. In the case that the elite set $\mathscr{E}$ is not yet full, then in lines 9 to 11 S is added to $\mathscr{E}$ if it is different from all elite set solutions. In the case that the elite set is full, an elite solution T is selected in line 13 and path-relinking is applied to the pair S,T in line 14, and finally, in line 15, the elite set $\mathscr{E}$ is updated, i.e. solution S is considered for inclusion in $\mathscr{E}$ and if accepted, it will replace some existing solution in $\mathscr{E}$. In line 18, the GRASP with path-relinking procedure returns the best-quality solution S^* among all elite solutions.

GRASP with path-relinking maintains a elite set of diverse good-quality solutions found during the search. Periodically evolutionary path-relinking can be applied to the elite set with the objective of improving the quality of some of the elite set solutions. The pseudo-code in Figure 5.8 shows how to modify GRASP with path-relinking in order to obtain GRASP with evolutionary path-relinking. If a criterion for evolutionary path-relinking is triggered (line 3) then evolutionary path-relinking is applied to the current elite set in line 4. This criterion is usually a number of iterations since the last call to evolutionary path-relinking. Since the same pair of elite solutions may be relinked several times (in different calls to evolutionary path-relinking), evolutionary path-relinking is usually implemented in the inner loop (line 4) using the greedy randomized adaptive path-relinking operator. That way if a pair is relinked more than once, a different solution can result from the path-relinking operator. Finally, at the conclusion of the GRASP iterations, evolutionary path-relinking is applied a final time in line 20 to possibly improve

Fig. 5.7 Pseudo-code of a GRASP with path-relinking

```
begin GRASP+PR
1  ℰ ← ∅;
2  while stopping criterion not satisfied do
3      S ← RandomizedGreedy(·);
4      if S is not feasible then
5          S ← Repair(S);
6      end-if
7      S ← LocalSearch(S);
8      if ℰ is not full then
9          if Δ(S,T) ≠ 0, for all T ∈ ℰ then
10             ℰ ← ℰ ∪ {S};
11         end-if
12     else
13         Select T ∈ ℰ;
14         S ← PathRelinking(S,T);
15         ℰ ← UpdateElite(ℰ,S);
16     end-if
17 end-while
18 return S* = argmin{f(S) | S ∈ ℰ};
end
```

the elite set and allow the algorithm to output a potentially better solution S^* in line 21.

In a paper on GRASP with path-relinking for the three-index assignment problem, Aiex et al. [2] applied path-relinking between all pairs of the elite set as search intensification and as post-processing. Resende and Werneck [61, 62] applied evolutionary path-relinking in a post-processing phase in GRASP with path-relinking heuristics for the p-median and uncapacitated facility location problems. Andrade and Resende [2] applied evolutionary path-relinking between the two best elite solutions and all other elite solutions as a search intensification in a GRASP with path-relinking for a network migration problem. Resende et al. [56] showed through experimental results that a GRASP with evolutionary path-relinking for a max-min diversity problem could outperform heuristics based on pure GRASP with path-relinking, simulated annealing, and tabu search.

5.5 Hybrid GRASP Lagrangean Heuristic

Pessoa et al. [51, 52] proposed LAGRASP, a hybrid heuristic combining GRASP with path-relinking and subgradient optimization to solve the set k-covering problem . Their algorithm extends the Lagrangean heuristic for set covering of Beasley [11] to the case of set k-covering. In addition, instead of following Beasley and using a simple greedy heuristic as the primal heuristic, Pessoa et al. use a GRASP with path-relinking heuristic in which Lagrangean reduced costs are used in place of the original costs.

```
begin GRASP+evPR
1  E ← ∅;
2  while stopping criterion not satisfied do
3      if evPR criterion triggered then
4          E ← evPathRelinking(E);
5      S ← RandomizedGreedy(·);
6      if S is not feasible then
7          S ← Repair(S);
8      end-if
9      S ← LocalSearch(S);
10     if E is not full then
11         if Δ(S,T) ≠ 0, for all T ∈ E then
12             E ← E ∪ {S};
13         end-if
14     else
15         Select T ∈ E;
16         S ← PathRelinking(S,T);
17         E ← UpdateElite(E,S);
18     end-if
19 end-while
20 E ← evPathRelinking(E);
21 return S* = argmin{f(S) | S ∈ E};
end
```

Fig. 5.8 Pseudo-code of a GRASP with evolutionary path-relinking

The comparison of LAGRASP with pure GRASP with path-relinking showed that LAGRASP was able to find much better quality solutions than the pure GRASP with path-relinking. Furthermore, the comparison of different variants of LAGRASP showed that, by properly tuning its parameters, it is possible to obtain a good trade-off between solution quality and running time. Extensive experiments on 135 instances showed that LAGRASP can take advantage of randomization to make better use of dual information provided by subgradient optimization than Beasley's algorithm. As a consequence, LAGRASP is able to discover better solutions and to escape from locally optimal solutions after the stabilization of the lower bounds, whereas the greedy Lagrangean heuristic of Beasley [11] fails to find new improving solutions.

5.6 Parallel GRASP with Path-Relinking

Multiple-walk independent-thread parallel implementations distribute the GRASP with path-relinking iterations over the processors. Each thread performs $i_{\max}/p$ iterations, where $i_{\max}$ is the total number of iterations and p is the number of processors. As opposed to pure GRASP, were linear speedup is usually observed, multiple-walk independent-thread parallel implementations of GRASP with path-relinking have had mixed results. For example, Aiex et al. [2] showed linear speedups for the 3-index assignment problem whereas for the job-shop scheduling problem, Aiex et al. [1] showed sublinear speedups.

In this section, we focus on multiple-walk cooperative-thread schemes for implementing GRASP with path-relinking in parallel. In multiple-walk cooperative-thread schemes superlinear speedups have been observed (see, e.g. [1, 2, 3]). Two basic mechanisms have be used to implement multiple-walk cooperative-thread GRASP with path-relinking heuristics.

In *distributed strategies* [1, 3], each thread maintains its own pool of elite solutions. Each iteration of each thread consists initially of a GRASP construction, followed by local search. Then, the local optimum is combined with a randomly selected element of the thread's pool using path-relinking. The output of path-relinking is then tested for insertion into the pool. If accepted, the solution is sent to the other threads, where it is tested for insertion into the other pools. Collaboration takes place at this point. Though there may be some communication overhead in the early iterations, this tends to ease up as pool insertions become less frequent.

The second mechanism is the one used in *centralized strategies* [45, 64, 65], in which a single pool of elite solution is used. As before, each GRASP iteration performed at each thread starts by the construction and local search phases. Next, an elite solution is requested and received from the centralized pool. Once path-relinking is performed, the solution obtained as the output is sent to the pool and tested for insertion. Collaboration takes place when elite solutions are sent from the pool to other processors different from the one that originally computed it.

In both the distributed and the centralized strategies each processor has a copy of the sequential algorithm and a copy of the data. One processor acts as the master, reading and distributing the problem data, generating the seeds which will be used by the pseudo-random number generators at each processor, distributing the iterations, and collecting the best solution found by each processor. In the case of a distributed strategy, each processor has its own pool of elite solutions and all available processors perform GRASP iterations. In the case of a centralized strategy, one processor does not perform GRASP iterations and is used exclusively to store the pool and to handle all operations involving communication requests between the pool and the slaves.

5.7 Concluding Remarks

This chapter reviewed the hybridization of greedy randomized adaptive search procedures (GRASP) and path-relinking. As originally proposed in Feo and Resende [25, 26], GRASP does not make use of any memory structures. The hybridization of path-relinking with GRASP, proposed in Laguna and Martí [41], introduced memory structures in GRASP. Though path-relinking adds extra work to each iteration of GRASP (maintenance of the elite set and the path-relinking operation itself), the total number of iterations required to find a solution of a given quality more than compensates for this additional work, resulting in a higher probability that a target solution will be found in a given amount of search time. Figure 5.9 shows runtime distributions (time to target plots [4]) comparing implementations of pure GRASP

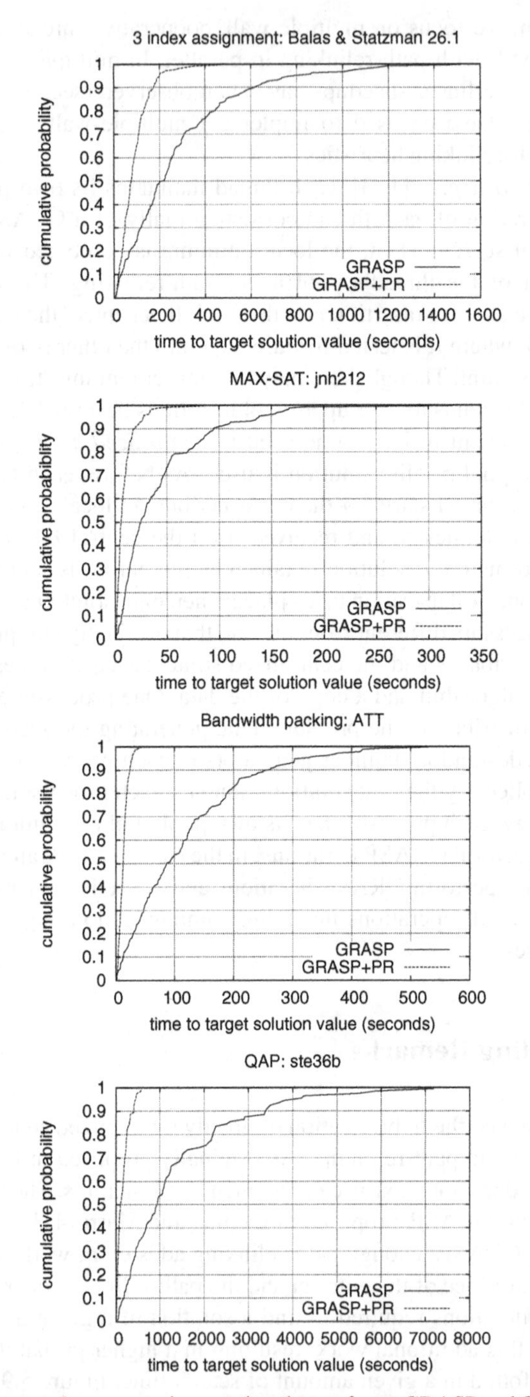

Fig. 5.9 Time to target plots comparing running times of pure GRASP and GRASP with path-relinking on four instances of distinct problem types: three index assignment [2], maximum satisfiability [27], bandwidth packing [57], and quadratic assignment [48].

and GRASP with path-relinking on four instances of distinct problem types: three index assignment [2], maximum satisfiability [27], bandwidth packing [57], and quadratic assignment [48]. The four plots are typical in the comparison of GRASP and GRASP with path-relinking in that:

- For a fixed running time, the probability that GRASP with path-relinking finds a solution at least as good as the target value is greater than the probability that pure GRASP will;
- For a fixed probability, the running time for GRASP with path-relinking to find a solution at least as good as the target value with that probability is smaller than the running time need for pure GRASP to find such a solution with the same probability.

Hybridization with path-relinking is now the standard approach to implementing GRASP.

We conclude this chapter with a list of applications of GRASP with path-relinking (which we do not intend to be exhaustive):

- Graph drawing [41];
- Job-shop scheduling [1], PBX migration scheduling [6], broadcast scheduling [17], network migration scheduling [7], machine scheduling [37], flowshop scheduling [70];
- Two-path network design [64], rural road network design [71], capacitated minimum spanning tree [73];
- Bandwidth packing [57], matrix bandwidth minimization [53], antibandwidth [22];
- Quadratic assignment [48], generalized quadratic assignment [46], three-index assignment [2], SONET ring assignment [10];
- Max-SAT [28], max-cut [29];
- *p*-median [61], uncapacitated facility location [62], health care facility location [50], capacitated clustering [20];
- Capacitated arc routing with time windows [55], traveling salesman problem [43];
- Production-distribution planning [14], assembly line sequencing [5], capacitated lot sizing [47];
- Maximum diversity [8], max-min diversity [56].

References

1. Aiex, R.M., Binato, S., Resende, M.G.C.: Parallel GRASP with path-relinking for job shop scheduling. Parallel Computing 29, 393–430 (2003)
2. Aiex, R.M., Pardalos, P.M., Resende, M.G.C., Toraldo, G.: GRASP with path-relinking for three-index assignment. INFORMS J. on Computing 17, 224–247 (2005)
3. Aiex, R.M., Resende, M.G.C.: Parallel strategies for GRASP with path-relinking. In: Ibaraki, T., Nonobe, K., Yagiura, M. (eds.) Metaheuristics: Progress as Real Problem Solvers, pp. 301–331. Springer (2005)

4. Aiex, R.M., Resende, M.G.C., Ribeiro, C.C.: TTTPLOTS: A perl program to create time-to-target plots. Optimization Letters 1, 355–366 (2007)
5. Alpay, S.: GRASP with path relinking for a multiple objective sequencing problem for a mixed-model assembly line. International J. of Production Research 47, 6001–6017 (2009)
6. Andrade, D.V., Resende, M.G.C.: A GRASP for PBX telephone migration scheduling. In: Proceedings of The Eighth INFORMS Telecommunications Conference (2006)
7. Andrade, D.V., Resende, M.G.C.: GRASP with path-relinking for network migration scheduling. In: Proceedings of the International Network Optimization Conference (2007)
8. de Andrade, M.R.Q., de Andrade, P.M.F., Martins, S.L., Plastino, A.: GRASP with Path-Relinking for the Maximum Diversity Problem. In: Nikoletseas, S.E. (ed.) WEA 2005. LNCS, vol. 3503, pp. 558–569. Springer, Heidelberg (2005)
9. Bartkutė, V., Felinskas, G., Sakalauskas, L.: Optimality testing in stochastic and heuristic algorithms. Technological and Economic Development of Economy 12(1), 4–10 (2006)
10. Bastos, L.O., Ochi, L.S., Macambira, E.M.: GRASP with path-relinking for the SONET ring assignment problem. In: International Conference on Hybrid Intelligent Systems, Los Alamitos, CA, USA, pp. 239–244. IEEE Computer Society (2005)
11. Beasley, J.E.: A Lagrangian heuristic for set-covering problems. Naval Research Logistics 37, 151–164 (1990)
12. Binato, S., Faria Jr., H., Resende, M.G.C.: Greedy randomized adaptive path relinking. In: Sousa, J.P. (ed.) Proceedings of the IV Metaheuristics International Conference, pp. 393–397 (2001)
13. Boender, C.G.E., Rinnooy Kan, A.H.G.: Bayesian stopping rules for multistart global optimization methods. Mathematical Programming 37, 59–80 (1987)
14. Boudia, M., Louly, M.A.O., Prins, C.: A reactive GRASP and path relinking for a combined production-distribution problem. Computers and Operations Research 34, 3402–3419 (2007)
15. Canuto, S.A., Resende, M.G.C., Ribeiro, C.C.: Local search with perturbations for the prize-collecting Steiner tree problem in graphs. Networks 38, 50–58 (2001)
16. Colomé, R., Serra, D.: Consumer choice in competitive location models: Formulations and heuristics. Papers in Regional Science 80, 439–464 (2001)
17. Commander, C.W., Butenko, S.I., Pardalos, P.M., Oliveira, C.A.S.: Reactive GRASP with path relinking for the broadcast scheduling problem. In: Proceedings of the 40th Annual International Telemetry Conference, pp. 792–800 (2004)
18. de la Peña, M.G.B.: Heuristics and metaheuristics approaches used to solve the rural postman problem: A comparative case study. In: Proceedings of the Fourth International ICSC Symposium on Engineering of Intelligent Systems, EIS 2004 (2004), `http://www.x-cd.com/eis04/22.pdf`
19. Delmaire, H., Díaz, J.A., Fernández, E., Ortega, M.: Reactive GRASP and tabu search based heuristics for the single source capacitated plant location problem. INFOR 37, 194–225 (1999)
20. Deng, Y., Bard, J.F.: A reactive GRASP with path relinking for capacitated clustering. J. of Heuristics 17, 119–152 (2011)
21. Dorea, C.C.Y.: Stopping rules for a random optimization method. SIAM J. on Control and Optimization 28, 841 (1990)
22. Duarte, A., Martí, R., Resende, M.G.C., Silva, R.M.A.: GRASP with path relinking heuristics for the antibandwidth problem. Networks 58, 171–189 (2011)
23. Ribeiro, C.C., Vianna, D.S.: A GRASP/VND heuristic for the phylogeny problem using a new neighborhood structure. International Transactions in Operational Research 12, 325–338 (2005)
24. Faria Jr., H., Binato, S., Resende, M.G.C., Falcão, D.J.: Transmission network design by a greedy randomized adaptive path relinking approach. IEEE Transactions on Power Systems 20, 43–49 (2005)

25. Feo, T.A., Resende, M.G.C.: A probabilistic heuristic for a computationally difficult set covering problem. Operations Research Letters 8, 67–71 (1989)
26. Feo, T.A., Resende, M.G.C.: Greedy randomized adaptive search procedures. J. of Global Optimization 6, 109–133 (1995)
27. Festa, P., Pardalos, P.M., Pitsoulis, L.S., Resende, M.G.C.: GRASP with path-relinking for the weighted MAXSAT problem. ACM J. of Experimental Algorithmics 11, 1–16 (2006)
28. Festa, P., Pardalos, P.M., Pitsoulis, L.S., Resende, M.G.C.: GRASP with path relinking for the weighted MAXSAT problem. J. of Experimental Algorithmics 11(2.4) (2007)
29. Festa, P., Pardalos, P.M., Resende, M.G.C., Ribeiro, C.C.: Randomized heuristics for the MAX-CUT problem. Optimization Methods and Software 7, 1033–1058 (2002)
30. Festa, P., Resende, M.G.C.: An annotated bibliography of GRASP, Part I: Algorithms. International Transactions in Operational Research 16, 1–24 (2009)
31. Festa, P., Resende, M.G.C.: An annotated bibliography of GRASP, Part II: Applications. International Transactions in Operational Research 16, 131–172 (2009)
32. Garey, M.R., Johnson, D.S.: Computers and intractability - A guide to the theory of NP-completeness. W.H. Freeman and Company (1979)
33. Gendreau, M., Potvin, J.-Y. (eds.): Handbook of Metaheuristics, 2nd edn. International Series in Operations Research & Management Science, vol. 146. Springer (2010)
34. Geng, Y., Li, Y., Lim, A.: A very large-scale neighborhood search approach to capacitated warehouse routing problem. In: 17th IEEE International Conference on Tools with Artificial Intelligence (ICTAI 2005), pp. 58–65 (2005)
35. Glover, F.: Tabu search and adaptive memory programming – Advances, applications and challenges. In: Barr, R.S., Helgason, R.V., Kennington, J.L. (eds.) Interfaces in Computer Science and Operations Research, pp. 1–75. Kluwer Academic Publishers (1996)
36. Glover, F., Kochenberger, G. (eds.): Handbook of Metaheuristics. Kluwer Academic Publishers (2003)
37. Gupta, S.R., Smith, J.S.: Algorithms for single machine total tardiness scheduling with sequence dependent setups. European J. of Operational Research 175, 722–739 (2006)
38. Hart, J.P., Shogan, A.W.: Semi-greedy heuristics: An empirical study. Operations Research Letters 6, 107–114 (1987)
39. Hart, W.E.: Sequential stopping rules for random optimization methods with applications to multistart local search. SIAM J. on Optimization, 270–290 (1998)
40. Laguna, M., González-Velarde, J.L.: A search heuristic for just-in-time scheduling in parallel machines. J. of Intelligent Manufacturing 2, 253–260 (1991)
41. Laguna, M., Martí, R.: GRASP and path relinking for 2-layer straight line crossing minimization. INFORMS J. on Computing 11, 44–52 (1999)
42. Liu, X., Pardalos, P.M., Rajasekaran, S., Resende, M.G.C.: A GRASP for frequency assignment in mobile radio networks. In: Badrinath, B.R., Hsu, F., Pardalos, P.M., Rajasejaran, S. (eds.) Mobile Networks and Computing. DIMACS Series on Discrete Mathematics and Theoretical Computer Science, vol. 52, pp. 195–201. American Mathematical Society (2000)
43. Marinakis, Y., Migdalas, A., Pardalos, P.M.: Multiple phase neighborhood search – GRASP based on Lagrangean relaxation, random backtracking Lin–Kernighan and path relinking for the TSP. J. of Combinatorial Optimization 17, 134–156 (2009)
44. Martins, S.L., Pardalos, P.M., Resende, M.G.C., Ribeiro, C.C.: Greedy randomized adaptive search procedures for the Steiner problem in graphs. In: Pardalos, P.M., Rajasejaran, S., Rolim, J. (eds.) Randomization Methods in Algorithmic Design. DIMACS Series on Discrete Mathematics and Theoretical Computer Science, vol. 43, pp. 133–145. American Mathematical Society (1999)
45. Martins, S.L., Ribeiro, C.C., Rosseti, I.: Applications and Parallel Implementations of Metaheuristics in Network Design and Routing. In: Manandhar, S., Austin, J., Desai, U., Oyanagi, Y., Talukder, A.K. (eds.) AACC 2004. LNCS, vol. 3285, pp. 205–213. Springer, Heidelberg (2004)

46. Mateus, G.R., Resende, M.G.C., Silva, R.M.A.: GRASP with path-relinking for the generalized quadratic assignment problem. J. of Heuristics 17, 527–565 (2011)
47. Nascimento, M.C.V., Resende, M.G.C., Toledo, F.M.B.: GRASP with path-relinking for the multi-plant capacitated plot sizing problem. In: European J. of Operational Research (2008) (to appear)
48. Oliveira, C.A.S., Pardalos, P.M., Resende, M.G.C.: GRASP with Path-Relinking for the Quadratic Assignment Problem. In: Ribeiro, C.C., Martins, S.L. (eds.) WEA 2004. LNCS, vol. 3059, pp. 356–368. Springer, Heidelberg (2004)
49. Orsenigo, C., Vercellis, C.: Bayesian stopping rules for greedy randomized procedures. J. of Global Optimization 36(3), 365–377 (2006)
50. Pacheco, J.A., Casado, S.: Solving two location models with few facilities by using a hybrid heuristic: A real health resources case. Computers and Operations Research 32, 3075–3091 (2005)
51. Pessoa, L.S., Resende, M.G.C., Ribeiro, C.C.: A hybrid Lagrangean heuristic with GRASP and path-relinking for set k-covering. Technical report. AT&T Labs Research, Shannon Laboratory, Florham Park, NJ 07932 (2010)
52. Pessoa, L.S., Resende, M.G.C., Ribeiro, C.C.: Experiments with LAGRASP heuristic for set k-covering. Optimization Letters 5, 407–419 (2011)
53. Pinana, E., Plana, I., Campos, V., Martí, R.: GRASP and path relinking for the matrix bandwidth minimization. European J. of Operational Research 153, 200–210 (2004)
54. Prais, M., Ribeiro, C.C.: Reactive GRASP: An application to a matrix decomposition problem in TDMA traffic assignment. INFORMS J. on Computing 12, 164–176 (2000)
55. Reghioui, M., Prins, C., Labadi, N.: GRASP with Path Relinking for the Capacitated arc Routing Problem with Time Windows. In: Giacobini, M. (ed.) EvoWorkshops 2007. LNCS, vol. 4448, pp. 722–731. Springer, Heidelberg (2007)
56. Resende, M.G.C., Martí, R., Gallego, M., Duarte, A.: GRASP and path relinking for the max-min diversity problem. Computers and Operations Research 37, 498–508 (2010)
57. Resende, M.G.C., Ribeiro, C.C.: A GRASP with path-relinking for private virtual circuit routing. Networks 41, 104–114 (2003)
58. Resende, M.G.C., Ribeiro, C.C.: GRASP with path-relinking: Recent advances and applications. In: Ibaraki, T., Nonobe, K., Yagiura, M. (eds.) Metaheuristics: Progress as Real Problem Solvers, pp. 29–63. Springer (2005)
59. Resende, M.G.C., Ribeiro, C.C.: Greedy randomized adaptive search procedures: Advances and applications. In: Gendreau, M., Potvin, J.-Y. (eds.) Handbook of Metaheuristics, 2nd edn. International Series in Operations Research & Management Science, vol. 146, pp. 281–317. Springer (2010)
60. Resende, M.G.C., Ribeiro, C.C., Glover, F., Martí, R.: Scatter search and path-relinking: Fundamentals, advances, and applications. In: Gendreau, M., Potvin, J.-Y. (eds.) Handbook of Metaheuristics, 2nd edn. International Series in Operations Research & Management Science, vol. 146, pp. 87–107. Springer (2010)
61. Resende, M.G.C., Werneck, R.F.: A hybrid heuristic for the p-median problem. J. of Heuristics 10, 59–88 (2004)
62. Resende, M.G.C., Werneck, R.F.: A hybrid multistart heuristic for the uncapacitated facility location problem. European J. of Operational Research 174, 54–68 (2006)
63. Ribeiro, C.C., Resende, M.G.C.: Path-relinking intensification methods for stochastic local search algorithms. J. of Heuristics (2011) (to appear)
64. Ribeiro, C.C., Rosseti, I.: A parallel GRASP for the 2-path network design problem CICLing 2001. LNCS, vol. 2004, pp. 922–926 (2002)
65. Ribeiro, C.C., Rosseti, I.: Efficient parallel cooperative implementations of GRASP heuristics. Parallel Computing 33, 21–35 (2007)
66. Ribeiro, C.C., Rosseti, I., Souza, R.C.: Effective Probabilistic Stopping Rules for Randomized Metaheuristics: GRASP Implementations. In: Coello, C.A.C. (ed.) LION 2011. LNCS, vol. 6683, pp. 146–160. Springer, Heidelberg (2011)
67. Ribeiro, C.C., Souza, M.C.: Variable neighborhood search for the degree constrained minimum spanning tree problem. Discrete Applied Mathematics 118, 43–54 (2002)

68. Ribeiro, C.C., Uchoa, E., Werneck, R.F.: A hybrid GRASP with perturbations for the Steiner problem in graphs. INFORMS J. on Computing 14, 228–246 (2002)
69. Ribeiro, C.C., Urrutia, S.: Heuristics for the mirrored traveling tournament problem. European J. of Operational Research 179, 775–787 (2007)
70. Ronconi, D.P., Henriques, L.R.S.: Some heuristic algorithms for total tardiness minimization in a flowshop with blocking. Omega 37, 272–281 (2009)
71. Scaparra, M., Church, R.: A GRASP and path relinking heuristic for rural road network development. J. of Heuristics 11, 89–108 (2005)
72. Schrijver, A.: Theory of linear and integer programming. John Wiley & Sons, Ltd., West Sussex (1996)
73. Souza, M.C., Duhamel, C., Ribeiro, C.C.: A GRASP heuristic for the capacitated minimum spanning tree problem using a memory-based local search strategy. In: Resende, M.G.C., de Sousa, J.P. (eds.) Metaheuristics: Computer Decision-Making, pp. 627–658. Kluwer Academic Publisher (2004)
74. Vazirani, V.V.: Approximation algorithms. Springer, Berlin (2001)
75. Williamson, D.P., Shmoys, D.B.: The design of approximation algorithms. Cambridge University Press, New York (2011)
76. Wolsey, L.A., Nemhauser, G.L.: Integer and combinatorial optimization. John Wiley & Sons, Inc., New York (1999)

Chapter 6
Hybrid Metaheuristics for the Graph Partitioning Problem

Una Benlic and Jin-Kao Hao

Abstract. The Graph Partitioning Problem (GPP) is one of the most studied NP-complete problems notable for its broad spectrum of applicability such as in VLSI design, data mining, image segmentation, etc. Due to its high computational complexity, a large number of approximate approaches have been reported in the literature. Hybrid algorithms that are based on adaptations of popular metaheuristic techniques have shown to provide outstanding performance in terms of partition quality. In particular, it is the hybrids between well-known metaheuristics and multilevel strategies that report partitions of the minimal cut-size value. However, metaheuristic hybrids generally require more computing time than those based on greedy heuristics which can generate partitions of acceptable quality in a matter of seconds even for very large graphs. This chapter is dedicated to a review on some representative hybrid metaheuristic approaches including genetic local search, basic multilevel search and recent development on hybrid multilevel search.

6.1 Introduction

The Graph Partitioning Problem (GPP) is one of the fundamental combinatorial optimization problems which is notable for its applicability to a wide range of domains, such as VLSI design [1, 43], data mining [49], image segmentation [42], etc. Since the general GPP is NP-complete, approximate methods constitute a natural and useful approach to address this problem. In the past several decades, many efforts have been made in devising a number of heuristic approaches such as graph growing and greedy algorithms, spectral methods, multilevel approaches, as well as algorithms based on well-known metaheuristics like tabu search, ant colony, simulated annealing, genetic and memetic algorithms. However, the application of these partitioning algorithms depends on several factors. An important factor is the trade-off between

Una Benlic · Jin-Kao Hao
LERIA, University of Angers, 2 Bd Lavoisier, 49045 Angers Cedex 01, France
e-mail: {benlic,hao}@info.univ-angers.fr

E.-G. Talbi (Ed.): Hybrid Metaheuristics, SCI 434, pp. 157–185.
springerlink.com

computation time and solution quality. Some algorithms run fast but deliver solutions of medium quality while others require significantly longer time but produce excellent quality partitions and even other that can be tuned between both extremes. This preference of time vs. quality is problem dependent. For instance, in the context of network layout or VLSI design, even a slight improvement of partition quality can be of significant importance. For these applications, it is worthwhile to employ a partition algorithm able to obtain excellent quality solutions even if the algorithm is computationally intensive. On the other hand, in other cases like sparse matrix-vector multiplication, a very fast algorithm is indispensable since the computing time required for the partitioning task has to be less than the time needed by a fast vector multiplication algorithm. Another factor that has to be considered when designing an appropriate partitioning algorithm is the partition balance. While some applications require partitions of perfect balance, others tolerate imbalance up to a certain degree in order to obtain a partition of better cut-size. All of these imply that there is no single best algorithm for all the cases, and that each one of them has its applications.

Hybrid metaheuristic approaches for GPP have shown to provide excellent performance in terms of solution quality. In particular, it is the hybrids between classical metaheuristics and multilevel methods that report partitions of the minimal cut-size value. However, these hybrids are generally more time consuming than those based on greedy iterative methods, which can produce partitions even for very large graphs in a matter of seconds.

This chapter is devoted to an overview of the most popular and effective hybrid metaheuristic approaches proposed in the literature for the k-way graph partitioning problem. We first provide a general definition of the graph partitioning problem and an overview of the benchmark instances, followed by a brief review on the most common heuristic approaches for GPP. In Section 6.4, we first describe the Kernighan-Lin (KL) heuristic and its improvement by Fiduccia and Mattheyses, and then review some of the most popular hybrids between KL-like algorithms and population-based methods. Before concluding, the multillevel paradigm for GP is detailed in Section 6.5 and some of the best multilevel metaheuristic hybrids are presented. Moreover, we try to provide an answer, based on a landscape analysis performed on a number of GP instances, to what makes these algorithms so effective.

6.2 Problem Definition and Bechmark Instances

6.2.1 *Problem Description and Notations*

The nature of a partitioning problem can greatly vary depending on the intended application. In this chapter we only focus on the partitioning approaches devised to tackle the general multi-way graph partitioning problem (also called k-way partitioning).

Consider an undirected graph $G = (V,E)$ consisting of a set of vertices (i.e. nodes) $V = \{v_1, v_2, ..., v_n\}$ and a set of edges E, such that each vertex and edge is associated with a non-negative weight that we denote by $w(v)$ and $w(e)$ respectively. Then, a k-partition of G can then be defined as a mapping (partition function) $\pi : V \rightarrow \{1,2,...,k\}$ that distributes the vertices of V among k disjoint subsets $S_1 \cup S_2 \cup ... \cup S_k = V$. The particular partitioning case when k is set to two is known as the Graph Bisection Problem (GBP).

Let $\{S_1, S_2, ..., S_k\}$ be a partition of V obtained by π, E^c the set of all the cutting edges of G induced by π, i.e., $E^c = \{\{x,y\} \in E \mid x \in S_i \text{ and } y \in S_j \text{ and } i \neq j\}$, and let φ be the set of all the partition functions of G. The k-way graph partitioning problem consists in determining $\pi^* \in \varphi$ such that the partition $\{S_1, S_2, ..., S_k\}$ given by π^* minimizes the sum of weight of edges in E^c, while ensuring that each S_i, $i \in \{1,2,...,k\}$ is of roughly equal weight. Here, the weight of a subset S_i is equal to the sum of weights of the vertices in S_i, $W(S_i) = \sum_{v \in S_i} w(v)$.

For some applications, perfect partition balance is required. On the other hand, some applications tolerate partition imbalance up to a certain limit, since allowing more imbalance may lead to partitions of better quality in terms of the total weight of edges in the cut. This notion of partition balance is defined as follows. Let $W_{opt} = \lceil |V|/k \rceil$ be the optimal subset weight, where $\lceil x \rceil$ represents the first integer $\geq x$, then the quantity $\varepsilon = max_{i \in \{1..k\}} W(S_i)/W_{opt}$ defines the degree of imbalance among the k subsets of a partition $\{S_1, S_2, ..., S_k\}$. $\varepsilon = 1$ means that the partition is perfectly balanced while $\varepsilon > 1$ indicates an imbalanced partition with larger ε corresponding to greater imbalance.

6.2.2 Benchmark Instances

In the context of various existing research studies on the graph partitioning problem defined in Section 6.2.1, a large number of benchmark instances have been used.

A well known source of graph instances that has been frequently used to compare and evaluate algorithm performance is proposed by Johnson et al. [25] and by Bui and Moon [14]. These benchmark graphs are classified into five following types where the first two classes are from [25]:

1. *Gn.d*: A random graph with n vertices, where an edge e is placed between any two vertices with probability p, such that p is chosen so that the expected vertex degree, $p(n-1)$, is d.
2. *Un.d*: A random geometric graph with n vertices uniformly distributed in a unit square. An edge is placed between two vertices if their Euclidean distance is less than or equal to $\sqrt{d/(n\pi)}$, where d is the expected vertex degree.
3. *breg.n.b*: A random regular graph with n vertices of degree 3, whose optimal bisection size is b with probability $1 - o(1)$.

4. *cat.n*: A caterpillar graph with n vertices. Starting with a spine in which every vertex is of degree 2 (except the two ending vertices), each vertex of the spine is connected to six new vertices. Given an even number of vertices in the spine, the optimal bisection size is 1. Another group of caterpillar graphs are denoted with *rcat.n* where each vertex on the spine is connected to $\sqrt{n}$ new vertices.
5. *grid.n.b*: A grid graph with n vertices whose optimal bisection size is known to be b. The same grid graph but with the boundaries wrapped around is denoted by *w-grid.n.b*.

The graph instances proposed in [25] and [14] are of rather small size (up to 1000 and 5252 vertices respectively), and as such do not represent a real challenge for recent algorithms that are designed to tackle large partitioning problems steaming from real-life applications. Another important source of GP instances is provided by the Walshaw's Graph Partitioning Archive (`http://staffweb.cms.gre.ac.uk/~c.walshaw/partition/`). These benchmark graphs represent samples of small to medium scale problems arising in different applications. Compared to the graphs provided in [25] and [14], these are of significantly larger dimensions with the biggest graph *auto* comprising 448695 vertices and 3314611 edges.

Since circuit partitioning is one of the most important applications of the GPP, the ISPD circuit benchmark suites are also used to evaluate the performance of partitioning algorithms. The circuit benchmark suite is regularly being updated with new circuits that are directly derived from real industrial designs, and that represent today's mixed-size physical design constraints in terms of size and complexity. The most recent circuit instances are presented at the ISPD-2011 placement contest [46]. These instances are large with the total number of nodes ranging from 483452 to 1293433 and the total number of nets from 468918 to 1293436. Their format can easily be converted into the standard format used by the current state-of-art k-way graph partitioning packages.

6.3 Classical Approaches for the Graph Partitioning Problem

Many different approaches have been proposed in the literature for the GPP. Some of these algorithms only take a local view of the graph and try to ameliorate the given partition, while others consider the problem globally. Some are purely deterministic always producing the same partition, while others rely on random decisions. Some operate on the graph itself, while others use some mathematical representations of it. Some are very time consuming, while others can find a partition even of very large graphs in a matter of seconds. In this section, we provide a brief review of the most common heuristic approaches applied to GPP.

Greedy graph partitioning methods are quite simple. The basic idea of these deterministic approaches is to accumulate in some way vertices into subsets, one subset at a time or alternating between subsets. Battiti and Bertossi propose two popular greedy procedures for graph bisection, the Min-Max-Greedy [7] and the Diff-Greedy [6], that assign one randomly chosen seed vertex to each bisection subset and subsequently add vertices to them alternatively. While the Min-Max-Greedy [7] method adds each time a vertex that will produce the smallest increase in the cut size of the partition, the vertex selection criterion is slightly modified in the Diff-Greedy [6] to choose vertices that have the minimum difference between the number of new external and internal edges. Another kind of greedy partitioning algorithm is proposed by Ciarlet and Lamour [15]. All of these algorithms are very fast, and have an additional advantage of being able to directly divide a graph into the desired number of subsets. This avoids applying recursive bisection that can give arbitrarily worse results than direct k-way partitioning [41]. On the other hand, the quality of partitions in terms of cut-size is not always great. Therefore, greedy partitioning algorithms are often used to generate an initial partition which is then refined with a local improvement algorithm.

The Kernighan-Lin (KL) algorithm [29] is one of the earliest and most popular local improvement heuristic for graph partitioning. It improves iteratively the quality of an existing partition obtained by other partitioning approaches. Originally, the KL procedure was intended to be applied several times starting from a different random partition. While this produces reasonable results on small graphs, it is quite ineffective on larger problem instances. Nowadays, the KL heuristic is used to complement algorithms that have a more global view of the problem but are likely to ignore local characteristics. Numerous improvements and adaptations of the basic KL procedure have been proposed in the literature. The most important improvement of the KL algorithm for graph bisection is the one proposed by Fiduccia and Mattheyses [21], which reduces the time per KL pass to linear. Both KL and Fiduccia-Mattheyses (FM) algorithms are devised to tackle only the graph bisection problem. Several adaptations of the KL and FM procedures have been proposed in the literature for the k-way partitioning. Among these is an extended FM algorithm by Sanchis [38, 39], and an adaptation of FM proposed by Hendrickson and Leland [23]. A description of the KL procedure and its modification by Fiduccia and Mattheyses is provided in Section 6.4.1.

More sophisticated vertex move based approaches for graph partitioning rely on well-known metaheuristics such as tabu search, simulated annealing, ant colony algorithms and evolutionary approaches. Moreover, Chardaire et al. [15] apply to the partitioning problem the Population Reinforced Optimization Based Exploration (PROBE) heuristic, which has been presented as a new metaheuristic [3] inspired by genetic algorithms. Although adaptations of metaheuristic algorithms are generally more time consuming than greedy iterative methods, they often yield an improvement on solution quality. In particular, the best performing approaches for the graph partitioning problem are often hybrids between a classical metaheuristic and a multilevel method. Indeed, a great number of the current best balanced partitions for the set of benchmark graphs from Walshaw's Graph Partitioning Archive are obtained with two multilevel hybrid approaches proposed by Benlic and Hao, based on an iterative tabu search algorithm [9] and a memetic algorithm [10, 11] respectively. Section 6.4 and 6.5 review respectively some of the most effective hybrid evolutionary approaches and multilevel metaheuristic hybrids.

Spectral heuristic approaches have also been extensively used. An advantage of these approaches is the possibility of defining lower bounds for the objective function of the partitioning problem. Spectral methods are based on computing eigenvalues and eigenvectors of the Laplacian matrix associated with the graph in order to construct various geometric representations of the graph. A fundamental spectral algorithm is the spectral bipartitioning (SB) based on the linear ordering representation. The SB uses the second eigenvector of the Laplacian (called the Fiedler vector), which contains important directional information about the graph G. The components of the Fiedler vector are weights associated to vertices of G, such that the differences between the components provide information about the distances between the vertices. A bisection with SB is obtained by sorting the vertices according to the sizes of the components of the Fiedler vector, and then distributing half of the vertices to each subset of the bisection. Spectral algorithms for k-way graph partitioning can be classified according to two approaches: recursive spectral bisection algorithm and direct spectral k-way partitioning algorithm. The former consists in finding the Fiedler eigenvector of a Laplacian matrix of graph G, and recursively partitioning G until the desired number of partitions is obtained. The latter uses $p \geq k$ eigenvectors and directly partitions G into k subsets with some heuristic. For a recent survey on algorithms based on these two approaches see [34]. Since spectral partitioning is a computationally intensive process, the first papers on multilevel methods for graph partitioning [4] propose multilevel implementations of spectral algorithms to simplify the calculation for a spectral method. Spectral methods are global approaches for partitioning graphs. Therefore, it is useful to improve the obtained partition with a local optimization algorithm. These are often called partition refinement algorithms. The algorithms for refinement of partitions found by spectral methods are most often of Kernighan-Lin type.

To handle very large graphs, multilevel algorithms prove to be quite useful. Various adaptations of the general multilevel technique have been tried on a number of combinatorial optimization problems including the traveling salesman, graph coloring, and the vehicle routing problem [47]. For graph partitioning, the multilevel approach has been very successful. The multilevel method was initially proposed to accelerate the performance of existing partitioning approaches. However, it was shortly recognized to be extremely effective and to have a more global vision of a graph than standard refinement procedures. The multilevel approach thus imposed itself as a global strategy using local partitioning algorithms. The basic idea of a multilevel graph partitioning approach is to successively create a sequence of progressively smaller graphs by grouping vertices into clusters. A partition of the coarsest graph is generated and then successively projected back towards the original graph followed by partition refinement. Hybrid algorithms that combine a multilevel method with a metaheuristic approach shortly became very popular for solving the partitioning problem. Section 6.5 is dedicated to the multilevel schemes for partitioning as well as to the best metaheuristic algorithms for multilevel partition refinement.

Although the current partitioning approaches are able to produce high quality partitions in reasonable time, performing partitioning in parallel is very important and has received a lot of attention. A great deal of work has been focused on parallelizing geometric graph partitioning and spectral bisection algorithms. Parallel formulations of multilevel graph partitioning schemes have also been proposed in the literature, although their development is quite challenging. Moreover, parallel versions of the three well known partition packages Jostle [48], Metis [40] and Scotch [36], based on the multilevel paradigm, have also been developed. Perhaps the fastest available parallel code is the parallel version of Metis (parMatis). It produces partitions which are worse that those obtained with the sequential version of Metis (kMetis). In general, parallelization of graph partitioning algorithms induces some penalty in terms of solution quality. However, Holtgrewe et al. [24] demonstrate in their recent work that high quality graph partitioning can be obtained in parallel in a scalable way. Moreover, their parallelization approach even seems to ameliorate partition quality, and in some cases improves the best-known partitions reported in the literature.

6.4 Evolutionary Hybrids for Graph Partitioning

Many hybrid evolutionary algorithms have been proposed in the literature for the graph partitioning problem. The success of these approaches lies in combining advantages of both recombination operator that discovers unexplored promising regions of the search space, and local search that finds good solutions by concentrating the search around these regions. Most of the popular population-based graph partitioning algorithms use the well-known Fiduccia-Mattheyses (FM) improvement of the Kernighan-Lin (KL) algorithm (or some slight modification of it) for fast iterative local improvement of partitions created in the recombination process. Before reviewing the current state-of-art hybrid population-based approaches, we thus describe the KL heuristic and its FM modification.

6.4.1 Kernighan-Lin Bisection Algorithm, Improvement and Adaptation

The Kenighan-Lin (KL) heuristic [29] improves upon a given initial bisection by exchanging two equal-size vertex subsets of the bisection. Let (A,B) be a graph bisection, i.e., $A \cup B = V$ and $A \cap B = \emptyset$. We denote by $g(a,b)$ the reduction in the cut size when two vertices $a \in A$ and $b \in B$ exchange their subsets, and by $g(v)$ the reduction when vertex v is moved to the opposite subset. The gain $g(a,b)$ can then be computed as

$$g(a,b) = g_a + g_b - 2\delta(a,b),$$

where

$$\delta(a,b) = \begin{cases} 1 & \text{if } (a,b) \in E \\ 0 & \text{otherwise.} \end{cases}$$

The KL algorithm selects a pair of vertices (a,b) which maximizes gain $g(a,b)$. Once a and b are selected, they are not considered any more for further exchange. This process is repeated to form a sequence of pairs $(a_1,b_1),...,(a_{n/2-1},b_{n/2-1})$. The algorithm then exchanges vertices in $X = \{a_1,...,a_k\}$ from one bisection subset with vertices in $Y = \{b_1,...,b_k\}$ from another bisection subset, such that $\sum_{i=1}^{k} g(a_i,b_i)$ is maximized. The above constitutes one KL pass of complexity $O(n^3)$. This process is repeated until no improvement on the bisection is possible.

To reduce the total running time per pass, Kernighan and Lin [29] suggest considering only several highest gain vertices in each subset and then selecting the pair with the maximum gain among all the combinations. This reduces the running time per pass to $O(n^2)$ and introduces only a slight degradation in solution quality.

Fiduccia and Mattheyses [21] modify the KL bisection heuristic by suggesting to move one vertex at a time instead of exchanging two vertices. Moreover, the authors propose an effective bucket data structure that reduces the time per pass to linear $O(|E|)$ by avoiding unnecessary search for the highest gain vertex and by minimizing the time needed for updating the gains of vertices affected by each move. The idea of the bucket data structure consists in placing all vertices with the same gain g in a bucket that is ranked g. Finding a vertex with the maximum gain simply consists in finding the non-empty bucket with the highest rank, and selecting a vertex from the bucket. After a vertex v has been moved to another subset, the bucket structure is updated by recomputing gains of vertex v and its neighbours, and transferring these vertices to appropriate buckets.

The bucket data structure consists of two arrays of buckets, one for each subset of a bisection, where each bucket of an array is represented by a doubly linked list. The arrays are indexed by the possible gain values for a move, ranging from g_{max} to g_{min}. A special pointer *maxgain* points to the highest index in the array whose bucket is not empty. The structure also keeps an additional array of vertices where each element (vertex) points to its corresponding vertex in the doubly linked lists. This enables direct access to the vertices in buckets and their transfer from one bucket to another in constant time. An example of the bucket data structure is illustrated in Fig 6.1.

Both KL and FM heuristics are devised only for the Graph Bisection Problem (GBP). Several adaptations of the FM algorithm have been proposed for the k-way partitioning. In [38], Sanchis proposes maintaining $k(k-1)$ previously described bucket structures, one for each of the $k(k-1)$ possible directions to move a cell (i.e., vertex) between partition subsets. The author also adopts the notion of level gain [31] that enables the algorithm to better distinguish between cells whose first level gains (regular gains) are the same. However, Sanchis suggests a more space efficient way to maintain level gains than that proposed in [31]. Moreover, making k^2 comparisons to determine the next legal cell move (i.e., move which preserves partition balance) is avoided by keeping a sorted list of the *maxgain* pointers corresponding to legal move directions. This is done by using a binary heap whose entries are *maxgain*

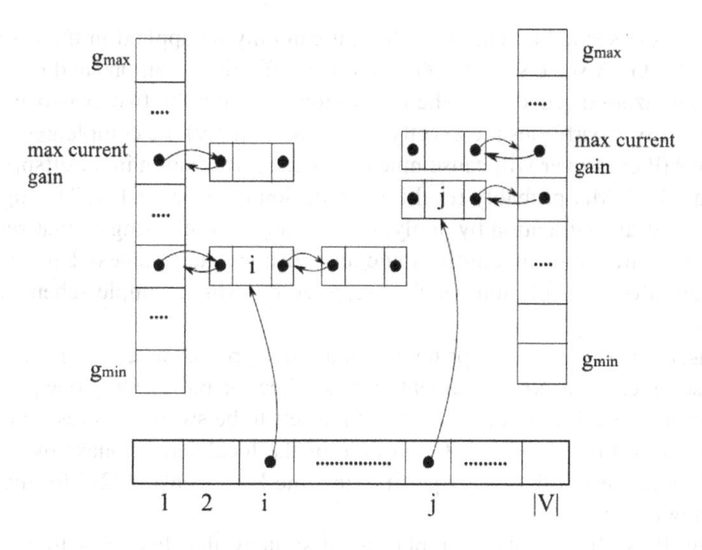

Fig. 6.1 The bucket sorting data structure [21] for graph bisection

pointers for the currently legal move directions. The pointer with the highest value is located at the root of the tree, and an array indexed by move directions is maintained holding pointers to the elements in the heap. The complexity involved in maintaining the heap is $O(lk|E|logk)$, where l is the number of gain levels.

6.4.2 Selected Evolutionary Approaches for Graph Partitioning

6.4.2.1 Hybrid Genetic Algorithm by Bui and Moon

A classical approach for GBP is the Breadth First Search Genetic Bisection Algorithm (BFS-GBA), proposed by Bui and Moon [14], that can easily be adapted for the k-way partitioning problem .

The approach uses a standard solution encoding that represents a partition as a bit string (or integer string in case of k-way partitioning), where the ith element in the string indicates the partition subset of vertex i. Although this solution representation is widely used for the GPP, it is not the most suitable because of the high redundancy which grows exponentially with the number of partition subsets k. That is, each solution can be represented in $k!$ possible ways which deteriorates the performance of traditional crossovers by introducing severe inconsistencies in an offspring partition.

In each iteration, BFS-GBA picks from the population two parents for recombination, such that the probability of selecting the best individual for recombination is four times as high as the probability of selecting the worst parent. Two offspring partitions are then created by recombining the selected parents using the standard multi-point (MP) crossover and a variation of MP (call it VMP) respectively. The VMP crossover is the same as MP except that it copies to the child the complement values from the second parent. Therefore, it can only be applied in the case of bisection. BFS-GBA selects the better of the two offspring solutions and passes it to a local optimization procedure. The motivation for using the two crossovers is the following. If two partitions are exactly (or almost exactly) the complement of each other, the MP crossover will cause much stronger perturbation in an offspring partition than the VMP, probably resulting a partition of poor quality. The algorithm reinforces the diversification by applying a mutation to offspring I^0 that selects at random m positions in the chromosome and changes their values. This generally leads to an infeasible solution which is repaired by using a simple scheme of flipping bits.

For fast improvement of offspring, Bui and Moon propose a variation of Fiduccia–Mattheyses linear time KL implementation (see Section 6.4.1). Only one pass of the local optimizer is allowed, and the size of the sets to be swapped is restricted. This decreases even more the computation time of the local serach phase by about an order of 10. In case of the k-way partitioning, the KL extension [29] for the k-way partitioning is used.

Finally, BFS-GBA applies a replacement strategy that first tries to replace I^0 with the more similar parent based on the Hamming distance measure, and if it fails, it tries to replace the other parent (replacement is carried out only when I^0 is better than one of the parents). Although this scheme preserves longer a diversified population, it is very time consuming since with large diversity the algorithm takes more time to converge. For this reason, the BFS-GBA scheme replaces offspring I^0 with the worst individual from the population in case when I^0 is worse than both of the parents.

The algorithm stops when 80% of the population is occupied by solutions of the same quality, which are not necessarily identical solutions.

An important component of this hybrid genetic algorithm is the preprocessing phase that can dramatically improve the performance on some types of graphs (geometric and caterpillar graphs) at very little cost in time. The rational behind this preprocessing scheme is to reorder vertices on the chromosome in an attempt to ensure that clusters of highly connected vertices are included in short schemas that have more chance to survive in a crossover. It consists in performing a breadth first search (BFS) on the input graph starting at a random vertex. The order in which vertices are visited by the BFS is used to reorder vertices on the chromosome. That is, the ith vertex in the BFS ordering takes the position i in the chromosome. This preprocessing phase is carried out only once before the start of the hybrid genetic algorithm.

The performance of BFS-GBA was tested on the set of graphs from [25], as well as on a number of instances specially designed for this evaluation [14]. In this study, BFS-GBA competes very favourably with the multi-start KL algorithm and the simulated annealing proposed by Johnson et al [25]. Moreover, it is considerably faster then the simulated annealing approach [25]. Unfortunately, the performance of BFS-GBA has not been demonstrated on larger instances ($|V| > 10000$).

6.4.2.2 A Memetic Algorithm by Merz and Freisleben

Merz and Freisleben [32] propose a memetic algorithm for GBP, which is based on the observations made from an exhaustive landscape analysis on a set of local optima sampled with the KL and greedy heuristics respectively for the instances proposed in [14, 25]. The results of this analysis showed that the landscape of the GBP is highly dependent on the graph structure and that some landscapes have slightly more correlated local optima than others. However, all these analysed graphs generally have a rather structured landscape (fitness-distance correlation coefficient $\rho_{fdc} > 0.15$), and local optima that are concentrated within a limited region of the search space (see Section 6.5.4.1). Therefore, the authors propose a memetic algorithm (MA) that attempts to exploit the landscape structure of the problem.

The algorithm uses the same solution encoding as BFS-GBA. For the local search phase, MA employs the standard KL algorithm which runs in $O(|E|)$ instead of $O(n^2)$ time by means of the bucket data structure proposed by Fidducia and Mattheyses.

Instead of generating the initial population randomly, the algorithm uses the randomized Diff-Greedy heuristic by Battiti and Bertossi [6] since it is one of the best constructive heuristics for the GBP and is able to generate a wide range of high quality solutions. The idea of the Diff-Greedy algorithm consists in generating a partition by adding vertices alternatively to partition subsets in a greedy way. Let S^0 and S^1 be two subsets of the bisection. At each stage, the vertex selected to enter a subset, say S^0, is the vertex for which the number of neighbour vertices in S^0 minus the number of neighbour vertices in S^1 is maximized. The rationale behind this selection criterion is that a bisection that minimizes the cut size maximizes at the same time the number of internal edges.

The Diff-Greedy heuristic thus exploits the structure of the search space that has shown to be very effective for instances of the GPP. The authors therefore propose a new crossover called greedy crossover (GC) which is based on the same idea as the Diff-Greedy heuristic. In the first phase of the GC, all vertices that are contained in the same partition subset in both parents are placed in the same subset in the offspring. Then, the rest of vertices is assigned to both subsets according to the selection strategy used in the Diff-Greedy algorithm. If $|S_0| < |S_1|$ a vertex is added to subset S_0, else to S_1.

Selection for recombination in MA is done uniformly at random without bias to fitter individuals, while selection for survival is performed by choosing the best individuals from the union of parents and children by taking care that there is no duplicate in the population.

Due to the computing time required by the local search phase, the population size is kept very small (up to 40) compared to genetic algorithms. This leads to premature convergence, especially in the absence of mutation. To overcome this problem, MA triggers a restart mechanism that has shown to be very effective for combinatorial optimization problems, including the QAP and the TSP. Upon convergence (the average Hamming distance has dropped below a threshold (d = 10) or there was no changes in the population for more than 30 generations), the whole population except the best individual is mutated by exchanging subsets of randomly chosen pairs of vertices (v_1, v_2) such that $v_1 \in S$ and $v_2 \notin S$. After mutation, each individual is improved with the KL local search and MA proceeds with performing the crossover as usual.

This memetic algorithm shows to be effective, scalable and very robust on different types of graphs, and is able to produce better average cut size than any previous heuristic search method including tabu search, simulated annealing, and hybrid genetic algorithms.

6.4.2.3 A PROBE Based Heuristic by Chardaire et al.

In [15], the authors propose an adaptation of a new population-based metaheuristic technique named PROBE (Population Reinforced Optimization Based Exploration) for the GBP. The PROBE method is conceptually much simpler than genetic algorithms as it does not include selection, replacement, and mutation procedures. The basic idea of PROBE is to find optimized solutions by exploring promising search subspaces, starting from solutions in which common characteristics found in both parents are preserved. These optimized solutions then constitute a new population in the next generation.

As in [14], the PROBE bisection algorithm (PROBE-BA) uses the bit string solution encoding where the ith bit indicates the subset of vertex i. Although this scheme is quite intuitive, the authors note that a less redundant encoding might improve the partition quality.

PROBE-BA uses standard graph partitioning approaches for exploration and exploitation of the search space, namely the Diff-Greedy heuristic by Battiti and Bertossi [6] (see Section 6.4.2.2) and the variation of the KL algorithm by Bui and Moon [14] (see Section 6.4.2.1) . Given a population $POP = \{I_q^1, I_q^2, ..., I_q^{|POP|}\}$ of feasible solutions at generation q, the next generation of solutions is obtained as follows. For each $i = 1, ..., |POP|$, a partial bisection I_{q+1}^i is computed from the pairs (I_q^i, I_q^{i+1}) (where the superscript is taken modulo $POP + 1$) by fixing the vertices corresponding to the bits shared by the two parent solutions I_q^i and I_q^{i+1}. This partial solution is then used as input to the Diff-Greedy algorithm to obtain a complete bisection of the given graph. Note that this recombination process is exactly the same as with the previously described greedy crossover devised by Merz and Freisleben [32], which tries to exploit the landscape structure. Once solution I_{q+1}^i has been constructed, its quality is improved with the fast KL local optimizer designed by Bui and Moon [14].

The performance of PROBE-BA has extensively been evaluated on a large number of graphs of different sizes (the largest graph *auto* has $|V| = 448695$ and $|E| = 331461$). The results show that PROBE-BA can compete with other population-based algorithms, reactive tabu search, or more specialized multilevel partitioning approaches. Moreover, it was able to improve, in reasonable time, the previous best cut values for a number of real world instances.

6.5 Multilevel Graph Partitioning

As illustrated in [47], the multilevel paradigm is a useful approach to solving combinatorial optimization problems that even appears to impart a 'global' quality to local search heuristics. Basically, the approach allows one to approximate the initial problem by approximating successively smaller (and easier) problems. Moreover, the coarsening helps filter the solution space by placing restrictions on which solutions the refinement algorithm can visit. We dedicate this section to some of the best performing and most popular multilevel hybrids that combine a refinement algorithm based on a metaheuristic approach. After a formal definition of the general multilevel procedure, we provide a review on two basic types of multilevel schemes and on the most effective adaptations of metaheuristic techniques that have been proposed for partition refinement of coarsened graphs.

6.5.1 Formal Definition of the Multilevel Paradigm

Let $G_0 = (V_0, E_0)$ be the initial graph, and let k denote the number of partition subsets. The multilevel paradigm can be summarized by the following steps.

1. Coarsening phase: The initial graph G_0 is transformed into a sequence of smaller graphs $G_1, G_2, ..., G_m$ such that $|V_0| > |V_1| > |V_2| > ... > |V_m|$. Each coarse graph represents the original problem, but with fewer degrees of freedom. Coarsening stops when $|V_m|$ reaches a fixed threshold (coarsening threshold).
2. Initial partitioning phase: A k-partition P_m of the coarsest graph $G_m = (V_m, E_m)$ is generated. It allows to get the first approximation of the problem.
3. Uncoarsening phase: Partition P_m is progressively projected back to each intermediate G_i $(i = m-1, m-2, ..., 0)$. Before each projection, the partition is first refined (improved) by a refinement algorithm.

This process leads thus to a sequence of partitions $P_m, P_{m-1}, P_{m-2}, ...P_0$. The last one, i.e., P_0 is returned as the final partition of the original graph G_0.

In Figure 6.2, we illustrate this multilevel procedure for the 4-way partitioning problem.

6.5.2 Multilevel Schemes

Two main classes of multilevel schemes have been proposed in the graph partitioning literature. In general, any coarsening can be defined as a process of aggregation of graph vertices to form the vertices of the next coarser graph. In the following, we describe the strict aggregation scheme, as well as the more effective weighted aggregation which has recently been proposed for the partition problem. While the former scheme makes hardened local decisions at each graph level, the latter

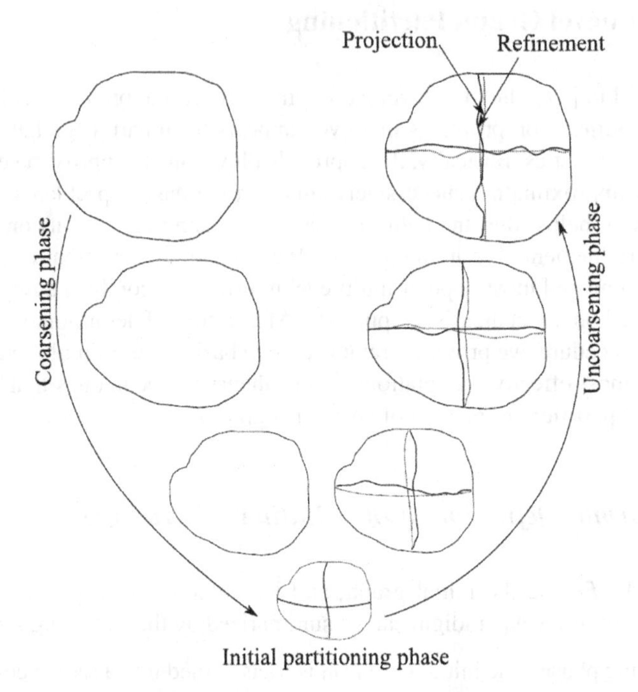

Fig. 6.2 The coarsening, initial partitioning and the uncoarsening (projection/refinement) phases for the 4-way partitioning problem. During the uncoarsening phase, the dashed lines represent projected partitions and the dark ones indicate partitions refined after projection.

introduces more freedom in solving the coarser levels and avoids making local decision before gathering the pertinent global information.

6.5.2.1 Strict Aggregation Scheme (SAG)

Strict aggregation (SAG) [11, 23, 28, 44], also called edge contraction or matching of vertices, is employed by most multilevel partitioning algorithms. The idea of the SAG is to form a new vertex $v \in V_{i+1}$ of a coarser graph G_{i+1} by merging a subset of vertices $V_i^c \subset V_i$ of G_i that (usually) have a strong local coupling (i.e., connectivity).

The weights of the resulting vertices and edges of the coarsened graph $G_{i+1} = (V_{i+1}, E_{i+1})$ are set accordingly. The weight of the new vertex $v \in G_{i+1}$ becomes equal to the sum of weights of the vertices that are aggregated to form v. Similarly, let $v_a, v_b \in V_{i+1}$ be two vertices formed by collapsing $\{v_1, v_2, v_3\} \in V_i^a$ and $\{v_4, v_5, v_6\} \in V_i^b$. All the edges incident to $\{v_1, v_2, v_3\}$ and $\{v_4, v_5, v_6\}$ are merged to form a new edge $\{v_a, v_b\} \in E_{i+1}$ with a weight that is set equal to the sum of weights of the edges incident to $\{v_1, v_2, v_3\}$ and $\{v_4, v_5, v_6\}$. Updating of vertex and edge weights is illustrated in Fig 6.3. For simplicity, the cardinality of the vertex subset that is merged to form a new vertex of a coarser graph is set to two.

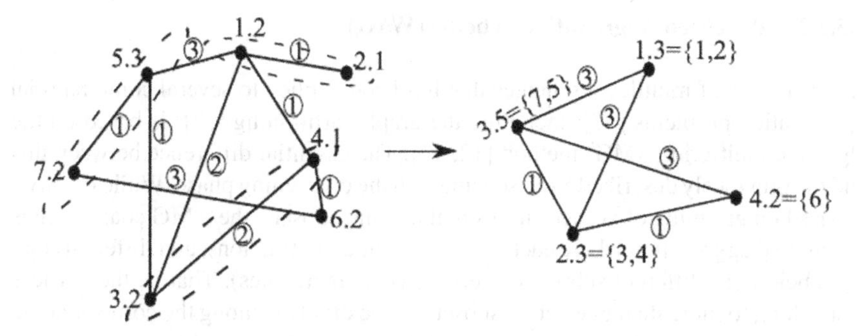

Fig. 6.3 An example of updating weights of new vertices and edges formed after vertex aggregation. For each vertex v, we indicate its corresponding weight w separated by a dot (e.g., $v.w$). For each newly formed vertex v_n, we also indicate the subset of vertices that are aggregated to form v_n.

Two main SAG schemes have been proposed for coarsening graphs. The first scheme is based on finding an independent subset of edges (matching) $\Gamma \subset E_i$, and then collapsing the two vertices of each edge in Γ to form a new vertex of a coarsened graph $G = (V_{i+1}, E_{i+1})$ [11, 27, 28, 44]. Any vertex that is not part of Γ is simply copied over to G_{i+1}. The second scheme is based on finding c-cliques for $c > 2$, and then collapsing the vertices of the cliques to form new vertices of the coarsened graph [19, 22]. We concentrate on the matching scheme, since it is the most commonly used coarsening method.

The key issue for the matching scheme is the selection of the independent subset of graph edges Γ to be collapsed at each step of the coarsening phase. This can be achieved by finding a maximal matching of the graph [35]. That is, the objective is to find the maximal number of edges no two which are incident on the same vertex. There exist polynomial time algorithms for tackling this problem, with running time of at least $O(|V|^{2.5})$. Unfortunately, this is too slow to be applicable to the partitioning problem. Therefore, several fast heuristic approaches have been proposed to compute an approximate maximal matching such as the heavy-edge matching heuristic (HEM), which has $O(|E|)$ time complexity [27]. The HEM considers vertices in random order, matching each unmatched vertex v with its unmatched neighbor u, if any, such that the weight of edge $\{u, v\}$ is maximal among all the edges incident to v. Other similar heuristics for computing an approximate maximal matching of a graph include Random Matching (RM), Light Edge Matching (LEM), and Heavy Clique Matching (HCM). If time is not a issue, metaheuristic algorithms might be used to approximate maximal matchings of a graph at the expense of running time. In [30], the authors propose a genetic algorithm to determine an approximate maximal matching that has shown to provide a significant improvement on solution quality.

The projection in case of the SAG is a trivial process. If a vertex $v \in V_i$ is in subset S_m, then the matched pair of vertices $v_1, v_2 \in V_{i-1}$ which represents vertex $v \in V_i$ will also be in subset S_m.

6.5.2.2 Weighted Aggregation Scheme (WAG)

Another class of multilevel schemes that has been applied to several combinatorial optimization problems [37] (including the graph partitioning [17]) is based on the algebraic multigrid (AMG) method [12, 13]. The essential difference between this and the previously described SAG scheme is in the coarsening phase. While the SAG is based on grouping of vertices into small disjoint subsets, the AMG coarsening is a weighted aggregation where each vertex is divided in fractions, and different fractions belong to different subsets of vertices (i.e., aggregates). That is, the vertices that belong to more than one vertex subset will be divided among the corresponding aggregates. Like with the SAG scheme, all the vertices from a subset are merged to form a vertex of the coarser level, where they will be blocked into larger aggregates, forming vertices of a still coarser graph, and so on. Weighted instead of strict aggregations is important to express a likelihood of vertices to belong to the same subset. These likelihoods are accumulated at the coarser levels, indicating tendencies of vertices to be associated together. In that way, weighted coarsening avoids making hardened local decisions, such as edge contractions, before gathering important global information.

The WAG coarsening scheme for graph partitioning first starts by selecting a set $C \subset V_i$ of seed vertices of the finer graph G_i, that will constitute the vertices of the coarser graph G_{i+1}. This process is guided by the principal that each vertex from $F = V \backslash C$ should have a strong dominant connection to C. Then, starting from $C = \emptyset$ and $F = V$, vertices are being transferred from F to C until all the remaining vertices from F satisfy the following condition:

$$\sum_{j \in C} w_{ij} / \sum_{j \in V} w_{ij} \geq Q,$$

where Q is a parameter (usually $Q \approx 0.5$).

Each vertex from C becomes a seed of one aggregate that will form one vertex of the coarser graph. Define for each $i \in F$ a coarse neighborhood $N_i = \{j \in C, w_{ij} \geq \alpha_i\}$, where α_i is a parameter that limits the nighborhood size in order to control complexity. The AMG interpolation matrix is then defined by

$$P_{ij} = \begin{cases} w_{ij} / \sum_{k \in N_i} w_{ik} & \text{for } i \in F, j \in N_i \\ 1 & \text{for } i \in C, j = i \\ 0 & \text{otherwise,} \end{cases}$$

where each entry P_{ij} of the matrix represents the likelihood of vertex i to belong to the jth aggregate. Let $I(k)$ be the order number in the coarser graph G_{i+1} of a vertex that constitutes an aggregate around a seed vertex whose order number in the finer graph G_i is k. The weight of an edge that connects two aggregates $p = I(i)$ and $q = I(j)$ in G_{i+1} becomes equal to $w_{pq} = \sum_{k \neq l} P_{ki} w_{kl} P_{lj}$. As in the case of the SAG scheme, the total sum of vertex weights is conserved throughout each graph level.

The projection process of the WAG scheme is more complex than in the case of the SAG. One of the simpler ways to project a partition consists in computing the probability that a fine vertex belongs to a particular subset of the partition. In the case of a graph bisection, only the probability of being part of subset 0 (or 1) matters. Then, the probability that vertex i belongs to subset 0 can be determined with the following relation:

$$P_0(i) = \sum_{k \in N, I(k) \in S_0} P_{iI(k)}.$$

With this strategy, vertex i is assigned to subset 0 if the probability $P_0(i) \geq 0.5$, and to subset 1 otherwise.

Although the SAG scheme has been used by most multilevel partitioning approaches, experimental evaluations have shown that the usage of WAG can significantly improve the quality of a partition. For a comparison between the two types of schemes for the graph partition problem, the reader is referred to a recent paper by Chevalier and Safro [17].

6.5.3 *Effective Refinement Strategies Based on Metaheuristics*

Partition refinement approaches that are used in conjunction with the multilevel paradigm are most often based on the KL linear-time complexity improvement by Fiduccia and Mattheyses. The KL heuristic has shown to be efficient in finding locally optimal partitions when it starts with a fine initial partition. Since the projected partition is already of fairly good quality, the KL considerably decreases the cut-size within a small number of iterations [27]. For that reason, four out of five public-domain graph partitioning packages (Chaco, Jostle, Metis, and Scotch), whose aim is to find reasonably good partitions in very short computing time, use a multilevel KL hybrid as the default setting. However, it has been shown that, given a longer computing time, other multilevel refinement algorithms that are based on well-known metaheuristic techniques, such as tabu search or evolutionary approaches, are able to largely improve on the solution quality in terms of cut-size.

We next review some of the best performing metaheuristic refinement approaches for the k-way partitioning. Note that each refinement procedure of a multilevel strategy can solely be applied to solve the partitioning problem. The multilevel paradigm is integrated since it can often either accelerate the convergence of the local serach or even improve the asymptotic convergence in solution quality [47].

6.5.3.1 Perturbation-Based Tabu Search by Benlic and Hao

A recently proposed perturbation-based iterated tabu search (ITS) procedure [9], combined with the multilevel matching scheme (see Section 6.5.2.1), has shown to be extremely effective in finding balanced ($\varepsilon = 1.0$) k-way partitions. Perfect partition balance for this algorithm is thus imposed as a constraint and is progressively established during the search, while the objective is to minimize the total sum of cutting edge weights.

The ITS employs two neighborhood relations (call them N_1 and N_2) which are explored in a token-ring way. That is, one neighborhood search is repeatedly applied to the best local optimum produced by the other neighborhood. Given a subset S_i of a k-partition $p = \{S_1, S_2, ..., S_k\}$, the basic idea of the neighborhood relations N_1 and N_2 is to move a vertex v from its current subset to subset S_i, under the constraint that v must be a border vertex relative to S_i, i.e., $v \notin S_i$ has at least one adjacent vertex in S_i. Note that in this way, the size of the neighborhoods is limited, since the set of border vertices relative to S_i is generally of small size. In addition, such a neighborhood allows the search to concentrate around these critical vertices.

Let $I = \{S_1, S_2, ..., S_k\}$ be a k-partition, $V(S_i)$ the set of border vertices relative to subset S_i, and $S_{max} = \{S_i | max_{i \in \{1..k\}} \{W(S_i)\}\}$ the subset with the maximum vertex weight. The neighborhood relations N_1 and N_2 can be explained by the two move operators given below.

Move 1: Move one highest gain vertex v_m. Choose randomly a subset $S_m \in \{S_1, S_2, ..., S_k\} - \{S_{max}\}$. Then, select the *highest gain* vertex $v_m \in V(S_m)$ whose current subset is S_c, such that $S_c \in \{S \in I | W(S) > W(S_m)\}$. Move the selected vertex v_m to subset S_m.

Move 2: Move two highest gain vertices v_m and v_n. Choose vertex v_m and its new subset S_m as with the first move operator. Choose randomly a new subset $S_n \in \{S_1, S_2, ..., S_k\} - \{S_{max}, S_m\}$. Then, select vertex $v_n \in V(S_n)$ whose current subset is S_c, such that $S_c \in \{S \in I | S \neq S_n\}$. Move v_m to S_m, and v_n to S_n.

As defined in Section 6.4.1, the gain for moving vertex v to subset S_m is the reduction in the cut size. The selection of the vertex with the highest gain, as well as the updates needed after each move, are achieved efficiently by using a new adaptation of Fiduccia-Mattheyses bucket sorting [21] for the k-way partitioning that maintains k arrays of buckets, one for each partition subset.

It is important to note that these move operators progressively lead the search toward a balanced partition since they basically constraint (partially with Move 2) vertex migration from heavy weight subsets to light weight subsets.

Let $V_{cand} \subset V(S_m)$ be the set of the highest gain vertices which are considered for migration to subset S_m. The selection of vertex v, which is moved to S_m, is based on several pieces of history information. This selection strategy is first conditioned by the tabu status. It also employs two additional criteria which are based on *vertex move frequency* and *vertex weight*. The move frequency is a long term memory that records, for each vertex v, the number of times v has been moved to a different subset. It gives priority to moves that have been applied less often. If there is more than one vertex with the same move frequency in the set V_{cand}, the second criterion is used to distinguish them and prefer a vertex v which, when moved to subset S_m, minimizes the weight difference between the target subset S_m and the original subset S_c.

Each time a vertex v is moved from a subset S_c to another subset S_m, it is forbidden to move v back to its original subset S_c for the next tt iterations (tabu tenure), where tt is tunded adaptively.

The described TS procedure applies a very aggressive search procedure since it focuses only around border (critical) vertices. Therefore, to avoid getting trapped in a local optimum, the algorithm periodically triggers a simple perturbation which consists in moving a fixed number of vertices γ, including non-border ones, such that the partition balance is not degraded.

This multilevel ITS algorithm (MITS) is designed to produce excellent quality partitions with the possibility to generate solutions of various qualities depending on the amount of computing time allowed. Indeed, experimental studies on a set of graphs from Walshaw's graph partitioning archive have shown that partitions generated with MITS within short computation time (from 1 second up to several minutes for a graph with $|V| = 143437$ and $|E| = 409593$) are generally far better than those produced by the current public-domain partitioning packages. When the running time is prolonged up to one hour, the described algorithm often outperforms the existing state-of-art graph partitioning algorithms in terms of solution quality.

6.5.3.2 An Evolutionary Approach by Soper et al.

A popular approach for generating high quality graph partitions, proposed by Soper et al. [44], is a combination of an evolutionary search algorithm and a multilevel partitioner.

The employed multilevel partitioner, known as JOSTLE, is based on the matching scheme with the HEM heuristic (see Section 6.5.2.1) and the linear-time KL improvement by Fiduccia and Mattheyses that the authors extend for use with non-integer gains by integer scaling. The fitness function used by the evolutionary approach is defined to be $-f\lambda$, where f is the number of edges in the cut and λ the degree of imbalance. The partition imbalance is thus not considered as a constraint, but induces a heavy penalty in case of greater imbalance. In this way, partitions within the balance constraint eventually dominate the population as the search progresses.

The basic idea of the approach is to assign a bias (≥ 0) to each vertex, and a weight to each edge that is equal to one plus the sum of the biases of its incident vertices. When applying JOSTLE to a graph with biased vertex and edge weights, vertices with a small bias are more likely to appear as boundary vertices than those with a larger one, and edges of a lower weight have higher probability to be cut. In this way, JOSTLE concentrates its search to a rather limited region of the search space just like the ITS from Section 6.5.3.1.

Each new offspring is obtained with a crossover or a mutation operator by determining a set of biased values from one or more parents from the population and than applying JOSTLE to generate a partition. The crossover creates a new set of biases from a given number of partitions in the following way. For each vertex v in the graph, check whether v appears as a border vertex (ends a cut edge) in two or more of the parent partitions. If so, assign to v a bias value selected uniformly at random from the range [0, 0.1]. Otherwise, assign to vertex v a bias value of 0.1 plus a random number chosen in the same range.

The mutation operator generates a new set of biases by considering information from only one parent in the following way. For each vertex v in the graph, check whether v is a border vertex, the neighbour of a border vertex, or the neighbour of a neighbour of a border vertex. If so, assign to v a bias value selected uniformly at random from the range [0, 0.1]. Otherwise, assign to vertex v a bias value of 2.0 plus a random number chosen in the same range.

After an offspring partition has been created, the associated biased values are removed.

The population size $|POP|$ is kept quite small (around 50) due to size of graphs and the time required to execute JOSTLE. Each new generation is produced as follows. $|POP|$ new offspring are created by either crossover or mutation at a given ratio. Mating groups of individuals for crossover and candidates for mutation are randomly selected from the current generation, such that each individual participates in at least one trial. The union of parent and offspring individuals are ranked by fitness, and the best $|POP|$ individuals are then selected to form the new generation. The proposed algorithm is thus a simplified version of the CHC Adaptive Search Algorithm [18] that lacks incest prevention and restarts.

Each run of this evolutionary approach consist of 50,000 calls to JOSTLE, and therefore requires very long execution time of hours and even days for large graphs. As expected, it thus provides higher quality partitions than any of the existing package that usually take less than a minute (and often less than a second) to generate a partition.

6.5.3.3 The Memetic Algorithm by Benlic and Hao

In [11], Benlic and Hao extend their MITS algorithm for balanced k-way partitioning from Section 6.5.3.1 to a multilevel memetic approach (MMA) by integrating a dedicated multi-parent crossover operator based on the notion of backbone and a distance-preserving pool updating strategy that maintains a healthily diversified population. To avoid high solution redundancy introduced by the standard string solution encoding, an individual $I = \{S_1, ..., S_k\}$ for MMA corresponds to a partition of V into k disjoint groups or subsets, such that each subset S_j, $j \in \{1, ..., k\}$ is composed of vertices that are assigned to the jth subset.

The success of the MMA partly lies in the dedicated backbone-based multi-parent crossover operator (BBC) that exploits an existing structure of a problem by preserving the elements which hopefully belong to the optimal partition, while permitting limited perturbations within offspring solutions. It thus provides high quality partitions for instances with exploitable global structure and search landscapes with highly correlated local optima.

Given the set $P = \{I^1, ..., I^p\}$ of p parent individuals chosen with the well-known tournament selection strategy, the BBC constructs the offspring $I^0 = \{S_1^0, ..., S_k^0\}$ in k passes (one for each subset of the partition). In each pass μ it performs the following steps:

1. Select a subset S_j^i of I^i such that the weight $W(S_j^i)$ is maximal across the subsets $j \in \{1..k\}$ of each individual $I^i \in P$, i.e. $max_{i\in\{1..p\},j\in\{1..k\}}\{W(S_j^i)\}$, with the constraint that at most $\lceil k/p \rceil$ subsets can be chosen from each individual $I^i \in P$.
2. Given I^i and S_j^i determined in Step 1, for *each* individual $I^t \in P$ $(t \neq i)$, let $\prod_t$ contain the largest number of vertices that are shared by the subset S_j^i of I^i and a subset S_η^t of I^t, i.e. $\prod_t = \{S_j^i \cap S_\eta^t | max_{\eta\in\{1..k\}} |S_j^i \cap S_\eta^t|\}$. Then, $\prod = \{\prod_1, .., \prod_{p-1}\}$ forms a set of these vertex subsets.
3. Set $S_\mu^0 = \prod_1 \cap \prod_2 \cap ... \cap \prod_{p-1}$. S_μ^0 is the largest subset of vertices that are shared by all the parent individuals. For each vertex $v \in S_j^i$ and $v \neq S_\mu^0$, v is assigned to subset S_μ^0 of I^0 if $c(v)/p - 1$ is greater than or equal to some random real number in the range $[0, 1]$, where $c(v)$ is the number of subsets of $\prod$ in which v occurs.
4. When a vertex v is assigned to subset S_μ^0 of I^0 in the μth pass, v is removed from all the parent individual subsets in which it occurs, and the weights of these subsets are adjusted accordingly.

After the previous four steps, the last step handles the unassigned vertices. Any vertex v missing from I^0 is placed at random to a subset S_r of I^0 such that $W(S_r \cup \{v\}) \leq W_{opt}$, where W_{opt} is defined in Section 6.2.1. This step introduces a degree of diversification in the crossover process.

Notice that the proposed BBC operator never degrades the balance with respect to the set of parent individuals P, since given a subset S_j^i of individual I^i which is chosen in the μth pass, at most $|S_j^i|$ vertices can be transmitted to the subset S_μ^0 of offspring I^0. In addition, an unassigned vertex v in I^0 is assigned to a subset S_r^0 only if adding v to S_r^0 does not exceed the expected optimal subset weight W_{opt}. An example of this crossover with three parent individuals $(p = 3)$ for $k = 3$ is provided in Figure 6.4.

After offspring I^0 has been generated with the BBC operator, it is improved with the ITS from Section 6.5.3.1. The MMA then decides whether I^0 should be inserted into the population by considering both the solution quality and the set-theoretic partition distance [20] (call it d) between individuals from the population. Offspring I^0 is inserted into *POP* if it is of the best quality relative to the population, or if the minimum distance between I^0 and any other individual in the population is greater than the minimum distance between any two individuals in the population. To determine the individual that is to be replaced by I^0, the authors adopt a strategy proposed in [33] that uses the following quality-and-distance scoring function H to rank the individuals of the population:

$$H_{i,POP} = f(I^i) + \beta / D_{i,POP}$$

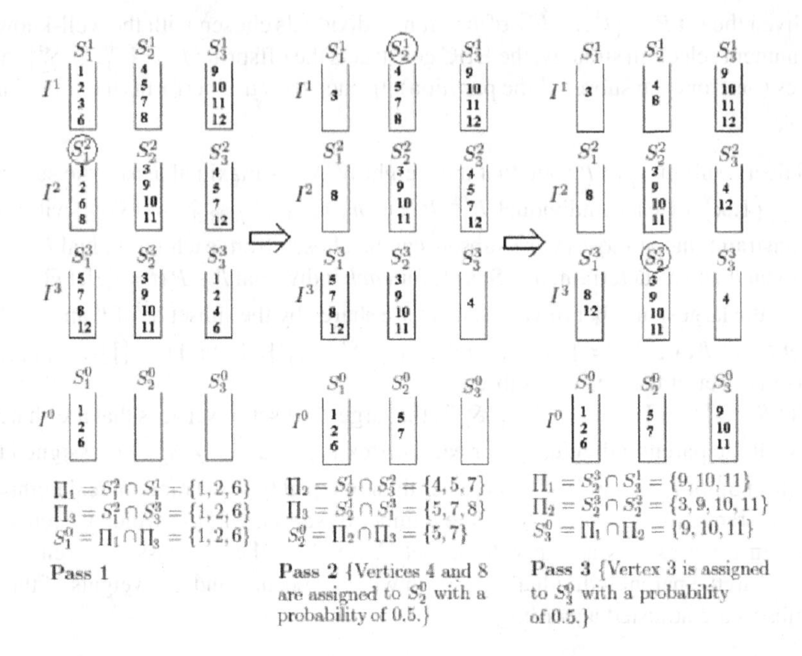

Fig. 6.4 An illustration of the BBC crossover with three parents taken from [11]. A circled subset of a parent corresponds to the subset chosen in the μth pass, i.e. the subset of maximal weight across all the parent individuals with the constraint that at most $\lceil k/p \rceil$ subsets can be chosen from each individual.

where f is the objective function (i.e., the sum of cutting edge weights), β a parameter, and $D_{i,POP}$ the minimum distance between individual I^i and any other individual from the population.

An extensive experimental evaluation of MMA has been performed on a set of benchmark instances from Walshaw's archive. It has been shown that the MMA can provide even better partitions in terms of solution quality than the ITS from Section 6.5.3.1. Moreover, the authors compare two version of MMAs integrating respectively the BBC and a standard uniform crossover where the diversification is further reinforced by a random mutation operator. The results show that there is no significant statistical difference between the solution sets generated by the two MMA versions for lower values of k, i.e., $k \in \{4,8\}$. However, as k increases, the BBC operator visibly outperforms the uniform crossover in almost each case for $k \in \{16,32\}$. One explanation is that intuitively, given the semantics of the BBC crossover, it favours the preservation of backbone information for larger k whereas the number of parts has a weak influence for the uniform crossover operator as to backbone preservation.

6.5.3.4 Other Partition Refinement Approaches

Beside the aforementioned multilevel refinement partitioning procedures, some other approaches are also worth mentioning.

In [8], Battiti et al. propose a multilevel algorithm for the balanced bipartitioning which integrates the Diff-Greedy algorithm [6], used in the initial partitioning phase, and a tabu search algorithm [7] employed as a partition refinement procedure. The TS algorithm is related to the KL heuristic [29]. However, the main differences are that each selected move is applied immediately to the current solution, and that worsening moves are also accepted. The authors consider two alternative choices to adjust the prohibition parameter T (i.e., the tabu tenure) of the TS algorithm. The first choice is to maintain T fixed during the search with a value that is selected by a preliminary off-line tuning phase for different types of graphs (FIXED-TS or FTS). The other choice is to determine the right value of T in a dynamic and on-line way depending on the past search history (Reactive Randomized Tabu Search RRTS). In this way, the tedious task of tuning by the user is avoided and the T value can automatically change during the search depending on the properties of a specific task.

A refinement procedure based on a mixture of simulated annealing and tabu search algorithm (RLrMSATS) is presented by Baños et al [5]. The idea of this hybrid approach is to employ the simulated annealing procedure to escape from local optima, while preventing the occurrence of cycles by means of a tabu search mechanism. A move with the proposed approach consist in moving a vertex v from its current to another partition subset. To jump from a local optimum, RMSATS accepts worsening moves as in simulated annealing. Once a move increasing the evaluation function cost is accepted, it is forbidden to apply the reverse move during a certain number of iterations as in tabu search in order to avoid cycling. A similar hybrid refinement approach inspired by RMSATS is proposed in [45].

6.5.4 The Key to Effectiveness of Partition Refinement Procedures

The success of a multilevel algorithm is greatly dependent on its two main components: the coarsening scheme and the solution refinement procedure. Since vertex aggregation filters the solution space by putting restrictions on which solutions the algorithm can visit, it is obvious that the way coarsening decisions are made is of an extreme importance for the quality of resulting solutions. As pointed out previously, most of the graph partitioning approaches are based on the same or very similar coarsening schemes. However, given the same amount of computing time, some of these algorithms perform better than others highlighting the importance of a refinement procedure.

In the previous sections, we described the three best performing partition refinement procedures in terms of partition quality. A common characteristic of these refinement algorithms is that they are based on stronger intensification and concentrate the search only around a limited region of the search space. In ITS [9], this is done by performing (most of the time) moves with only border (critical) vertices according to a selection strategy. A similar idea is also used in [44], where a smaller bias is assigned to vertices that appear as border vertices in one or more parent partitions, which results higher probability that these vertices will remain border in successive generations. In [11], besides the ITS, the BBC crossover also limits the region of explored search space by preserving vertex groupings that are common to a number of population individuals. Moreover, the intensification plays a major role in other popular graph partitioning approaches.

We next provide an explanation, based on observation made from a landscape analysis [11, 32], to why a more pronounced intensification mechanism constitutes a highly effective search in case of the graph partitioning problem.

6.5.4.1 Landscape Analysis for the Graph Partitioning Instances

The performance of a stohastic algorithm crucially depends on the characteristics of search landscape like the average distance between local optima and the relative distance of local optima to the nearest global optimum. The fitness distance correlation (FDC) coefficient ρ_{fdc} [26] is a well-known tool for landscape analysis and can provide useful indications about the problem hardness, even if such an analysis has some known shortcomings and limits. FDC estimates how closely related are the fitness and distance to the nearest optimum. For a minimization problem, if the fitness of a solution decreases with the decrease of distance from the optimum, then it would be easy to reach the target optimum for an algorithm that concentrates around the best candidate solutions found so far, since there is a "path" to the optimum via solutions with decreasing (better) fitness. A value of $\rho_{fdc} = 1$ indicates perfect correlation between fitness and distance to the optimum. For correlation of $\rho_{fdc} = -1$, the fitness function is completely misleading. FDC can also be visualized with the FD plot, where the same data used for estimating ρ_{fdc} is displayed graphically.

A landscape analysis for the GPP has been performed in two works. In [32], Merz and Freisleben provide a thorough analysis of the landscape for the GBP on a set of instances introduced in [14, 25], and perform a fitness distance correlation analysis (FDA) [26] based on solutions samples with the KL [29] and Diff-Greedy [6] heuristics respectively. In [11], Benlic and Hao make a FDA for the k-way partitioning problem (for $k \in \{4,8,16,32\}$) on a set of graphs from the Walshaw's graph partitioning archive (used for performance evaluation in [9, 11, 44]), based on a sample of local optima obtained after 1500 independent runs of the ITS [9]. While Merz and Freisleben use the Hamming distance, Benlic and Hao use the set-theoretic distance to perform the landscape analysis.

Tables 6.1 and 6.2 show the results from [32] and [11] respectively. Column 'ρ_{fdc}' of the two tables reports FDC coefficients ρ_{fdc} for the analysed graphs. For illustrative purpose, FD plots of only two graphs (*3elt* and *vibrobox*) are given in Figure 6.5 for $k \in \{4,8,16,32\}$. As it can be seen from Tables 6.1 and 6.2, there is a signification fitness distance correlation in many cases. However, the FDA

Table 6.1 Analytical results for graph bisection, taken from [32], for graph partitioning instances provided in [14, 25]. Columns 'd_{lo}' and 'd_{go}' report respectively the average distance between local optima and the average distance of local optima from the best local optimum, expressed as a percentage of $|V|$. Column 'ρ_{fdc}' shows the correlation coefficients with respect to fitness and distance.

	KL heuristic			Diff-Greedy heuristic		
Graph	*avg* d_{lo}	*avg* d_{go}	ρ_{fdc}	*avg* d_{lo}	*avg* d_{go}	ρ_{fdc}
G1000.0025	22.53	21.29	0.37	22.09	20.86	0.34
G1000.005	22.79	21.97	0.22	22.36	21.68	0.18
G1000.01	22.6	21.54	0.37	22.21	21.36	0.29
G1000.02	22.47	21.24	0.47	22.2	20.86	0.41
U1000.05	22.47	21.62	0.28	14.44	11.15	0.63
U1000.10	20.65	19.16	0.36	15.39	12.81	0.58
U1000.20	15.76	12.94	0.63	14.94	13.53	0.58
U1000.40	13.02	9.45	0.82	13.6	10.47	0.66
Breg5000.16	3.95	2.08	0.99	17.9	11.98	0.99
Cat.5252	24.14	23.91	0.02	14.58	11.52	0.21
Rcat.5114	22.79	22.36	0.07	14.46	11.45	0.7
Grid5000.50	4.3	2.4	0.91	15.07	12.09	0.7
W-grid5000.100	14.43	13.56	0.66	14.43	13.55	0.7

Table 6.2 Analytical results, taken from [11], for seven graph partitioning instances from Walshaw's graph partitioning archive when $k \in \{4,8,16,32\}$. Columns 'd_{lo}' and 'd_{go}' report respectively the average distance between local optima and the average distance of local optima from the best local optimum, expressed as a percentage of $|V|$. Column 'ρ_{fdc}' shows the correlation coefficients with respect to fitness and distance.

	k=4			k=8			k=16			k=32		
Graph	*avg* d_{lo}	*avg* d_{go}	ρ_{fdc}	*avg* d_{lo}	*avg* d_{go}	ρ_{fdc}	*avg* d_{lo}	*avg* d_{go}	ρ_{fdc}	*avg* d_{lo}	*avg* d_{go}	ρ_{fdc}
data	30.5	34.8	0.57	17.8	16.0	0.68	22.5	23.7	0.08	24.9	23.1	0.6
3elt	19.1	18.7	0.7	17.2	14.6	0.53	14.0	12.1	0.75	20.5	17.1	0.53
uk	18	14.3	0.61	26.3	25.7	0.24	26.9	25.1	0.33	27.4	24.9	0.44
crack	3.5	2.2	0.89	22.5	19.6	0.51	27.7	22.9	0.74	28.1	26.3	0.58
wing-nodal	26.1	21.6	0.81	17.1	13.6	0.91	31.0	27.3	0.56	37.5	35.6	0.4
fe-4elt2	9.8	6.7	0.74	26.0	24.4	0.68	16.4	14.7	0.51	28.7	25.5	0.51
vibrobox	40.1	41.4	-0.02	22.4	19.7	0.03	41.5	45.5	0.65	49.7	46.8	0.21

analysis also reveals the existence of several cases among the selected instances for which there is virtually no correlation between fitness and distance, i.e., cases where $\rho_{fdc} < 0.15$. Indeed, from plots in Figure 6.5, it is clear that there is practically no correlation for 'vibrobox' when $k \in \{4,8\}$. On the other hand, the plots indicate the strongest correlation for the graph '3elt' when $k \in \{4,16\}$ and 'vibrobox' when $k = 16$.

The existence of a strong correlation between solution quality and its distance to the nearest global optimum, as observed from the FDC analysis in [11, 32], is often refered to as a big valley structure of the landscape. Intuitively, in this structure a global optimum is surrounded by local optima with evaluation values that deteriorate with the increase of distance to the global optimum. In case of landscapes with a big valley structure, stronger intensification leads to algorithms of better performance.

Additionally, tables 6.1 and 6.2 report the average distance between local optima (column '*avg* d_{lo}') and the average distance of local optima from the best local optimum (column '*avg* d_{go}'), expressed as a percentage of $|V|$. Given that the maximum

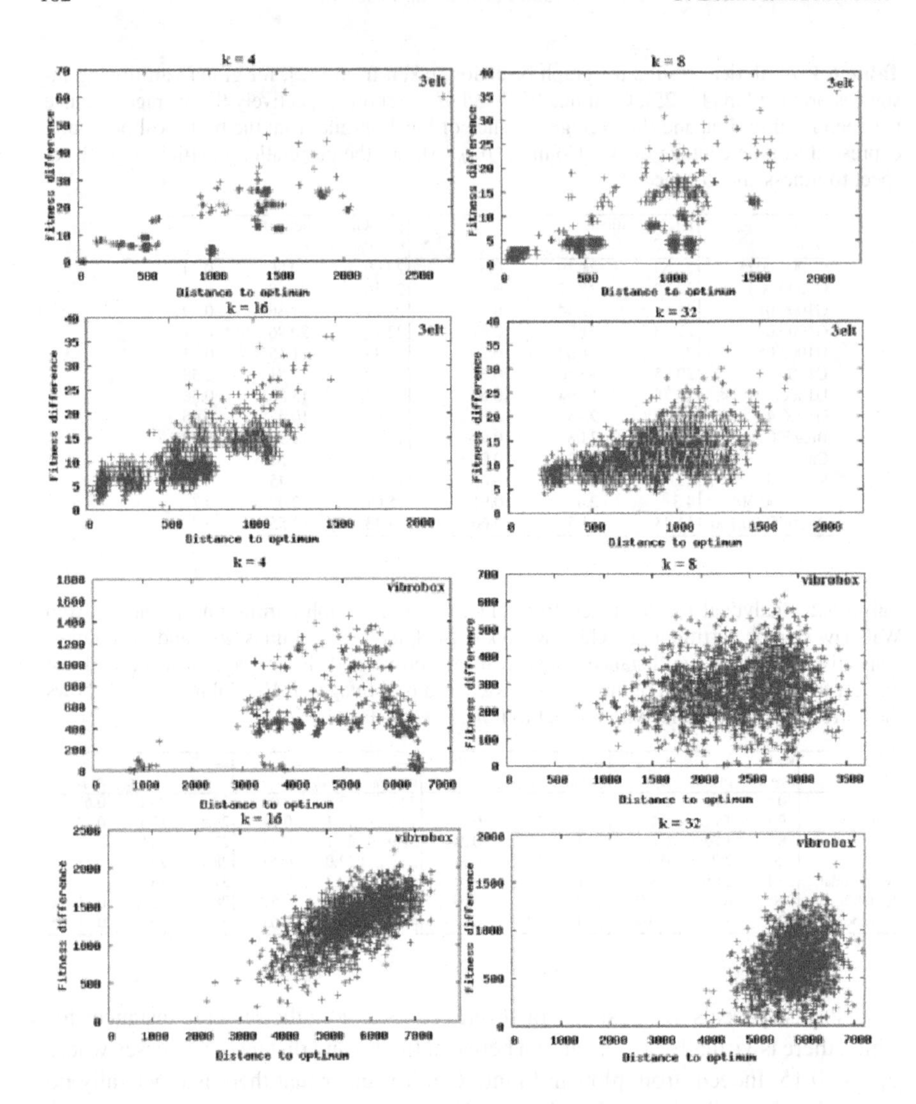

Fig. 6.5 FD correlation plots with respect to the normalized solution fitness and distance to the optimum for *3elt* and *vibrobox* when $k \in \{4, 8, 16, 32\}$. The first four plots are related to the *elt* graph, while the last four are related to the *vibrobox*. The plots are taken from [11].

distance between any two solutions is $|V|$, these results imply that local optima are not uniformly distributed, but are rather concentrated within a limited number of regions in the search space.

These observations constitute an explanation to why algorithms that perform a stronger intesification do so well on these GP instances.

6.6 Conclusion

In this chapter, we provided a review on hybrid metaheuristics for solving the well-known k-way partitioning problem (GPP). GPP is an NP-complete problem with a broad spectrum of applicability. Therefore, many efforts have been made in devising a number of different heuristic approaches such as spectral methods, graph growing and greedy heuristics, multilevel approaches, as well as algorithms based on popular metaheuristics. The application of these methods depends on several factors including time vs. quality and the degree of imbalance. The most popular graph bisection heuristic is the linear time implementation of the Kernighan-Lin algorithm by Fiduccia and Mattheyses, which improves iteratively the quality of an existing partition. Different adaptations and modifications of its basic procedure have been proposed in the literature. These KL-like algorithms are often hybridized with other approaches such as multilevel and genetic algorithms. The current best performing GPP algorithms in terms of solution quality are hybrids between classical metaheuristic techniques and multilevel methods. Indeed, the three most effective algorithms reviewed in this chapter, that were able to produce state-of-art partitions, are hybrids between multilevel methods and adaptations of well-known metaheuristics. We noted that a common characteristic of these approaches is that they are based on a strong intensification mechanism which seems to work well on most GP instances whose landscapes generally display the big valley structure.

References

1. Alpert, J.C., Kahng, B.A.: Recent directions in netlist partitioning: A survey. Integration, the VLSI Journal 19(12), 1–81 (1995)
2. Alpert, J.C., Hagen, W.L., Kahng, B.A.: A hybrid multilevel/genetic approach for circuit partitioning. In: Proceedings of the IEEE Asia Pacific Conference on Circuits and Systems, pp. 298–301 (1996)
3. Barake, M., Chardaire, P., McKeown, G.P.: The PROBE metaheuristic for the multiconstraint knapsack problem. In: Resende, M.G.C., de Sousa, J.P. (eds.) Metaheuritics, pp. 19–36. Springer (2004)
4. Barnard, T.S., Simon, D.H.: A Fast Multilevel Implementation of Recursive Spectral Bisection for Partitioning Unstructured Problems. In: Proceedings of the 6th SIAM Conference on Parallel Processing for Scientific Computing, pp. 711–718 (1993)
5. Baños, R., Gil, C., Ortega, J., Montoya, F.G.: Multilevel Heuristic Algorithm for Graph Partitioning. In: Raidl, G.R., Cagnoni, S., Cardalda, J.J.R., Corne, D.W., Gottlieb, J., Guillot, A., Hart, E., Johnson, C.G., Marchiori, E., Meyer, J.-A., Middendorf, M. (eds.) EvoIASP 2003, EvoWorkshops 2003, EvoSTIM 2003, EvoROB/EvoRobot 2003, EvoCOP 2003, EvoBIO 2003, and EvoMUSART 2003. LNCS, vol. 2611, pp. 143–153. Springer, Heidelberg (2003)
6. Battiti, R., Bertossi, A.: Differential Greedy for the 0-1 Equicut Problem. In: Proceedings of the DIMACS Workshop on Network Design: Connectivity and Facilities Location, pp. 3–21 (1997)
7. Battiti, R., Bertossi, A.: Greedy, prohibition, and reactive heuristics for graph partitioning. IEEE Transactions on Computers 48(4), 361–385 (1999)

8. Battiti, R., Bertossi, A., Cappelletti, A.: Multilevel reactive tabu search for graph partitioning. Preprint UTM 554. Dip. Mat., University Trento, Italy (1999)
9. Benlic, U., Hao, J.K.: An effective multilevel tabu search approach for balanced graph partitioning. Computers and Operations Research 38(7), 1066–1075 (2010)
10. Benlic, U., Hao, J.K.: An Effective Multilevel Memetic Algorithm for Balanced Graph Partitioning. In: ICTAI, vol. (1), pp. 121–128 (2010)
11. Benlic, U., Hao, J.K.: A multilevel memetic approach for improving graph k-partitions. To appear in IEEE Transactions on Evolutionary Computation (2011)
12. Brandt, A., McCormick, S., Ruge, J.: Algebraic multigrid (AMG) for sparse matrix equations. In: Evans, D.J. (ed.) Sparsity and its Applications, pp. 257–284 (1984)
13. Brandt, A.: Algebraic multigrid theory: The symmetric case. In: Preliminary Proceedings of the International Multigrid Congerence, vol. 19, pp. 23–56 (1986)
14. Bui, T.N., Moon, B.R.: Genetic Algorithm and Graph Partitioning. IEEE Transactions on Computers 45(7), 841–855 (1996)
15. Chardaire, P., Barake, M., McKeown, G.P.: A PROBE-based heuristic for graph partitioning. IEEE Transactions on Computers 56(12), 1707–1720 (2007)
16. Ciarlet, P., Lamour, F.: On the validity of a front-oriented approach to partitioning large sparse graphs with a connectivity. Numerical Algorithms 12(1), 193–214 (1996)
17. Chevalier, C., Safro, I.: Comparison of Coarsening Schemes for Multilevel Graph Partitioning. In: Stützle, T. (ed.) LION 3. LNCS, vol. 5851, pp. 191–205. Springer, Heidelberg (2009)
18. Eshelman, L.J.: The CHC adaptive search algorithm: How to have a safe search when engaging in non-traditional genetic recombination. In: Rawlings, G.J.E. (ed.) Foundations of Genetic Algorithms, pp. 265–283 (1991)
19. Garbers, J., Prome, H.J., Steger, A.: Finding clusters in VLSI circuits. In: Proceedings of IEEE International Conference on Computer Aided Design, pp. 520–523 (1990)
20. Gusfield, D.: Partition-Distance: A Problem and Class of Perfect Graphs Arising in Clustering. Information Processing Letters 82(3), 159–164 (2002)
21. Fiduccia, C., Mattheyses, R.: A linear-time heuristics for improving network partitions. In: Proceedings of the 19th Design Automation Conference, pp. 171–185 (1982)
22. Hagen, L., Kahng, A.: A new approach to effective circuit clustering. In: Proceedings of IEEE International Conference on Computer Aided Design, pp. 422–427 (1992)
23. Hendrickson, B., Leland, R.: A multilevel algorithm for partitioning graphs. In: Proceedings of Supercomputing, CDROM (1995)
24. Holtgrewe, M., Sanders, P., Schulz, C.: Engineering a scalable high quality graph partitioner. In: Proceedings of IEEE International Parallel & and Distributed Processing Symposium, pp. 1–12 (2010)
25. Johnson, D.S., Aragon, C.R., Mcgeoch, L.A., Schevon, C.: Optimization by Simulated Annealing: An Experimental Evaluation; Part-I, Graph Partitioning. Operations Research 37, 865–892 (1989)
26. Jones, T., Forrest, S.: Fitness Distance Correlation as a Measure of Problem Difficulty for Genetic Algorithms. In: Proceedings of the 6th International Conference on Genetic Algorithms, pp. 184–192. Morgan Kaufmann (1995)
27. Karypis, G., Kumar, V.: A Fast and High Quality Multilevel Scheme for Partitioning Irregular Graphs. SIAM Journal on Scientific Computing 20(1), 359–392 (1998)
28. Karypis, G., Kumar, V.: Multilevel k-way Partitioning Scheme for Irregular Graphs. Journal of Parallel and Distributed Computing 48(1), 96–129 (1998)
29. Kernighan, B.W., Lin, S.: An efficient heuristic procedure for partitioning graphs. Bell System Technical Journal 49, 291–307 (1970)
30. Küçükpetek, S., Polat, F., Oğuztüzün, H.: Multilevel graph partitioning: an evolutionary approach. Journal of the Operational Research Society 56, 549–562 (2005)
31. Krishnarnurthy, B.: An Improved Min-Cut Algorithm for Partitioning VLSI Networks. IEEE Transactions on Computers 33, 438–446 (1984)
32. Merz, P., Freisleben, B.: Fitness Landscapes, Memetic Algorithms. and Greedy Operators for Graph Bipartitioning. Journal of Evolutionary Computation 8(1), 61–91 (2000)

33. Lü, Z., Hao, J.K.: A Memetic Algorithm for Graph Coloring. European Journal of Operational Research 203(1), 241–250 (2010)
34. Nascimento, M., de Carvalho, A.: Spectral methods for graph clustering: A survey. European Journal of Operational Research 211(2011), 221–231 (2010)
35. Papadimitriou, C., Steiglitz, K.: Combinatorial Optimization: Algorithms and Complexity. Prentice-Hall (1982)
36. Pellegrini, F.: Scotch home page, `http://www.labri.fr/pelegrin/scotch`
37. Safro, I., Dorit, R., Brandt, A.: Multilevel algorithms for linear ordering problems. Journal of Experimental Algorithmics 13, 1–14 (2008)
38. Sanchis, L.: Multiple-Way Network Partitioning. IEEE Transactions on Computers 38(1), 62–81 (1989)
39. Sanchis, L.: Multiple-Way Network Partitioning with Different Cost Functions. IEEE Transactions on Computers 42(12), 1500–1504 (1993)
40. Schloegel, K., Karypis, G., Kumar, V.: Graph partitioning for high performance scientific simulations. In: Dongarra, J., et al. (eds.) CRPC Parallel Computing Handbook. Morgan Kaufmann (2000)
41. Simon, H., Teng, S.H.: How good is recursive bisection. SIAM J. Sci. Comput. 18(5), 1436–1445 (1997)
42. Shi, J., Malik, J.: Normalized Cuts and Image Segmentation. In: Proceedings of the IEEE Computer Society Conference on Computer Vision and Pattern Recognition, pp. 731–737 (1997)
43. Słowik, A., Białko, M.: Partitioning of VLSI Circuits on Subcircuits with Minimal Number of Connections Using Evolutionary Algorithm. In: Rutkowski, L., Tadeusiewicz, R., Zadeh, L.A., Żurada, J.M. (eds.) ICAISC 2006. LNCS (LNAI), vol. 4029, pp. 470–478. Springer, Heidelberg (2006)
44. Soper, A.J., Walshaw, C., Cross, M.: A Combined Evolutionary Search and Multilevel Optimisation Approach to Graph-partitioning. Journal of Global Optimization 29(2), 225–241 (2004)
45. Sun, L., Leng, M.: An Effective Multi-Level Algorithm Based on Simulated Annealing for Bisecting Graph. In: Yuille, A.L., Zhu, S.-C., Cremers, D., Wang, Y. (eds.) EMMCVPR 2007. LNCS, vol. 4679, pp. 1–12. Springer, Heidelberg (2007)
46. Viswanathan, N., Alpert, C.J., Sze, C., Li, Z., Nam, G.J., Roy, J.A.: The ISPD-2011 Routability-Driven Placement Contest and Benchmark Suite. In: Proc. ACM International Symposium on Physical Design, pp. 141–146 (2011)
47. Walshaw, C.: Multilevel refinement for combinatorial optimisation problems. Annals of Operations Research 131, 325–372 (2004)
48. Walshaw, C., Cross, M.: JOSTLE: Parallel Multilevel Graph-Partitioning Software – An Overview. In: Magoules, F. (ed.) Mesh Partitioning Techniques and Domain Decomposition Techniques, pp. 27–58 (2007)
49. Zha, H., He, X., Ding, C., Simon, H., Gu, M.: Bipartite Graph Partitioning and Data Clustering. In: Proceedings of the ACM 10th International Conference on Information and Knowledge, pp. 25–31 (2001)

Chapter 7
Hybrid Metaheuristics for Medical Data Classification

Sarab Al-Muhaideb and Mohamed El Bachir Menai

Abstract. Medical data exhibit certain features that make their classification stand out as a distinct field of research. Several medical classification tasks exist, among which medical diagnosis and prognosis are most common. Deriving a medical classification is a complex task. In particular, the rule–discovery problem is NP-hard. Identifying the most suitable strategy for a particular medical classification problem along with its optimal parameters is no less difficult. Heuristics and meta-heuristics are normally applied to approximate its solution. This chapter reviews hybrid meta-heuristics for medical data classification task, particularly diagnosis and prognosis, and their application to model selection, including parameter optimization and feature subset selection.

Keywords: Medical data classification, medical data complexity, evolutionary computation, swarm intelligence, model selection, model optimization, hybrid meta-heuristics, artificial neural networks.

7.1 Introduction

Modern clinical information systems store extensive amount of data in medical databases. This encourages the extraction of useful knowledge from these databases providing valuable insight for medical decision support. A branch of data mining, known as medical data mining, is currently considered one of the most popular research subjects in the data mining community [68]. This, in part, is due to the

Sarab Al-Muhaideb · Mohamed El-Bachir Menai
Department of Computer Science, College of Computer and Information Sciences, King Saud University, P.O. Box 51178, Riyadh 11543, Saudi Arabia
e-mail: salmuhaideb@acm.org,
menai@ksu.edu.sa

E.-G. Talbi (Ed.): Hybrid Metaheuristics, SCI 434, pp. 187–217.
springerlink.com

societal significance of the subject and also to the computational challenge it presents. Normally, there exist a dataset of historic data describing a particular medical disorder. Such datasets consist of records of patients' data relating to demographic, clinical and pathological data, along with results of particular investigations that were collected for the diagnosis and prognosis of a particular medical disorder. These medical datasets are typically incomplete, noisy, imbalanced and inexact [55]. Developing a computational diagnostic or a prognostic system is thus a challenging task.

This chapter is not intended to present a review of medical data classification techniques, but rather to introduce a snapshot of data mining techniques used to aid medical decision making. Several computational techniques have been proposed including machine learning, evolutionary computation and statistical techniques. Since each of these techniques have their own advantages and drawbacks, they are commonly hybridized in search of a more robust solution. Metaheuristics can be effective and efficient tools. They are well known for solving various optimization problems, for their adaptation ability and for their ability to produce good solutions in reasonable time and memory requirements. The chapter starts with a brief introduction of the classification problem in general, followed by medical data classification in particular. Next, features and challenges of medical datasets that make their classification stand out as a separate domain are explored. Based on that, the computational complexity of medical data classification is analyzed. Next, light is shed on some state-of-the-art solutions for medical data classification, in particular, hybrid meta-heuristics. It is possible to classify the hybrid metaheuristic techniques used for medical data classification into two broad categories according to their purpose:

1. Model learning and optimization; where the objective is to learn the classification hypothesis.
2. Model selection; that is selecting the model that best describes a dataset. This may include parameter and hyper-parameter optimization, neural network weight optimization, or feature subset selection, etc.

Each of these categories is illustrated by published work.

7.2 The Classification Problem

Classification aims at capturing hidden regulations and/or relations between the attributes (predictor features) in a set of class-labeled instances. These relations and/or regulations are modeled producing a general hypothesis. The resulting hypothesis is next applied to unseen future instances, with known predictor features and unknown class labels.The goal is to automatically make predictions about the class of those future instances [49]. Formally, given a set of training instances $\mathbf{D}_n$ with the form $\{(x_1,y_1),(x_2,y_2),\ldots,(x_n,y_n)\}$, the task is to approximate or project a function; $f(x)$, where $x \in \Re^m$ is a vector of attributes or predictor features of the form $\langle x_{i1}, x_{i2}, \ldots, x_{im} \rangle$, and y is the expected output (i.e. class) for the given x vector. Normally, y is drawn from a discrete set of classes [71]. The discovered model can be represented in different forms. Production rules in the form of $(IF\langle condition\rangle THEN\langle class\rangle)$ are often used. Other forms include decision trees (DTs) and artificial neural networks (ANNs).

Errors in classification may be in one of three cases [68]. Type-I error (false-positive) occurs when the system erroneously classifies a case as positive when in fact, it is not. For example, in a diagnosis scenario, a patient is wrongly labeled with a certain disease. Type-II error (false-negative), on the other hand, describes missing an existent positive. For example, a patient who is affected by a certain disease is diagnosed as disease-free. Usually, improving one type of error comes on the expense of the other [16]. In practice, the significance of these error costs vary with the application itself. For example, in life threatening medical conditions that require prompt intervention, missing a diagnosis (a false-negative) might result in a waste of time that may lead to life losses or at least cases that are not treated properly. On the other hand, a false-positive may result in unnecessary procedures, anxiety and financial costs [68]. The last type of error is the unclassifiable error. In this case, the system is unable to classify a case, possibly due to the lack of historic data.

There are many approaches to estimate the expected error of the classification model. Computing the error on the training set itself is an optimistic estimator of the true error [54]. In the training–testing method, the data set is normally split into two partitions called training and testing sets respectively. The most common technique is called the k-fold cross-validation [83]. Here, the whole data set $\mathbf{D}_n$ is partitioned into k disjoint folds, each of size k/n. Cross-validation is done k times each using $k-1$ folds for training the model and the one fold left out of the training phase is used as a test set. Each time a different fold is used as a test set. Results are then averaged over the k iterations.

Different performance metrics are used to measure the effectiveness of a classifier with respect to a given data set. The prior and posterior probabilities, also known as the Sensitivity (Sn), Specificity (Sp) and Precision (P) [71] are among the most commonly used.

Let the number of positive instances correctly classified be denoted TP, the number of positive instances incorrectly classified into negative FN. Similarly, the number of negative instances correctly classified as negative TN and those falsely classified into positive as FP. Sensitivity measures the proportion of positive samples being correctly classified as positive (7.1).

$$Sn = \frac{TP}{TP + FN} \tag{7.1}$$

Specificity, on the other hand, measures the proportion of negative samples being correctly recognized as negative (7.2).

$$Sp = \frac{TN}{TN + FP} \tag{7.2}$$

Precision (classification accuracy [26, 55] measures the proportion of samples being correctly classified (7.3).

$$P = \frac{TP + TN}{TP + TN + FP + FN} \tag{7.3}$$

Using Bayes theorem , it can be shown that P is entirely dependent on the values of Sp and Sn only if the data set is balanced [71]. A tradeoff between the hit rate (Sn, plotted on the Y-axis) and false alarm rate ($1 - Sp$, plotted on the X-axis) can be illustrated by the receiver operating characteristic curve. Each classification algorithm has a parameter, for instance, a threshold of decision, which can be fine-tuned to balance the tradeoff between hit rates and false alarms. Increasing the hit rate leads to an increase in false alarms as well. Different applications exhibit different significance levels of these two factors leading to the selection of a different point on the curve. Another performance measure used by classification algorithms is the area under the ROC curve (AUC). AUC index values range from 0.5 (random behavior) to 1.0 (perfect classification performance). For more detail see [12, 45].

7.3 Features and Challenges of Medical Data Classification

Several medical classification tasks exist, among which diagnosis and prognosis are most common. Other medical classification tasks include medical imaging, signal processing and scheduling [65]. In a diagnosis process, the patient's information is selectively collected and interpreted based on previous knowledge as evidence for or against the existence or nonexistence of disorders [58]. In the case of prognosis, the patient's information is selectively gathered and analyzed to predict the "course and outcome of disease process" [59]. Prognosis is considered an important instrument for medical management [59, 65]. For example, in the case of cancer prognosis, the intention is to predict cancer susceptibility, recurrence or survivability [19].

Medical diagnosis and prognosis can be modeled as classification problems. An instance is a patient's case. The predictor features are the patient's medical data. These might include demographic, clinical and pathological data. The class in case of diagnosis is the medical disorder. In case of prognosis, the class is the course and outcome of disease process. Production rules and decision trees are particularly attractive representation forms for the classification model in the medical field due to their comprehensibility. Using these forms, extracted models can be verified by medical experts and can enhance understanding the problem in-hand [84]. For example, in [47], a consultant pathologist in the domain of primary breast cancer evaluated the resulting rules for primary breast cancer diagnosis and classified them into three types; interesting new knowledge that could be further investigated, rules that are useful for the diagnosis and confirm medical knowledge, and those that contradict existing medical knowledge.

A physician relies on medical knowledge and personal experience to perform the desired classification task (diagnosis and/or prognosis). In many cases, physicians find difficulty in deciding the correct diagnosis or prognosis of a patient [11]. Patient presentation of disease varies significantly. It is a qualitative perception of symptoms that is difficult to quantify. The fluid representation is also perceived by qualitative receptors, i.e. physicians. These two factors; patient presentation and physician reception, are usually variable which participate significantly in understanding the medical case. The subjective interpretation results in variable output in terms of diagnosis and/or prognosis. For example, heart attack may be represented with pain in both arms; that can be interpreted as different diagnosis, some of which are not cardiac [76]. In addition, medical field experts are scarce and do not cooperate in converting their unique knowledge and art into a practical decision tool [65]. Also, the medical literature grows at a speed the physicians cannot cope with. Computer-aided diagnosis and/or prognosis systems bridge the knowledge gap in the era of evidence-based medicine [76]. The development of an adaptive model that learns from experience is more desirable than a best-fit solution for inherently complex and non-linear systems like the human body [92].

There are difficulties associated with medical data as well. Medical data includes demographic data, clinical observations, laboratory tests and radiology exams. Medical decisions are based on patient's medical records. Health care institutions are maintaining permanent patient medical records. Modern medical screening and diagnostic methods generate high volume of heterogeneous data. This data is continually accumulating. Mining such data requires intelligent methods [65, 88].

In addition to the high dimensionality, medical data exhibit unique features including noise resulting from human as well as systematic errors, missing values and even sparseness [88]. To illustrate, Table 7.1 presents medical data set examples. Most of these datasets are obtained from the UCI repository of machine learning databases, University of California-Irvine, Department of Information and Computer Science[1]. For example, some datasets like Dermatology, consist of different types of attributes. The high dimensionality is a feature of the Ovarian 8-7-02 dataset. Thyroid dataset contains more than 7000 instances. The Hepatitis dataset is imbalanced. The percentage of missing values in the Hungarian Heart dataset exceeds 20%. Finally, the Chest Pain dataset exhibits the multiclass problem featuring 12 different classes. Due to this nature, Tanwani et al. [88] calls for the classification of medical data as a separate domain, of which is currently considered one of the most popular research subjects in the data mining community [68]. This, in part, is due to the societal significance of the subject and also to the computational challenge it possess.

Tanwani and Farooq [85, 86, 87] performed an extensive study to present the challenges associated with biomedical data and approximate the classification potential of a biomedical dataset using qualitative measure of this complexity. The

[1] Address="`http://archive.ics.uci.edu/ml/datasets.html`"

complexity of biomedical datasets was found to be highly associated with a new factor; the correlation-based feature selection subset merit. This factor measures the quality of attributes in terms of how much they are correlated with the outcome class and not correlated with each other. Several empirical studies involving various evolutionary computing and machine learning classification algorithms were performed on UCI biomedical datasets. The classification accuracy was found to be dependent on the complexity of the biomedical dataset - not on the classifier choice. The two main effectors are noise and correlation-based feature selection subset merit. Second, the number and type of attributes has no noticeable effect on the classification accuracy as compared to the quality of the attributes. It is shown that biomedical datasets are noisy and that noise is the dominant factor that affects the resulting classification accuracy. Only high percentages of missing values severely degrade the classification accuracy. Third, evolutionary algorithms tend to overfit for small-sized datasets and are not much affected by the imbalanced classes' problem. A meta-study was performed consisting of the complexity measures as attributes. Using a decision tree and rule learner classifiers, the datasets were categorized into having good, satisfactory, or bad classification potential, according to their complexity factors. An equation is presented to find the classification potential of a dataset based on the level of its' noise and correlation-based feature selection subset merit.

Table 7.1 Example medical data sets and their associated complexity

Data set	Source	No. Instances	No. Attributes	No. Classes	Missing Values	Input Data Type
Chest Pain	[10]	138	165	12	No	Binary
Hungarian Heart	UCI; [88]	294	13	5	20.46%	3 Binary, 10 real
Dermatology	UCI; [88]	366	34	6	0.06%	1 Categorical, 1 binary, 32 integer
Wisconsin breast cancer (WDBC)	UCI; [84]	569	32	2[a]	No	Real
Hepatitis	UCI; [84]; [88]	155	19	2[b]	5.67%	13 Integer,6 real
Ovarian 8-7-02	CCR[c]; [88]	253	15, 154	2	No	Real
Thyroid	UCI; [88]	7200	21	3	No	15 Binary, 6 real

[a]Benign (62.7%)/Malignant (37.3%)
[b]Live (79.35)/Die (20.65%)
[c]Ovarian cancer studies, Center for Cancer Research, National Cancer Institute, USA, address="http://home.ccr.cancer.gov/ncifdaproteomics/ppatterns.asp"

In light of all of this, deriving a medical classification is a complex task [11, 65]. In particular, the rule-discovery problem is NP-hard [18]. This task involves searching for the hypothesis that models the diagnosis and/or prognosis concept, over all possible patient instances, in the space of all possible hypotheses. Penã-Reyes and Sipper [65] state "the medical search space is usually very large and complex" The Chest pain dataset [10] is a simple example to show the complexity of the search space. It consists of 165 binary attributes. The instance space $|X|$ contains exactly $2^{165} = 4.6768E49$ distinct instances. Therefore, the target hypothesis space includes $2^{|X|}$ possible hypothesis. That is, the target space includes $2^{|4.6768E49|}$ possible hypothesis. Execution time and memory demands grow rapidly with the number of instances and attributes of the problem at hand. Exact methods cannot be applied in this case.

Various classification paradigms exist, each with a related decision surface that decides the type of problems the classifier is suitable for. Machine learning algorithms like decision trees (DTs) suffer from trapping in local optima for a problem with a large number of attributes [18]. The back propagation algorithm (BP) [74] for training ANNs exhibit local search ability and can similarly get trapped into the nearest local optima [43]. A single run of BP is normally unrepeatable, unreliable and suboptimal, particularly on multi-local optima decision surfaces [43]. The main problem with machine learning methods is scalability especially when dealing with huge data [75]. In this respect, Provost and Kolluri [70] present a survey of methods for scaling up these algorithms. Statistical methods such as logistic regression (LR) and linear discriminant analysis (LDA) are widely used for classification. However, they do not produce accurate models when the relationship between the inputs and outputs of the dataset are non-linear and/or complex [25]. There exists no best classifier over all possible problem types [54]. Each technique has its own set of capabilities and limitations.

One way to deal with this shortage is to combine the properties of intelligent techniques so that each technique complements the capabilities and covers the limitations of the other. Combining or hybridizing various methods including heuristics and metaheuristics such as soft computing methods can significantly improve an analysis in terms of tractability, robustness, solution cost, and accuracy [25]. Metaheuristics in particular such as genetic algorithms (GA) [36], tabu search (TS) [29, 30], memetic algorithms (MA) [62], and simulated annealing (SA) [48], perform heuristic local search rather than exhaustive search producing good solutions within reasonable time and memory requirements [18]. Early hybrid systems, like evolutionary–neural hybrid systems [65] appeared in the early 90's. For instance, GAs were used to select predictor variables for the neural network [63] used to predict patient's response to Warafin. GAs were also used to optimize weights of ANNs in the prognosis for ICU patients [23]. Penã-Reyes and Sipper [64] used an evolutionary–fuzzy hybrid system for breast cancer diagnosis. In this study, a rule-based classifier that uses fuzzy logic called a 'fuzzy inference system' is used for the medical classification model learning. GAs are used to search for the parameters of the fuzzy inference system. A similar evolutionary–fuzzy hybrid was used by Jain et al. [44] for the diagnosis of coronary artery disease and breast cancer.

The next sections provide a snapshot of the state-of-the-art approaches in medical data classification. Section 7.4 demonstrates a sample of the literature that applies hybrid Metaheuristics for the problem of learning and optimizing medical data classification models. Section 7.5 illustrates the use of hybrid metaheuristics for model selection in medical data classification.

7.4 Hybrid Metaheuristics for Model Learning and Optimization in Medical Data Classification

This section starts with the use of learning classifier systems and their variants for model learning [5, 28, 38, 39, 40, 41, 47, 66, 77, 78, 84, 89, 91, 94, 97]. Other hybrid systems for model learning are next exemplified including the combination of genetic programming (GP) [51] with genetic algorithms [84], the blending of self-organized maps (SOMs) with ANNs and sUpervised Classifier Systems (UCSs) [73], the combination of TS with SA [18], and MAs [9]. Finally, two examples illustrate the use of metaheuristics for enhancing classifier accuracy as in the use of GA to enhance the classifier model generated by a decision tree classifier [75] and the use of homogeneity-based algorithm (HGA) [67] for optimizing the classifier models generated by support vector machines (SVMs), DTs and ANNs [68]. Table 7.2 presents a summary of these systems.

7.4.1 Learning Classifier Systems

Learning Classifier Systems (LCSs) [37, 95, 96] represent the merger of different fields of research including evolutionary computing and machine learning (reinforcement and supervised learning). They are adaptive systems that learn rules to direct their performance in a certain environment. In these rule-based systems, evolutionary methods (mainly GAs) are used to search the solution space while the reinforcement part from machine learning is used to guide the search to improved results. Their first appearance, Cognitive System One (CS-1) [37] seemed to be "complex and difficult to realize" [14]. The mid-1990s witnessed the birth of new models and new applications which revived this area. The 'zeroth-level' classifier system, ZCS [95] is a striped-down version of Holland's LCS that has better performance and comprehensibility. Wilson's ZCS was parameter-sensitive but has demonstrated optimal performance on several well-known test problems [13]. Not much later, Wilson introduced a variant of LCS with a new fitness measure, XCS [96]. Wilson's XCS has obtained more success and acceptance in the LCS community [14]. Stolzmann introduced a new line in the LCS research that stems from the theory of anticipatory behavioral control and cognitive psychology; Anticipatory Classifier Systems (ACSs) [82]. Rules in ACS aim at predicting action consequences in all possible cases in an environment. The models evolved in ACS direct the system to the most promoted action and also provide anticipation on what will happen next. On the other extreme, sUpervised Classifier Systems (UCSs) [8] replace the reinforcement learning component that was basic in all previous systems by supervised learning. That is, immediate reward system is used as the correct action is known in advance.

There are two styles of learning classifier systems. The first follows Holland's original model [37] developed at the University of Michigan and is thus termed 'Michigan-Style'. The solution is represented by the whole population. Rules

Table 7.2 Hybrid Metaheuristics for Model Learning and Optimization

System	Medical Purpose	Method	Rival Algorithms	Performance Metric
		Learning Classifier Systems		
EpiCS [38, 40, 41]	Epidemiologic surveillance	LCS	C4.5, LR	S_n, S_p, P, AUC, ... etc.
EpiXCS [39]	Epidemiologic surveillance	XCS	See5 DT	S_n, S_p, P, AUC, ... etc.
ClaDia [94]	Breast cancer diagnosis	LCS (fuzzy)	—	P
[47]	Breast cancer diagnosis	XCS	Bayesian, SVM, C4.5	P, medical expert
[77, 78]	EEG signal classification	XCS	NB, SMO, k-NN, PART	S_n, S_p, P
XCSI [97]	Breast cancer diagnosis	XCSI	Best on UCI cite, other published work	P, rule quality
LCSE [28]	Diabetes classification	LCS ensemble	LCS, DT, ANN	P
[91]	Mixed	ACS	XCS, XCSL, C4.5	P, no. rules
ZCS-DM [89]	Mixed	ZCS	DT, C4.5, XCS, HIDAR	P
[66]	Mixed	Pitt-style LCS	—	P, rule quality
		Other Hybrid Metaheuristics		
[57]	Breast cancer diagnosis	ensemble: SVM, AdaBoost, and GA	—	AUC
[3]	Mixed	Hybrid BN–k-NN–GA	Bayesian (EM)	P
[9]	Cancer cell diagnosis	MA-optimized cell graph coloring	—	No. colors (cell graph)
[18]	Mixed	Hybrid TS–SA	Ant Miner, CN2	$S_n \times S_p$, P, no. rules
[73]	Mixed	SOANN, SOUCS	UCS, ANN, other published work	P, computational time
[84]	Mixed	Hybrid GA–GP	C4.5, PART, NB, other published work	P
		Enhancing Classification Accuracy		
[75]	Mixed	GA	UCS, C4.5	P
[68]	Mixed	HBA and GA	DT, SVM, ANN, other published work	P

compete under GA which operates at the individual rule level. In Smith's Pitt-style [79] developed at the University of Pittsburg, each individual in the population represents a complete solution. In Pitt-style, GA operates at the rule-set level. Both styles have their own advantages and shortcomings. However, since entire solutions are simultaneously being evolved and compared in pitt-style, it is computationally heavier than Michigan-style LCS. This favors Michigan-style LCS in terms of popularity in the LCS community [92]. For interested readers, the survey by Urbanowicz and Moore [92] is recommended.

Learning Classifier Systems exhibit several attractive features. First of all, their rule-based nature leads to comprehensible hypothesis, as opposed to black-box solutions presented by ANNs for example. This implies that physicians can validate if the resulting classification hypothesis is clinically plausible. This also means that there is room for discovering new interesting relations. Second, LCSs tackle complex learning problems [42], and this is particularly important when dealing with the medical domain. LCSs are also on-line learners that avoid local minima due to the EC component [6]. Different kinds of representation can be used for LCSs [6, 52]. Other advantages include adaptability, robustness [6], and good generalization ability [96]. These features are especially interesting when dealing with medical data.

The main weaknesses of LCSs include overfitting for small data [5, 28] and difficulty with imbalanced classes [6]; as they tend to bias towards the majority class.

The use of learning classifier systems and their variants for the purpose of model learning in medical data classification has been well established. LCSs were applied with considerable results in medical data classification field. For example, learning classifier systems for epidemiologic surveillance EpiCS [38, 40, 41], EpiXCS [39],LCS with fuzzy rule representation [94], XCS [5, 47, 77, 78, 97], learning classifier system ensembles [28], ACS [91], ZCS for data mining (ZCS-DM) [89], and Pitt-style LCS [66]. Below a summary is presented for medical data classification solutions that are based on learning classifier systems.

EpiCS [38] was the first specialized LCS in the medical field; a learning classifier system for epidemiologic surveillance data. EpiCS predicts risk of disease; the probability of developing a disease. The estimate is given by the proportion of the matching classifiers that classify the case as positive. Using synthetic epidemiologic data generated such that one variable is associated with the outcome, EpiCS was compared to logistic regression-derived probability of disease and has shown significant advantage in terms of classification performance measured using the area under the receiver-operating characteristic curve (*AUC*).

The study by Holmes et al. [40, 41] was performed on epidemiologic surveillance data obtained from the Partners for Child Passenger Safety (PCPS). The aim of the study was reducing child automobile crash-associated morbidity and mortality through discovering patterns associated with head injury (head-injury/no-head injury classification task) [40], with inappropriate child restraint (appropriate/inappropriate child restraint classification), and the associated risk analysis [41]. Epidemiologic data are characterized by their large size and number of features that may result in huge number of relations. These relations can be modeled using the IF-THEN format. 47 numeric features [40] were selected out of over 500 available variables. Data were equally partitioned into testing and training sets with positive and negative classes equally distributed. Missing data were treated as don't-cares. Performance was evaluated in terms of sensitivity, specificity and *AUC*. All of these evaluation metrics were modified by the indeterminate rate (IR); cases that the model could not classify. EpiCS significantly outperformed the decision tree classifier algorithm C4.5 and LR in terms of *AUC* (0.97%) [40]. The number of rules produced by C4.5 was significantly lower. Based on that, the authors suggest that the use of C4.5 to initialize the EpiCS population might be advantageous. The authors also point out the need to improve LCS in terms of macrostate reduction, dealing with numeric data in native form and dealing with noisy data. The paper addresses the limitations of decision trees and linear regression models with respect to clinical and epidemiologic data.

In 2005, Holmes and Sager [39] introduced a new LCS for the epidemiologic community, EpiXCS. EpiXCS is an XCS classifier application tailored to the needs of epidemiologic research. The main feature is an interface workbench that allows researchers to set different parameters in a user-friendly manner. Using EpiXCS, researchers can watch the performance in terms of parameters like sensitivity, specificity, *AUC*, learning rate, and indeterminate rate. These parameters are updated in frames of 100 iterations. In addition, EpiXCS views resulting rules both in textual (IF–THEN format) and graphical forms. The graphical rule display option enables researchers to see possible clustering of features and their values forming a certain outcome. EpiXCS was compared to the See5 decision tree classifier in forming rules that discover the features associated with teenage automobile fatality in the census of all fatal United States and Puerto Rico automobile crashes ;FARS database. Results show that while classification accuracy of both classifiers was comparable, EpiXCS produced far fewer rules making the analysis much more manageable. Also, EpiXCS has discovered several features that were missed by See5.

For the diagnosis of breast cancer, Walter and Mohan present a classifier system for disease diagnosis, ClaDia [94]. A fuzzy rule representation was used where the attribute values were mapped to the ranges (low, medium and high). Instance-rule match degree correspond to the median membership degree of the instance's constituent attributes. Rule fitness was computed as the difference between the number of correctly classified instances and those incorrectly classified. Rule fitness was later reinforced by correct classifications and penalized otherwise. Niching was applied such that recombination is only allowed among individuals in the same niche (benign/malignant). Unlike the original LCS, mutation is performed on rule antecedent as well as consequent. That is, the rule consequent of weak rules may be mutated (reversed) as these may result in good rules for the opposite class. ClaDia was applied to Wisconsin Breast Cancer (WBC) database from the UCI repository and achieved over 90% accuracy.

Bacardit and Butz [5] compared the performance and generality level of two LCS classifiers, namely the on-line XCS and the off-line GAssist [4]. The comparison is done over thirteen different data sets. While GAssist is a Pitt-style classifier, XCS is a Michigan-style classifier that basis fitness on rule accuracy and applies GA selection to the currently active classifier subsets. Six types of problem difficulty are considered in this study. These include the input data volume, size and type of the search space, concept complexity, input noise and missing data in addition to the overfitting problem. The goal is to achieve a maximum level of generality. Results show that while both systems perform well on all data sets, the produced solutions are quite different. XCS has a weaker strategy in handling missing data and tends to over-fit training data especially in small data sets. XCS thus requires a large training set. GAssist on the other hand tends to ignore additional complexity and struggles when facing problems with multiple classes or those featuring large search spaces. Two conclusions are drawn: XCS needs to address its generality difficulty and GAssist needs to address its problem handling data sets with multiple classes.

In Kharbat et al. [47] primary breast cancer data from the Franchay Breast Cancer (FBC) data set is mined using XCS. Results are compared to other classifying techniques including Bayesian network classifier, SVM and C4.5 decision trees. As a preprocessing step, numeric values were normalized and data in nominal and Boolean attributes were decoded. The imbalance problem is handled by random over-sampling. Missing data were treated with Wild-to-Wild method; in which missing values are replaced with don't-cares for nominal data and general intervals for numerical data. Results showed that XCS outperformed other methods. The number of rules produced with XCS was much more than those produced by the C4.5 algorithm. However, these rules were described by a medical expert to be more informative and useful. Clustering and rule compaction were applied on the resulting rules.

Skinner et al. [77, 78] also use XCS but for EEG signal classification. EEG signals are characterized by their high dimensionality and noisy nature. This study investigates the efficacy of XCS in the classification of mental tasks based on human multi-channel EEG signals. In particular, the binary classification of four diverse mental tasks for three individuals. The significance of this investigation lies in the potential to use EEG classification results to control wheelchairs or similar devices for paralyzed individuals. The novelty of the approach is in the investigation of using XCS to process large and noisy condition strings. EEG signals were preprocessed to reduce the number of channels and their associated frequencies. Data was then segmented. The results were compared with four ML classification methods; naïve Bayes (NB), SMO, *k*-nearest-neighbor (*k*-NN with *k*=3), and PART which combines the learning strategies of decision trees and rule learners. Results were compared in terms of classification accuracy and showed that XCS significantly outperformed PART and *k*-NN. XCS was comparable to the SMO but inferior to naïve Bayes.

The study by Skinner et al. [78] investigates the effect of different migration policies on distributed and parallel XCS classifier population with different topologies and parameters. The study was performed on the single-step classification for human EEG signals associated with two mental tasks; Mental Counting and Figure Rotation for two persons. Three topologies were examined; fully connected, and uni- and bi-directional rings with different number of demes (2, 4 and 8). Migration policies are based on the selection and replacement criteria for the immigrant rules; based on their fitness, numerosity, or random. The study concludes that lower migration frequencies and rates produce better classification performance. High degree of connectivity speeds up the learning process. All policies result in a significant classification accuracy improvement with respect to XCS alone. Random immigrant selection results in a slower learning. Also, fitness-based migration selection increases the selection pressure and thus degrades the classification performance. As for population size, it is entirely dependent on the immigrant selection policy. Fitness-based immigrant selection policy gives better results for the fully connected topology while random-based immigrant selection is more beneficial for the uni- and bi-directional rings.

Also in the field of breast cancer is the study by Wilson [97]. The classifier predicates of XCS describe logical problems by defining hyper-rectangles in the decision space. In [97], XCS was modified to handle integer input spaces. The modified version, XCSI, was tested on oblique data. The study started with a simple

2-dimensional synthetic oblique data for which XCS achieved 100% (training) classification accuracy. A second experiment was conducted also with synthetic oblique data that resembles the UCI WBC dataset in terms of the number and type of attributes, their data ranges and the number of instances. Again 100% (training) performance was achieved although slightly slower. The final experiment used the UCI WBC dataset with a 10-fold cross-validation technique. Accuracies averaging 95.56% were reached. Results further show that the hyper-rectangles modeled by XCS predicates were good at approximating the oblique discrimination surface of the data. Also, results suggest the presence of logical patterns in WBC dataset that is evident by the presence of several accurate classifiers showing logical dependencies on one or a few attributes. Classifiers describing regions close to the discrimination surface feature a match set with strong evidence on both directions that can be used for risk of disease analysis instead of a concrete diagnosis.

The first LCS ensemble was introduced by Gao et al. [28]. The Learning Classifier System Ensemble (LCSE) [28] is an extension of LCS that aims at achieving better generality through using several sub-LCSs. Diabetes data input is distributed over these sub-LCSs. Each sub-LCS may then produce different rules even for the same input data. Results are then aggregated by means of a popularity voting method. Overfitting problem is managed with a 10-fold cross-validation approach. Results of LCSE outperform LCS, DTs and ANNs as well. Experiments also show that the accuracy of results increases with the number of sub-LCSs.

Unold and Tuszynski [91] applied ACS to three data mining data sets from the UCI repository; Monks, Voting-record and WBC. Results show that ACS achieve results no less than 97% except for the Monk's 2 data set, were the accuracy was limited to 75%. A comparison with XCS, XCS with s-expression (XCSL) [52, 53] and C4.5 shows that overall; XCS and XCSL achieved best results. XCSL have succeeded in producing the least rule set size. C4.5 was far behind. Future research aims at developing ACS to enable the handling of attributes of continuous type.

A modified version of ZCS for data mining applications named ZCS-DM is presented and applied to several UCI repository datasets including WBC, Hepatitis, Pima Indiana Diabetes and Bupa Liver Disorder benchmark sets [89]. The main changes include evolving the action part as well as the condition part and using user-tunable reward/penalty for the different class combinations (predicted and actual classes). In their model, users decide the number of individuals in the population and whether the action selection mechanism is deterministic or stochastic. A preprocessing step involves removing all duplicate rules after adjusting their strength. Rules are ordered based on their strength and either the first matching rule is selected or a voting scheme is employed. Missing values were treated by setting their corresponding predicates in the rule's condition part to true. Experiments were done using a 10-fold cross-validation and results show the classification accuracy advantage of the proposed model in 11 out of 12 UCI datasets over the DT algorithm C4.5, XCS and HIDAR [2]; which is a hierarchical decision rule that uses a sequential covering GA. C4.5 was the fastest algorithm.

An improved Pitt-style LCS is introduced by Peroumalnaik and Enee [66]. The training set is equally partitioned among the individual classifiers following the divide-and-conquer approach. Since prediction is performed by the whole population, and GA combines the genetic material of different individuals, the authors argue that the partitioning would eventually lead to a segmentation of the cognitive space. The proposed algorithm was tested on four medical UCI data sets with various parameters for the population size, number of rules per individual and rule selection strategy (random, most general or most specific). The proposed method has produced good results for the Wisconsin Diagnostic Breast Cancer (WDBC) dataset using a 10-fold cross-validation. No comparison with other methods was performed.

7.4.2 Other Hybrid Metaheuristics

Applying multiple classifiers is analogues to consulting a team of specialists. Each specialist considers the problem from a different perspective thus allowing the exploration of different regions of the search space. Multiple classifiers usually result in higher accuracy compared to a single classifier [88]. This is because the strengths of one method are utilized to complement the weaknesses of another [49]. The use of multiple classifiers is particularly useful for imbalanced data sets [88]. In addition, using multiple classifiers usually feature a strong generalization ability [28]. As classification model learners, there are endless possibilities for hybridizing metaheuristics together or with other classification methods in seek for a better classifier. However, using multiple classifiers comes in one of two forms. The first form is the ensemble classifier [28, 57]. Classifier ensembles achieve model diversification by using different subsets of training data with a single learning method, different training parameters with a single learning method, or using different learning methods [49].The second form for using multiple classifiers is employing a hybridization of metaheuristics with machine learning and/or evolutionary computing methods [3, 9, 18, 73, 84].

Like employing a team of specialists, the cost of using multiple classifiers is more than that of a single classifier. First, since all component classifiers need to be stored after training, the storage requirement increases accordingly. Second, all component classifiers need to be processed adding to the computational cost. Finally, it is more difficult when using multiple classifiers to comprehend the underlying reasoning and conclude a classification, particularly for non-experts [49].

An ensemble method for the detection of breast cancer from x-ray images is investigated by the authors in Lo et al. [57]. The proposed classifier was chosen as the joint winner in KDD Cup 2008. The data set is characterized by being highly imbalanced. The ratio of positive samples to negative samples is 163. Each patient is represented by a set of data points. The evaluation criterion was to minimize the *AUC* per patient rather than per data point. This was intended to minimize overfitting. The ensemble consisted of four classifiers; AdaBoost, Class-based SVM (CB-SVM), Patient-based SVM (PB-SVM), and GA. In CB-SVM, the intention was to balance the positive and negative classes. A class-sensitive loss function was employed where the weight of positive samples was 163 times more than that of negative ones. The problem faced was that patients with fewer positive instances were more difficult to identify than those with more positive instances. To resolve this problem, PB-SVM was designed such that the sum of weights of positive samples was equal to the sum of weights of negative ones. In addition, for each patient i, the sum of positive sample weights is equal to that of patient j. A slight improvement was obtained over the CB-SVM. AdaBoost was based on 50 weak learners, were Classification and Regression Tree (CART) was chosen as a weak learner. As for GA, the fitness was based on the *AUC* itself and resulted in better recognition of patients with fewer positive instances. The best ensemble outcome was obtained by averaging the two best classifiers.

The study by [3] focuses on randomly generating data sets based on the observed data and that will maximize classification accuracy. This technique is particularly useful in cases featuring missing data, small training data set size or noisy data. The study suggests an iterative hybrid model that starts with applying the Bayesian method based on the expected maximization algorithm (EM). Misclassifications are recorded. Next, a new data set twice as large as the observed data is randomly generated. A *k*-NN classifier is trained on this data and tested on the observed data set. This process is repeated until a lower misclassification rate is observed. Then, GA is used to further improve the generated instances. Bayesian classification based on EM is applied on the resulting data and new data generations are evolved and tested until an improved misclassification number is obtained. The algorithm was applied to five UCI data sets including Iris, Breast Cancer, Wine, Yeast and Glass. Results were compared against using the Bayesian classification based on EM alone. Improvements up to about 75% were recorded for the Breast cancer data set. On the other hand, Wine dataset resulted in a slight retrogression. The algorithm involves several iteration cycles resulting in an increased computational time. Also, results were not compared to other algorithms that are not based on data set generation.

Bhattacharyya et al. [9] present an introductory work for the diagnosis of cancerous cells from human-extracted low-resolution biopsy BMP images. Currently the diagnosis is based on the subjective pathologist evaluation of the tissue sample. The authors introduce a new automated diagnostic method that is based on the generic organizational structure of tissue cells. Two phases are implemented. The first is constructing the Cell Graph. This step transforms the BMP image into a monochrome graph; where nodes correspond to cells or cell clusters depending on the resolution used. Edges are assigned a probability based on the Euclidian distance between the nodes. The second phase is graph coloring using the minimum number of colors such that nodes within the same range of Euclidian distance obtain the same color. Memetic algorithms (MAs) are used to optimize graph coloring. In this work, MAs are composed of a heuristic search; sequential graph coloring algorithm, and a genetic algorithm with a modified mutation operator. The output of the program was the number of colors used for the sample image. It was not clarified how cancerous cell diagnosis can be derived from this information. However, the work provides more formalism about the density/organizational characteristics of the tissue cells that aid in the diagnosis process.

A hybrid tabu search–simulated annealing rule induction algorithm for classification tasks is presented by Chorbev et al. [18]. Continuous attributes were discretized. Classification rules are created incrementally and pruned for better readability and higher predictive value. The probability of an addition of a term to a classification rule depends on the entropy of the attribute's value as opposed to entropy in decision trees that is computed for attributes as a whole. Tabu timeouts aim at reducing the probability that a particular attribute value is selected twice; therefore increasing the search diversity. The quality of a rule computed as the product of sensitivity and specificity serves as the energy parameter for SA. An initial high temperature in SA allows low quality solutions to be accepted in the beginning for better exploration of the search space. As the search proceeds and the temperature cools down, only high quality rules will be accepted into the final rule list thus intensifying the search in promising areas. The proposed classification algorithm was compared against Ant Miner and the rule induction algorithm CN2 on four UCI medical datasets. In terms of predictive accuracy, Ant Miner was in the lead. However, in terms of the number of rules and terms per rule, SA Tabu Miner achieved good results that outperformed CN2 and was highly comparable to that for Ant Miner.

Rojanavasu et al. [73] present the use of self-organized maps (SOMs) as a pregate. That is, the SOM is used to cluster the data on-line and thus decomposing the search space into smaller sub-problems that are conceptually simpler. Class labels for the data are masked in this phase. Separate classifiers are then used to learn each sub-problem. The paper investigates the utility of connecting the pregate to two different classifies; a set of sUpervised Classifier Systems (UCSs) thus forming Self-Organized UCS or SOUCS; and an artificial neural network (ANN) thus forming SOANN. ANN layout is fixed for each dataset. The authors experiment with three data sets; the first is a group of five synthetic problems of increasing complexity. The second is a set of UCI datasets, and the third is a large and complex Forest-Cover-Type dataset from the Roosevelt National Forest in northern Colorado. Experiments have been applied with varying number of SOM sizes $(2 \times 2,\ 3 \times 3, and\ 4 \times 4)$; which implies a different number of UCS classifiers, and varying the number of individual UCS populations. All experiments are done using 10-fold cross-validation. In comparison with UCS alone and other published work, SOUCS showed an equivalent or better results in terms of classification accuracy except for the Forest-Cover-Type data set. The complexity of this dataset was not properly addressed by the smaller population sizes in the individual constituent UCSs. The SOUCS was superior in terms of computational time. The reason is the smaller population size in each UCS. Experiments also show the high sensitivity of the outcome to the population size and number of SOM cells as well as the problem type. This was suggested as a future research line. The ANN/SOANN environment obtained better results for the Forest-Cover-Type dataset but no better in the rest.

Finally, in Tan et al. [84], a two-phase strategy is presented as follows: the first phase uses a hybrid Michigan GA and GP, to produce per-class single rule poles in the form of: $\langle IF\ X_1\ and\ X_2\ and\ \cdots\ X_n\ THEN\ class = Y \rangle$. Michigan GA is applied to numeric data while the GP is applied to nominal data sets. The second phase involves applying Pittsburgh GA to find an optimal combination of the resulting rules with the OR operator. The population is divided into a number of subpopulations evolving simultaneously and corresponding to different number of rules in a single solution (rule set). Results of applying the algorithm to hepatitis prognosis (live/die) and breast cancer diagnosis (benign/malignant) were very encouraging and outperformed other classification methods like the DT algorithm C4.5 and trained neural networks. However, the system is computationally expensive and is suited for off-line classification.

7.4.3 Hybrid Metaheuristics for Enhancing Classification Accuracy in Medical Data Classification

Metaheuristics like GA that are well known for solving complex optimization problems can be used to optimize classification models obtained using other data mining techniques. The idea is to evolve a population of individuals (classification models or classification model components) that compete on the basis of their fitness. This 'fitness' measure can be defined in terms of their classification accuracy, compactness, computational complexity, or some other similar or compound measure. For example, a two-stage hybrid machine learning classifier approach is proposed by [75]. The first phase involves creating an initial classification rule set by the C4.5 decision tree classifier. 3-Fold cross-validation is used and the best accuracy generated rule set out of the three is used for the second stage. In the second stage, a genetic algorithm with one-point cross-over and optional mutation is applied to improve the generated rule set. Rules having invalid class type and resulting from crossover operation are omitted. Invalid attribute values in rules resulting also from crossover operator are replaced with don't-cares. In comparison with accuracy scores of the C4.5 alone and the accuracy-based learning classifier UCL, the proposed approach produced better classification results over most of the eight UCI data sets used in the study.

The second example on the use of hybrid metaheuristics for enhancing classification accuracy in medical data classification is the work by Pham and Triantaphyllou [68]. While most studies assign equal weight to the different types of error for a classifier; FP, FN, and UC (un-classifiable), the study by Pham and Triantaphyllou [68] focuses on the optimization problem of the penalty costs for those three error types. The study investigated the use of three traditional machine learning classification algorithms; DT, SVM and ANN in combination with a metaheuristic termed Homogeneity Based Algorithm (HBA) [67], to optimize the penalty costs of the three error types. HBA works on defragmenting the decision surface space resulting from the classifiers into homogeneous regions according to their density. In addition, GA is also applied to optimize the parameters for HBA. The proposed hybrid system was tested on five medical datasets from the UCI repository. Results were compared to using the three classification algorithms alone and with other published studies. It is shown that using HBA significantly improves the classification accuracy for all five datasets. The shortage of HBA is its high complexity as it cannot deal with datasets with a high number of attributes (greater than 10). This is currently the research focus of the authors.

7.5 Hybrid Metaheuristics for Model Selection in Medical Data Classification

There is a recent trend to use optimization techniques, including mainly EC methods, for parameter selection, feature subset selection, class representative selection and even for preprocessing and classifier selection. These factors highly affect the quality of the resulting classifier. For example, there is no rule of thumb on how to guide parameter setting for ANNs. Usually these are determined experimentally for each problem. Also, the high dimensionality of the data set not only slows down the classification process, but also confuses the classification algorithm and may lead to poor results [33]. The main benefit is the achievement of competitive classifiers without using background knowledge, careful data analysis, long experimental trials, or even knowledge about the classification model being employed [24].

Model selection is defined as "estimating the performance of different models in order to choose the best one" in describing a dataset [35]. There are many interpretations for model selection, these include parameter selection and optimization [15, 17, 25, 60], feature subset selection [90, 93], artificial neural network modeling including learning the weights of the neural nets [61, 80, 81], or optimizing the architecture of the neural net [16, 43]. Reference [24] have extended this definition and introduced the so called full model selection (FMS). In this system, a pool of preprocessing methods, feature subset selection and learning algorithms is introduced and the task is to select the best combination that would yield the lowest classification error for a given problem. In addition, the parameters for these methods are being selected as well. Stochastic optimization algorithms are well suited for dealing with the vast search space introduced by such problems. This section samples published work that utilize hybrid metaheuristics for model selection in medical data classification. Section 7.5.1 focuses on their use for feature subset selection. Section 7.5.2 illustrates the use of hybrid metaheuristics for ANN model selection in medical data classification. Finally, sect. 7.5.3 exemplifies their use for FMS. Table 7.3 presents a summary of these systems.

The work by Candelieri [15] is derived from the author's PhD thesis. The paper investigates and compares the hybridization of several metaheuristics including GA, TS and ACO to perform model selection for an SVM classifier both as a single classifier and as an ensemble. In this framework, SVM is used for learning the classification model. In the case of single classifier, the metaheuristic is either used to search for the best performing kernel function (linear, Normalized Polynomial, or Radial Basis Function) and its associated parameter(s); of which the author calls Model

Table 7.3 Hybrid Metaheuristics for Model Selection

System	Medical Purpose	Method	Rival Algorithms	Performance Metric
[15]	Mixed	SVM hyberdized with GA, TA, and ACO	—	Balanced classification accuracy
[17]	Mixed	Ensemble of BP-ANN, SVM, C4.5, parameter and FSS by SS	Other published work, individual ensemble components	P
[25]	Mixed	Clustering by CBR then fuzzy DT	*k*-NN, NB, SVM, fuzzy DT, other published work	P
[60]	Mixed	Class representatives and, parameters chosen by DE	*k*-NN, SVM, discriminant analysis, ANN	P
		Hybrid Metaheuristics for Feature Subset Selection		
[90]	Mixed	LR with PSO for FSS	Exhaustive search, TS, SS, random subset generation	P, computational time
[93]	Mixed	Fuzzy rule-based classifier and ACO for FSS	Fuzzy rule-based classifier, other published work	P, min no. features
		Hybrid Metaheuristics for Artificial Neural Network Model Selection		
[61]	Mixed	EDA for training ANN	BP-ANN, LM-ANN	CEP
[80]	Colorectal cancer prognosis	Step-wise regression, clustering, ANN ensemble with LS + BP training	—	P
[81]	Mixed	ANN trained by ACOR, ACOR–BP, and ACOR–LM	ANN trained with BP, LM, GA, GA–BP, GA–LM	CEP
[16]	Mixed	MG-Prob for MOP of MLP ensemble	—	minimize(FP, FN, network size)
[43]	Mixed	BP–ANN, PSO-trained ANN	BP vs. PSO, other published work	CEP, MSE, computational time
		Hybrid Metaheuristics for Full Model Selection		
PSMS [24]	Mixed	PSO, PS, other systems	BER	

Selection. The second feature, Multiple Kernel Learning, is to use the metaheuristic to optimize the n kernels along with their related parameters and coefficients. In the case of Ensemble Learning, the metaheuristic is used to find the best m SVM classifiers and their associated weights for combination. ACO was only applied to Multiple Kernel Learning. The framework was tested on 8 datasets of which several were medical. Balanced classification accuracy using 10-fold cross-validation was used for evaluation. Results show that the three models were highly competitive. GA was generally faster and more effective than TS. As for ACO, results were comparable to the others and promising as a new application.

In a study by Chen et al. [17]; classifier parameters and data features are stochastically chosen and evolved independently using scatter search [31]. The parameters were for an ensemble of three classifiers; SVM, BP-ANNs and DT (C4.5). Data instances with missing values were removed from all classifiers except for the DT. Each classifier was run three times and a majority voting was obtained to combine the 9 runs. Experiments were conducted on 18 UCI datasets and were compared to four similar studies of which some involve using ensembles of a larger number of classifiers. The proposed approach achieved the highest classification accuracy. Also, in comparison with results of individual component classifiers, the average performance of the ensemble outperformed the single classifiers.

Fan et al. [25] introduce a four-stage model for medical data classification that utilizes data preprocessing and clustering techniques for improved classification accuracy. Comprehensibility of the generated model was a main objective. First stage involved feature subset selection using step-wise regression. Selected features are then weighted by the gradient method. The next step involves case-based reasoning (CBR) clustering of the input data. Next, the fuzzy Triangle member ship function is applied to discretize the data and ID3 decision tree is applied to build a classification tree for each cluster. The last step involves evolving the fuzzy terms used in the decision tree by means of genetic algorithms to further improve the classification accuracy. The model was applied to two UCI datasets; WDBC and liver disorder. Comparisons against several ML classification methods including k-NN, naïve Bayes, SVM, fuzzy decision tree, and to other similar studies were done. Results show the consistent advantage of the proposed model. Average accuracy rate achieved was 99.5% and 85% for breast cancer and liver disorder respectively. The paper does not show clearly how GA encoding is done to allow for different number of fuzzy terms for each feature.

Also to demonstrate the advantage of using evolutionary computation methods to guide the parameter and class representatives' choice is the study by Luukka and Lampinen [60]. This study assess the effect on classification accuracy of adding noise to dataset features, adding extra noisy variables, and adding all two-component variables . A simple minimum distance classifier (instance based) was applied to four UCI data sets; New Thyroid, Hungarian Heart, Heart–Statlog and Lenses. Minkowsky distance metric parameter and individual class representatives were stochastically chosen and evolved by using differential evolution (DE) [69]. Three sets of experiments were conducted to study the effect of adding noise directly to data set feature values or as independent variables as well as adding all two-component terms on the classification accuracy. Noise variance and the number of noisy variables added were varied. Experimental results show improvements in classification accuracy up to 8% in the case of adding all component terms. Performance was degraded in the other cases. However, a comparison with k-NN, SVM, Discriminant analysis and BP-ANNs showed that the DE-enhanced classifier obtained higher classification accuracy in all experiments. Particularly, BP-ANNs and the DE-enhanced classifier showed the best results in the extra noise parameters and the two-component terms experiments.

7.5.1 Hybrid Metaheuristics for Feature Subset Selection in Medical Data Classification

Many of the feature attributes in a typical medical dataset are collected for reasons other than data classification. Some of the features are redundant while others are irrelevant adding more noise to the dataset. The Feature Subset Selection problem (FSS) consists in selecting the minimum subset of feature that represents the dataset without loss in classification accuracy [93]. FSS not only reduces storage and computational complexity, but also enhances comprehensibility and classification accuracy particularly in small sample size datasets. FSS also reduce the overfitting effect [50]. In medical diagnosis, it is desirable to select the clinical tests that have the least cost and risk and that are significantly important in determining the class of the disease. There are two approaches for solving the FSS problem. In the filter approach, features are selected independently of the classifier. In wrapper approach, a classifier is used to test each feature subset candidate and thus is classifier-dependent and computationally heavier than the filter approach. FSS problem is NP-hard [50]. Using exhaustive search to find all the possible feature subsets is computationally impractical, even for a medium sized feature set. This requires the use of heuristics and meta-heuristics. A recent survey of feature selection problem for machine learning classification can be found in [50]. What is needed is an algorithm with good global and local search abilities, that can converge to a near optimal solution in reasonable time, and that is computationally efficient [50]. Nature inspired methods like Particle Swarm Optimization (PSO) [46] Ant Colony Optimization (ACO) [22] have been successfully applied for many combinatorial optimization problems including FSS. These swarm intelligence search algorithms are based on the collective behavior of intelligent agents that use both direct and indirect interaction.

For example, in Unler and Murat [90], a Discrete PSO method is applied as wrapper feature subset selection methodology for a (binary) classification problem. Linear regression was chosen as the learning algorithm. Features are considered on an individual basis and the decision for inclusion combines the feature's predictive contribution, independent likelihood as well as the stochastic factor. The feature is added to the so-far collected feature subset if it results in improved classification accuracy. Computational complexity is managed by restricting the number of features considered in every iteration. PSO parameters are based on earlier empirical and theoretical research. Experiments were conducted on 10 UCI datasets and comparisons were made to exhaustive search and Random Subset Generation. Results show that PSO produced identical or near identical accuracy to exhaustive search accompanied with a significant time advantage. Results further show that PSO accuracy was competitive to Random Subset Generation. Other comparisons were done with other wrapper feature subset selection methods introduced in published research and using the same learning algorithm and data sets. These studies include tabu search and scatter search. Results show the superiority of the proposed algorithm in terms of both classification accuracy and computational cost.

The ACO metaheuristic was used by Vieira et al. [93] in conjunction with fuzzy rule-based classifiers. Fuzzy rule-based classifiers perform model learning while the ACO selects feature subsets. This specialized feature selection ACO is termed (AFS). It consists of two colonies; the first is assigned the determination of the number of features to be selected. The second, selects the features themselves. Choosing the cardinality of features is based on Fisher discriminant criterion. The fitness function is based on minimizing the number of selected features and the classification accuracy of the obtained model as evaluated using the fuzzy rule-based classifier. Experiments were conducted using 5 UCI benchmark datasets among which 2 are medical. Results for the medical data sets show a significant improvement over fuzzy rule-based classifiers alone and also over other published studies that use PSO and rough set-based feature selection.

7.5.2 Hybrid Metaheuristics for Artificial Neural Network Model Selection in Medical Data Classification

ANNs are capable of learning complex, non-linear decision surfaces with multiple classes [43]. A recent survey about machine learning in cancer prediction and prognosis [19] shows that more than half of the surveyed papers were using or referring to ANNs. The idea was derived from human biological neural system where multiple neurons are interconnected to each other. The basic unit is called a neuron. The simplest ANN is called a perceptron and is able to do a binary classification task that has a linear discriminate function. Neurons are organized in layers; producing a structure called a multilayer perceptron (MLP). The first layer is connected to the inputs. Each input predictor is normally connected to an input neuron. The last layer produces the output(s). There is usually one output neuron per class in the dataset [81]. With a certain precision, a two layer MLP can approximate any classification region [54]. In feed-forward ANNs, the most popular ANNs, there are no backward connections and no loops. Each node in a hidden layer has connections coming from the nodes in the previous layer, and others going to nodes in the next layer. Assuming that the weights of neuron connections are available, the features of an instance are fed to the input layer, and is propagated through the hidden layer(s), until it reaches the output layer. Each output neuron is associated with a class. The output neuron that generates the highest signal wins in determining the class of that instance.

There is no universal ANN architecture. The architectural design of ANNs should be optimized for each application [19]. In order to generate an ANN classifier, the weights of the network's connections needs to be determined. As these weights are real-valued, the problem of determining ANN weights (the optimization of the network training error, or in short training ANNs [61]) can be casted as a continuous optimization problem [81]. Back-propagation (BP) optimization algorithm [74] is normally used for tuning the values of the set of weights. It follows a gradient-decent technique on the error surface and exhibits a local search ability that causes it to get trapped into the nearest local minima [43].

The problem of simultaneous optimization of the network's training error and its architecture can be modeled as a multi-objective optimization problem [1]. Abbas [1] found that combining back-propagation algorithm with an evolutionary multi-objective optimization algorithm leads to a considerable drop in computational cost. In the following, an illustration is presented of literature that substitutes or combines traditional ANN training algorithms with hybrid metaheuristics. The last two examples use hybrid metaheuristics for the task of optimizing both ANN architecture and training error.

The use of Estimation of Distribution Algorithms (EDAs) [7] is investigated by Madera and Dorronsoro [61] for the training of ANNs. The ANNs are used for the medical classification task of four PROBIN1 benchmark datasets. EDAs are evolutionary stochastic search techniques that base the construction of a new generation on estimations of the probability distribution of current population, rather than by means of variation operators. In this study, six different EDAs are being tested. These algorithms cover discrete and continuous search spaces and in each type of search space, three different correlation types of input variables are tested: those without dependencies, with bivariate dependencies and multiple dependencies. Training–testing method was used rather than the cross-validation. The performance was compared against other famous ANN training techniques; namely BP and Levenberg–Marquardt (LM) [32] algorithm. Performance was also compared to few ANN training techniques based on EAs; including those using GAs, MAs, or evolutionary programming (EP) [26, 27]. In general, EDAs performance was comparable to the others. This initial result is promising as further parameter tuning might likely improve the findings. EDAs for discrete domains were generally slower and less accurate than those for continuous domains. This is due to the large search space resulting from the discretization of the input variables. In addition, EDAs based on higher degrees of dependencies are better suited for the more complex problems as they exhibit slow convergence.

The study by Smithies et al. [80] aims at predicting the recurrence of colorectal cancer. In particular, the study tests the efficacy of a type of chemotherapy termed FUFA in preventing cancer recurrence. The data set obtained from NHS hospitals in the UK features different types of attributes and a considerable amount of missing data. The classification process started with a relaxed linear regression stage to remove irrelevant attributes and those with markedly missing data. Next, data is clustered to better deal with the different data types and allow a better inference mechanism for missing data rather than statistical methods. A new clustering methodology is introduced that is tailored to data with mixed attribute types. Each cluster is then fed to a three-layer feed-forward ANN ensemble. The members of each ensemble differ only in the number of hidden nodes. For training the neural nets, local search is combined with a modified form of back propagation algorithm known as the batched error back propagation with an enhanced Resilient Propagation for learning rate adaptation (iRPROP). The combination of gradient-independent local search with the gradient-based enhanced iRPROP has shown to enhance the classification performance. In addition, the proposed search algorithm utilizes the forbidden neighborhood region idea from tabu search. Given that previous studies focus on statistical models, the 66% of patients being correctly predicted forms a promising result and encourages further enhancements.

Reference [81] extends the ACO algorithm to tackle unconstrained continuous optimization problems; ACOR. It is then used to optimize the weights for a feed-forward ANN used in a medical classification task. Classification is applied on three PROBIN benchmark medical datasets. The resulting performance was compared against two neural network training algorithms; BP and LM. Training–testing method was used rather than the cross-validation. As a general optimization algorithm and unlike BP and LM, ACOR does not require that the neuron function is known and is differentiable. However, ACOR does not exploit additional information, such as the gradient information. Results show that the performance of ACOR was inferior to the other two. A hybrid approach was also tested where ACOR was combined with BP (ACOR-BP) and with LM (ACOR-LM). In the hybrid approach, each solution of ACOR is enhanced by running a single iteration of the BP or LM methods respectively. The performance of the hybrid approached was comparable to BP and LM and in some cases outperformed them. The proposed algorithms where also compared to GA and its' hybrids (GA-BP and GA-LM) on the same data sets, and have significantly outperformed them.

Castillo et al. [16] and Ince et al. [43] investigate the optimization of ANN architecture and training error. Castillo et al. [16] used a multi-objective evolutionary algorithm called MG-Prob for the simultaneous optimization of three objectives; the reduction of type-I and type-II errors as well as minimizing the artificial neural network's size. MG-Prob is based on the Single Front Genetic Algorithm (FSGA) [20] that builds on the Pareto optimality principle. Elite set represents the non-dominated individuals and in FSGA only a diverse part of this set that is spread across the search space is copied into the next generation. Individuals are multi-layer perceptrons (MLPs). The resulting non-dominated individuals in the population are used as an ensemble to perform the classification. Three methods were used to combine the ensemble results; voting, average and largest activation among all outputs. Experiments on breast cancer dataset from the UCI repository demonstrate the effectiveness of this method as compared to other methods obtaining slightly better classification error with a minimal difference between the two type errors, and smaller network size for individual MLPs.

PSO was used by Ince et al. [43] to set the parameters for feed-forward fully-connected ANNs and compared the results with those obtained by the traditional BP training algorithm for different training depths (deep/shallow). Experiments were conducted on three UCI Probin1 medical datasets; breast cancer, heart disease and diabetes. Classification Error Percentage (CEP) was used to measure performance were $CEP = 1 - P$. Other performance measures were used including mean square error (MSE) and average processing time (ms). The study concluded that PSO has better generalization ability and more stable performance with respect to changing network architecture. On the other hand, BP resulted in better classification accuracy for smaller networks. The accuracy for PSO and BP is otherwise comparable. BP training however was consistently superior in terms of computational complexity.

7.5.3 Hybrid Metaheuristics for Full Model Selection in Medical Data Classification

PSO is used to perform a full model selection (FMS) for a classification task [24]. No background knowledge about the problem is required. Full model selection involves choosing and chaining preprocessing methods (zero or more), feature subset selection method (zero or one), a learning algorithm and a post processing method (zero or one). FMS includes the choice of all the associated parameters as well as the order of preprocessing and feature subset selection (i.e. to perform FSS first or else preprocessing first). The choice of these methods is made from objects available at the CLOP machine learning package. This package includes three preprocessing methods, twelve feature selection methods, ten ML classification algorithms and a single post processing method. Each individual is encoded such that it represents a definition to all the previous stages. Fitness is evaluated in terms of balanced error rate (BER). The advantage of this evaluation criterion is that it considers classification errors in both classes and thus avoids rewarding an algorithm that favors the majority class. Computational complexity is reduced by means of sub-sampling heuristic. The proposed PSO-based FMS was compared to another FMS method that is based on a simple direct search and optimization algorithm termed pattern search (PS) [21]. Results show that the PSO alternative consistently outperformed the PS-based FMS. The proposed algorithm was also challenged against other models that use background knowledge or are based on model selection for a single learning algorithm in the framework of a model selection competition named Agnostic Learning vs. Prior Knowledge Challenge. The proposed model has demonstrated comparable results.

7.6 Conclusion

Medical data classification is a new field of research that will improve the cost, accessibility, and quality of health care. The complexity associated with medical data classification prohibits the use of exact methods. This chapter overviews the state-of-the-art approaches in medical data classification. Studies suggest the use of fuzzy and hybrid meta-heuristic methods for model learning, selection, and optimization. Hybridizing different approximation algorithms, including metaheuristics, is a promising approach. However, most of the studies choose their system components arbitrarily, for example due to their success in other fields of study. Model comprehensibility is an important factor in the selection of these components. There exists a need for a meta-study that focuses on the basis of choosing hybrid system components for medical data classification. Perhaps the work by Tanwani and his team [85, 86, 87, 88] on formalizing medical data complexity forms a gateway to this area.

The use of transparent comprehensible models enables physicians to validate the clinical plausibility of the resulting classification hypothesis and allows discovering new interesting relations. In some cases, obtaining explanations and conclusions that enlighten and convince medical experts is more important than suggesting a particular class. XCS particularly showed comprehensible results with high classification accuracies. Several studies employing XCS were presented in this document. However, there is still room for research and improvement. For example, the use of XCS ensembles is an interesting line of research to be investigated.

Most of the studies highlighted in this document aim at the design of general classification systems rather than systems that are geared towards a specific medical disorder. It is true that medical data have many common characteristics. However, each dataset has its own character (for example, see Table 7.1). Therefore, classification models that work well with one dataset may not exhibit the same level of performance on another. Medical data classification may benefit more with focused research.

Another issue is the evaluation metric. Most studies limit the evaluation metric to precision (P). As this may be satisfactory in the machine learning and data mining community, it may be not for the medical community. The inclusion of other evaluation metrics that are routinely used in the medical field would certainly give more value and depth to the results. For example, Holmes EpiXCS tailor XCS to the epidemiologic community mainly by considering evaluation metrics needed by the epidemiologic field.

Due to the societal significance of the subject and also to the computational challenge it presents, more research in the field of medical data classification is needed. The papers introduced in this chapter only represent the tip of the iceberg in the medical data classification field. Despite this wealth in literature, very few systems are put into practical use. The inclusion of a medical professional in the study team is a valuable asset that is most often ignored. Several authors have addressed the clinical approval of intelligent systems before [34, 56, 72] and so they need to be considered seriously.

Acknowledgements. The authors would like to thank Dr. Bryan Bergeron, a member of the Affiliate Faculty, Harvard-MIT Division of Health Sciences and Technology, for his valuable input in reviewing this document.

References

1. Abbas, H.A.: Speeding up Back-Propagation using Multiobjective Evolutionary Algorithms. Neural Computation 15(11), 2705–2726 (2003)
2. Aguilar-Ruiz, J., Riquelme, J., Toro, M.: Evolutionary Learning of Hierarchical Decision Rules. IEEE Transactions on Systems, Man, and Cybernetics, Part B 33(2), 324–331 (2003)
3. Aci, M., Inan, C., Avci, M.: A Hybrid Classification Method of K-Nearest Neighbor, Bayesian Methods and Genetic Algorithm. Expert Systems with Applications 30, 5061–5067 (2010)
4. Bacardit, J.: Pittsburgh Genetics-Based Machine Learning in the Data Mining era: Representations, generalization, and run-time. Ramon Llull University, Dissertation (2004)
5. Bacardit, J., Butz, M.V.: Data Mining in Learning Classifier Systems: Comparing XCS with GAssist. In: Kovacs, T., Llorà, X., Takadama, K., Lanzi, P.L., Stolzmann, W., Wilson, S.W. (eds.) IWLCS 2003. LNCS (LNAI), vol. 4399, pp. 282–290. Springer, Heidelberg (2007)
6. Bacardit, J., Bernadó-Mansilla, E., Butz, M.V.: Learning Classifier Systems: Looking Back and Glimpsing Ahead. In: Bacardit, J., Bernadó-Mansilla, E., Butz, M.V., Kovacs, T., Llorà, X., Takadama, K. (eds.) IWLCS 2006 and IWLCS 2007. LNCS (LNAI), vol. 4998, pp. 1–21. Springer, Heidelberg (2008)
7. Back, T., Fogel, D., Michalewicz, Z.: Handbook of Evolutionary Computation. Oxford University Press, London (1997)
8. Bernadó-Mansilla, E., Garrell-Guiu, J.M.: Accuracy based learning classifier systems: models, analysis and applications to classification tasks. Evolutionary Computation 11(3), 209–238 (2003)
9. Bhattacharyya, D., Pal, A.J., Kim, T.: Cell-graph coloring for cancerous tissue modeling and classification. Multimedia Tools and Applications (2011), doi:10.1007/s11042-011-0797-y
10. Bojarczuk, C., Lopes, H., Freitas, A.: Genetic Programming for Knowledge Discovery in Chest Pain Diagnosis. IEEE Engineering in Medicine and Biology Magazine 19(4), 38–44 (2000)
11. Bojarczuk, C., Lopes, H., Freitas, A., Michaliewicz, E.: A Constrained-syntax Genetic Programming System for Discovering Classification Rules: Application to Medical Data Sets. Artificial Intelligence in Medicine 30(1), 27–48 (2004)
12. Bradley, A.P.: The Use of the Area under the ROC Curve in the Evaluation of Machine Learning Algorithms. Pattern Recognition 30, 1145–1159 (1997)
13. Bull, L., Hurst, J.: ZCS Redux. Evolutionary Computation 10(2), 185–205 (2002)
14. Bull, L., Bernadó-Mansilla, E., Holmes, J.: Learning Classifier Systems in Data Mining: An Introduction. In: Bull, L., Bernadó-Mansilla, E., Holmes, J. (eds.) Learning Classifier Systems in Data Mining: Studies in Computational Intelligence, pp. 1–16. Springer (2008)
15. Candelieri, A.: A hyper-solution framework for classification problems via metaheuristic approaches. 4OR-Q J Oper Res (2010), doi:10.1007/s10288-011-0166-8
16. Castillo, P.A., Arenas, M., Merelo, J.J., Rivas, V.M., Romero, G.: Multiobjective Optimization of Ensembles of Multilayer Perceptrons for Pattern Classification. In: Runarsson, T.P., Beyer, H.-G., Burke, E.K., Merelo-Guervós, J.J., Whitley, L.D., Yao, X. (eds.) PPSN 2006. LNCS, vol. 4193, pp. 453–462. Springer, Heidelberg (2006)
17. Chen, S., Lin, S., Chou, S.: Enhancing the Classification Accuracy by Scatter Search-Based ensemble Approach. Applied Soft Computing 11, 1021–1028 (2011)
18. Chorbev, I., Mihajlov, D., Jolevski, I.: Web Based Medical Expert System with Self Training Heuristic Rule Induction Algorithm. In: Proceedings of the First International Conference on Advances in Databases, Knowledge, and Data Application (DBKDA 2009), pp. 143–148. IEEE Computer Society, Washington, DC (2009), doi:10.1109/DBKDA.2009.21

19. Cruz, J., Wishart, D.: Applications of Machine Learning in Cancer Prediction and Prognosis. Cancer Informatics 2006(2), 59–77 (2006)
20. de Toro, F., Ortega, J., Fernandez, J., Diaz, A.: Parallel genetic algorithm for multiobjective optimization. In: Proceedings of the 10th Euromicro Workshop on Parallel, Distributed and Network-based Processing, pp. 384–391. IEEE Computer Society (2002)
21. Dennis, J., Torczon, V.: Derivative-free pattern search methods for multidisciplinary design problems. In: Proceedings of the 5th AIAA/ USAF/NASA/ISSMO Symposium on Multidisciplinary Analysis and Optimization, Panama City, FL, pp. 922–932 (1994)
22. Dorigo, M.: Optimization, Learning and Natural Algorithms, Dissertation, Politecnico di Milano, Italie (1992)
23. Dybowski, R., Weller, P., Chang, R.: Gant V Prediction of outcome in critically ill patients using artificial neural network synthesized by genetic algorithm. Lancet 347(9009), 1146–1150 (1996)
24. Escalante, H., Montes, M., Sucar, L.: Particle Swarm Model Selection. J. Mach. Learn. Res. 10, 405–440 (2009)
25. Fan, C., Chang, P., Hsieh, J.L.: A Hybrid Model Combining Case-based Reasoning and Fuzzy Decision Tree for Medical Data Classification. Applied Soft Computing 11, 632–644 (2011)
26. Fogel, L.: Evolutionary Programming in Perspective: the Top-Down View. In: Zurada, J., Marks II, R., Robinson, C. (eds.) Computational Intelligence: Imitating Life, pp. 135–146. IEEE Press (1994)
27. Fogel, L., Owens, A., Walsh, M.: Artificial Intelligence through a Simulation of Evolution. In: Callahan, A., Maxfield, M., Fogel, L.J. (eds.) Biophysics and Cybernetic Systems, pp. 131–155. Spartan, Washington DC (1965)
28. Gao, Y., Huang, J., Rong, H.: Learning Classifier System Ensemble for Data Mining. In: Proceedings of the 2005 Genetic and Evolutionary Computation Conference IWLCS, pp. 63–66 (2005)
29. Glover, F.: Tabu Search - Part I. ORSA Journal on Computing 1(3), 190–206 (1989)
30. Glover, F.: Tabu Search - Part II. ORSA Journal on Computing 2(1), 4–32 (1990)
31. Glover, F.: A Template for Scatter Search and Path Relinking. In: Hao, J.-K., Lutton, E., Ronald, E., Schoenauer, M., Snyers, D. (eds.) AE 1997. LNCS, vol. 1363, p. 13. Springer, Heidelberg (1998)
32. Hagan, M.T., Menhaj, M.B.: Training Feedforward Networks with the Marquardt Algorithm. IEEE Transactions on Neural Networks 5, 989–999 (1994)
33. Han, J., Kamber, M.: Data Mining: Concepts and Techniques, 2nd edn., Jim Gray. The Morgan Kaufmann Series in Data Management Systems. Series Editor Morgan Kaufmann Publishers (2006) ISBN 1-55860-901-6
34. Hanson, C.W.: Marshall BEArtificial intelligence applications in the intensive care unit. Crit. Care Med. 29(2), 427–435 (2001)
35. Hastie, T., Tibshirani, R., Friedman, J.: The Elements of Statistical Learning: Data Mining, Inference, and Prediction, 2nd edn. Springer (2009) ISBN: 9780387848570
36. Holland, J.H.: Adaptation in Natural and Artificial Systems, 1st edn. The University of Michigan Press, Ann Arbor (1975); MIT Press, Cambridge, MA (1992)
37. Holland, J., Reitman, J.: Cognetive Systems based on Adaptive Agents. In: Waterman, D.A., Inand, F. (eds.) Pattern-Directed Inference Systems, Hayes-Roth (1978)
38. Holmes, J.: Discovering Risk of Disease with a Learning Classifier System. In: Proceedings of the 7th International Conference on Genetic Algorithms (ICGA 1997), pp. 426–433 (1997)
39. Holmes, J., Sager, J.: Rule Discovery in Epidemiologic Surveillance Data Using EpiXCS: An Evolutionary Computation Approach. In: Miksch, S., Hunter, J., Keravnou, E.T. (eds.) AIME 2005. LNCS (LNAI), vol. 3581, pp. 444–452. Springer, Heidelberg (2005)
40. Holmes, J., Durbin, D., Winston, F.: Discovery of Predictive Models in an Injury Surveillance Database: An Application of Data Mining in Clinical Research. In: Proceedings of AMIA Symposium, pp. 359–363 (2000a)

41. Holmes, J., Durbin, D., Winston, F.: The learning classifier system: an evolutionary computation approach to knowledge discovery in epidemiologic surveillance. Artificial Intelligence in Medicine 19, 53–74 (2000b)
42. Holmes, J., Lanzi, P., Stolzmann, W., Wilson, S.: Learning Classifier Systems: New Models, Successful Applications. Information Processing Letters archive 82(1), 23–30 (2002)
43. Ince, T., Kiranyaz, S., Pulkkinen, J., Gabbouj, M.: Evaluation of global and local training techniques over feed-forward neural network architecture spaces for computer-aided medical diagnosis. Expert Systems with Applications 37, 8450–8461 (2010)
44. Jain, R., Mazumdar, J., Moran, W.: Application of fuzzy-classifier system to coronary artery disease and breast cancer. Australas Phys. Eng. Sci. Med. 21(3), 141–147 (1998)
45. Jiang, Y., Metz, C.E., Nishikawa, R.M.: A Receiver Operating Characteristic Partial Area Index for Highly Sensitive Diagnostic Tests. Radiology 201, 745–750 (1996)
46. Kennedy, J., Eberhart, R.C.: Particle swarm optimization. In: Proceedings of IEEE International Conference on Neural Networks, Piscataway, NJ, vol. 4, pp. 1942–1948 (1995)
47. Kharbat, F., Bull, L., Odeh, M.: Mining breast cancer data with XCS. In: Proceedings of the 9th Annual Conference on Genetic and Evolutionary Computation, pp. 2066–2073 (2007)
48. Kirkpatrick, S., Gelatt, C.D., Vecchi, M.P.: Optimization by Simulated Annealing. Science. New Series 220(4598), 671–680 (1983)
49. Kotsiantis, S.B.: Supervised Machine Learning: A Review of Classification. Informatica Journal 31, 249–268 (2007)
50. Kotsiantis, S.B.: Feature selection for machine learning classification problems: a recent overview. Artif. Intell. Rev (2011), doi:10.1007/s10462-011-9230-1
51. Koza, J.R.: Genetic Programming. MIT Press, Cambridge (1992)
52. Lanzi, P.: Extending the Representations of Classifier Conditions. Part II: From Messy coding to S-expression. In: Banzhaf, W., et al. (eds.) Proceedings of the Genetic and Proceedings of the Genetic and Evolutionary Computation Conference, vol. 1, pp. 345–352 (1999)
53. Lanzi, P.: Mining interesting knowledge from data with the XCS classifier system. In: Proceedings of the Genetic and Evolutionary Computation Conference (GECCO 2001), pp. 958–965. Morgan Kaufmann, San Francisco (2001)
54. Larrañaga, P., Calvo, B., Santana, R., et al.: Machine Learnig in Bioinformatics. Brief Bioinform. 7(1), 86–112 (2006)
55. Lavrac, N.: Selected Techniques for Data Mining in Medicine. Artificial Intelligence in Medicine 16(1), 3–23 (1999)
56. Lisboa, P.J.: Taktak AFG The use of artificial neural networks in decision support in cancer: A systematic review. Neural Networks 19(4), 408–415 (2006)
57. Lo, H.-Y., Chang, C.-M., Chiang, T.-H., et al.: Learning to Improve Area-Under-FROC for Imbalanced Medical Data Classification Using an Ensemble Method. In: ACM SIGKDD Explorations Newsletter, vol. 10(2), ACM, New York (2008), doi:10.1145/1540276.1540290
58. Lucas, F.: Analysis of Notions of Diagnosis. Artificial Intelligence 105(12), 295–343 (1998)
59. Lucas, F., Abu-Hanna, A.: Prognosis Methods in Medicine. Artificial Intelligence in Medicine 15(2), 105–119 (1998)
60. Luukka, P., Lampinen, J.: Differential Evolutionary Classifier in Noisy Settings with Interactive Variables. Applied Soft Computing 1, 891–899 (2011)
61. Madera, J., Dorronsoro, B.: Estimation of Distribution Algorithms. In: Metaheuristic Procedures for Training Neutral Networks Operations Research/Computer Science Interfaces Series, Part III, vol. 36, pp. 87–108 (2006)
62. Moscato, P.: On Evolution, Search, Optimization, Genetic Algorithms and Martial Arts: Towards Memetic Algorithms - Caltech Concurrent Computation Program, C3P Report (1989)
63. Narayanan, M.N., Lucas, S.B.: A genetic algorithm to improve a neural network to predict a patient's response to warfarin. Methods Inf. Med. 32(1), 55–58 (1993)

64. Penã-Reyes, C., Sipper, M.: A fuzzy-genetic approach to breast cancer diagnosis. Artif. Intell. Med. 17(2), 131–135 (1999)
65. Penã-Reyes, C., Sipper, M.: Evolutionary Computation in Medicine: an Overview. Artif. Intell. Med. 19(1), 1–23 (2000)
66. Peroumalnaik, M., Enee, G.: Prediction using Pittsburgh Learning Classifier Systems: APCS use case. In: Proceedings of the 12th Annual Conference Companion on Genetic and Evolutionary Computation GECCO 2010, pp. 1901–1907 (2010)
67. Pham, H.N.A., Triantaphyllou, E.: The impact of overfitting and overgeneralization on the classification accuracy in data mining. In: Maimon, O., Rokach, L. (eds.) Soft Computing for Knowledge Discovery and Data Mining, Part 4, pp. 391–431. Springer, New York (2007)
68. Pham, H.N., Triantaphyllou, E.: An application of a new meta-heuristic for optimizing the classification accuracy when analyzing some medical datasets. Expert Systems with Applications 36, 9240–9249 (2009)
69. Price, K., Storn, R., Lampinen, J.: Differential Evolution - A Practical Approach to Global Optimization. Springer (2005)
70. Provost, F., Kolluri, V.: A Survey of Methods for Scaling up inductive Algorithms. Datamining and Knowledge Discovery 3(2), 131–169 (1999)
71. Ranawana, R., Palade, V.: Optimized Precision, A New Measure for Classifier Performance Evaluation. In: Proceedings of the IEEE Congress on Evolutionary Computation, Vancouver, Canada, pp. 2254–2261 (2006)
72. Rao, R.B., Bi, J., Fung, G., et al.: LungCAD: A Clinically Approved, Machine Learning System for Lung Cancer Detection. In: Proceedings of the 13th ACM SIGKDD International Conference on Knowledge Discovery and Data Mining, ACM, New York (2008), doi:10.1145/1281192.1281306
73. Rojanavasu, P., Dam, H., Abbass, H., Lokan, C., Pinngern, O.: A Self-Organized, Distributed, and Adaptive Rule-Based Induction System. IEEE Transctions on Neural Networks 20(3), 446–495 (2009)
74. Rumelhart, D., Hinton, G., Williams, R.: Learning Representations by Backpropagation Errors. Nature 323, 533–536 (1986)
75. Sarkar, B.K., Sana, S.S.: A Hybrid Approach to Design Efficient Learning Classifiers. Computers and Mathematics with Applications 58, 65–73 (2009)
76. Shortliffe, E., Cimino, J.: Biomedical Informatics: Computer Applications in Health Care and Biomedicine. Springer, New York (2006)
77. Skinner, B., Nguyen, H., Liu, D.: Classification of EEG Signals Using a Genetic-Based Machine Learning Classifier. In: Proceedings of the 29th Annual International Conference of the IEEE EMBS, Lyon, France, pp. 3120–3123 (2007a)
78. Skinner, B., Nguyen, H., Liu, D.: Distributed Classifier Migration in XCS for Classification of Electroencephalographic Signals. In: Proceedings of the IEEE Congress on Evolutionary Computation CEC 2007, Singapore, pp. 2829–2836 (2007b)
79. Smith, S.: A learning system based on genetic adaptive algorithms. Dissertation, University of Pittsburgh, Pittsburgh (1980)
80. Smithies, R.G., Salhi, S., Queen, N.M.: Predicting colorectal cancer recurrence: A hybrid neural networks-based approach. In: Ibaraki, T., Nonobe, K., Yagiura, M. (eds.) Metaheuristics: Progress as Real Problem Solvers. Series: Operations Research/Computer Science Interfaces Series, vol. 32, pp. 259–285 (2005)
81. Socha, K., Blum, C.: Ant Colony Optimization. In: Metaheuristic Procedures for Training Neutral Networks, Part IV. Operations Research/Computer Science Interfaces Series, vol. 36, pp. 153–180 (2006)
82. Stolzmann, W.: Anticipatory Classifier Systems. In: Proceedings of the 3rd Annual Genetic Programming Conference, pp. 658–664 (1998)
83. Stone, M.: Cross-validatory choice and assessment of statistical predictions. Journal of the Royal Statistical Society Series B 36, 111–147 (1974)
84. Tan, K., Yu, Q., Heng, C., Lee, T.: Evolutionary Computing for Knowledge Discovery in Medical Diagnosis. Artificial Intelligence in Medicine 27(2), 129–154 (2003)

85. Tanwani, A., Farooq, M.: Performance Evaluation of Evolutionary Algorithms in Classification of Biomedical Datasets. In: Proceedings of the 11th Annual Conference Companion on Genetic and Evolutionary Computation: Late Breaking Papers, GECCO 2009, Canada, pp. 2617–2624 (2009a)
86. Tanwani, A., Farooq, M.: The Role of Biomedical Dataset in Classification. In: Combi, C., Shahar, Y., Abu-Hanna, A. (eds.) AIME 2009. Lecture Notes in Computer Science (LNAI), vol. 5651, pp. 370–374. Springer, Heidelberg (2009b)
87. Tanwani, A.K., Farooq, M.: Classification Potential vs. Classification Accuracy: A Comprehensive Study of Evolutionary Algorithms with Biomedical Datasets. In: Bacardit, J., Browne, W., Drugowitsch, J., Bernadó-Mansilla, E., Butz, M.V. (eds.) IWLCS 2008/2009. Lecture Notes in Computer Science (LNAI), vol. 6471, pp. 127–144. Springer, Heidelberg (2010)
88. Tanwani, A., Afridi, J., Shafiq, M., Farooq, M.: Guidelines to Select Machine Learning Scheme for Classification of Biomedical Datasets. In: Pizzuti, C., Ritchie, M.D., Giacobini, M. (eds.) EvoBIO 2009. LNCS, vol. 5483, pp. 128–139. Springer, Heidelberg (2009)
89. Tzima, F., Mitkas, P.: ZCS Revisited: Zeroth-level Classifier Systems for Data Mining. In: Proceedings of the 2008 IEEE International Conference on Data Mining Workshops, pp. 700–709 (2008)
90. Unler, A., Murat, A.: Discrete Optimization: A discrete particle swarm optimization method for feature selection in binary classification problems. European Journal of Operational Research 206(3), 528–539 (2010)
91. Unold, O., Tuszynski, K.: Mining Knowledge from Data using Anticipatory Classifier Systems. Knowledge-Based Systems 21(5), 363–370 (2008)
92. Urbanowicz, R., Moore, J.: Review Article: Learning Classifier Systems: A Complete Introduction, Review and Roadmap. Journal of Artificial Evolution and Applications, 1–25 (2009)
93. Vieira, S.M., Sousa, J., Runkler, T.A.: Multi-Criteria Ant Feature Selection Using Fuzzy Classifiers. In: Swarm Intelligence for Multi-objective Problems in Data Mining: Studies in Computational Intelligence, vol. 242, pp. 19–36 (2009)
94. Walter, D., Mohan, C.: ClaDia: A Fuzzy Classifier System for Disease Diagnosis. In: Proceedings of the 2000 Congress on Evolutionary Computation, CA, USA, vol. 2, pp. 1429–1435 (2000)
95. Wilson, S.W.: ZCS: A Zeroth-Level Learning Classifier System. Evolutionary Computation 2(1), 1–18 (1994)
96. Wilson, S.W.: Classifier Fitness Based on Accuracy. Evolutionary Computation 3(2), 149–175 (1995)
97. Wilson, S.W.: Mining Oblique Data with XCS. In: Lanzi, P.L., Stolzmann, W., Wilson, S.W. (eds.) IWLCS 2000. LNCS (LNAI), vol. 1996, pp. 158–290. Springer, Heidelberg (2001)

Chapter 8
HydroCM: A Hybrid Parallel Search Model for Heterogeneous Platforms

Julián Domínguez and Enrique Alba

Abstract. Here we present HydroCM (HydroCarbon inspired Metaheuristic), a parallel metaheuristic model specifically designed for its execution on heterogeneous hardware environments. With HydroCM we actually propose a schema for describing a family of parallel hybrid metaheuristics inspired by the structure of hydrocarbons in Nature, establishing a resemblance between atoms and computers, and between chemical bonds and communication links. Our goal is to gracefully match computers of different computing power to algorithms of different behavior (GA and SA in this study), all them collaborating to solve the same problem. The analysis will show that our proposal, though simple, can solve search problems in a faster and more robust way than well-known panmictic and distributed algorithms very popular in the literature.

8.1 Introduction

Metaheuristics are an important branch of research since they provide a fast an efficient way for solving problems. In many cases, parallelism is necessary, not only to reduce the computation time, but to enhance the quality of the solutions obtained. Many parallel models exist, both for local search methods (LSMs) and evolutionary algorithms (EAs), and even parallel hybrid models combining both methods are present in the literature [4] [6].

In a modern lab, it is very common the coexistence of many different hardware architectures. It has been proven that such heterogeneous resources can also be used efficiently to solve optimization problems with standard parallel algorithms [7] [20]

Julián Domínguez · Enrique Alba
Universidad de Málaga, Spain
e-mail: `{julian,eat}@lcc.uma.es`

E.-G. Talbi (Ed.): Hybrid Metaheuristics, SCI 434, pp. 219–235.
springerlink.com

[21], but there exist few works about the design of specific parallel models for an heterogeneous environment.

Here we present HydroCM, a hybrid parallel metaheuristic model. With this work we propose a general model for describing a family of hybrid metaheuristics specifically designed for their execution in heterogeneous hardware environments, being inspired in the structure of the hydrocarbons that can be found in Nature.

Our contribution is not only methodological, but we also have carried out an analysis in order to study the behavior of our proposal. For our analysis, we have implemented two versions of the model making use of two well-known metaheuristics: steady state Genetic Algorithm (ssGA) and Simulated Annealing(SA). We have compared our proposal against the panmictic versions of these algorithms and against a unidirectional ring of ssGA islands executed on the same hardware infrastructure. Our results show that the running times of our proposal are faster in some cases and more robust in the rest than the reference ssGA ring.

We will here present an overview of the proposed model as well as the results of the analysis of the implemented algorithms. Previously, we will start with a brief review on the background concepts used in this chapter.

8.2 Decentralized, Heterogeneous and Hybrid Parallel Metaheuristics

In this section we include a quick review on the existing implementations of decentralized and parallel metaheuristics, as well as on heterogeneity. We also include a description of the metaheuristics used in our hybrid algorithm and how they classify as hybrid metaheuristics.

Many parallel implementations exist for different groups of metaheuristics. We will focus in two of the more common families of metaheuristics: Evolutionary Algorithms (EAs) and Local Search Metaheuristics (LSMs). On the one hand, EAs are population based methods, where a random population is created and further enhanced through a Nature-like evolution process. On the other hand, only one candidate solution is used in LSMs, and it is enhanced by moving through its neighborhood replacing the candidate solution by another one, usually one with a better quality (fitness) value. EAs commonly provide a good exploration of the search space, so they are also called exploration-oriented methods. On the contrary, LSMs allow to find a local optima solution and subsequently they are called exploitation-oriented methods. Many different parallel models have been proposed for each method, and here we present the more representative ones.

8.2.1 Parallel EA Models

A panmictic EA applies its stochastic operators over a single population, which makes them easily parallelizable. A first strategy for its parallelization is the use of

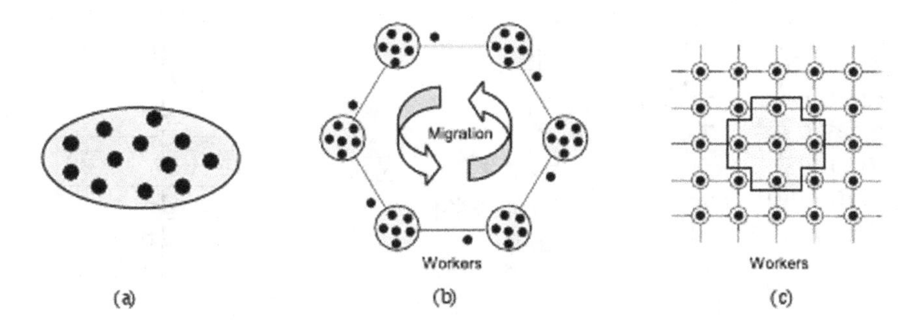

Fig. 8.1 A panmictic EA (a), and two structured EAs: distributed (b) and cellular (c)

a master-slave approach where evaluations are performed in parallel but the population, unless divided, is treated as a whole, maintaining its panmictic behavior. It could be interesting for many tasks, but it does not offer the benefits of a structured population. Therefore, we are going to focus in structured populations, which leads to a distinction: cellular versus distributed EAs [3] (Figure 8.1).

- **Distributed EAs (dEA):** In the case of distributed EAs, the population is divided into a number of islands that run an isolated instance of the EA (Figure 8.1b). Although there is not a single population the sub-populations are not completely isolated: some individuals are sent from one population to another following a migration scheme. It is common that in this model there only exist a few sub-algorithms, loosely coupled among them.
- **Cellular EAs (cEA):** In the cellular model, there exists only one population which is structured into neighborhoods, so that an individual can only interact with the individuals inside its neighborhood (Figure 8.1c). Different neighborhood structures can lead to a different behavior. With the cellular model there exists a large number of sub-algorithms and they are tightly coupled [5].

8.2.2 Parallel LSM Models

Many different parallel models have been proposed for LSMs, but there exist three models that are widely extended in the literature: parallel multistart model, parallel moves model, and move acceleration model (Figure 8.2).

- **Parallel multistart model:** In this model, several independent instances of the LSM are launched simultaneously (Figure 8.2a). They can exchange individuals following a migration scheme. This model can usually compute better and more robust solutions than the panmictic version.
- **Parallel moves model:** This model is a kind of master-slave model where the master runs a sequential LSM but, at the beginning of each iteration, the current solution is distributed among all the slaves (Figure 8.2b). The slaves perform a

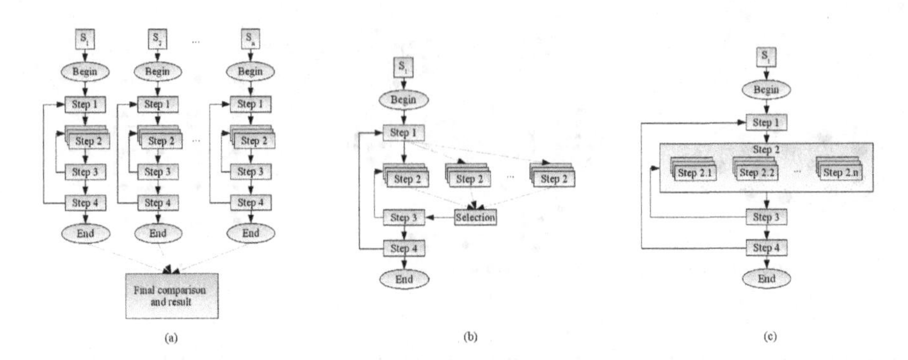

Fig. 8.2 Parallel multistart model (a), parallel moves model (b), and move acceleration model (c)

move and return the candidate solution to the master, which selects one of them. This model does not alter the behavior of the sequential algorithm.

- **Move acceleration model:** The quality of each candidate solution is evaluated in a parallel centralized way (Figure 8.2c). It is useful when the evaluation function can be itself parallelized. The move acceleration model does not alter the behavior of the sequential algorithm.

In both, EAs and LSMs parallel models, each sub-algorithm includes a phase for communication with a neighborhood according to some topology. This communication can be carried out in a synchronous or asynchronous manner. Many works have found advantages in using an asynchronous execution model [8] [11]. Additionally, asynchronism is essential in our study because of the heterogeneous hardware, which could easily produce bottlenecks, so our communications will be carried out in an asynchronous way.

8.2.3 Being Heterogeneous

In the models presented above, all the sub-algorithms share the same search features. But we could modify the behavior of a parallel metaheuristic by changing the search features between sub-algorithms, obtaining a globally heterogeneous hybrid metaheuristic. Also the hardware being used to run the algorithm can be homogeneous or heterogeneous, so we have not to be confused between the hardware platform heterogeneity and the heterogeneous software model. Parallel heterogeneous metaheuristics can be classified into four levels depending on the source of heterogeneity [3]:

- *Parameter level*: At this level, the same algorithm is used in each node, but the configuration parameters are different in one or more of them.
- *Operator level*: At operator level, heterogeneity is achieved by using different mechanisms for exploring the search space, such as different operators.

- *Solution level*: Heterogeneity is obtained using a different encoding for the solutions in each component.
- *Algorithm level*: At this level, each component can run a different algorithm. This level is the most widely used.

Here we present an algorithm level parallel heterogeneous metaheuristic which is later run in heterogeneous hardware. This solver is based in two different methods. We have chosen one method of each of the two well-known families, LSMs and EAs, in order to obtain a good balance between exploitation and exploration. The used methods are a Genetic Algorithm (GA) and a Simulated Annealing (SA).

GAs are one of the more popular EAs present in the literature. In Algorithm 8 we can see an outline of a panmictic GA. A GA starts by randomly generating an initial population $P(0)$, with each individual encoding a candidate solution for the problem and its associated fitness value. At each iteration, a new population $P'''(t)$ is generated using simple stochastic operators, leading the population towards regions with better fitness values.

Algorithm 8. Standard Genetic Algorithm

```
Generate(P(0));
Evaluate(P(0));
t := 0;
while not stop_condition(P(t)) do
   P'(t) := Selection(P(t));
   P''(t) := Recombination(P'(t));
   P''(t) := Mutation(P''(t));
   Evaluate(P'''(t));
   P(t+1) := Replace(P(t),P'''(t));
   t := t+1;
end while
```

In our algorithm, we have actually used a special variant of the generic GA called steady state Genetic Algorithm (ssGA) [22]. The difference between a common generational GA and a ssGA is the replacement policy: while in a generational GA a full new population replaces de old one, in a ssGA only a few individuals, usually one here, are generated at each iteration and merged with the existing population.

Because of its easy utilization SA has become one of the most popular LSMs. SA is an stochastic algorithm which explores the search space using a hill-climbing process. A panmictic SA is outlined in Algorithm 9. SA starts with a randomly generated solution S. At each step, a new candidate solution S' is generated. If the fitness value of S' is better or equal than the old value, S' is accepted and replaces S. As the temperature T_k decreases, the probability of accepting a lower quality solution S' decays exponentially towards zero according to the Boltzmann probability distribution. The temperature is progressively decreased following an annealing schedule.

Based on the classic SA, many different versions have been implemented by using a different annealing schedule. In our algorithm we have used the New Simulated Annealing (NSA) [26], which uses a *very fast* annealing schedule.

Algorithm 9. Standard Simulated Annealing

```
Generate(S);
Evaluate(S);
Initialize(T_0);
k := 0;
while not stop_condition(S) do
   S' := Generate(S,T_k);
   if Accept(S,S',T_k) then
      S := S';
   end if
   T_{k+1} := Update(T_k);
   k := k+1;
end while
```

8.2.4 *Classifying Hybrid Metaheuristics*

Attending to the classification proposed by E.-G. Talbi [23] (Figure 8.3), we can classify a hybrid metaheuristic attending to its structure (hierarchical) or to the features of the algorithms involved in the hybrid (flat). Four classes are derived from the hierarchical taxonomy:

- *LRH (Low-level Relay Hybrid)*. This class of hybrids represents algorithms in which a given metaheuristic is embedded into a single-solution algorithm. We can find some examples of LRH in the literature [1] [19].
- *LTH (Low-level Teamwork Hybrid)*. This class comprises combinations of metaheuristics with strong exploring capabilities (like most EAs) with exploitation-oriented metaheuristics (most single-solution metaheuristics). Usually, exploitation-oriented methods replace or extend genetic operator such as mutation or crossover. There are numerous examples of this strategy, for example [17] [14] [10].
- *HRH (High-level Relay Hybrid)*. In this class of algorithms, self-contained metaheuristics are executed in a sequence. In HRH, an algorithm is used for improving the results obtained by another one. Many authors have used this idea [24] [18].
- *HTH (High-level Teamwork Hybrid)*. Self-contained algorithms perform a search in parallel, and cooperating to find an optimum. This model has been widely used in the literature [12] [25].

As to the flat classification, we can distinguish between:

- *Homogeneous/heterogeneous*. In homogeneous hybrids, all the combined algorithms use the same metaheuristic, while in heterogeneous algorithms different metaheuristics are used.
- *Global/partial*. In global hybrids, all the algorithms search in the whole search space. However, the search space is decomposed into subspaces in the partial hybrids.
- *Specialist/general*. In a general hybrid, all the algorithms solve the same problem, while specialist hybrids combine algorithms which solve different problems.

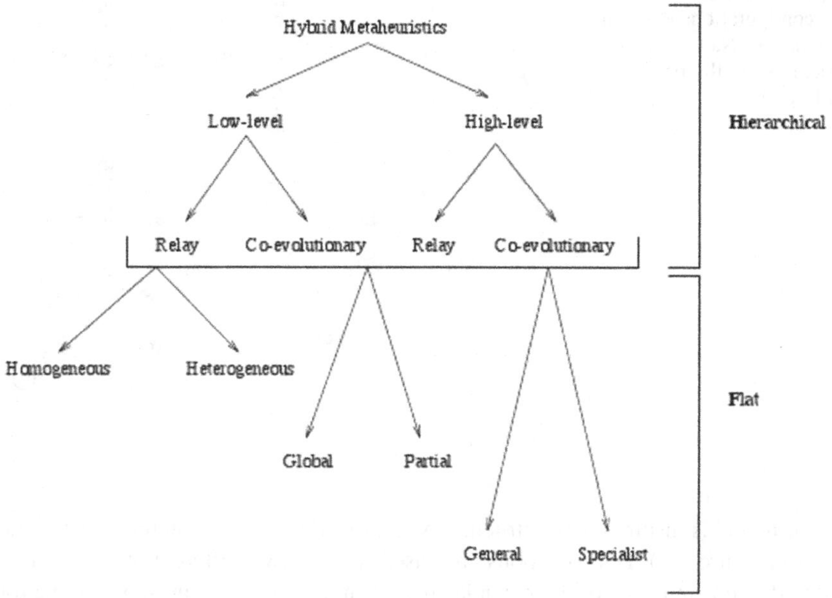

Fig. 8.3 Talbi's classification of hybrid metaheuristics

Attending to this taxonomy, our model can be classified as a High-level Teamwork Hybrid metaheuristic, while several self-contained algorithms cooperate in order to find a solution. HydroCM can be classified as well as heterogeneous, global and general, because two different metaheuristics search in the whole search space trying to solve the same problem.

8.3 Description of Our Proposal

In this section we present the particularities of HydroCM, as well as we briefly outline the algorithm that we have implemented in our tests, which has been called Ethane [13].

8.3.1 An Overview of HydroCM

In this work, we present a generic model for a complete family of parallel hybrid metaheuristics. The goal of the model is to provide a schema for the islands and communications of the parallel algorithm to efficiently perform a search over heterogeneous hardware architectures.

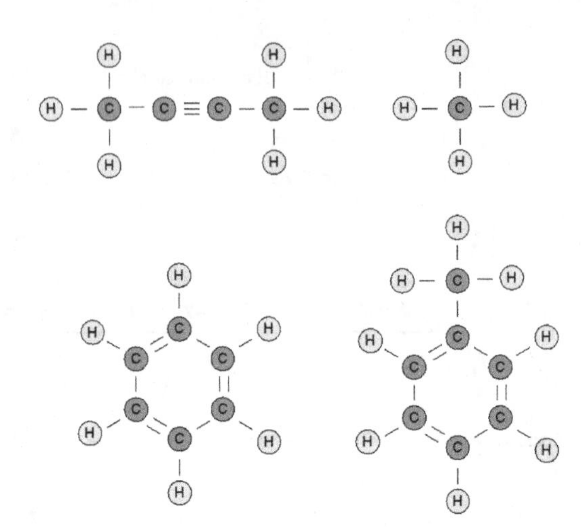

Fig. 8.4 Different hydrocarbon configurations that can be found in Nature; their structures are the basis of HydroCM

Our model is inspired in the structure of hydrocarbons as we can find them in the Nature (Figure 8.4). Hydrocarbons are based in only two different atoms, carbon and hydrogen, and each of them can keep a given number of bounds, being one for hydrogen and four for carbon.

In our model, we establish a resemblance between computers and atoms in the hydrocarbon. The bonds between atoms have a correspondence to communication channels, and double or triple bonds can be modeled as the amount of information being migrated (intensity of the interaction) or, in the case of non-population based algorithms, a higher migration rate. In our model, the fastest machines are associated with central carbon atoms (because of the higher computational effort caused by the migrations) and the slowest ones are associated with hydrogen atoms.

This model provides us with plenty of different schemes for designing a parallel heterogeneous algorithm because of the amount of hydrocarbons present in Nature and their different architectures: linear, ring, branches... obtaining a huge amount of different combinations depending on the number of fast and slow available computers and the topology of the network.

Ethane [13] can be viewed as an instance of HydroCM for an environment composed of eight nodes, where two of them are more powerful than the rest, and making use of ssGA and SA as the composing atoms. As well as Ethane is such an instance, we could instantiate many different algorithms depending on the underlying hardware architecture following the model proposed by HydroCM.

8.3.2 *Ethane*

With Ethane we propose an instance of HydroCM model, based in the chemical compound of the same name. The chemical compound called ethane consists of two carbon atoms and six hydrogen atoms, joined together with single chemical

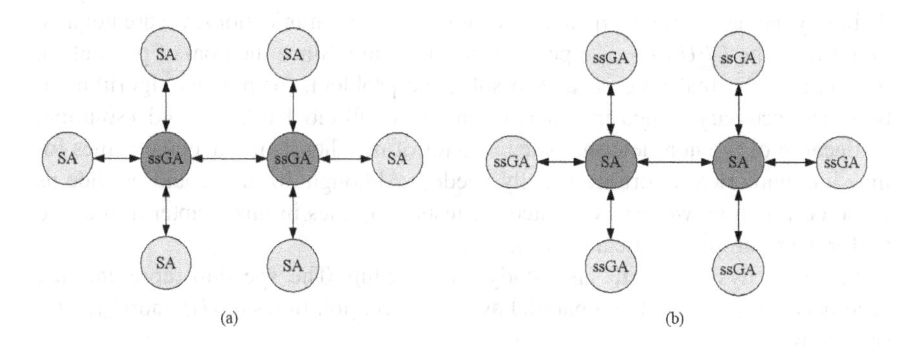

Fig. 8.5 Communication schema for Ethane G (a) and S (b)

bonds. In ethane, each carbon atom is bonded to three hydrogen atoms, and there is another bond between both carbon atoms. In our Ethane algorithm, we propose the same schema, using two basic algorithms resembling different atoms, and migration channels resembling bonds.

For our study, we have implemented two different versions of the algorithm. In Figure 8.5 we show the schema for the two instances of Ethane studied in this chapter. Ethane G (Figure 8.5a) assigns a ssGA sub-algorithm to the central nodes, and a SA sub-algorithm to the *slave* nodes. On the contrary, Ethane S (Figure 8.5b) allocates a SA sub-algorithm in each one of the central carbon nodes, and a ssGA sub-algorithm in the *slave* nodes. With this schema, the most of the communication load falls on the *master* nodes, which are provided with the best hardware, moving some of the load out of the slowest nodes.

8.4 Performance Measures and Speedup

In this section we present the performance measures used for assessing the performance of the studied algorithms. The measures that are going to be used are the numerical effort, the total run time, and the speedup.

A widely accepted way of measuring the performance of a parallel metaheuristic is to check the number of evaluations of the fitness function needed to locate an optimum. This performance measure is called numerical effort. Numerical effort is widely used in the field of metaheuristics because it removes the effects of the implementation and the platform, but it could be misleading in many cases for parallel methods. Furthermore, the goal of the parallelism is not the reduction of the number of evaluations (this is a goal for decentralized algorithms) but the reduction of the running time.

The most significant performance measure for a parallel algorithm is the total run time needed to locate a solution. In a non-parallel algorithm, the use of the *CPU time* is a common performance measure. While parallelizing an algorithm should definitely include some overhead, for example for communications, we are not able to use only the *CPU time* as a performance measure. Since the goal of parallelism is to reduce the real time needed to solve the problem, for parallel algorithms it becomes necessary to measure the real run time (wall-clock time) to find a solution.

Because of the non-deterministic behavior of metaheuristics, average values for time and numerical effort are usually needed. Although 30 runs could provide us a good estimation, we have executed the tests 100 times in this chapter in order to perform a rigurous statistical analysis.

In our analysis we will also study the speedup. The speedup represents the ratio between sequential and parallel average execution times ($E[T_1]$ and $E[T_m]$ respectively).

$$s_m = \frac{E[T_1]}{E[T_m]} \tag{8.1}$$

For the speedup to be a meaningful metric, we have to take care of many aspects for its analysis. Because of the aforementioned non-deterministic behavior of metaheuristics it is necessary to use average times, being these times the wall-clock times. The algorithms run in the single and multiprocessor platform must be exactly the same, thus panmictic algorithms can not be used for the analysis. The algorithms have to be executed until they found the solution or a solution of the same quality [2]. Since in our study we are working over a heterogeneous platform, our reference point is the execution time of the program on the fastest single processor.

8.5 Problems, Parameters, and Platform

In this section we include the basic information necessary to reproduce the experiments that have been carried out for this work. First we will present the set of benchmark problems used for assessing the performance of our proposal. Second we will briefly explain the parameters used within the sub-algorithms, and then the underlying hardware and software platform.

8.5.1 Benchmark Problems

In order to assess the performance of our algorithms, we have used two problems in the analysis: the Subset Sum Problem (SSP) [16] and the Massively Multimodal Deceptive Problem (MMDP) with 6 bits [15].

The SSP problem consists in finding a subset of values $V \subseteq W$ from a set of integers $W = \{w_1, w_2, ..., w_n\}$, such that the subset sum approaches a constant C without exceeding it. We have chosen an instance with 2048 random integer numbers in the

Table 8.1 Bipolar deception (6 bits) sub-function value

#ONES	sub-function value
0	1.000000
1	0.000000
2	0.360384
3	0.640576
4	0.360384
5	0.000000
6	1.000000

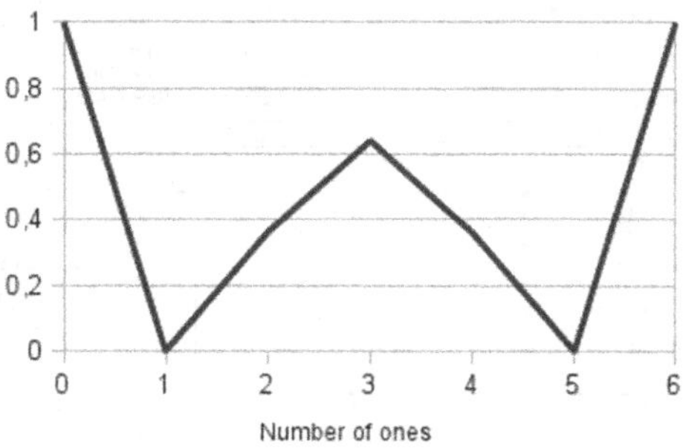

range $[0..10^4]$ following a Gaussian distribution, being the value of the sum for the optimum 3256234.

MMDP is one of so called deceptive problems. Deceptive problems are specifically designed to make the algorithm converge to wrong regions of the search space, decorrelating the relationship between the fitness of a string and its genotype. In MMDP a binary string encodes k 6-bit sub-problems which contribute with a partial fitness depending on its number of 1's (unitation) following Table 8.1. We have used an instance with strings of 150 bits so that the global optimum is $k = 25$.

8.5.2 *Parameters of the Algorithms and Platform*

The parameters used in every ssGA sub-population are: a population size of 64 individuals, a crossover probability of 0.8 and a mutation probability of 4.0 divided by the chromosome length. The genetic operators are a single point crossover and a bit flip mutation. For the SA, we used the same mutation probability. For the SSP the chromosome length is 2048 an in the case of MMDP its length is 150 for both algorithms. In the case of the panmictic ssGA, the population size has been set to 64 individuals because larger populations have performed much worse than smaller ones for the proposed problems in our tests, and they have been not able to find the solution of the benchmark problems in a reasonable time.

We have chosen a migration frequency of 50 iterations for all the configurations after several initial preliminary experiments. The number of individuals migrated are 1 in all cases. For the ssGA, the emigrant is randomly selected and the immigrant always replaces the worst individual of the population. In the SA, the immigrant is treated as a new move.

The hardware infrastructure used in our analysis 8.6 consists of 8 different machines: 2 of them have an Intel Core 2 Quad Q9400 @ 2.66GHz processor and 4GB of RAM (namely Type A, fast), the other 6 computers have an Intel Pentium 4 @ 2.4GHz processor and 1GB of RAM (namely Type B, slow). All the computers are managed by a GNU/Linux distribution, being Debian 5.0 for Type A, and SuSE 8.1, Debian 3.1 and Ubuntu 6.10 for Type B. The computers are connected

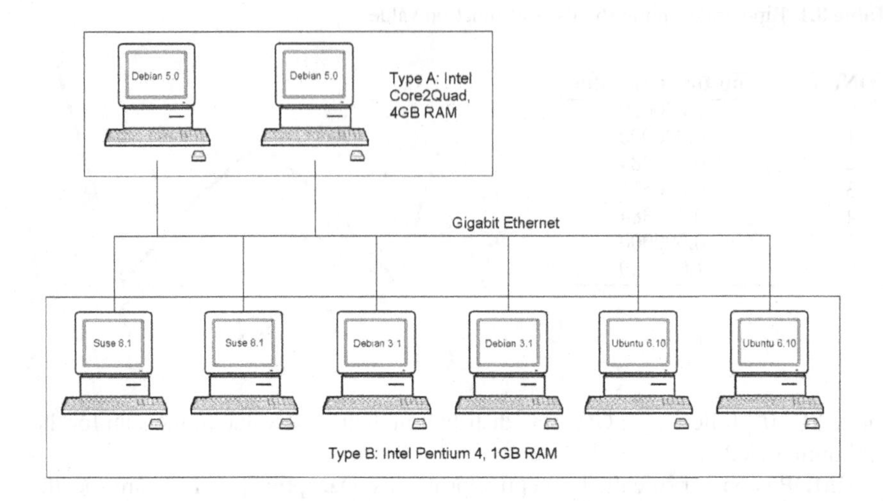

Fig. 8.6 Schema of the hardware infrastructure

by a Gigabit Ethernet Network. The algorithms have been implemented in Java in order to support both hardware and software heterogeneity. For the purpose of the analysis the version 1.6.0_01 of the Java Virtual Machine (JVM) is used in all the nodes.

8.6 Tests and Analysis

In this section we analyze the behavior of Ethane, and compare it with the well-known ssGA unidirectional ring. We have analyzed the aforementioned performance measures, being numerical effort, total run time and speedup, as well as the evolution of the fitness.

We have implemented two different algorithms based on Ethane. For the first one, Ethane G, we have provided the Type A computers with a central ssGA island, and Type B computers with a SA island. For the second algorithm, Ethane S, the fastest machines run central SA islands and the slowest ones run ssGA. As we mentioned above, the migration scheme resembles a molecule of ethane as represented in Figure 8.5. In the parallel ssGA used as reference, the islands have been distributed over a unidirectional ring, placing the most powerful computers in the first and fourth place in a sort of MaxSumSort [9]. As we do not know the statistical distribution of the data, they have been statistically compared with Mann-Whitney U test.

Table 8.2 Number of evaluations for the tested models and panmictic algorithms

Algorithm	Subset Sum		MMDP6	
	Average	Std. Deviation	Average	Std. Deviation
Ethane G	**146418**	**174433**	1572735	919691
Ethane S	202815	198696	**708231**	**430353**
ssGA Ring	214824	239125	786583	805837
Panm. ssGA	179792	175177	*	*
Panm. SA	81737	93627	*	*

8.6.1 Numerical Effort

In Table 8.2 we show the numerical effort needed to find the optimum for each algorithm. It can be seen that our proposals performed better than the panmictic algorithms for both problems (in the case of MMPD, panmictic algorithms where not even able to find the optimum in a reasonable time). For the SSP, both Ethane versions performed numerically better than the reference ssGA ring, and one of the instances (Ethane S) performed better even for the MMDP.

From the point of view of numerical effort, all the differences are statistically significant according to the Mann-Whitney U test . Note that also the standard deviation is better in our algorithms, so that its behavior is more robust. We can see how the panmictic SA has reached the solution with less numerical effort because SA is a fast converging trajectory method, but as we will see in the forthcoming analysis of the run time, the time needed to find a solution is worse than for the studied parallel models.

Since the objective of our model is the reduction of the total execution time let us begin with the study of a more meaningful performance metric, the total run time.

8.6.2 Total Run Time

Table 8.3 shows the average execution time of each algorithm for each problem until global optimum is reached. As we can see, our proposals performed clearly better than the panmictic algorithms for both problems (remember that the panmictic algorithms where not able to find the optimum for the MMDP in a reasonable time) as well as better than the ssGA ring does.

As we can see in Table 8.3, Ethane G was the best performing algorithm for the SSP problem. The Mann-Whitney U test gives a p-value of 0.0412 for the Ethane G compared to the ssGA ring, so the difference is statistically significant. The average time needed for Ethane G to find a solution is more than 30% better than for ssGA ring.

Ethane S was the best algorithm solving the MMDP problem, with an average time slightly better than the ssGA ring, but with a much lower standard deviation, as Mann-Whitney U test confirms with a p-value of 0.007. The standard deviation

Table 8.3 Time - ms - for the tested models and panmictic algorithms

Algorithm	Subset Sum		MMDP6	
	Average	Std. Deviation	Average	Std. Deviation
Ethane G	**5318**	**6226**	9195	4942
Ethane S	7155	6922	**3052**	**1546**
ssGA Ring	7453	8107	3194	3380
Panm. ssGA	30008	29387	*	*
Panm. SA	13300	15443	*	*

of ssGA is more than twice the standard deviation of Ethane S. This means that the two representative instances of the Ethane family evaluated in this chapter can be both more efficient and more robust/stable than standard sequential and distributed popular algorithms.

8.6.3 Speedup

In Table 8.4, we can see a summary of the execution time of the studied algorithms within a single processor and its speedup with respect to the execution in the eight processors heterogeneous platform. As we can see, both versions of Ethane have obtained a better speedup than the ssGA for the SSP, but only Ethane S has achieved a better speedup for the MMDP.

As it is shown in Table 8.4, Ethane G has performed better than the reference ssGA ring even in a single processor in the case of SSP. Even when its performance over a single processor is still better, its speedup is the best of the three models. However, in general, the value for the speedup is not good for any of the algorithms for this problem, being the value for Ethane G a small 3×.

Ethane S still performed slightly better than the ssGA ring for a single processor for both problems. Even the speedup is better in both cases, being the best of the studied algorithms for the MMDP with a value of 6.76×. In the case of MMDP the speedup of the three algorithms was quite good although linear speedup was not reached.

In the case of SSP, Ethane G and S have not showed a very good speedup, and ssGA has showed even a worse speedup. This fact could be explained by the huge difference among the computational power of the different hardware configurations

Table 8.4 Time - ms - for the tested models in a single processor and its speedup

Algorithm	Subset Sum		MMDP6	
	Avg. time	Speedup	Avg. time	Speedup
Ethane G	**15995**	**3.00**×	41943	4.56×
Ethane S	17817	2.49×	**20627**	**6.76**×
ssGA Ring	18137	2.43×	21227	6.64×

used (remember that the reference point for speedup is the best performing processor). Heterogeneous hardware might not be expected of very high speedup as homogeneous hardware.

8.6.4 Evolution of the Fitness

Figures 8.7 and 8.8 are showing, for each algorithm and each problem, the execution whose value for the run time is the median of the results.

In the case of SSP, the Figure shows that the two Ethane versions clearly outperform the ssGA ring, converging quite faster. We can see how Ethane G performs even better than Ethane S for this problem.

For the MMDP, Ethane S performed clearly better than Ethane G as we can see in Figure 8.8. Ethane S outperformed the ssGA ring, but the difference is not as large as with the SSP.

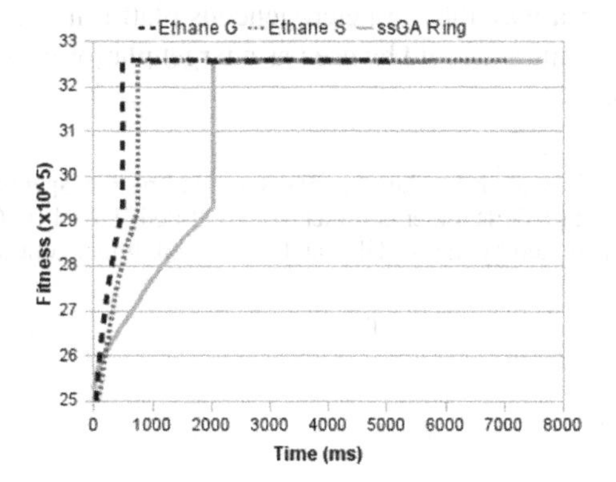

Fig. 8.7 Evolution of the fitness (time (ms) vs. fitness) for SSP

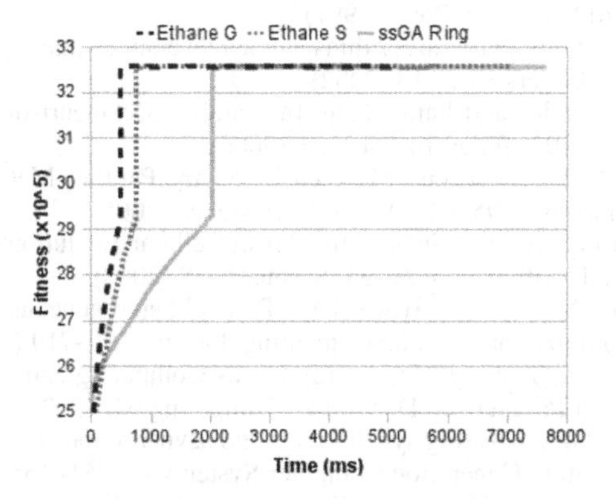

Fig. 8.8 Evolution of the fitness (time (ms) vs. fitness) for MMP

8.7 Conclusions

In this chapter we have presented a general model for designing hybrid algorithms depending on the underlying heterogeneous platform, inspired in the structures of the hydrocarbons present in Nature. We have also analyzed an instance of HydroCM: Ethane, a hybrid parallel search algorithm based on the structure of ethane.

We have performed a set of tests in order to assess the performance of our proposal, and compared it with a well-known state-of-the-art model, the ssGA unidirectional ring, and two well-known algorithms: SA and ssGA. Our tests have shown that the hybrid model can perform better in terms of time and numerical effort than the reference model, and Ethane is even able to find the solutions in a more robust/stable manner. Also the speedup of the proposed models is competitive with that of the reference model, obtaining quite good values even with the huge differences between the performance of the computers of the heterogeneous platform.

With HydroCM, our goal is to offer a hybrid general model for gracefully matching computers of different powers to run different algorithms for efficiently solve the same problem, in a way that an heterogeneous platform does not constitute a problem but, on the contrary, could be used as a target platform for specialized new parallel algorithms.

Acknowledgements. This work has been partially funded by the Spanish Ministry of Science and Innovation and FEDER under contract TIN2008-06491-C04-01 (the M* project). It has also been partially funded by the Andalusian Government under contract P07-TIC-03044 (DIRICOM project).

References

1. Aarts, E.H.L., Verhoeven, M.G.A.: Genetic local search for the traveling salesman problem. In: Handbook of Evolutionary Computation, pp. G9.5:1–7. Institute of Physics Publishing and Oxford University Press (1997)
2. Alba, E.: Parallel evolutionary algorithms can achieve super-lineal performance. Information Processing Letters 82, 7–13 (2002)
3. Alba, E.: Metaheuristics and Parallelism. In: Parallel Metaheuristics: A new Class of Algorithms, pp. 79–103. Wiley-Interscience (2005)
4. Alba, E.: Parallel Heterogeneous Metaheuristics. In: Parallel Metaheuristics: A new Class of Algorithms, pp. 395–422. Wiley-Interscience (2005)
5. Alba, E., Dorronsoro, B.: The State of the Art in Cellular Evolutionary Algorithms. In: Cellular Genetic Algorithms, pp. 21–34. Springer, US (2008)
6. Alba, E., Luna, F., Nebro, A.J., Troya, J.M.: Parallel heterogeneous genetic algorithms for continuous optimization. Parallel Computing 30(5-6), 699–719 (2004)
7. Alba, E., Nebro, A.J., Troya, J.M.: Heterogeneous Computing and Parallel Genetic Algorithms. Journal of Parallel and Distributed Computing 62, 1362–1385 (2002)
8. Alba, E., Troya, J.M.: Analyzing synchronous and asynchronous parallel distributed genetic algorithms. Future Generation Computer Systems 17, 451–465 (2001)
9. Branke, J., Kamper, A., Schmeck, H.: Distribution of Evolutionary Algorithms in Heterogeneous Networks. In: Deb, K., et al. (eds.) GECCO 2004. LNCS, vol. 3102, pp. 923–934. Springer, Heidelberg (2004)

10. Chen, H., Flann, N.S.: Parallel Simulated Annealing and Genetic Algorithms: A Space of Hybrid Methods. In: Davidor, Y., Männer, R., Schwefel, H.-P. (eds.) PPSN 1994. LNCS, vol. 866, Springer, Heidelberg (1994)
11. Crainic, T.G., Toulouse, M.: Parallel strategies for meta-heuristics. In: Handbook of Metaheuristics, pp. 474–513. Kluwer (2003)
12. De Falco, I., Del Balio, R., Tarantino, E., Vaccaro, R.: Improving search by incorporating evolution principles in parallel tabu search. In: Int. Conf. on Machine Learning, pp. 823–828 (1994)
13. Domínguez, J., Alba, E.: Ethane: A Heterogeneous Parallel Search Algorithm for Heterogeneous Platforms. In: DECIE (2011), doi:arXiv:1105.5900v2
14. Fleurant, C., Ferland, J.A.: Genetic and hybrid algorithms for graph coloring. Annals of Operations Research 63, 437–461 (1996)
15. Goldberg, D.E., Deb, K., Horn, J.: Massively multimodality, deception and genetic algorithms. Parallel Problem Solving from Nature 2, 37–46 (1992)
16. Jelasity, M.: A wave analysis of the subset sum problem. In: Proceedings of the Seventh International Conference on Genetic Algorithms, San Francisco, CA, pp. 89–96 (1997)
17. Lozano, M., Herrera, F., Krasnogor, N., Molina, D.: Real-coded memetic algorithms with crossover hill-climbing. Evolutionary Computation 12(3), 273–302 (2004)
18. Mahfoud, S.W., Goldberg, D.E.: Parallel recombinative simulated annealing: A genetic algorithm. Parallel Computing 21, 1–28 (1995)
19. Martin, O.C., Otto, S.W., Felten, E.W.: Large-step markov chains for the TSP: Incorporating local search heuristics. Operation Research Letters 11, 219–224 (1992)
20. Salto, C., Alba, E.: Designing Heterogeneous Distributed GAs by Efficient Self-Adapting the Migration Period. Applied Intelligence (2011), doi:10.1007/s10489-011-0297-9
21. Salto, C., Alba, E., Luna, F.: Using Landscape Measures for the Online Tuning of Heterogeneous Distributed GAs. In: Proceedings of the GECCO 2011, pp. 691–694 (2011)
22. Syswerda, G.: A study of reproduction in generational and steady-state genetic algorithms. In: Foundations of Genetic Algorithms, pp. 94–101. Morgan Kauffman (1991)
23. Talbi, E.-G.: A taxonomy of hybrid metaheuristics. Journal of Heuristics 8(5), 541–564 (2002)
24. Talbi, E.-G., Muntean, T., Samarandache, I.: Hybridation des algorithmes génétiques aveq la recherche tabou. In: Evolution Artificielle, EA 1994 (1994)
25. Voigt, H.-M., Born, J., Santibanez-Koref, I.: Modeling and simulation of distributed evolutionary search processes for function optimization. In: Schwefel, H.-P., Männer, R. (eds.) PPSN 1990. LNCS, vol. 496, pp. 373–380. Springer, Heidelberg (1991)
26. Yao, X.: A new Simulated Annealing Algorithm. International Journal of Computer Mathematics 56, 161–168 (1995)

Chapter 9
A Multi-thread GRASPxELS for the Heterogeneous Capacitated Vehicle Routing Problem

Christophe Duhamel, Christophe Gouinaud,
Philippe Lacomme, and Caroline Prodhon

Abstract. This chapter focuses on the definition of an efficient parallel metaheuristic which takes advantage of the multi-core design of recent processors. The approach is designed as a Greedy Randomized Adaptive Search Procedure (GRASP) hybridized with a multi-threaded version of an Evolutionary Local Search (ELS) metaheuristic scheme. Our approach is evaluated on an extension of the Vehicle Routing Problem where a heterogeneous fleet of vehicles is available to service a set of customers. The objective consists in designing a set of trips for a limited heterogeneous fleet of vehicles located at a depot node which minimizes the total transportation cost. Each type of vehicles is defined by a capacity and by the number of available vehicles. The efficiency of the parallel approach is evaluated on a new set of real-life instances built out of data from the French districts. A fair comparative study, using a same implementation, is done to evaluate the impact of the number of threads on the convergence rate. Thus, a better trade-off between solution quality and computational time can be reached. The numerical experiments show that the hybrid GRASPxparallel ELS outperforms the classical iterative version and provides numerous new best solutions.

9.1 Introduction

The design of parallel implementations of metaheuritics has received a considerable amount of attention in the last two decades with the development of new hardware

Christophe Gouinaud · Philippe Lacomme
Laboratoire d'Informatique (LIMOS, UMR CNRS 6158), Campus des Cézeaux,
63177 Aubière Cedex, France
e-mail: {christophe.gouinaud,placomme}@isima.fr

Christophe Duhamel · Caroline Prodhon
Institut Charles Delaunay (LOSI) and STMR (UMR CNRS 6279),
Université de Technologie de Troyes, BP 2060, 10010 Troyes Cedex, France
e-mail: christophe.duhamel@isima.fr, caroline.prodhon@utt.fr

E.-G. Talbi (Ed.): Hybrid Metaheuristics, SCI 434, pp. 237–269.
springerlink.com

technologies which provide several efficient and cheap opportunities. This chapter focuses on the definition of a Greedy Randomized Adaptive Search Procedure (GRASP) hybridized with a multi-threaded version of an Evolutionary Local Search (ELS) metaheuristic scheme which takes advantage of the multi-core design of the recent processors. Our approach is tested on the Heterogeneous Vehicle Routing Problem (HVRP), an extension of the classical Vehicle Routing Problem (VRP) where a heterogeneous fleet of vehicles is available to service a set of customers.

A brief reminder of the recent technological advances in parallel implementations of algorithms is presented in the next Section, with a state of the art on parallel metaheuristics. Then, Section 9.3 introduces the problem under consideration in this chapter, the HVRP. The parallel hybrid framework is stated in Section 9.4. Its implementation for the HVRP is explained in Section 9.5. Then, the details of the components of the method are exposed in Section 9.6. Finally, a computational evaluation of the method is proposed through Section 9.7 before concluding remarks.

9.2 Parallel Metaheuristics

Before presenting our hybrid metaheuristic based on a GRASP and a parallel implementation of an ELS scheme, this first section gives an overview of the available technologies and of the state of the art to justify our choice of a multi-thread implementation to solve the HVRP.

9.2.1 Technologies

Parallel computating environment offers various ways to implement parallelization. More specifically, three main trends can be identified.

Historically, the first trend is based on computer cluster architectures and software environment. It relies on communication frameworks as MPI (Message Passing Interface) and it usually requires a large communication bandwidth in order to stay efficient. Such an approach provides a master-slave message passing paradigm but it also requires a huge financial investment in order to build the cluster and the high speed communication network.

The second trend is more recent. It consists in taking advantage of the capabilities of the graphics processors included in modern computers (Graphics Processing Unit - GPU). Even if they were first developed to afford the computing requirements of the visual effects in modern 3D games, it has been lately evidenced they could be used for other purposes. For instance, NVIDIA provides a CUDA (Compute Unified Device Architecture) library with a user-friendly C++ interface in order to simplify the development of parallel algorithms. Their GPU can handle a huge number of threads to be used but they are unfortunately gathered in blocks with a slow shared memory. Depending on the GPU capability, the dedicated memory is limited to several kilobytes. This is a strong limitation, especially when one needs to implement Operations Research algorithms in which the amount of required data is large. Note also that GPU frequency is quite low when comparing to modern CPU (Central Processing Unit). In the same trend, a new generation of parallel computing architectures (as in Tesla workstations), designed by NVIDIA as well, delivers fast data-parallel processing. It is however limited to dedicated and professional complex supercomputing problems due to the price of such workstations.

The third trend consists in using the multi-core technologies available on every recent processor to address the parallelization challenge. Those CPU typically offer less parallel threads (from 2-cores for basic processors to 8-cores for high-end processors, and even more on specific machines) than NVIDIA GPU. However, it allows a high speed memory and higher frequencies. In addition, the large data accesses are convenient for Operations Research algorithms. These reasons prompt us to develop a framework based on this trend even if the parallelization is limited compared with the number of parallel threads offered in NVIDIA technology.

9.2.2 State of the Art for Operations Research Problems

Various optimization problems have been addressed for years using parallel metaheuristics. Literature is scarce and also confusing as papers address a wide range of problems, propose different approaches and different criteria are used to analyze the parallelization efficiency.

Experiments in parallelizing metaheuristic for Operations Research problems encompass both population-based metaheuristics, such as Genetic Algorithms, and single solution-based metaheuristics including Tabu Search and GRASP (Greedy Randomized Adaptive Search Procedure) for instance. The interest mainly comes from many real-life applications, from scheduling to vehicle routing, when size is too large to get high quality solutions in a reasonable time for sequential approaches. Fiechter in 1994 [15] developed an efficient parallel Tabu Search algorithm for the Traveling Salesman Problem (TSP), implemented on a network of transputers in OCCAM language (concurrent programming language working on Communicating Sequential Processes - CSP). The author provided a thorough analysis of the speed-up, *i.e.* the ratio of the average CPU time using one transputer and p transputers. He also introduced a careful description of the algorithmic key points which includes, among others, a diversification and an intensification strategy. Ten years later, Berger and Barkaoui [24] introduced a parallel Hybrid Genetic Algorithm (HGA) for the Vehicle Routing Problem with Time Windows (VRPTW) based on the simultaneous evolution of two populations of solutions focusing on separate objectives and subject to time windows constraints relaxation. The implementation uses the PVM (Parallel Virtual Machine) library and the parallel procedure has been implemented in C++ on a cluster of 19 computers. The master process manages the execution of the algorithm at a high level. It synchronizes atomic genetic operations and handles the parent selection process. The slave processes concurrently execute the reproduction and the mutation operators. Numerical experiments are based on the medium-scale instances of Solomon. They include 56 instances with up to 100 nodes. The authors provide a performance analysis of the parallel HGA and a comparison with previous published methods. Their approach is proven to be time efficient, cost effective and thus it is competitive as it matches the performance of best-known heuristic routing procedures. Moreover RHGA provides six new best-known solutions.

The Vehicle Routing Problem (VRP) is also addressed by Le Bouthilier and Crainic [2] using a solution warehouse strategy, in which several search threads cooperate by asynchronously exchanging information on the best solutions found so far. The exchanges are performed through a mechanism, denoted solution warehouse, which holds and manages a pool of solutions. The asynchronous communication ensures independence of the individual search processes. Each search process implements a different metaheuristic including an Evolutionary Algorithm and a Tabu Search procedure. The results obtained on an extended set of test problems show that the parallel procedure achieves linear accelerations. This approach identifies solutions of comparable quality to those obtained by the best sequential methods in the literature. Caricato et al. [36] propose a parallel Tabu Search for the pickup and delivery problem, *i.e.* a special VRP where each request (product) has to be transported from an origin node to a destination node by means of vehicles. The authors introduce two sequential heuristics and a parallel Tabu Search. The method is run on a cluster of 4 PC, each one including two Pentium III processors. The code is implemented in C using the MPI library. The Periodic VRP (PVRP) is addressed by Drummond et al. [32] with a parallel Genetic Algorithm (PGA) and local search heuristics. The PGA relies on the Island model and includes periodic migration between islands (local pools) with low frequencies. Experiments are done on a workstation with 4 RISC System/6000 processors and the methods are implemented in C using MPI for parallelism. Numerical experiments let authors conclude that the parallel implementation provides significant advantages not only in terms of running time but also in terms of search quality. The Heterogeneous VRP (HVRP) has been addressed in 1998 by Ochi et al. [33] using a parallel Evolutionary Algorithm. The authors combine a parallel Genetic Algorithm with a Scatter Search and they apply decomposition procedures (petals decomposition).

Bortfeld et al. [1] introduce a parallel Tabu Search Algorithm for the container loading problem. They propose a distributed parallel approach based on the concept of multi-search threads developed by Toulouse et al. [35]. Several search paths are explored in a concurrent way: cooperation is achieved by the exchange of solutions at the end of the search phases. The parallel search processes are run on LAN (Local Area Network) workstations. The efficiency of the parallel Tabu Search Algorithm (TSA) is evidenced by an extensive comparative test including classical instances. According to their test, the container volumes are already loaded at a high rate in the solutions provided by the sequential TSA. Only slight improvements are achieved by the parallelization. The authors also note that the communication between the TSA processes weakly contribute to this effect, which corroborates previous similar results. For instance, Crainic et al. [47] proposed the parallelization of a TSA for solving a warehouse location problem. They show that the best results were obtained without communication between the concurrent processes. Previous experiences in the field make them consider the parallelization of a method as a relevant methodical extension, especially if other concepts for the improvement of a sequential method are already exhausted. They conclude that such an approach provides only limited enhancements in terms of solution quality.

Parallel hybrid metaheuristics have been proposed for the flexible job shop problem by [26] including two major modules: the machine selection module is sequential while the operation scheduling module is run in parallel. Numerical experiments are provided using a NVIDIA GPU Tesla C870 GPU (512 GFLOPS) with 128 streaming processors cores. They use benchmark problems from literature. The Hybrid Flow-Shop problem with multiprocessors tasks scheduling has recently been addressed by [26] using a parallel greedy algorithm approach. The algorithm consists on two phases: a destruction phase and a construction one. A similar problem has been recently addressed by Bozejko [51].

The sequential ordering problem is addressed in a parallel environment by Guerriero and Mancini in 2003 [20]. They use a parallel version of the rollout algorithm first proposed by [17]. It is based on a multi-thread parallelization strategy where different regions of the solution space are explored by different threads. Information exchange is done periodicaly. The implementation is done on a cluster of PC and the authors conclude that parallelization both speeds up the convergence and the solution quality. Lately, in 2009, the quadratic assignment problem has been addressed by James et al. [45] using a Cooperative Parallel Tabu Search (CPTS). The numerical experiments demonstrate the benefits that can be obtained with parallel computing in terms of convergence rate and of solution quality. CPTS provides several new best solutions. A parallel Scatter Search metaheuristic for solving the feature subset selection problem in classification is introduced by [22]. Two methods are proposed to combine solutions in the Scatter Search (SS). The parallelization is done by simultaneously running the two combination methods. The parallel Scatter Search achieves values similar to both sequential SS algorithms, but uses a smaller subset of features. Moreover, the parallel algorithm is more accurate than sequential algorithms.

Speedups can strongly vary since they depend on the problem and on the metaheuristic used for parallelization. Sometimes a slight improvement in terms of solution quality can be achieved while the computational time remains stable. However there are many ways to perform a relevant parallelization. For instance, Ribeiro and Rosseti [13] investigate parallel strategies for GRASP metaheuristics. The experiments they did for the 2-path Network Design Problem show that a speedup about 17 is obtained with the 32-processor cluster. Our aim is to provide a parallel method that can be run on computers earned in any compagny, *i.e.* without investing in expensive machines. Thus our choice goes to a multi-threaded strategy that takes advantage of the multi-core processors available on any recent computer.

Independently to the parallel technology used and whatever the problem and the metaheuristic used, the state of the art enable us to name key points for efficient parallelization. They include the way to perform diversification and intensification, and the cooperation mode between threads. These three points are considered in our approach as summarized in Table 1 from Section 9.4.3. Before detailing them, next Section exposes the problem used to attest the performance of our GRASP hybridized with a multi-threaded ELS.

9.3 Heterogeneous Capacitated Vehicle Routing Problem

The Capacitated Vehicle Routing Problem (CVRP) is a standard NP-hard node routing problem which has received considerable attention in the last decades, see [40, 39, 12] for instance. It consists in optimizing the delivery of goods required by a set of customers. It can be fully defined by considering a depot and a set of n customers which correspond to the nodes of a complete graph $G = (V;E)$. V is the set of $n+1$ nodes, 0 being the depot and $1...n$ being the customers. Each edge $e \in E$ has a finite nonnegative routing cost $c_e \geq 0$ and each node $v \in V \setminus \{0\}$ has a demand $d_v \geq 0$. A fleet of homogenous vehicles of limited capacity Q is based at the depot. The objective is to design a set of trips of minimal total cost to service all customers. A trip is a circuit performed by one vehicle, starting at the depot, sequentially visiting a subset of nodes and returning to the depot. The total trip load must not exceed the vehicle capacity Q. Since split deliveries are not allowed, each customer is serviced by exactly one vehicle. As stressed in [26], efficient resolution of medium to large CVRP instances is currently limited to metaheuristics.

In its basic formulation, the CVRP is not realistic enough to accomodate real features coming from companies. In this context, many extensions have been considered during the last decades. Thus, keeping track of CVRP development is really difficult as node routing problems transcends several academic disciplines. In 2008, Eksioglu et al. [5] have provided a methodology to classify the literature of the VRP, *i.e.* a taxonomic framework. Their proposal extends the previous proposal of Current and Marsh in 1993 [29]. Additional VRP constraints can be classified into three sets: scenario characteristic, physical characteristics and information characteristics. Scenario characteristics include, for instance, customer service demand quantity (deterministic, stochastic), load splitting constraints (splitting allowed or not), time windows (soft time windows, strict time windows), time horizon (single period, multi period) or customer types (linehaul, backhaul, transfer). Physical characteristics include the number of origin points (single or multiple origins), the time window types (restrictions on customers, on roads), the number of vehicles (exact number of vehicles, limited number of vehicles, and unlimited number of vehicles). Information characteristics include the evolution of information (static or partially dynamic) or the quality of information (stochastic, deterministic).

In this chapter, we consider the deterministic case in which a company owns a limited heterogeneous fleet of vehicles. This is an extension of the Vehicle Fleet Mix Problem (VFMP) for which the fleet is composed of an unlimited number of vehicles of several types. In VFMP, each type k is defined by a vehicle capacity Q_k, a fixed cost f_k, a variable cost v_k and a travel cost between pairs of nodes that can be defined by a distance matrix (VFMP-F), by a variable (VFMP-V) or by a mix of both (VFMP-FV). A trip of length L performed by a vehicle of type k has a cost $f_k + L.v_k$. The goal is to compute a set of trips and to assign vehicles to trips to minimize the total cost. The VFMP typically appears in situations where the fleet is not yet purchased: it combines tactical decisions (selecting the number of vehicles to be acquired) and operational decisions (computing the trips and the vehicles assigned to them). The Heterogeneous Fleet VRP (HFVRP or HVRP) shares the same structure as the VFMP with a limited availability a_k for vehicles of type k. Both the VFMP and the HVRP are NP-hard, since they generalize the VRP.

The fleet composition induces a trip assignment to vehicle type and a limitation on resources for the HVRP case. Thus, the latter is more restrictive. Published methods are, in most of the cases, dedicated to only one of the three VFMP versions or to the HVRP. For instance, two articles have recently considered the HVRP only: Li et al. [21] introduce a Memory Programming metaheuristic and Brandão [4] proposes a Tabu Search algorithm. Prins [12] is the first to propose a single approach to handle all the non-homogeneous fleet VRP versions. It is based on a Memetic Algorithm specially tuned for each version of either the VFMP or the HVRP. Besides, he presents a thorough review of the literature on the history of VFMP and HVRP. Duhamel et al. [9] also handle both the VFMP and the HVRP. Besides, they propose 96 new realistic instances based on data from the real French counties.

Results provided by state-of-the-art methods on "classical" HVRP instances (Taillard's instances) are usually very close to the best-known solutions, and they are obtained within a reasonable amount of time. Thus, these instances cannot be used to compare the methods anymore. The situation is different on the recent large instances proposed by Duhamel et al. [9][1]. To the best of our knowledge, they are the only recent large instances available. The aim of this chapter is to emphasize the interest of a new parallel implementation of ELS to achieve good solutions and to evaluate the performance on such large instances.

9.4 Hybrid GRASP x Parallel ELS

In this chapter, we propose a hybrid metaheuristic that focuses on the third parallelization trend, *i.e.* taking advantage of the multi-thread capabilities of the modern multi-core processors. This choice is motivated by two characteristics:

- first, algorithms dedicated to Operations Research problems typically require large amount of data and can benefit from shared-memory machines;
- second, our goal is to propose a threaded cooperative search strategy to better explore the solution space through a functional parallelism. In this context, fine-grained application is not much useful and parallel programming languages are

[1] `http://www.isima.fr/~duhamel`, `http://www.isima.fr/~lacomme/hvrp/hvrp.html` or `http://prodhonc.free.fr/`

not well-suited. An information synchronization done at a low or medium frequency (medium- or coarse-grained application) is sufficient.

To achieve an efficient cooperative search, we propose a clever parallelization of ELS embedded within a GRASP.

9.4.1 GRASPxELS Principle

ELS (Evolutionary Local Search) is an extension of ILS (Iterated Local Search). ILS was introduced by Lourenço et al. [16]. It consists in applying a perturbation to the current solution S before improving the resulting solution S' through a local search. The new solution becomes the new current solution. Note that the perturbation is similar to the mutation operator found in Genetic Algorithms.

ELS was enhanced by Prins in 2009 [11] for routing problems. It is similar to ILS except that nd "children" are generated from S at each iteration. Thus, each child is a copy of S on which a mutation, then a local search, are applied. The best child is selected to become the new ongoing solution S. The purpose of ELS is to better explore the current local optimum neighborhood before leaving it.

In addition, to ensure some diversity along the search space exploration, a multi-start scheme can be implemented. Using any randomized process ensuring diversity, a set of initial solutions is built. ELS is applied to each initial solution generated in this set. In this chapter, such a scheme is applied with some diversification brought by a metaheuristic called GRASP (Greedy Randomized Adaptive Search Procedure) developed by Feo and Resende [46].

A classical GRASP generates a solution thanks to a greedy randomized heuristic. This solution is improved by a local search and the process is re-iterated until a stopping criteria is reach. The best solution encountered during the iterations is saved and returned at the end. In the hybrid version, the local search is replaced by ELS, leading to the hybrid GRASP×ELS [11]. Both GRASP×ELS and multi-start ELS have proven to be highly efficient on some optimization problems, including but not limited to, VRP [11] and LRP [8].

9.4.2 Parallelization

The proposed hybrid metaheuristic is based on the GRASPxELS scheme. It is parallelized on the ELS part to bring several concurrent searches in the solution space attempting to explore even wider regions of the search space. Thus, it belongs to type 3 according to Guerriero's classification [20]: several search threads that simultaneously perform an ELS with a given degree of synchronization and cooperation. The whole framework we propose is illustrated in Figure 9.1 and the details of the components are exposed in Section 9.6.

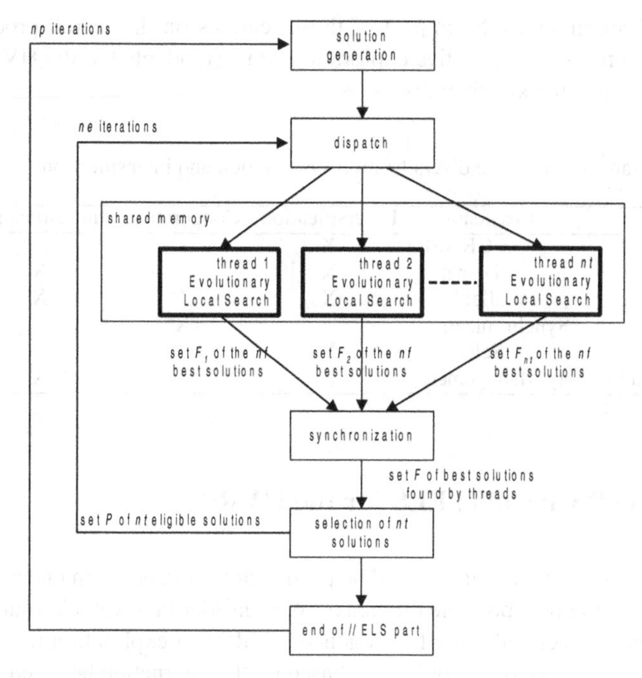

Fig. 9.1 Multi-start parallel ELS principle

9.4.3 *Key-Features of the GRASP x Parallel ELS*

Increasing the number of threads allows more evaluations to be performed. However, this may not help improving the performances if no further bright strategy is implemented. Our goal is first directed towards achieving a good speed-up while keeping a natural and reasonable parallel implementation effort. Thus, the proposed parallelization relies on an asynchronous cooperative search thread strategy (as seen in Figure 9.1). On the other hand, a wrong equilibrium between diversification and intensification may also lead to a premature convergence and trap the method on a local optimum.

An equilibrium between diversification, intensification and cooperation is important and it is ensured by all the key-features we promote within our GRASP x parallel ELS, which are:

- the greedy randomized heuristic through the GRASP scheme;
- the assignment of an initial solution to each thread through a *Dispatch* procedure;
- the parallel ELS scheme;
- the synchronization of the threads;
- the selection of the solutions to be dispatched;
- the shared memory management.

Table 9.1 summarizes the impact of those features on the search process. The next Section presents an effective exploration strategy, adapted to the HVRP. Then Section 9.6 details the key features

Table 9.1 Parameter impact in diversification, cooperation and intensification

Key features	Diversification	Cooperation	Intensification
GRASP	X		
Dispatch	X		X
Parallel ELS	X		X
Synchronization		X	
Selection	X		
Shared Memory Management	X	X	X

9.5 GRASP x Parallel ELS for the HVRP

The solution space exploration should be proficient for the problem under consideration to ensure the best possible efficiency. We consider here vehicle routing problems, and more specifically the heterogeneous VRP. An exploration that has been proven effective for routing problems is based on the alternation between two solution representations: solutions encoded as giant tours (TSP tours on the n customers) and solutions encoded as sets of trips (see Figure 9.2).

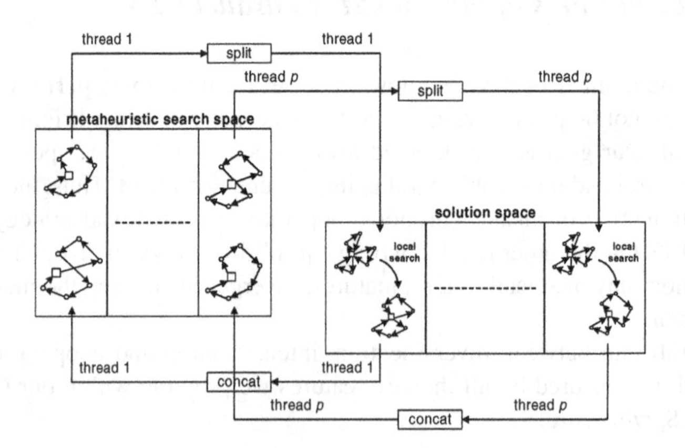

Fig. 9.2 Combination of the two search spaces

The alternation between solution representations is based on the following principle:

- **Split:** a giant tour T, made of the n customers to visit, is transformed by a splitting procedure into a HVRP solution S with respect to the given sequence and

the vehicle fleet composition. This solution S can then be improved by a local search. It can also be perturbed or left unchanged.
- **Concat:** a procedure transforms S into a giant tour T' by concatenating its trips.

The giant tour T' can be split again in order to get a new HVRP solution. This process allows alternation between the giant tours space and the HVRP space.

Splitting was originally proposed by Beasley in 1983 [27] as the second phase of a route-first cluster-second heuristic to solve the VRP. It has been successfully applied to many routing problems within general optimization frameworks. A state of the art [9] about Split approaches in routing problems surveys more than 40 recent publications on the topic. It shows this strategy has proved its efficiency and received a considerable amount of attention especially since 2001. It covers a large area including node routing problems, arc routing problems and more complex routing problems with additional constraints as the Capacitated Arc Routing Problem (CARP [38]), the Vehicle Routing Problem (VRP [10]) and the Location Routing Problem (LRP [7, 8]). Moreover, the high quality solutions obtained by Prins [10] are a suitable indication of the interest of such an approach. Our GRASP x parallel ELS belongs to this line of research.

The transformation from T to S through the Split procedure proposed by Prins [10] consists first in building an auxiliary acyclic graph H based on the sequence of customers given by T. The auxiliary graph is composed of $n+1$ nodes numbered from 0 to n, 0 being the depot and $T(i)$ being the customer at position i in the sequence. An arc from node i to j corresponds to a subsequence $u_{ij} = (T(i+1),\ldots,T(j))$ of T. The quantity to collect along u_{ij} is denoted as $Q(u_{ij})$. Thus, the trip u_{ij} consists in a route starting from a depot and going to node $T(i+1)$, from $T(i+1)$ to $T(i+2)$ and so on until $T(j)$ before coming back to the departure depot node. Moreover, all the additional problem constraints have to be satisfied by u_{ij} for the arc (i,j) to belong to the graph. Then, optimally splitting T into routes corresponds to computing the min-cost path from node 0 to node n in H. Since the HVRP includes some resource constraints as the vehicle types (to each type is given a limited number of available vehicles), several labels per node have to be handled to compute the shortest path in H [34]. Thus, several strategies to perform the label propagation can be used. Since the graph is acyclic, the original strategy consists in iteratively considering the nodes from 0 to $n-1$ to generate the labels in a greedy way. A recent variation [9] uses a depth first search exploration of the graph to reach the final node as fast as possible in order to get sharp bounds. The second strategy is used in this chapter.

When dealing with resource constraints, the number of labels per node generated by Split in H can be large [9]. A solution consists in limiting both the maximal number of labels per node ($mlpn$) and the total number of labels generated during the whole split procedure (mtl). The Split procedure becomes suboptimal since it may discard the optimal label, but its computational time is greatly reduced. Thus, it becomes compatible with an iterative metaheuristic scheme. The Split procedure depends on these two parameters.

The alternation between solution representations as giant tours and HVRP solutions is applied in the core of the ELS we propose. At each iteration of the ELS, a ***mutation*** operator is applied. It aims at modifying the solution to explore its neighborhood. The obtained children are then improved by a ***local search***.

For the HVRP, the ***mutation*** is defined on giant tour to generate new customer sequences. A child is obtained by swapping two randomly selected nodes in the giant tour. The resulting sequence is evaluated by the Split procedure and the local search is called. Such a customer exchange would mostly be ineffective on HLRP solutions, since the local search can cancel it in most cases. It is far more effective on giant tours, since a basic customer exchange may force Split to radically modify several trips.

The ***local search*** we implemented works on the HVRP solutions. It uses a first improvement strategy combining several classical VRP neighborhoods: 2-Opt within a trip, 2-Opt between two trips, Swap within a trip and Swap between two trips. This kind of approach is commonly used in routing problems including the CARP [37], the VRP [10], the HVRP [12] and the LRP [8].

9.6 Detailed Components of the GRASP x Parallel ELS

A detailed description of the key-features of the proposed method to solve the HVRP is given in this Section. They include:

- the greedy randomized heuristic through the GRASP scheme;
- the assignment of an initial solution to each thread through a *Dispatch* procedure;
- the parallel ELS scheme;
- the synchronization of the threads;
- the selection of the solutions to be dispatched;
- the management of the shared memory.

9.6.1 Initial Solutions of the GRASP

The ***greedy randomized heuristic*** is the first step of the method. It is called at the beginning of each iteration of the GRASP. It creates an initial solution S by generating a giant tour through a greedy randomized heuristic and splitting it into HVRP trips. Then, the resulting solution is improved by local search. Such generation was promoted by Prins in its genetic algorithm for the HVRP [12]: the initial population is made of random trips and of a limited number of high-quality solutions computed by randomly generating a giant tour and performing a local search on it.

Our greedy randomized heuristic is based on the nearest neighbor algorithm. From the current node i (initially the depot), the nearest customer j is identified, with c_{ij} the distance between them. Then, the randomized version looks for the set of nodes k such that $c_{ik} \leq frac * c_{ij}$, $frac$ being a given percentage. The procedure chooses at random a customer s from this set and adds it to the giant tour. s becomes the new current node and the algorithm goes on until it remains no customer.

Once the giant tour is completed, Split is called to evaluate the corresponding HVRP solution S. However, due to resource constraints on the vehicle fleet, the procedure might fail to compute a feasible solution. In this case, a new call to the greedy randomized heuristic is called to generate a new giant tour. If after *max_try* attempts no feasible solution is built, another procedure is called to built a giant tour T at random until the splitting of T provides a suitable solution S. Such a strategy seems to be a suitable trade-off since generalization of well-know heuristics (including Clarke and Wright [23] heuristic and Golden et al. [6] heuristic) is not easy due to the limited heterogeneous fleet of vehicles.

The solution S is then improved by the local search detailed in Section 9.5 and saved nt times in a pool P.

9.6.2 Dispatching: Threads Management

The ***Dispatch()*** procedure assigns an initial solution to each thread. These initial solutions are selected from the pool P of eligible solutions, according to a dispatching strategy R. Then, each ELS runs asynchronously on the threads.

It is possible to consider several dispatching strategies in order to promote diversification or, at the opposite, intensification, *i.e.* to focus on specific areas of the solution space. For instance, one can consider the following strategies:

- *BF* (Best First) consists in assigning the best solution in P to every thread: this dispatching strategy favours intensification;
- *DF* (Diversity First) consists in assigning one different solution to each processor: this dispatching strategy favours diversification.
- *CBF* (Couple Best First) consists in assigning the best solution to threads 1 and 2, the second best solution to threads 3 and 4 etc. Thus, this is a trade-of between *BF* and *DF*.

9.6.3 Parallel ELS Scheme

The algorithm below describes the ***Parallel_ELS()*** procedure in which each of the nt thread performs simultaneously one ELS.

Each thread starts an ELS with its assigned solution S. At the first iteration of the parallel part ($j = 1$) P contains only one solution. Thus, ***Dispatch()*** assigns the same initial solution to each thread. Then, ns ELS iterations, or nr unproductive ELS iterations, *i.e.* iterations without improving the incumbent solution, are performed (*while* loop lines 2-29).

The children generation (mutations and local search) is done lines 5-19. It produces nd children and updates the incumbent pair (S,T) accordingly. From T, the

Algorithm 10. Start_ELS_Thread

```
global parameters :
ns: maximum number of iterations per ELS
nd: number of children (mutations)
nf: number of solutions to save
nr: maximum number of iterations without improvement per ELS
nc_max: maximal number of unproductive iterations during mutation
input parameters :
T: initial solution
i: thread identification
output parameters:
F_i: vector storing the nb best solutions of the ELS in this thread i
r := 0, nc:=0, S := call Split(T)
while (u < ns ) and (r < nr ) do
    u := u+1; f" := ∞ ; j := 1
    // mutation loop
    while (j < nd) and (nc < nc_max) do
        T' := call Mutation(T)
        S' := call Split(T')
        if S' is a solution then
            // HVRP feasible solution
            S' := call Local_Search(S')
            if (S' is not a clone) then
                T' := call Concat(S')
                if (f(S') < f") then
                    f" := f(S'); T" := T'; S" := S'
                j := j+1
            else
                nc := nc + 1
        else
            nc := nc+1
    // check the improvment
    if f" ≥ f(S) then
        r := r+1 // update the number iterations without improvement
    else
        Save T' into F_i //only keep the best nf solutions
    // if a new best solution
    if f" < f* then
        S* := S" // update S*
        f* := f"
    T := T" // the best ELS solution becomes the new initial solution
```

mutation produces a new giant tour T'. If splitting T' leads to a feasible solution S', a local search is applied on it line 10. Then, an efficient clone detection (detailed in Section9.6.5) is included to avoid thread search to be done on the same solution space area. Thus, if the S' is not a clone (line 11), it is considered for the next ELS iteration. Lines 13-15 are classical ELS steps: the best solution is saved into S''. Note the value *nc* is increased if S' is unfeasible or if it is a clone. The value *r* is increased line 22 each time an ELS iteration does not improve the incumbent solution. Otherwise S is considered to enter F_i, the ordered set keeping the nf best solutions found during the ELS. Lines 26-28 consist in updating S^* whenever required. Line 29 assigns T'' to T, providing the new starting giant tour required at line 2.

9.6.4 Synchronization and Selection Procedure

Parallel_ELS() simultaneously runs on the *nt* threads. However, each ELS can stop asynchronously. A **synchronization** is necessary before beginning a next parallel iteration. At this point, the ***Selection()*** procedure recover information from the threads by retrieving the set F_i of best solutions saved during each ***Parallel_ELS()*** and selecting a new pool P of *nt* solutions by considering both diversification and quality of solutions.

Diversity is important to avoid premature convergence as it favours exploration of new search space regions. Managing diversity requires the definition of a distance between solutions in an approach similar to what was defined in MAPM (Memetic Algorithm with Population Management) approach [31]. The distance depends on the representation of a solution. For binary problems, the Hamming distance can be used and several distances have been proposed for permutation problems [42, 43, 50]. The Wagner and Fischer's distance (the edit distance) is commonly used in many combinatorial problems [41, 30]. This way, a solution is stated as eligible if and only if it is far enough to the best found solution, or to a set of high quality solutions previously visited.

One can also consider that solutions of identical cost are quite similar in the sense they are at the same distance from the optimal solution, if one considers a cost-based distance. Thus, the cost-base distance has been considered sufficient to check if a solution must be kept into the set F of nf best solutions.

9.6.5 Shared Memory Management

The shared memory management is the last key feature of the framework. In fact, an efficient search strongly depends on the coordination between threads. The ***Dispatch()*** procedure does most of it, but one need another feature: the **clone detection** to avoid as much as possible costly and unprofitable solution evaluation. A clone can be defined as a solution that has already been visited.

Clones weaken the effectiveness of the parallel search and an option is to forbid clones to be considered during the search. Exact clone detection might be time consuming and even a basic comparison between two giant tours runs in $O(n)$. However, an efficient management be done by using hashing techniques [48]. Indeed, by associating each solution to a key which depends on the solution cost and on the number of trips ($key(S) = (S.Cost \times S.nbtrip) mod K$, with K a large integer), an approximate and fast system can be implemented. Each key is stored in an array *Clone* such that $Clone[x]$ gives the number of visited solutions whose key is x. Thus the clone detection can be done in $O(1)$. Using this array, one can control how many time the parallel ELS search are allowed to consider a previously visited solution.

However, this is not sufficient as several threads may consider the same solution. Thus, another policy has to be defined, leading to two ways to perform the detection:

- only consider solutions investigated by the thread;
- consider the whole set of solutions over the threads.

The latter policy increases the amount of communication between threads and favours diversification while the former favours intensification. These policies either impose one array *Clone* for each thread or a global *Clone* array over all the threads. For convenience, these strategies are called:

- *LCD* (Local Clone Detection) when the clone detection is restricted to solutions of the current thread;
- *GCD* (Global Clone Detection) when the clone detection is done over all the solutions investigated by the threads.

Moreover, whatever the policy used (*GCD* or *LCD*), two memory management strategies are defined:

- a Long Term Memory Strategy (*LTMS*) in which the memory is only initialized at the beginning of the GRASP x parallel ELS. Such strategy favors diversification by preventing identical solution investigation during two different ELS iterations.
- a Short Term Memory Strategy (*STMS*) in which the memory is initialized at the end of each ELS loop. This favors intensification since two ELS iterations can lead to the exploration of similar regions.

Depending on the policy (*LCD / GCD*) and on the strategy (*LTMS / STMS*), the shared memory management creates a cooperation by indirectly controlling the behaviour of the search threads during optimization. It also helps balancing between diversification and intensification.

9.7 Computational Evaluation

As stressed before, parallelizing the search space investigation in combinatorial optimization problems should be made in order to both widen the search and enforce the convergence, in addition to speed up the algorithm with parallel calculation. In this ethic, our GRASP x parallel ELS takes advantage of multi-core processors trying to make the threads collaborate. This section provides a computational evaluation attesting whether the search strategy we promote does not only allow solving larger problems but also finding improved solutions (with respect to the sequential counterpart), and leading to a better convergence.

9.7.1 *Settings and Benchmarks*

Our method has been implemented in Pascal using Delphi 6.0 package. The numerical experiments were run on a 2.3 GHz AMD Opteron quad-cores computer running Windows 2003 with 256 Go of memory. Several sets of instances have been used. They can be divided into two main categories: classical HVRP instances and new HVRP instances.

9.7.1.1 Classical HVRP Instances

A classical set of HVRP instances has been used to assess the performance of the proposed parallel approach. It has been proposed by Taillard [19]. It is made of 8 small instances with 50 to 100 nodes. They correspond to the VFMP-V transformed into HVRP instances by limiting the vehicles availability.

We compare our results with 5 methods from the literature on these classical HVRP instances. The first approach is a heuristic column generation [19] proposed by the author [19]. Tarantilis et al. [14] designed a threshold accepting algorithm (TA). Li et al. [21] published a record-to-record (RTR) travel metaheuristic. Prins [12] proposed a memetic algorithm (SMA-D2). Finally, Brandão [4] developed a tabu search algorithm (TSA).

To perform a fair comparison between methods that have been implemented on different processors, Dongarra [25] provides scaling factors depending on the processor speed. Thus we scale all the times with respect to the computer used in this chapter. For each of the 5 previously mentioned methods, Table 9.2 gives the processor used, the processor speed in Mflops and the scale factor applied in the next Tables.

Table 9.2 Scale factors in numerical experiments

Publications	Computers	Mflops	Scale factor
Taillard [19]	50MHz Sun Sparc 10	27	0.006
Prins [12]	1.8GHz Pentium 4M	1564	0.347
Tarantilis et al. [14]	400Mhz Pentium II	262	0.058
Li et al. [21]	1GHz Athlon	1168	0.259
Brandão [4]	1.4GHz Pentium 4M	1216	0.270
Our work	2.1 GHz Opteron	4500	1.000

9.7.1.2 New Realistic HVRP Instances (DLP_HVRP)

To better evaluate the proposed method, we also use large size instances, denoted DLP_HVRP and available at http://www.isima.fr/ lacomme/hvrp/hvrp.html and at `http://prodhonc.free.fr/`. To the best of our knowledge, these instances are the first ones based on real life district data.

Using the GIS software developed by Bajart and Charles [49], one instance is built for each of the 96 French districts, and contains from 60 to more than 250 nodes. Nodes correspond to cities larger than 100 or 500 citizens. Costs between nodes are computed as the shortest paths using the Google web service and thus they correspond to the real distance in kilometers between cities. The fleet composition contains up to 10 types of vehicles and has been randomly generated. As a consequence, some node demand can exceed the capacity of some types of vehicles and it may also happen that the total fleet capacity is close to the total demand to service. In addition, both fixed and variable vehicle costs are not dependant of the vehicles capacity. Thus, for some instances the smallest vehicles have smallest fixed cost while for some others the smallest vehicles are the most expensive in terms of variable and/or fixed costs. These characteristics might lead to very hard instances and be challenging for the metaheuristics.

9.7.1.3 Parameters

To have a fair comparison, two sets of parameters are defined: one for the classical instances and another for the DLP_HVRP instances. The parameters are detailed in Tables 9.3 and 9.4.

Table 9.3 Parameter Definition

Parameter	Definition
mtl	Maximal labels generated in split
mlpn	Maximal labels saved on each node in split
Nls	Maximal number of iterations during local search
Ns	Number of ELS iterations
Nr	Maximal number of ELS without improvement
nd	Number of diversifications
Nc_max	Maximal number of unproductive iterations during mutation
Ne	Number of parallel ELS iterations
Nt	Number of threads
Np	Number of GRASP iterations
Nf	Number of solutions to save by each thread in *Fi*
Strategy	Dispatch strategy
Share Memory Policy	Clone detection strategy and memory management strategy

The main difference from the two settings is the maximal number of labels generated in split, set to 50 000 for the DLP instances since those instances are larger than the classical ones. For a similar reason, the number of labels per node is increased from 3 to 100.

Table 9.4 Parameter settings

Parameter	Classical HVRP instances	DLP_HVRP instances
mtl	10 000	50 000
mlpn	3	100
Nls	500	100
Ns	3	3
Nr	3	3
nd	10	10
Nc_max	10	10
Ne	15	15
Nt	from 2 to 32	from 2 to 32
Np	50	50
Nf	1.15	1.15
Strategy	DF	DF
Share Memory Policy	LCD + STMS	LCD + STMS

9.7.2 Evaluation of the Communication Time

Before focusing on the results obtained on the sets of instances, it would be interesting to check the communication time of our machine. Ideally, when increasing the number of threads from 1 to k in our framework, the total running time should remain the same (except small variations due to the synchronization feature). Indeed, the number of iterations does not depend on the number of threads. However, as already explained, if k threads, assigned to k cores, imply the use of a greater number of processors than for p threads, then a slowdown occurs. It is due to communication through buses. In our experiments with quad-core processors, this is observed with a factor r between 1 and 2 processors, $2r$ between 2 and 4 processors and $4r$ between 4 and 8 processors, with $r = 1.55$. If an extra convergence rate is obtained one can state it is induced by an effective dispatch, synchronization and shared memory management.

Thus, in the next Tables, Columns TT provide the average total computational time in seconds on the instances. The time in seconds to the find the best solution is given by T^* and this value can be corrected (*Corrected* T^*) as explained above by the value that T^* would have taken if the threads were associated to cores belonging to a same processor (no extra time due to communication through buses).

9.7.3 Results on Classical HVRP Instances

Table 9.5 provides a comparative study of the last publications using the set from Taillard. *BKS* refers to the value of the best-known solution for each instance. *Avg. dev* corresponds to the average gap from a considered method to BKS in percent. *Avg. time* is the average CPU time of the methods in seconds, and *Scaled time* provides the corresponding time that would be achieved on our computer using the scaling factors from Table 9.2. The line *#Best* expresses the number of BKS achieved by the method. Asterisks indicate when the BKS is an optimal cost.

The best published method so far has been proposed by Li et al. [21] with a deviation to the best-known solutions lower than 0.04%. This method provides 7 times the best solutions. The last two publications of Brandão [4] and Prins [12] have a higher average deviation to the best-known solutions (about 0.08%) but they get 6 BKS (vs. 7 for Li et al.) and they are twice faster (scaled).

Table 9.5 Performances of the last published methods on the HVRP instances from Taillard

	n	BKS	Taillard [19]	Tarantilis [14]	Li [21]	Prins [12]	Brandão [4]
			Cost	Cost	Cost	Cost	Cost
13	50	1517.84*	1518.05	1519.96	**1517.84**	**1517.84**	**1517.84**
14	50	607.53*	615.64	611.39	**607.53**	**607.53**	**607.53**
15	50	1015.29*	1016.86	**1015.29**	**1015.29**	**1015.29**	**1015.29**
16	50	1144.94*	1154.05	1145.52	**1144.94**	**1144.94**	**1144.94**
17	75	1061.96*	1071.79	1071.01	**1061.96**	1065.85	**1061.96**
18	75	1823.58*	1870.16	1846.35	**1823.58**	**1823.58**	1831.36
19	100	1117.51	**1117.51**	1123.83	1120.34	1120.34	1120.34
20	100	1534.17*	1559.77	1556.35	**1534.17**	**1534.17**	**1534.17**
Avg. dev			0.9310	0.6172	0.0317	0.0774	0.0850
Avg. time			2011.12	607.12	285.75	94.74	151.66
Scaled time			12.06	35.34	74.16	32.92	40.98
# Best			1	1	7	6	6

Tables 9.6 reports the performance of the GRASP x parallel ELS with 1 to 32 threads with the same statistical indicators as in Table 9.5, except the times that are given in Table 9.7 providing the average CPU times to achieve the best found solution and the corrected time due to the communication between processors.

While the parameters have not been especially tuned on these instances, the initial performance (single thread) is at about 0.53% of the best-known solutions, which is already better than both Taillard and Tarantilis et al. proposals. However, the single thread performance is slower and it does not compete with Li et al., Brandão or Prins methods.

The conclusion is totally different when the number of threads increases. On these small instances, it clearly appears that going from 1 to 2 threads significantly improves the results. The average deviation drops to 0.05%, becoming 10 times lower than with a single thread, and the average time to get the best solution is halved. Thus our proposed approach with 2 threads is only outperformed by Li et al. in terms of solution cost. A first sign of a better convergence seems to appear.

Until 8 threads, no better solution is found (6 instances over 8 are already solved to optimality). Then, with 16 and 32 threads, the methods provides results of the same quality as the best published method of Li et al. with an average deviation of 0.0317%. Increasing the number of threads allows to enforce the search on the solution space thanks to the balanced diversification, intensification and cooperation.

Table 9.6 Performances of the GRASP x parallel ELS on the classical HVRP instances from Taillard

	n	BKS	1 Thread Cost	2 Threads Cost	4 Threads Cost	8 Threads Cost	16 Threads Cost	32 Threads Cost
13	50	1517.84*	**1517.84**	**1517.84**	**1517.84**	**1517.84**	**1517.84**	**1517.84**
14	50	607.53*	609.17	**607.53**	**607.53**	**607.53**	**607.53**	**607.53**
15	50	1015.29*	**1015.29**	**1015.29**	**1015.29**	**1015.29**	**1015.29**	**1015.29**
16	50	1144.94*	**1144.94**	**1144.94**	**1144.94**	**1144.94**	**1144.94**	**1144.94**
17	75	1061.96*	1065.2	1064.07	1064.07	1064.07	**1061.96**	**1061.96**
18	75	1823.58*	**1823.58**	**1823.58**	**1823.58**	**1823.58**	**1823.58**	**1823.58**
19	100	1117.51	1120.34	1120.34	1120.34	1120.34	1120.34	1120.34
20	100	1534.17*	**1534.17**	**1534.17**	**1534.17**	**1534.17**	**1534.17**	**1534.17**
Avg. dev			0.5313	0.0565	0.0565	0.0565	0.0317	0.0317
# Best			5	6	6	6	7	7

Table 9.7 Times of the GRASP x parallel ELS on the classical HVRP instances from Taillard

	n	1 Thread Time	2 Threads Time	4 Threads Time	8 Threads Time	16 Threads Time	32 Threads Time
13	50	**60.7**	**80.7**	**5.8**	**14.3**	**17.9**	**38.7**
14	50	46.6	**353.1**	**153.9**	**230.7**	**79.0**	**111.7**
15	50	**17.7**	**62.0**	**18.4**	**10.7**	**10.9**	**22.9**
16	50	**109.0**	**2.1**	**1.1**	**2.5**	**3.8**	**7.7**
17	75	182.6	109.7	49.2	37.2	**123.8**	**482.0**
18	75	**287.8**	**68.3**	**72.3**	**32.9**	**25.5**	**38.9**
19	100	702.0	16.0	142.8	22.4	31.4	10.3
20	100	**297.7**	**190.3**	**434.2**	**116.8**	**27.5**	**249.5**
Avg. T^*		213.0	110.3	109.7	58.5	40.0	120.2
Avg. corrected T^*		213.0	110.3	109.7	37.7	12.9	19.4

However, to really attest the performance, both the solution cost and the computational time have to be observed to confirm the convergence brought by the key-features of the proposed method. Tables 9.7 helps us to focus on that aspect.

As already said, from 1 to 2 threads, both the solution cost and T^* decrease. The same observation is made between 8 and 16 threads, reinforcing our thought about a better convergence rate provided by a higher number of threads. Moreover with 8 and 16 threads, T^* reduces from around 110s to respectively 58s and 40s (the latter also coming with better results) instead of experiencing a slow-down.

The small increase of corrected CPU time happens with 32 threads. It can be explained by two reasons. The first one is the synchronization time that can be longer because of a large number of threads. The second one is the main ground. The management of the shared memory forbids the search on solution already visited among the threads. However, the solution space is not so large on small instances, and many solutions are discarded due to clone detection leading to many unproductive trials.

To conclude, our method is competitive with the literature on Taillard's set, even if this must be moderated since all the recent published methods provide less than

1% deviation. This study exposes the significant impact of number of threads on the convergence and therefore on the solution quality and the computational time. Next Section is dedicated to larger instances to confirm the behavior brought by the balanced diversification, intensification and cooperation between parallel calculations thanks to the key-features provided in our framework.

9.7.4 *Results on DLP_HVRP Instances*

These instances can be divided into the following subsets:

- DLP_HVRP_1: 15 small instances with less than 100 nodes;
- DLP_HVRP_2: 40 medium instances with 100 to 150 nodes;
- DLP_HVRP_3: 33 large instances with 150 to 200 nodes;
- DLP_HVRP_4: 11 very large instances with more than 200 nodes.

The two next Sections summarize the performance of our GRASP x parallel ELS. Detailed results per instance are provided in Appendix, in which Cost, $T*$ and TT respectively mean the solution cost, the time to achieve the best found solution and the total time of the method. The last line indicates the average values of the columns.

9.7.4.1 DLP_HVRP_1 Instances (Small Scale Instances with Less Than 100 Nodes)

This set encompasses 15 small-scale instances with less than 100 nodes. Table 9.8 summarizes the results obtained according to the number of threads. The meaning of the indicators are the same as in Tables from Section 9.7.3, and *% Best* expresses the percent number of BKS computed by the method.

Table 9.8 GRASP x parallel ELS performance with 1 to 32 threads on DLP_HVRP_1

# Threads	Avg. Cost	Avg. dev	# Best	% Best	Avg. T^*	Corrected Avg. T^*
1 thread	4391.37	0.11	7	46.7	109.1	109.1
2 threads	4389.96	0.10	7	46.7	92.6	92.6
4 threads	4390.30	0.10	8	53.3	105.9	105.9
8 threads	4387.30	0.05	9	60.0	182.6	117.8
16 threads	4386.14	0.03	11	73.3	193.4	62.4
32 threads	4386.10	0.01	13	86.7	409.0	66.0

The gap to the best-known solutions clearly decreases between 1 to 32 threads while the average T^* is halved. Both reductions on cost and on computational time illustrate an improvement on the convergence rate.

The quality of the solution is better stressed by an analysis of the percent of best-known solutions found as shown in column *% Best* of Table 9.8. Even if the reduction of the average cost is not large (from 4391.37 to 4386.10), the number of best-known solutions found is doubled between a single thread version of the method and the parallelization on 32 threads.

9.7.4.2 DLP_HVRP_2, DLP_HVRP_3, DLP_HVRP_4 Instances

Table 9.9 sums up the results obtained on the three subsets of medium and large-scale DLP_HVRP instances. Once again, the quality of the solution costs is clearly improved with the number of threads used while the computational time increases but not more than the expected extension due to the communication between processors.

Table 9.9 Average solutions depending on the number of threads

DLP_HVRP_2						
# Threads	Avg.Cost	Avg. dev	# Best	% Best	Avg. T^*	Corrected Avg. T^*
1	8712.516	0.52	4	10.5	427.1	427.1
2	8704.628	0.43	3	7.9	413.8	413.8
4	8696.673	0.35	6	15.8	429.8	429.8
8	8684.732	0.22	9	23.7	389.8	251.5
16	8675.003	0.13	13	34.2	498.4	160.8
32	8669.608	0.05	26	68.4	794.9	128.2
DLP_HVRP_3						
# Threads	Avg.Cost	Avg. dev	# Best	% Best	Avg. T^*	Corrected Avg. T^*
1	11995.19	1.34	0	0.0	663.5	663.5
2	11954.47	0.99	1	3.2	693.3	693.3
4	11913.26	0.66	3	9.7	704.3	704.3
8	11897.53	0.52	4	12.9	704.8	454.7
16	11877.95	0.32	9	29.0	773.1	249.4
32	11856.20	0.14	16	51.6	1042.9	168.2
DLP_HVRP_4						
# Threads	Avg.Cost	Avg. dev	# Best	% Best	Avg. T^*	Corrected Avg. T^*
1	14543.41	1.29	0	0.0	899.1	899.1
2	14509.72	1.02	0	0.0	1063.0	1063.0
4	14404.31	0.55	1	8.3	983.7	983.7
8	14389.63	0.36	1	8.3	1138.1	734.3
16	14398.82	0.37	5	41.7	1171.2	377.8
32	14365.49	0.23	6	50.0	1466.4	236.5

The full analysis of the DLP_HVRP_1 subset (see Section 9.7.4.1), which encompasses small-scale instances with less than 100 nodes, defines a general trend which is confirmed over the 3 last subsets. Depending on the instances set of interest the percent of best-known solutions obtained varies from (Table 9.9):

- 10% with 1 thread to 68% with 32 threads for the instance set DLP_HVRP_2;
- 0% with 1 thread to 51% with 32 threads for the instance set DLP_HVRP_3;
- 0% with 1 thread to 50% with 32 threads for the instance set DLP_HVRP_4.

For the small-scale instances, the difference from the single thread program and the 32 threads program is significant with a variation from 46% to 86% which correspond to a factor 2 (see Section 3.3.1). This ratio is about 6 for the DLP_HVRP_2 subset of instances (from 10% to 68%).

The last subsets (3 and 4) are interesting since the single thread version got no best-known solution. For these large-scale instances, the improvement in percent of the number of best-known solutions found is impressive when increasing the number of threads. On DLP_HVRP_3 instances, it steadily grows from 2 to 32 threads. On DLP_HVRP_4 instances, it first grows for 4 threads. Then it increases to 41% before stabilizing. This behavior points out the relationship between the size of the solution space and the benefit of extra threads to search a good solution.

On the other hand, the corrected average T^* significantly decreases with the number of threads. The double impact brought by the cooperative parallelization (improvement of the solution costs plus reduction of the computational times) attests the efficiency of proposed algorithm. The last Section of this experimental part details the convergence rate provided by the multi-threads.

9.7.5 Convergence Rate of the Parallel ELS

From the results presented in the previous Sections, it seems clear that the parallelization proposed in this chapter, with its key-features lending intensification, diversification and cooperation, leads to a convergence rate that is improved not only because of the number of threads but also thanks to the threads policy management during ELS synchronization, dispatch and selection (see Section 9.4).

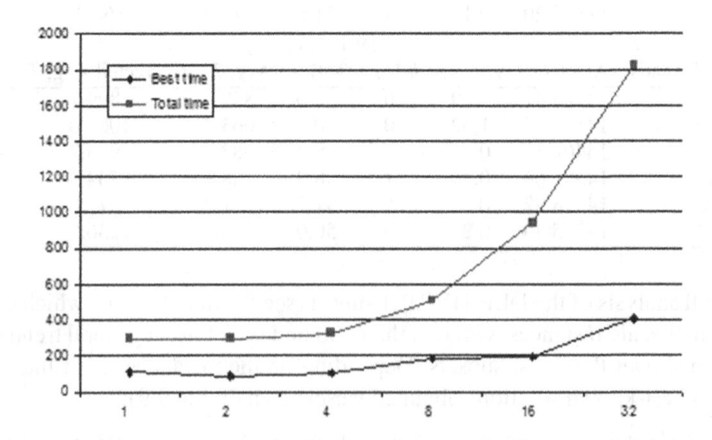

Fig. 9.3 Computational time and time to the best evolution depending on the number of threads - DLP_HVRP_1

The convergence rate is materialized by better results obtained within equivalent or shorter computational times (columns called Corrected Avg. T^*, see Tables 9.7, 9.8 and 9.9). Even when analysing the original value of T^*, one can note a quasi-linear evolution of the time to the best (T^*) and an exponential trend of the total time (see Figure 9.3).

The behavior of the convergence is also illustrated on Figure 9.4 representing the percentage of BKS found by the algorithm. The first set of instances (DLP_HVRP_1) seems to be easier to solve: the single thread approach is able to get 50% of the best-known solutions which can be compared with the 10% or 0% obtained for the last three sets. The general curve for the set 2, 3 and 4 seems to be piecewise linear especially for the last set.

We can now focus on the convergence through the evolution of the best found solution over the iterations. Comparative study is not straightforward since the

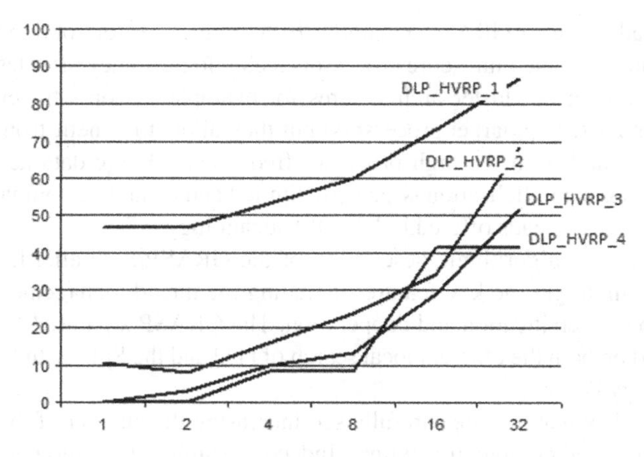

Fig. 9.4 Percents of best solutions depending on the number of threads - DLP_HVRP set

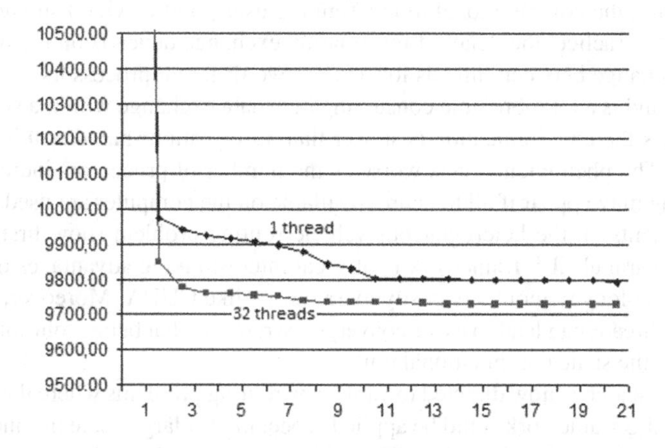

Fig. 9.5 Best solutions evolution with 1 and 32 threads

meta-heuristic is stochastic. Trying to analyze the relative impact of the number of threads, we introduce hereafter 5 runs with 1, 2, 4, 8, 16, 32 threads (for exactly the same parameters) for the largest example of the benchmark *i.e.* instance 18 with 256 nodes and 5 types of vehicles. Each run starts with the same initial solution at cost 13364.60.

After one iteration of the GRASP x parallel ELS, the best found solution varies from 10048 to 9849 with a strong advantage to the 32 threads run. Figure 9.5 gives the convergence curve over 20 iterations for 1 thread only and 32 threads. It tends to show that both the convergence rate and the solution quality are better when increasing the number of threads.

9.8 Concluding Remarks and Future Research

This article addresses a GRASP x parallel ELS framework for routing problems taking advantages of the multi-core processors technologies. They offer less parallel threads than computer cluster architectures for example (2 cores for entry point processors and 8 for upmarket processors) but they allow to benefit from the high speed memory and from the high processor frequencies. Large data accesses are convenient but the parallelisation is strongly limited and cannot be compared to the number of parallel threads offered in NVIDIA technology.

We introduce an algorithmic description of the GRASP x parallel ELS framework and we highlight the key-features including the thread management through intensification, diversification and cooperation. The GRASP x parallel ELS framework is based on both the efficient local search of [11] and the Split with Depth First Search Strategy [9].

Even if the key features are carefully set, increasing the number of threads may lead to an increased computational time. Indeed, assuming the same number of iterations per thread, the search should be enlarged proportionally to the number of threads while the computational time (from the user point of view) should remain the same. In practice, the delay of information exchange depends on the thread location. Exchange between threads located on two different processors requires the use of bus and is a lot more time consuming than data exchange within a single processor. Thus the communication is slower than information shared within a single processor. The phenomenon gets worse as the number of processors increases and overloading may appear if all the cores available on the computer are used.

Experiments on the heterogeneous vehicle routing problem show first that the GRASP x parallel ELS framework is of great interest taking advantages of a wide spread technology without any costly extra device like CUDA. Moreover, it shows that multi-threads can lead to extra-convergence rate and that better solutions can be obtained in the same computational time.

Future research is now directed towards new routing problems where the GRASP x parallel ELS framework could be applied especially for large-scale instances. This line of research could include the LRP with heterogonous fleet of vehicles.

References

1. Bortfeldt, A., Gehring, H., Mack, D.: A parallel tabu search algorithm for solving the container loading problem. Parallel Computing 29(5), 641–662 (2003)
2. Le Bouthillier, A., Crainic, T.G.: A cooperative parallel meta-heuristic for the vehicle routing problem with time windows. Computers & Operations Research 32(7), 1685–1708 (2005)
3. Tantar, A.-A., Melab, N., Talbi, E.-G., Parent, B., Horvath, D.: A parallel hybrid genetic algorithm for protein structure prediction on the computational grid. Future Generation Computer Systems 23(3), 398–409 (2007)
4. Brand, A.O.J.: A tabu search algorithm for the heterogeneous fixed fleet vehicle routing problem. Computers & Operations Research 38(1), 140–151 (2011)
5. Eksioglu, B., Vural, A.V., Reisman, A.: The vehicle routing problem: a taxonomic review. Computers and Industrial Engineering 57, 1472–1483 (2008)
6. Golden, B.L., DeArmon, J.S., Baker, E.K.: Computational experiments with algorithms for a class of routing problems. Computers & Operations Research 10(1), 47–59 (1983)
7. Duhamel, C., Lacomme, P., Prins, C., Prodhon, C.: A memetic approach for the capacitated location routing problem. In: EU-MEeting 2008, Troyes, France (2008)
8. Duhamel, C., Lacomme, P., Prins, C., Prodhon, C.: A GRASPxELS approach for the capacitated location-routing problem. Computers & Operations Research 37(11), 1912–1923 (2010)
9. Duhamel, C., Lacomme, P., Prodhon, C.: Efficient frameworks for greedy split and new depth first search procedures for routing problems. Computers & Operations Research 38(4), 723–739 (2010)
10. Prins, C.: A simple and effective evolutionary algorithm for the vehicle routing problem. Computers & Operations Research 31, 1985–2002 (2004)
11. Prins, C.: A GRASP x evolutionary local search hybrid for the vehicle routing problem. In: Pereira, F.B., Tavares, J. (eds.) Bio-Inspired Algorithms for the Vehicle Routing Problem. SCI, vol. 161, pp. 35–53. Springer (2009)
12. Prins, C.: Two memetic algorithms for heterogeneous fleet vehicle routing problems. Engineering Applications of Artificial Intelligence 22, 916–928 (2009)
13. Ribeiro, C.C., Rosseti, I.: Efficient parallel cooperative implementations of GRASP heuristics. Parallel Computing 33(1), 21–35 (2007)
14. Tarantilis, C.D., Kiranoudis, C.T., Vassiliadis, V.S.: A threshold accepting metaheuristic for the heterogeneous fixed fleet vehicle routing problem. European Journal of Operational Research 152, 148–158 (2004)
15. Fiechter, C.N.: A parallel tabu search algorithm for large traveling salesman problems. Discrete Applied Mathematics 51(3), 243–267 (1994)
16. Lourenço, H., Martin, O., Sttzle, T.: Iterated local search. In: Glover, F., Kochenberger, G. (eds.) Handbook of Metaheuristics, pp. 321–353. Kluwer, Dordrecht (2003)
17. Bertsekas, D.P., Tsitsiklis, J.N., Wu, C.: Rollout algorithms for combinatorial optimization problems. Journal of Heuristics 3, 245–262 (1997)
18. Talbi, E.-G., Cahon, S., Melab, N.: Designing cellular networks using a parallel hybrid metaheuristic on the computational grid. Computer Communications 30(4), 698–713 (2007)
19. Taillard, E.D.: A heuristic column generation method for the heterogeneous fleet VRP. RAIRO Operations Research 31(1), 1–14 (1999)
20. Guerriero, F., Mancini, M.: A cooperative parallel rollout algorithm for the sequential ordering problem. Parallel Computing 29(5), 663–677 (2003)
21. Li, F., Golden, B.L., Wasil, E.A.: A record-to-record travel algorithm for solving the heterogeneous fleet vehicle routing problems. Computers & Operations Research 34, 2734–2742 (2007)
22. López, F.G., Torres, M.G., Batista, B.M., Pérez, J.A.M., Moreno-Vega, J.M.: Solving feature subset selection problem by a parallel scatter search. European Journal of Operational Research 169(2), 477–489 (2006)

23. Clarke, G., Wright, J.W.: Scheduling of vehicles from a central depot to a number of delivery points. Operations Research 12, 568–581 (1964)
24. Berger, J., Barkaoui, M.: A parallel hybrid genetic algorithm for the vehicle routing problem with time windows. Computers & Operations Research 31, 2037–2053 (2004)
25. Dongarra, J.: Performance of various computers using standard linear equations software. Report CS-89-85. University of Tennessee (2009)
26. Cordeau, J.-F., Gendreau, M., Hertz, A., Laporte, G., Sormany, J.S.: New heuristics for the vehicle routing problem. In: Langevin, A., Riopel, D. (eds.) Logistics Systems - Design and Optimization, pp. 279–298. Springer (2005)
27. Beasley, J.E.: Route-first cluster-second methods for vehicle routing. Omega 11, 403–408 (1983)
28. Cadenas, J.M., Garrido, M.C., Muñoz, E.: Using machine learning in a cooperative hybrid parallel strategy of metaheuristics. Information Sciences 179(19), 3255–3267 (2009)
29. Current, J.R., Marsh, M.: Multiobjective transportation network design and routing problems: Taxonomy and annotation. European Journal of Operational Research 65, 4–19 (1993)
30. Sörensen, K.: Distance measures based on the edit distance for permutation-type representations. In: Barry, A. (ed.) Proceedings of the Workshop on Analysis and Design of Representations and Operators (ADoRo), GECCO Conference, pp. 15–21 (2003)
31. Sörensen, K., Sevaux, M.: *MA* | *PM*: memetic algorithms with population management. Computers & Operations Research 33(5), 1214–1225 (2006)
32. Drummond, L.M.A., Ochi, L.S., Vianna, D.S.: An asynchronous parallel metaheuristic for the period vehicle routing problem. Future Generation Computer Systems 17(4), 379–386 (2001)
33. Ochi, L.S., Vianna, D.S., Drummond, L.M.A., Victor, A.O.: A parallel evolutionary algorithm for the vehicle routing problem with heterogeneous fleet. Future Generation Computer System 14, 285–292 (1998)
34. Desrochers, M.: An algorithm for the shortest path problem with resource constraints. Technical report G-88-27. GERAD (1988)
35. Toulouse, M., Crainic, T.G., Thulasiraman, K.: Global optimization properties of parallel cooperative search algorithms: A simulation study. Parallel Computing 26(1), 91–112 (2000)
36. Caricato, P., Ghiani, G., Grieco, A., Guerriero, E.: Parallel tabu search for a pickup and delivery problem under track contention. Parallel Computing 29, 631–639 (2003)
37. Lacomme, P., Prins, C., Ramdane-Chérif, W.: A Genetic Algorithm for the Capacitated Arc Routing Problem and Its Extensions. In: Boers, E.J.W., Gottlieb, J., Lanzi, P.L., Smith, R.E., Cagnoni, S., Hart, E., Raidl, G.R., Tijink, H. (eds.) EvoIASP 2001, EvoWorkshops 2001, EvoFlight 2001, EvoSTIM 2001, EvoCOP 2001, and EvoLearn 2001. LNCS, vol. 2037, pp. 473–483. Springer, Heidelberg (2001)
38. Lacomme, P., Prins, C., Ramdane-Cherif, W.: Competitive memetic algorithms for arc routing problems. Annals of Operations Research 131, 159–185 (2004)
39. Toth, P., Vigo, D.: An overview of vehicle routing problems. In: The Vehicle Routing Problem, pp. 1–26. SIAM Monographs on Discrete Mathematics and Applications, Philadelphia (2002)
40. Baldacci, R., Battarra, M., Vigo, D.: Routing a heterogeneous fleet of vehicles. In: Wasil, E., Raghavan, S., Golden, B.L. (eds.) The Vehicle Routing Problem: Latest Advances and New Challenges. Operations Research/Computer Science Interfaces, vol. 43, pp. 3–27. Springer, Berlin (2008)
41. Wagner, R.A., Fischer, M.J.: The string-to-string correction problem. Journal of the Association for Computing Machinery 21, 168–173 (1974)
42. Ronald, S.: Distance functions for order-based encodings. In: Fogel, D. (ed.) Proceedings of the IEEE Conference on Evolutionary Computation, pp. 641–646 (1997)
43. Ronald, S.: More distance functions for order-based encodings. In: Proceedings of the IEEE Conference on Evolutionary Computation, pp. 558–563 (1998)

44. Porto, S.C.S., Kitajima, J.P.F.W., Ribeiro, C.C.: Efficient parallel cooperative implementations of GRASP heuristics performance evaluation of a parallel tabu search task scheduling algorithm. Parallel Computing 26(1), 73–90 (2000)
45. James, T., Rego, C., Glover, F.: A cooperative parallel tabu search algorithm for the quadratic assignment problem european. Journal of Operational Research 195(3), 810–826 (2009)
46. Feo, T.A., Resende, M.G.C.: Greedy randomized adaptive search procedures. Journal of Global Optimization 6, 109–133 (1995)
47. Crainic, T.G., Toulouse, M., Gendreau, M.: Parallel asynchronous tabu search for multicommodity location-allocation with balancing requirements. Annals of Operations Research 63, 277–299 (1996)
48. Cormen, T.H., Leiserson, C.L., Rivest, M.L.: Introduction to algorithms. MIT Press (1990)
49. Bajart, V., Charles, C.: Systèmes d'information géographique. 3rd Year Project Report. ISIMA (2009), http://www.isima.fr/~lacomme/students.html
50. Campos, V., Laguna, M., Martí, R.: Context-independent scatter and tabu search for permutation problems. INFORMS Journal on Computing 17, 111–122 (2005)
51. Bozejko, W.: Solving the flow shop problem by parallel programming. Journal of Parallel and Distributed Computing 69(5), 470–481 (2009)

Appendix

Table 9.10 DLP_HVRP_1 (small scale instances with less than 100 nodes)

	1 thread			2 threads			4 threads			8 threads			16 threads			32 threads		
	Cost	TT	T*	Cost	TT	T*	Cost	TT	T*	Cost	TT	T*	Cost	TT	T*	Cost	TT	T*
HVRP_DLP_01	9219.65	466.5	290.3	9219.65	501.5	122.3	9219.65	449.6	24.7	9210.72	576.5	558.3	9210.14	760.2	587.1	9219.65	1494.7	14.6
HVRP_DLP_08	4597.35	281.6	277.4	4596.52	296.4	79.0	4596.52	278.5	40.4	4596.52	404.2	93.4	4596.52	766.7	29.7	4591.75	1528.1	450.1
HVRP_DLP_10	2107.55	220.3	10.9	2107.55	215.1	8.2	2107.55	247.6	3.7	2107.55	420.9	7.8	2107.55	952.3	3.6	2107.55	1783.1	45.6
HVRP_DLP_11	3370.47	298.9	43.9	3370.52	320.9	177.4	3369.91	310.5	264.6	3369.91	455.5	176.4	3369.91	859.8	64.2	3367.41	1632.1	624.8
HVRP_DLP_36	5721.84	378.1	377.2	5723.62	377.8	226.4	5719.79	346.8	286.6	5695.68	452.9	433.8	5684.61	715.9	276.3	5688.37	1452.8	1379.9
HVRP_DLP_39	2934.55	319.0	14.2	2934.55	325.3	104.4	2934.55	299.4	15.6	2934.55	417.2	19.8	2929.94	754.5	680.8	2923.72	1538.2	1449.5
HVRP_DLP_43	8753.95	579.5	134.3	8737.34	531.6	425.1	8749.01	485.1	311.6	8737.02	541.4	499.8	8737.34	743.6	214.0	8737.02	1464.4	110.1
HVRP_DLP_52	4027.27	296.5	161.5	4029.42	311.5	1.8	4027.27	305.7	161.5	4027.27	471.7	105.8	4027.27	886.2	200.5	4027.27	1767.2	105.7
HVRP_DLP_55	10244.34	201.4	127.1	10244.34	200.8	100.9	10244.34	244.4	5.3	10244.34	435.2	34.3	10244.34	879.4	27.5	10244.34	1745.5	30.1
HVRP_DLP_70	6692.91	354.3	103.5	6685.24	356.9	67.4	6685.24	353.8	164.8	6685.24	439.2	357.6	6685.24	741.5	25.5	6685.24	1506.8	30.7
HVRP_DLP_75	452.85	185.1	0.1	452.85	214.2	0.1	452.85	406.9	0.1	452.85	850.0	0.1	452.85	1786.1	0.1	452.85	3558.3	0.1
HVRP_DLP_82	4768.21	424.7	38.7	4768.21	342.0	24.1	4768.21	425.8	255.2	4768.21	811.5	284.4	4766.74	1402.2	499.1	4766.74	1912.3	1783.7
HVRP_DLP_92	564.39	172.0	1.7	564.39	181.5	0.9	564.39	256.0	1.0	564.39	536.7	0.1	564.39	1061.1	0.2	564.39	2188.5	0.5
HVRP_DLP_93	1036.99	128.8	43.5	1036.99	137.8	47.8	1036.99	195.6	51.6	1036.99	401.7	162.3	1036.99	833.7	286.4	1036.99	1679.9	99.8
HVRP_DLP_94	1378.25	150.3	12.1	1378.25	156.7	2.7	1378.25	231.03	2.3	1378.25	478.6	5.3	1378.25	1017.6	5.4	1378.25	2041.4	9.9
Avg.	4391.37	297.1	109.1	4389.96	298.0	92.6	4390.30	322.4	105.9	4387.29	512.9	182.6	4386.13	944.1	193.4	4386.10	1819.5	409.0

Table 9.11 DLP_HVRP_2 (medium scale instances with a number of nodes from 100 to 150)

	1 thread			2 threads			4 threads			8 threads			16 threads			32 threads		
	Cost	TT	T*	Cost	TT	T*	Cost	TT	T*	Cost	TT	T*	Cost	TT	T*	Cost	TT	T*
HVRP_DLP_03	11418.57	657.2	505.6	11424.08	732.8	641.9	11366.57	715.7	46.6	11350.48	748.0	424.4	11189.12	887.6	711.9	11201.68	1503.9	1027.0
HVRP_DLP_05	10940.02	757.2	699.7	10913.93	712.1	691.6	10908.67	703.9	576.4	10908.67	820.6	462.1	10906.93	984.5	748.0	10906.93	1563.5	1225.8
HVRP_DLP_06	11723.21	563.2	540.8	11742.55	986.8	235.1	11731.39	1112.0	877.7	11692.85	1233.5	926.3	11698.22	1566.3	577.2	11698.22	1940.6	1436.3
HVRP_DLP_07	8139.18	431.9	170.9	8118.44	580.1	552.7	8103.98	474.2	360.9	8095.88	633.0	627.4	8108.82	821.0	507.3	8101.80	1525.4	249.8
HVRP_DLP_12	3543.99	451.0	58.2	3543.99	491.6	22.9	3543.99	483.4	66.6	3543.99	553.5	25.4	3543.99	837.6	21.6	3543.99	1677.4	188.6
HVRP_DLP_13	6710.35	495.4	253.1	6700.57	546.6	198.8	6697.58	479.2	98.0	6697.58	590.5	145.7	6697.58	780.6	94.3	6696.43	1550.6	747.6
HVRP_DLP_16	4156.97	602.7	14.0	4156.97	732.5	88.7	4156.97	584.7	15.2	4156.97	784.7	13.1	4156.97	1021.1	24.2	4156.97	1792.5	20.0
HVRP_DLP_17	5367.34	591.1	424.1	5365.98	600.4	580.0	5367.34	580.9	168.3	5362.83	644.8	287.1	5365.97	806.7	564.0	5366.61	1515.7	294.5
HVRP_DLP_2A	7835.61	591.3	493.2	7854.58	693.9	396.8	7845.38	592.1	586.6	7798.92	685.1	321.9	7843.21	804.4	796.6	7793.16	1490.0	1374.4
HVRP_DLP_2B	8478.96	537.7	462.4	8479.19	655.0	651.3	8485.69	531.8	437.6	8471.12	587.0	383.0	8464.69	767.3	750.2	8467.73	1467.3	1022.2
HVRP_DLP_21	5150.05	570.6	476.6	5144.09	545.8	407.0	5147.27	549.7	491.1	5149.27	626.7	409.6	5144.76	787.4	525.9	5141.49	1521.8	538.1
HVRP_DLP_25	7215.01	695.7	534.7	7208.39	796.3	428.8	7212.68	711.2	427.8	7209.29	819.0	587.3	7209.29	974.3	414.9	7206.64	1614.1	778.0
HVRP_DLP_26	6486.99	482.4	475.7	6460.00	541.9	520.8	6453.99	2472.9	452.1	6509.14	569.7	259.4	6459.53	742.1	417.7	6446.31	1466.5	850.1
HVRP_DLP_28	5535.18	559.9	486.5	5540.73	642.7	607.0	5551.37	565.8	376.7	5537.87	689.9	633.6	5537.71	809.3	293.7	5531.06	1516.1	959.2
HVRP_DLP_30	6345.48	401.7	284.0	6334.18	458.0	435.9	6321.50	425.2	392.6	6329.88	475.2	408.9	6321.55	758.6	751.6	6313.39	1534.6	1020.0
HVRP_DLP_31	4123.83	584.2	167.2	4113.03	585.0	481.7	4116.72	645.5	413.0	4105.53	636.8	236.9	4110.18	819.1	101.2	4091.52	1575.2	1311.7
HVRP_DLP_34	5814.64	643.2	590.5	5823.80	631.0	487.7	5826.49	662.0	658.2	5786.73	711.9	153.6	5758.89	846.8	524.5	5768.76	1497.0	1201.5
HVRP_DLP_40	11211.61	711.3	476.8	11249.86	898.4	821.3	11186.59	924.1	716.7	11174.83	906.2	442.1	11148.07	1066.7	1032.6	11133.50	2433.8	922.5
HVRP_DLP_41	7633.72	724.9	633.4	7667.16	693.0	311.2	7647.67	660.6	658.8	7630.47	795.4	421.5	7651.70	884.0	305.3	7616.17	1788.1	1186.0
HVRP_DLP_47	16224.87	557.0	506.7	16218.32	703.9	602.2	16234.39	562.0	560.7	16210.56	615.8	454.1	16214.37	817.2	417.2	16206.14	1518.6	1195.0
HVRP_DLP_48	21370.88	531.6	346.8	21396.99	580.9	281.8	21359.57	548.8	130.7	21327.82	581.4	95.7	21318.04	775.1	722.1	21329.71	1496.4	207.4
HVRP_DLP_51	7798.40	717.0	678.8	7786.78	632.8	408.5	7794.67	629.2	578.4	7785.21	700.5	406.5	7774.52	824.6	487.5	7721.47	1506.2	545.1
HVRP_DLP_53	6435.24	636.3	522.0	6447.03	696.6	562.4	6455.53	629.8	478.4	6434.83	725.6	228.1	6435.24	858.4	461.8	6434.83	1549.4	773.4
HVRP_DLP_60	17037.39	665.3	614.9	17059.83	668.7	382.2	17079.64	659.3	429.7	17045.33	724.2	419.5	17045.33	882.0	322.2	17085.82	1543.0	608.0
HVRP_DLP_61	18681.83	500.7	294.3	18636.30	504.6	173.4	18636.30	471.1	160.8	18709.70	585.7	437.0	18580.30	768.2	63.2	18580.30	1487.5	1019.1
HVRP_DLP_66	13586.86	435.4	233.9	13412.58	523.8	85.2	13442.50	529.4	354.4	13383.27	620.3	591.7	13414.20	802.1	659.2	13301.34	1487.6	1449.8
HVRP_DLP_68	9174.21	656.6	595.7	9098.51	668.1	570.3	9114.35	628.4	501.2	8976.53	691.1	529.2	8988.82	848.2	812.4	9001.50	1478.0	1391.6
HVRP_DLP_73	10219.30	765.9	544.5	10241.86	843.1	441.5	10229.44	775.2	503.8	10212.69	799.6	443.8	10208.72	946.8	317.6	10206.05	1494.4	1490.8
HVRP_DLP_74	11695.40	626.5	299.6	11685.95	650.5	129.0	11634.84	659.5	592.1	11633.64	686.7	650.1	11614.37	856.7	771.4	11608.54	1513.4	1426.0
HVRP_DLP_79	7345.99	725.0	668.2	7273.67	762.3	756.1	7277.81	781.2	676.3	7262.91	756.0	244.0	7262.91	916.7	67.3	7259.54	1529.2	599.4
HVRP_DLP_81	1824.76	560.4	226.2	1824.76	651.7	19.5	1824.76	697.8	439.8	1824.68	739.2	50.2	1824.68	1026.5	281.0	1819.11	1715.5	503.6
HVRP_DLP_83	10027.91	598.5	550.4	10042.64	613.8	589.9	10019.15	617.6	561.3	10021.57	683.6	537.5	10019.15	867.0	732.9	10019.15	1553.1	110.7
HVRP_DLP_84	7266.08	495.6	456.7	7266.08	592.3	62.0	7227.88	510.5	161.6	7229.48	1158.3	290.2	7227.88	1261.2	502.9	7227.88	2415.7	281.0
HVRP_DLP_85	8882.36	839.9	386.6	8864.67	950.7	551.9	8854.33	1243.5	947.1	8819.22	1564.9	859.9	8779.76	1180.3	1038.9	8815.20	6755.3	930.2
HVRP_DLP_87	3753.87	437.0	236.1	3753.87	469.4	20.5	3753.87	446.8	71.7	3753.87	487.8	38.1	3753.87	834.4	25.6	3753.87	1631.2	67.3
HVRP_DLP_88	12423.76	1025.7	951.0	12452.11	997.9	890.9	12402.85	949.8	896.4	12420.81	1069.0	482.2	12415.30	1255.0	986.5	12436.22	1723.3	110.8
HVRP_DLP_89	7136.36	680.5	344.1	7114.72	649.5	442.5	7112.05	642.8	364.4	7106.84	770.9	652.7	7109.06	894.5	729.1	7109.65	1488.4	800.2
HVRP_DLP_90	2359.42	344.6	25.1	2357.68	404.0	193.2	2347.81	368.9	101.8	2348.55	484.8	234.1	2346.43	870.5	378.3	2346.43	1674.1	344.7
Avg.	8712.51	601.4	427.1	8704.62	660.2	413.8	8696.67	637.5	429.8	8684.73	735.7	389.8	8675.00	906.6	498.4	8669.60	1750.9	794.9

Table 9.12 DLP_HVRP_3 (large scale instances with a number of nodes from 150 to 200)

	1 thread			2 threads			4 threads			8 threads			16 threads			32 threads		
	Cost	TT	T*	Cost	TT	T*	Cost	TT	T*	Cost	TT	T*	Cost	TT	T*	Cost	TT	T*
HVRP_DLP_02	12306.47	1029.9	177.1	12116.07	1002.7	442.8	12245.68	1109.8	731.3	12096.78	1232.0	825.5	11982.15	1362.8	1185.4	12089.60	1617.6	1145.5
HVRP_DLP_04	11313.77	1010.5	101.0	11295.90	1121.4	986.2	11084.51	1267.7	910.0	11267.20	1187.7	425.2	11104.18	1371.2	907.7	11074.77	1608.3	861.7
HVRP_DLP_09	7653.11	715.4	553.6	7705.06	740.41	335.3	7681.72	885.6	829.0	7656.16	847.6	669.5	7633.90	1090.1	263.9	7619.19	1675.2	1097.2
HVRP_DLP_14	5672.28	917.0	824.1	5693.09	771.8	744.2	5644.98	903.0	601.6	5648.92	941.5	661.9	5670.52	985.2	886.1	5658.26	1556.0	1333.6
HVRP_DLP_15	8296.74	1089.0	1044.8	8290.65	1010.8	934.6	8236.40	1135.5	1078.5	8249.72	1178.6	924.5	8246.44	1298.0	1147.0	8251.40	1658.3	1320.3
HVRP_DLP_24	9194.39	821.1	810.6	9155.78	828.8	776.7	9156.73	899.1	857.6	9146.54	910.7	901.9	9101.47	1086.8	1014.6	9103.69	1553.8	1539.3
HVRP_DLP_29	9180.78	679.2	663.2	9144.50	677.1	286.6	9144.11	711.3	377.4	9143.69	848.0	531.5	9147.23	951.6	497.0	9143.69	1506.3	402.7
HVRP_DLP_33	9483.53	1075.3	260.7	9443.30	1014.7	591.3	9488.53	1137.1	901.6	9517.93	1157.9	1080.7	9421.01	1178.3	697.4	9422.58	1620.7	1537.9
HVRP_DLP_35	9681.97	1102.1	1008.3	9623.06	1120.6	807.7	9622.31	1229.1	876.6	9641.28	1177.5	380.2	9601.03	2115.2	2005.4	9574.71	2061.7	1846.0
HVRP_DLP_37	6905.30	728.6	554.5	6904.01	729.1	632.9	6887.17	772.4	770.5	6882.27	807.7	644.3	6868.27	921.6	416.9	6858.23	1496.0	1221.8
HVRP_DLP_42	12092.69	782.5	741.5	11912.60	865.6	93.7	11809.31	883.8	655.6	11904.36	1007.5	619.8	11624.80	1140.7	1097.9	11658.94	1582.3	1511.2
HVRP_DLP_44	12480.47	891.8	597.4	12373.56	969.1	954.7	12350.71	1059.0	1012.4	12315.75	1061.9	583.2	12284.26	1177.9	394.5	12197.46	1654.4	1112.6
HVRP_DLP_45	10615.94	1106.8	1040.4	10587.68	1052.7	1036.3	10521.15	1238.2	1044.0	10537.58	1204.2	1055.4	10484.23	1371.7	661.0	10530.81	1586.3	838.1
HVRP_DLP_50	12453.04	722.6	518.1	12635.68	742.5	655.0	12481.23	830.0	816.5	12426.01	749.1	647.9	12459.49	928.0	846.2	12374.04	1528.2	589.0
HVRP_DLP_54	12505.92	544.7	239.4	11946.77	587.4	409.2	12030.81	633.9	172.7	11780.97	673.1	407.0	11809.07	908.1	463.7	11558.79	1517.5	916.1
HVRP_DLP_56	31228.55	770.0	741.5	31270.66	737.2	590.4	31152.02	768.4	616.6	31159.61	840.1	455.5	31129.76	968.6	519.4	31111.42	1500.1	602.4
HVRP_DLP_57	45103.90	901.8	869.9	45033.11	880.1	624.7	44837.58	1027.6	947.4	44818.52	1009.4	923.9	44818.18	1126.1	1123.5	44832.06	1691.7	1567.8
HVRP_DLP_59	14423.32	1100.3	750.8	14316.43	1038.5	895.3	14309.48	1230.5	366.5	14352.78	1211.5	1178.5	14374.43	1335.3	541.6	14378.29	1597.5	772.0
HVRP_DLP_63	20197.82	1020.7	1010.6	20136.76	1020.8	850.8	20150.83	981.6	660.4	20000.29	1072.7	1021.7	20075.70	1184.2	654.5	19951.76	1607.9	956.8
HVRP_DLP_64	19240.87	889.9	622.9	19232.11	791.3	377.9	19210.39	864.6	336.1	19202.47	865.3	469.1	19252.24	937.0	729.0	19202.47	1569.3	784.0
HVRP_DLP_67	11851.79	644.0	133.8	11906.52	713.5	509.9	11584.19	785.6	633.3	11555.73	828.2	268.1	11603.74	935.8	387.2	11483.25	1517.8	1297.1
HVRP_DLP_69	9309.46	908.4	712.5	9289.26	836.8	665.4	9222.10	897.4	676.6	9206.06	929.3	838.1	9162.78	1019.3	908.7	9198.22	1494.3	1189.2
HVRP_DLP_71	9936.25	994.6	817.8	9885.10	981.8	890.1	9891.70	1007.8	568.3	9892.86	1173.5	313.7	9870.22	1252.5	734.3	9873.41	1596.9	1557.6
HVRP_DLP_72	5989.28	1114.2	1096.7	5989.80	973.7	885.0	5975.63	1027.1	640.4	5923.25	1097.7	1004.6	5905.58	1183.9	1169.3	5920.67	1538.7	824.9
HVRP_DLP_76	12111.78	970.2	697.6	12132.31	920.5	789.2	12032.43	1037.1	564.2	12018.26	1044.6	727.5	12076.93	1179.5	914.9	12033.66	1538.1	371.6
HVRP_DLP_77	7005.73	925.2	676.6	6930.44	925.9	923.0	6960.52	957.4	943.8	6973.85	964.4	917.6	6952.56	1063.0	930.3	6947.52	1534.7	1127.1
HVRP_DLP_78	7054.8	962.4	792.6	7057.45	953.1	871.7	7058.02	980.8	705.1	7047.56	994.6	891.6	7040.79	1147.2	149.1	7035.01	1545.8	865.6
HVRP_DLP_80	6849.63	930.8	727.2	6864.41	887.7	885.1	6842.34	909.4	653.9	6839.92	939.1	593.7	6849.29	1127.1	811.6	6816.89	1568.4	478.7
HVRP_DLP_86	9065.30	714.6	578.9	9058.65	769.5	764.0	9060.11	740.6	326.5	9056.19	934.2	797.2	9051.72	1024.6	219.9	9030.68	1517.0	832.5
HVRP_DLP_91	6403.09	885.5	620.9	6413.77	934.1	900.8	6403.08	1034.0	775.3	6379.46	1040.4	817.4	6381.77	1224.3	677.9	6377.48	1649.4	1118.8
HVRP_DLP_95	6242.97	786.3	585.0	6244.13	734.5	381.5	6235.22	838.9	784.8	6181.60	819.1	273.2	6232.80	1018.0	1009.6	6233.21	1605.2	711.0
Avg.	11995.19	894.7	663.5	11954.47	881.7	693.3	11913.26	960.8	704.3	11897.53	991.9	704.8	11877.95	1148.8	773.1	11856.20	1590.2	1042.9

Table 9.13 DLP_HVRP_4 (strongly large scale instances with a number of nodes greater than 200)

	1 thread			2 threads			4 threads			8 threads			16 threads			32 threads		
	Cost	TT	T*	Cost	TT	T*	Cost	TT	T*	Cost	TT	T*	Cost	TT	T*	Cost	TT	T*
HVRP_DLP_18	9792.66	1569.2	1518.4	9752.84	1337.2	1141.7	9768.43	1483.3	1238.9	9736.63	1830.2	1710.7	9704.38	1653.0	1137.0	9702.75	1961.2	1903.1
HVRP_DLP_19	11895.84	1207.8	1141.3	11909.97	1087.0	907.0	11869.61	1207.5	1072.4	11750.93	1300.3	1156.4	11839.38	1560.3	1534.6	11702.77	1692.5	1582.5
HVRP_DLP_22	13227.74	1486.1	741.8	13182.24	1531.16	1427.8	13120.01	1645.5	642.9	13115.59	1626.4	1293.3	13068.03	1784.1	996.3	13172.25	2548.0	2526.1
HVRP_DLP_23	7786.12	1326.5	780.5	7786.54	1353.7	1171.4	7804.05	1424.1	1060.6	7782.44	1513.4	875.4	7750.27	1755.8	449.6	7763.12	1813.5	1628.6
HVRP_DLP_27	8575.58	917.0	685.4	8535.36	989.2	947.7	8511.74	1063.5	1049.8	8484.22	1128.1	535.2	8485.90	1352.9	1261.3	8469.19	1685.6	1592.5
HVRP_DLP_32	9510.72	1444.6	1199.0	9464.42	1485.8	1363.6	9493.27	1557.2	1096.9	9417.62	1565.6	1090.8	9478.74	1848.8	1340.9	9430.57	1981.7	1697.4
HVRP_DLP_38	11338.71	941.0	852.7	11275.69	1011.5	928.1	11253.47	1030.4	906.1	11327.79	1059.2	1026.6	11242.95	1200.0	1194.2	11299.01	1556.9	1424.3
HVRP_DLP_46	25299.34	877.2	836.5	25213.41	984.7	978.2	24734.47	1326.4	1094.9	24741.84	1389.3	1331.8	24809.99	1503.0	1333.2	24674.26	1808.5	1585.3
HVRP_DLP_49	16726.72	820.9	701.7	16596.06	1046.6	1023.8	16530.99	969.4	726.0	16435.91	1047.4	756.8	16377.69	1285.2	1172.9	16394.64	1621.9	1225.8
HVRP_DLP_58	23654.21	1524.6	509.1	23777.31	1448.6	1223.9	23397.76	1634.8	1338.9	23400.96	1643.4	1634.0	23538.34	1764.1	1718.9	23492.93	1805.2	894.4
HVRP_DLP_62	23595.34	1162.9	1125.4	23549.89	1372.8	1288.7	23268.30	1263.3	1018.0	23415.10	1525.4	1060.6	23433.62	1820.0	1161.8	23149.61	1692.4	1075.2
HVRP_DLP_65	13117.99	1076.8	697.4	13072.92	1121.2	354.0	13099.59	1136.3	559.1	13066.55	1210.3	1185.7	13056.53	1435.8	753.9	13134.81	1634.4	461.5
Avg.	14543.41	1196.2	899.1	14509.72	1230.8	1063.0	14404.31	1311.8	983.7	14389.63	1403.3	1138.1	14398.82	1580.2	1171.2	14365.49	1816.8	1466.4

Part III
Combining Metaheuristics with Exact Methods from Mathematical Programming Approaches

Chapter 10
The Heuristic (Dark) Side of MIP Solvers

Andrea Lodi

Abstract. The evolution of Mixed-Integer Linear Programming (MIP) solvers has reached a very stable and effective level in which solving real-world problems is possible. However, the computed solution is not always the optimal one also because optimality is often not of primary interest for day-by-day users. We show some structural characteristics of MIP solvers and of computation for MIP problems that reveal the heuristic nature of the solvers. Moreover, we discuss the key components of MIP solvers with special emphasis on the role of heuristic decisions within the solution process. Finally, we present MIP solvers as "open" frameworks whose flexibility can be exploited to devise sophisticated hybrid algorithms.

10.1 Introduction

We consider a general *Mixed Integer Linear Programming* problem (MIP) in the form

$$\min\{c^T x : Ax \geq b, x \geq 0,\ x_j \text{ integer},\ \forall j \in \mathscr{I}\} \tag{10.1}$$

where $\mathscr{I}$ is the set of integer-constrained variables.

We assume that matrix A does not have a "clean" special structure to be exploited through combinatorial algorithms, either exact or heuristic (and metaheuristic), but instead is a collection of heterogenous groups of constraints, as it is often the case in real-world applications. Of course, there are general-purpose heuristic approaches in the literature for these problems that are not based on mathematical programming techniques. However, in this paper we are interested in the case in which problem

Andrea Lodi
DEIS, University of Bologna, and
IBM-UniBo Mathematical Optimization Center of Excellence
Viale Risorgimento 2 – I-40136, Bologna, Italy
e-mail: andrea.lodi@unibo.it

E.-G. Talbi (Ed.): Hybrid Metaheuristics, SCI 434, pp. 273–284.
springerlink.com

(10.1) is solved through a general-purpose MIP solver , i.e., through branch and bound with bounds computed by iteratively solving the *Linear Programming* (LP) relaxations, via a general-purpose LP solver.

The evolution of MIP solvers has reached a very stable and effective level in which the solution of real-world problems is possible. The progress over the first 50 years has been impressive and it is discussed, for example, in [2, 15, 18]. For a recent survey on MIP software the reader is referred to Linderoth and Lodi [16].

This general-purpose MIP setting is, *apparently*, far away from what one thinks belongs to the context of (*i*) metaheuristic algorithms and (*ii*) hybrid approaches. Indeed, common sense says an MIP solver implements a pure and exact algorithm, in contrast to hybrid and heuristic ones. Thus, MIP solvers are mostly conceived as exact *black boxes*, especially within the Metaheuristic community. This might be the reason of a certain diffidence for the MIP solvers and, in general, for the MIP technology, which is reflected by many papers in the Metaheuristic literature.

The goal of this short paper is to confute this viewpoint by showing that

1. MIP solvers are used for a large portion as heuristics;
2. MIP solvers are heuristic in nature;
3. the computation for $\mathcal{NP}$-hard problems is intrinsically heuristic;
4. benchmarking is by design heuristic;
5. heuristic decisions and techniques are hidden everywhere in MIP solvers.

In other words, we aim at presenting MIP solvers as "open" frameworks whose flexibility can be exploited in many ways. On the one side, we will show that it is largely possible nowadays to develop all sorts of algorithms within an MIP solver framework. On the other hand, the intrinsic presence of heuristic components within the solvers suggests that ideas developed in contexts different from classical (exact) Mixed Integer Programming are already and could be more and more fruitfully incorporated.

The remainder of the paper is organized as follows. In Section 10.2 we will discuss some natural characteristics of MIP solvers and of computation for Mixed Integer Linear Programming problems that reveal the heuristic nature of the solvers. In Section 10.3 we will present the main components of MIP solvers and we will show that in a number of crucial points the implemented algorithms heavily rely on heuristic reasonings. Finally, in Section 10.4 we will outline two research directions: (*i*) the use of MIP solvers as open frameworks for algorithmic development, and (*ii*) the hybridization of the MIP solvers through ideas coming from different research areas.

10.2 The Heuristic Nature of MIP Solvers

As anticipated in the introduction, modern MIP solvers are brilliant, stable and effective pieces of code in which a flexible modeling is accompanied by powerful algorithms and sophisticated software engineering. Most of the techniques implemented in the solvers originated from academia, where the emphasis is mainly on theory and methodology. However, the move from theory to practice requires some "painful" steps that make the solvers to appear much less clean than expected. In the next two sections we will discuss a few facts, from trivial to slightly less so, which highlight the heuristic nature of the solvers.

10.2.1 Some Trivial Facts

The first trivial reason for MIP solvers being "heuristic" is actually associated with the way they are used in general. Indeed, the classical user of an MIP solver enforces *limits* to the computation. The most classical of those is the *time limit*, i.e., the user fixes a maximum amount of computing time he/she is willing to spend. Another (less frequent) limit can be in the number of branch-and-bound nodes, or the computation can be halted once the gap between the lower and the upper bounds[1] has decreased until a certain threshold or a prefixed number of feasible solutions (often only one) has been found within the enumeration tree. The reason for imposing these limits is that the computational resources are limited and sometimes solving the problems does not require/allow searching for the optimal solution. This is true in several circumstances two of which are.

- The overall decision system is highly complex and it has been split *a priori* into pieces. In such a case it is not necessarily true that the optimal solution of a specific piece if used as starting point for a subsequent optimization/simulation phase would guarantee to provide an "optimal" decision for the complete system. A classical example of complex systems of this type arises in public transportation where vehicle and crew scheduling must be solved, theoretically together but in practice one after the other [7].
- The problem must be solved in (extremely) short computing times as part of a quasi-online system like, for example, the transmission of data packets in wireless telecommunications [19].

The second trivial reason that makes MIP solvers somehow heuristic is that they work with *tolerances*. More precisely, essentially all MIP solvers are based on floating-point arithmetic and use tolerances to check solutions for feasibility and to decide on optimality, i.e., to fathom nodes in the branch-and-bound tree. On the one hand, feasibility is tested by MIP solvers with absolute tolerances for the integrality constraints and relative ones for linear constraints. On the other side, nodes with relative percentage gap between the incumbent solution and the lower bound smaller than a certain threshold value are disregarded. A commonly-used default value for such a threshold is 0.01%, a value that, for specific applications, might be far from acceptable.

[1] Note that for minimization problems of type (10.1), the lower bound is established by the minimum objective of the unexplored nodes in the branch-and-bound tree, while the upper bound, or incumbent solution value, is the value of the best (smallest) feasible solution encountered in the enumeration.

Of course, guaranteeing true feasibility/optimality comes with a price in terms of computing time and MIP solvers have to establish a reasonable compromise on the matter. The tolerance issue is deeply discussed in the work by Koch et al. [9], the recent paper providing the fourth version of the library of MIP instances commonly used for benchmarking. In addition to the library, the paper also provides, for the first time, scripts to run automated tests in a predefined environment and a solution checker to test the accuracy of the solutions using exact arithmetic. Examples in [9] show that solutions provided by MIP solvers might fail a simple test in which (*i*) one fixes all integer-constrained variables from a reported solution to the closest integer value[2] and (*ii*) recomputes the value of the continuous variables by solving the associated LP with exact arithmetic. This fail does not happen because the MIP solvers are mistaken. The reported solutions are (most likely) "feasible": not within the region described by the input file, but within that obtained by reading the problem and introducing tolerances.

Finally, to floating-point computation is associated a very small amount of error incurred by a single operation, error that can accumulate and propagate significantly for algorithms requiring millions of operations. This is deeply discussed in the survey by Goldberg [14].

10.2.2 *Less Trivial Facts*

It is well known that Mixed Integer Linear Programming is $\mathcal{NP}$-hard [6], thus, by definition, there is always a polynomial path to the optimal solution in the search tree, but unless $\mathcal{P} = \mathcal{NP}$, in the worst case, the path followed by any enumerative algorithm will be exponentially long. In other words, branch-and-bound algorithms *heuristically* explore the search tree and such an exploration can be unlucky. This is because of "bad" algorithmic decisions: an ineffective choice of the first branching variable potentially leads to a search tree twice as big. (The algorithmic side of MIP, i.e., the way MIP solvers explore the solution space taking those decisions will be discussed in Section 10.3.3.)

However, ineffective algorithmic decisions are not the only reason for the computational effort to become high. Namely, while running computational experiments, one often experiences a change in performance for the same problem (or problems in the same "family") created by a change in the solver or in the computational environment that seem performance neutral. This situation, quite commonly observed, has been formalized for the first time by Danna [4] who reported the following enlightening example.

Example 10.1. CPLEX 11 solves the well-known MIPLIB 2003 [21] instance "10teams" in 0 branch-and-bound nodes and 2,731 Simplex iterations in a computer equipped by the Linux Operating System. The same version of CPLEX, i.e., exactly

[2] Because of the mentioned tolerances, a solver might return a value for an integer-constrained variable differing from a true integer of, say, 10^{-8}.

the same algorithm and computing code, needs 1,426 branch-and-bound nodes and 122,948 Simplex iterations in a computer running the AIX Operating System. □

This phenomenon has been referred to as *performance variability* and has been systematically discussed in Section 5 of Koch et al. [9]. The variability observed in Example 10.1 is associated with the difference of the computing platform, which, however, does not explain it. This unstable behavior is intrinsically related to imperfect tie-breaking, i.e., to the fact that among a set of equivalent possibilities the algorithm has to take a sort of "arbitrary" decision. Sets of this type are computed throughout the entire solution process and vary in size and rank depending on apparently unrelated reasons, like the mentioned change in computing environment (due most probably to minor differences in floating-point computation) or the order in which pieces of information, e.g., cutting planes, are stored.

Although some of the sources of variability are known, there is no cure yet to avoid it. (An example will be discussed in Section 10.3.2.) Thus, it becomes clear that *benchmarking* MIP solvers is a very delicate activity. Specifically, the testbeds of MIP solvers, in particular the commercial ones, are composed of thousands of instances. These instances are classified into categories: from very easy, to very hard. Any new algorithmic idea is tested on the entire testbed and in order to "make it into the solver" it must (*i*) improve on the subset of problems to which the idea applies, and (*ii*) not deteriorate (too much) on the rest of the problems, generally the easy ones. Because of MIP $\mathscr{NP}$-hardness, it is theoretically difficult to recognize if a problem is "easy", and, more generally, deciding if an idea is useful for a particular class of instances. Then, such an algorithmic idea must be heuristically "weakened" to accomplish simultaneously the two given goals. Thus, benchmarking is complex and heuristic, and, of course, the discussed variability strongly increase the difficulty of the process.

10.3 Key Features of MIP Solvers

The current generation of MIP solvers incorporates key ideas developed continuously during the first 50 years of Integer Programming [8]. Precisely, in this section we will briefly present[3] the basic components of an MIP solver, namely, preprocessing, cutting plane generation, branching, and primal heuristics. The emphasis is on showing the key role of heuristic decisions within each of these components, often in contexts rather unexpected.

10.3.1 Preprocessing

In the preprocessing phase (often called presolve) an MIP solver tries to detect certain changes in the input that will probably lead to a better performance of the so-

[3] For more details, the reader is referred to Achterberg [1] and Lodi [18].

lution process. In general, this does not include removing feasible solutions, even if that might happen in some special cases, as for example due to symmetry breaking reductions, see, e.g., Margot [20].

Preprocessing is especially important for models originated from real-world applications and/or created by using modeling languages. Indeed, it is often the case that MIP models of this type can be improved with respect to their initial formulation by either removing redundant information (variables or constraints) or strengthening the variable bounds generally by exploiting the integrality of variables in $\mathscr{I}$.

Nowadays MIP solvers have the capability of "cleaning up" the models so as to create a presolved version of the original instance to which the components discussed in the next sections are then applied. Many of the techniques used to create such a prosolved instance are called in the *Constraint Programming* (CP) context *propagation* algorithms [22]. Specifically, CP models are built through so-called *global constraints*, i.e., combinatorial objects defining a portion of the feasible region and able to check feasibility of an assignment of values to variables. The intersection of all global constraints defines the feasible region as pictorially shown in Figure 10.1. Moreover, and more importantly here, a global constraint contains an algorithm that prunes (filters) values from the variable domains so as to reduce as much as possible the search space. In this way, large and combinatorially well-defined structures are effectively exploited.

MIP technology does not explicitly include global constraints, thus, some sort of propagation is applied by *locally* comparing and analyzing constraints and variables, a structurally heuristic process highly dependent on the way the model has been written. Indeed, random permutations of rows/columns of the MIP generally lead to a performance deterioration of the solvers, mostly because of reduced preprocessing effectiveness.

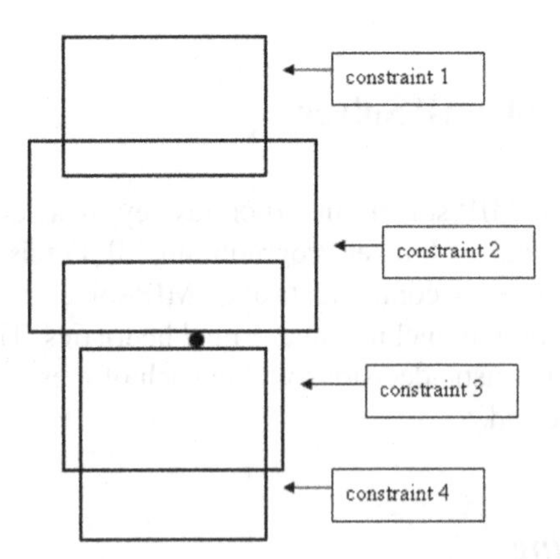

Fig. 10.1 Constraint Programming modeling through global constraints

10.3.2 Cutting Plane Generation

The chances of solving an MIP by a general-purpose solver are directly proportional to the quality of its LP relaxation, i.e., the problem obtained by removing (relaxing) the integrality requirements on the variables in $\mathscr{I}$. However, such a relaxation, especially for real-world problems, might be not tight with respect to the convex hull of mixed integer solutions. Therefore, MIP solvers apply a rather intense phase of strengthening based on cutting plane generation. A tighter relaxation can be obtained by adding new linear inequalities (cutting planes, or cuts, for short) obtained by solving the so-called *Separation problem*:

> Given a feasible solution x^* of the LP relaxation that is not feasible for the MIP (10.1), find a linear inequality $\alpha^T x \geq \alpha_0$ that is valid for (10.1), i.e., satisfied by all feasible solutions $\bar{x}$ of the system (10.1), while it is violated by x^*, i.e., $\alpha^T x^* < \alpha_0$.

The overall strengthening is achieved by (*i*) solving the LP relaxation, (*ii*) solving the separation problem, (*iii*) amending the LP relaxation with the obtained cut(s), and iterate. The effect of a cut in the LP relaxation of an MIP is shown in Figure 10.2, where the blue arrow indicates the optimal solution x^* of the LP relaxation at a given strengthening iteration, and the red line represents the cut. The role of cutting planes in MIP technology is crucial: without the iterative strengthening described, current enumerative schemes would fail in solving difficult MIP instances.

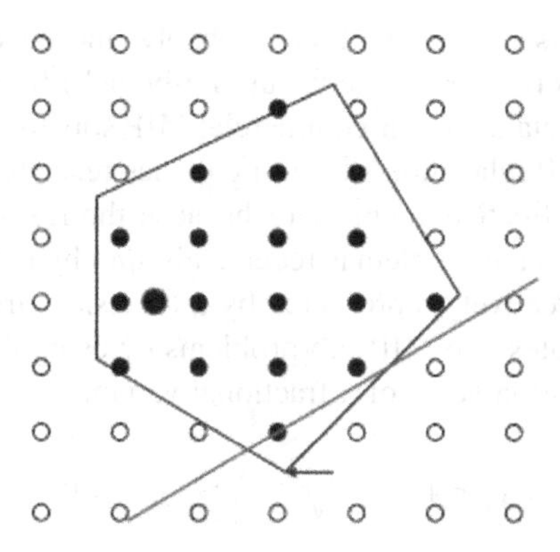

Fig. 10.2 Strengthening the LP relaxation by cutting planes

The arsenal of separation algorithms in the MIP solvers has been continuously extended over the years, and a large group[4] of cuts, namely, Chvátal-Gomory cuts, Gomory mixed integer cuts, mixed integer rounding cuts, $\{0, \frac{1}{2}\}$ cuts, lift-and-project cuts, and split cuts, are derived essentially in the same way. Precisely, all these inequalities are obtained in two steps:

[4] This group of cuts is presented in a brilliant and unified way by Cornuéjols [3].

1. *heuristically* aggregating the entire MIP into a mixed integer set of (only!) one row, often referred to as *base inequality*, and
2. applying a cut generation procedure to such a mixed integer set, based on a disjunctive argument that exploits integrality.

The heuristic aggregation of Step 1 above is sometimes an implicit byproduct of a sophisticated procedure, as in the case of Gomory mixed integer cuts, where the base inequalities are the rows of the optimal Simplex tableau. Nevertheless, the above procedure is intrinsically heuristic, and it is not the only heuristic decision in the context of cutting plane generation. One of the hardest to settle among these decisions is *cut selection*, i.e., the identification of which cuts, among all those separated within a cutting plane iteration, should be added to the next LP relaxation. Another very intriguing decision recently emerged in the context of performance variability (see Section 10.2.2), concerns the selection of a specific basis in the optimal face of any LP relaxation. More precisely, it has been observed [9] that a crucial source of variability is the basis to which the cutting plane phase is applied among all equivalent ones[5] in the optimal face. Currently, the Simplex algorithm returns one of them at random, and a natural question is about the characterization of a basis that, at least heuristically, would lead to a better cutting plane performance.

10.3.3 Sophisticated Branching Strategies

The branching phase is by construction (probably) the most delicate part of the branch-and-cut (or, more appropriately, cut-and-branch) framework currently implemented by commercial and non-commercial MIP solvers. Essentially, when the strengthening of the LP relaxation by cutting planes reaches a steady point, either because cuts are not effective anymore or because the resulting LP starts having numerical troubles, then the problem is recursively split by fixing variables, and implicit enumeration, effectively represented by a (decision) tree, is performed. The splitting generally creates two MIP sub-problems by using the rounding of the solution of the LP relaxation value of a fractional variable, say x_j, constrained to be integral ($j \in \mathscr{I}$)

$$x_j \le \lfloor x_j^* \rfloor \quad \bigvee \quad x_j \ge \lfloor x_j^* \rfloor + 1. \tag{10.2}$$

The idea is to split the solution space into pieces that are easier to explore because some of the decisions have already been taken. The selection of the variable x_j (whose domain must be split) among all those in $\mathscr{I}$ that are fractional at the current decision point is not an issue for the theoretical convergence of the enumerative algorithm[6]. However, in practice, decisions are dramatically *not equivalent*: branching

[5] The existence of several optimal bases, i.e., an optimal face, for an LP is likely to happen especially in case of degenerate LPs.

[6] Note that the convergence and correctness of the algorithm does not require branching on variables, but rather just branching on discrete expressions or objects in the model in a way that splits the domains in a mutually exclusive way.

decisions of type (10.2) are "good" if the created sub-MIPs are indeed easier, i.e., finding their optimal mixed integer solution is "fast". In other words, the only strict measure of the quality of a branching decision is the computational effort required to solve the sub-MIPs it originates, i.e., the size of the search sub-trees. This measure is *a posteriori*, and, because of the $\mathscr{N}\mathscr{P}$-hardness of MIP discussed in Section 10.2.2, there is very little hope it can be theoretically anticipated. Thus, heuristic approximations of that measure must be used and there is a flourishing literature in the area (see, e.g., [17, 18]).

Almost all sophisticated techniques used to compute an *a priori* proxy of the above quality measure of effectiveness of a branching decision are variations of the so-called *strong branching* (see, e.g., Linderoth and Savelsbergh [17]). In its heaviest form strong branching solves (to optimality) the two LP relaxations associated with the decision (10.2), for all possible $j \in \mathscr{I}$ such that x_j^* is fractional, and selects the best variable x_j to branch on according to (slightly) different scores. The underlining idea is that the variable for which the lower bound of both sub-MIPs has increased the most[7] with respect to that of the father (sub-)MIP is the most promising one, because the smaller gap between lower and upper bound in the sub-MIPs is likely to be closed "faster".

However, the computational effort required by the "full" strong branching would be way too high, and MIP solvers adopt two heuristics:

- only a subset of the variables is considered (candidate set), and
- each LP is not solved to optimality but within a given Simplex iteration limit.

In other words, not only heuristics are needed to overcome *theoretically* hard tasks like the lack of an *a priori* measure of the quality of a branching decision, but MIP solvers need to find a heuristic compromise for *practical* tasks too, specifically, in the way the ideas are implemented.

10.3.4 Primal Heuristics

This is clearly the area in which it is trivial to recognize the impact of heuristics within MIP solvers. The methods implemented in the solvers go from the so-called *rounding heuristics*, where a fractional solution obtained from the LP relaxation is converted into a mixed integer one trying to preserve the feasibility with respect to the linear constraints, to local search and metaheuristic techniques to improve feasible solutions (see, Fischetti and Lodi [12] for a recent survey).

From the one side, in the last decade we have seen a tremendous improvement in the capability of primal[8] heuristics to find very good (nearly-optimal) solutions

[7] Generally, the difference of the bounds in the two sub-MIPs is multiplied in the score associated with the variable.

[8] It might be useful to recall that the term "primal" is used in MIP context to indicate the pessimistic approximation of the optimal solution value (corresponding to a feasible solution), in opposition to the term "dual" used for the optimistic approximation.

early in the tree. On the other hand, a meaningful experiment by Achterberg [1] has shown that the knowledge of the optimal solution from the very beginning of the search improves the running time of an MIP solver, on average, *only* by a factor of 2.

In other words, heuristics dramatically impact on the user perception of the quality of a solver, although the current algorithmic framework, intrinsically dual, does not benefit much from accurate primal (i.e., feasible) solutions to prove optimality. Nevertheless, a solver is perceived to be weak for an instance if it does not "fill" the log column reporting the primal bound almost immediately, one reason of this perception being the sometimes minor need of the optimal solution discussed at the beginning of Section 10.2.1.

However, the most surprising impact of primal heuristics in the MIP solvers is on the so-called *MIPping* approach [10, 13], which consists in internally invoking within the algorithmic process the MIP solver itself to "solve" other MIPs, so as to find good feasible solutions [5] and/or to generate cutting planes [11].

The "surprise" comes from the fact that the MIPs solved internally in a black-box fashion, i.e., without exploiting any *a priori* knowledge, are theoretically as difficult as the original MIP for which one is "simply" trying to improve the incumbent solution or strengthen the LP relaxation. In other words, at first, MIPping appears way too expensive computationally to be effective. In fact, the MIP technology has reached such an effective and stable performance to allow MIPping (especially if used with a lot of care), mainly thanks to the integration in the solvers of sophisticated primal heuristics able to generally produce feasible solutions quickly. Indeed, none of those MIPs needs to be solved to optimality for providing useful pieces of information: they are constructed in such a way that any of their feasible solutions is either an improving primal solution or a cutting plane for the original instance. Finally, the solution process of those additional MIPs can be controlled safely, for example, by limiting the number of explored branch-and-bound nodes.

10.4 MIP Solvers, Metaheuristics and Hybrid Algorithms

We have shown that *not only* MIP solvers (*i*) are often used as heuristics[9], (*ii*) are developed and tested using heuristic criteria, and (*iii*) use heuristic decisions in each of their basic components, *but also* they (*iv*) incorporate heuristic algorithms that are not only used to find good solutions but also as building blocks of sophisticated algorithmic strategies like MIPping, and (*v*) benefit from ideas originated in different communities like propagation algorithms from Constraint Programming and neighborhood exploitation from Metaheuristics.

On the one side, this leads us to view MIP solvers as open frameworks for effective and sophisticated algorithmic development. In other words, we believe that it is

[9] MIP solvers are, in several important cases, among the best heuristic approaches even for combinatorial/structured problems, e.g., set covering type problems.

worthwhile trying to use more and more the MIP solvers also for the implementation of dedicated heuristic approaches. Of course, this goes far beyond the simple try to run an MIP solver on a naively constructed model. In fact, an MIP solver can be effectively fed with the knowledge of the problem at hand in many forms, like redundant constraints, symmetry breaking rules, priorities in the variable selection for branching, good neighborhoods to explore. This can be done quite smoothly in some of the MIP solvers, namely those providing the so-called *callback functions*, i.e., pieces of code that allow the flexibility of accommodating the user code for specific algorithmic tasks like cut generation, primal heuristics, branching strategies, etc. Moreover, several of these callbacks allow recovering information from the system in the solution phase so as to favor a better understanding of the algorithm evolution. However, what we are advocating here is the exploitation of an MIP solver (or, to some extent, of the MIP technology) at an intermediate depth, possibly disregarding some more theoretical features like polyhedral analysis or exhaustive enumeration, but working within a framework that allows the user unexpected flexibility and freedom. We believe the discussion so far has shown that such an intermediate level of depth exists, and that it is relatively natural to understand and use.

On the other hand, the previous sections have also shown that MIP solvers are open to ideas originated in different areas, from the classically related ones like Graph Theory to more recent connections in the contexts of Constraint Programming, Artificial Intelligence, Metaheuristics. A lot of potentially very good work remains to be done for effectively integrating those ideas in the solvers, thus leading to sophisticated *hybrid* algorithms.

Acknowledgements. The author is grateful to the current and past IBM ILOG CPLEX team members for endless discussions on many of the topics covered by this paper. A special thank goes to Ed Klotz for sharing with me his deep experience with MIP solvers, and to Sophie Parragh for carefully reading a preliminary version of this paper. I am grateful to the organizers and program committee members of HM 2009 and MIC 2011 (both in Udine, Italy), especially Andrea Schaerf, for inviting me to lecture, thus triggering the thoughts that led to this paper. Finally, thanks to the participants of the two conferences above for their positive and encouraging feedbacks.

References

1. Achterberg, T.: Constraint Integer Programming. PhD thesis, ZIB, Berlin (2007)
2. Bixby, R.E., Fenelon, M., Gu, Z., Rothberg, E., Wonderling, R.: Mixed-Integer Programming: A Progress Report. In: Grötschel, M. (ed.) The Sharpest Cut, pp. 309–326. SIAM (2004)
3. Cornuéjols, G.: Valid inequalities for mixed integer linear programs. Mathematical Programming 112, 3–44 (2008)
4. Danna, E.: Performance variability in mixed integer programming. Talk at MIP (2008), `http://coral.ie.lehigh.edu/~jeff/mip-2008/index.html`
5. Danna, E., Rothberg, E., Le Pape, C.: Exploiting relaxation induced neighborhoods to improve MIP solutions. Mathematical Programming 102, 71–90 (2005)

6. Garey, M.R., Johnson, D.S.: Computers and Intractability: A Guide to the Theory of NP-Completeness. W.H. Freeman (1979)
7. Groot, S.W., Huisman, D.: Vehicle and Crew Scheduling: Solving Large Real-World Instances with an Integrated Approach. In: Hickman, M., Mirchandani, P., Voss, S. (eds.) Computer-aided Systems in Public Transport. Lecture Notes in Economics and Mathematical Systems, pp. 43–56. Springer (2008)
8. Jünger, M., Liebling, T.M., Naddef, D., Nemhauser, G.L., Pulleyblank, W.R., Reinelt, G., Rinaldi, G., Wolsey, L.A. (eds.): 50 Years of Integer Programming 1958-2008. Springer (2009)
9. Koch, T., Achterberg, T., Andersen, E., Bastert, O., Berthold, T., Bixby, R.E., Danna, E., Gamrath, G., Gleixner, A.M., Heinz, S., Lodi, A., Mittelmann, H., Ralphs, T., Salvagnin, D., Steffy, D.E., Wolter, K.: MILPLIB 2010. Mathematical Programming Computation 3, 103–163 (2011)
10. Fischetti, M., Lodi, A.: Local Branching. Mathematical Programming 98, 23–47 (2003)
11. Fischetti, M., Lodi, A.: Optimizing over the first Chvátal closure. Mathematical Programming 110, 3–20 (2007)
12. Fischetti, M., Lodi, A.: Heuristics in Mixed Integer Programming. In: Cochran, J.J. (ed.) Wiley Encyclopedia of Operations Research and Management Science, vol. 3, pp. 2199–2204. Wiley (2011)
13. Fischetti, M., Lodi, A., Salvagnin, D.: Just MIP it! In: Maniezzo, V., Stützle, T., Voss, S. (eds.) MATHEURISTICS: Hybridizing Metaheuristics and Mathematical Programming. Operations Research/Computer Science Interfaces Series, pp. 39–70. Springer (2009)
14. Goldberg, D.: What every computer scientist should know about floating-point arithmetic. ACM Computing Surveys 23, 5–48 (1991)
15. Laundy, R., Perregaard, M., Tavares, G., Tipi, H., Vazacopoulos, A.: Solving hard mixed integer programming problems with Xpress-MP: a MIPLIB 2003 case study. Informs Journal of Computing 21, 304–319 (2009)
16. Linderoth, J.T., Lodi, A.: MILP Software. In: Cochran, J.J. (ed.) Wiley Encyclopedia of Operations Research and Management Science, vol. 5, pp. 3239–3248. Wiley (2011)
17. Linderoth, J.T., Savelsbergh, M.W.P.: A computational study of search strategies for mixed integer programming. Informs Journal on Computing 11, 173–187 (1999)
18. Lodi, A.: MIP computation. In: Jünger, M., Liebling, T.M., Naddef, D., Nemhauser, G.L., Pulleyblank, W.R., Reinelt, G., Rinaldi, G., Wolsey, L.A. (eds.) 50 Years of Integer Programming 1958-2008, pp. 619–645. Springer (2009)
19. Lodi, A., Martello, S., Monaci, M., Cicconetti, C., Lenzini, L., Mingozzi, E., Eklund, C., Moilanen, J.: Efficient two-dimensional packing algorithms for mobile WiMAX. Management Science (2011), doi:10.1287/mnsc.1110.1416
20. Margot, F.: Symmetry in Integer Linear Programming. In: Jünger, M., Liebling, T.M., Naddef, D., Nemhauser, G.L., Pulleyblank, W.R., Reinelt, G., Rinaldi, G., Wolsey, L.A. (eds.) 50 Years of Integer Programming 1958-2008, pp. 647–686. Springer (2009)
21. MIPLIB - Mixed Integer Problem Library, `http://miplib.zib.de`
22. Régin, J.-C.: Global Constraints: A Survey. In: Milano, M., Van Hentenryck, P. (eds.) Hybrid Optimization: the 10 Years of CPAIOR, pp. 169–190. Springer (2011)

Chapter 11
Combining Column Generation and Metaheuristics

Filipe Alvelos, Amaro de Sousa, and Dorabella Santos

Abstract. In this Chapter, we consider the hybridization of column generation (CG) with metaheuristics (MHs) for solving integer programming and combinatorial optimization problems. We describe a general framework entitled "metaheuristic search by column generation" (for short, SearchCol). CG is a decomposition approach in which one linear programming master problem interacts with subproblems to obtain an optimal solution to a relaxed version of a problem. The subproblems may be solved by problem-specific algorithms. After CG is applied, a set of subproblem's solutions, optimal primal and dual values of the master problem variables and a lower bound to the optimal value of the problem are available. In contrast with enumerative approaches (e.g, branch-and-price), in SearchCol the information provided by CG is used in a MH search. The search is based on representing a solution (to the overall problem) as being composed by one solution from each subproblem. After a search is conducted, a perturbation for CG is defined and a new iteration begins. The perturbation consists in forcing or forbidding attributes of the subproblem's solutions and, in general, leads to the generation of new subproblem's solutions and different optimal primal and dual values of the master problem variables. In this Chapter, we discuss (i) which models are suitable for decomposition approaches as SearchCol, (ii) different alternatives for generating initial solutions for the search (with different degrees of randomization, greediness and influence of CG) (iii) different search approaches based on local search, (iv) different alternatives for perturbing CG (influenced by CG, based on the incumbent, and based on the memory of the search).

Filipe Alvelos
Centro Algoritmi / DPS, Universidade do Minho, 4710-057 Braga, Portugal
e-mail: falvelos@dps.uminho.pt

Amaro de Sousa
Instituto de Telecomunicações / DETI, Universidade de Aveiro, 3810-193 Aveiro, Portugal
e-mail: asou@ua.pt

Dorabella Santos
Instituto de Telecomunicações, 3810-193 Aveiro, Portugal
e-mail: dorabella@av.it.pt

E.-G. Talbi (Ed.): Hybrid Metaheuristics, SCI 434, pp. 285–334.
springerlink.com

11.1 Introduction

We describe a framework for obtaining approximate solutions for mixed integer programming (MIP) and combinatorial optimization (CO) problems. The framework, named metaheuristic search by column generation , or *SearchCol* for short, relies on the combination of column generation (CG) and metaheuristic (MH) search. The aim of a SearchCol algorithm is to obtain high quality solutions in reasonable amounts of time for a wide range of problems.

The core ideas of this work were first proposed in [2] and have their roots in [4]. In this Chapter, we further develop some concepts, provide additional details on others, and extend the SearchCol framework to a more general setting.

SearchCol is a decomposition approach in which a solution to a problem is composed by several components, which are themselves solutions to smaller problems. Its central idea is the exchange of information between CG and a MH. Basically, CG provides subproblem solutions which define a search space for the MH. The MH provides an incumbent solution and information on desired or avoidable attributes of the subproblem solutions to CG. This exchange of information is repeated until a stopping criterion is met.

A particular MH is not specified in SearchCol but a set of algorithmic components are defined such that different (hybrid) MHs can be used in an actual SearchCol algorithm. Those components use CG information, randomness, greediness, neighborhoods, and memory for implementing different strategies of building and improving solutions.

SearchCol can be applied in the vast majority of problems for which approaches based on Dantzig-Wolfe decomposition [19] or Lagrangean relaxation [30] were devised (even when the model to be solved does not result directly from those reformulation methods). Dantzig-Wolfe decomposition and Lagrangean relaxation are dual equivalent of each other and the corresponding solution approaches, CG and Kelley's cutting plane method [44], are also dual equivalent of each other (see [28] for a clear explanation). Having present this equivalence, in the remainder of this Chapter, we use the primal perspective of CG and Dantzig-Wolfe decomposition.

In Dantzig-Wolfe decomposition , it is assumed that an original model exists. By defining a set of subproblems with some of the constraints of the original MIP model, a reformulated model, the decomposition model, is obtained. In many applications, a decomposition model can be formulated directly. In this Chapter, although we mention Dantzig-Wolfe decomposition when we think that this important particular case is worth to detail, we consider the more general decomposition model.

11.1.1 Motivations

Decomposition approaches are appealing for MIP and CO problems for several reasons:

- Decomposition models may capture the decomposable structure already present in the problem (which in fact, seems to exist in most pratical problems [49]) being the simpler and more direct way to model it.
- Decomposition is attractive for large-scale problems when a non-decomposition model is too large to be dealt with computationally.
- Decomposition approaches can be parallelized easier than direct approaches.
- A problem may involve complex constraints or objective functions which are not easily addressed by MIP (where linearity is assumed - disregarding the integrality constraints of the variables) but can be efficiently addressed by other modelling approaches such as dynamic programming [22] and constraint programming [35]. These approaches can be used in subproblems, but still framed by MIP / linear programming.
- Decomposition approaches can be faster than the other available approaches. There are two main reasons for this:
  - The Linear relaxation of a decomposition model may provide a better gap than other models. In particular, when a Dantzig-Wolfe decomposition is applied, if the subproblem does not possess the integrality property (i.e, not all its extreme points are integers), the lower bound (assuming a minimization problem, as will be done throughout the Chapter) provided by the linear relaxation of the decomposition model is greater than or equal to the one provided by the linear relaxation of the original model. Good quality lower bounds are a major factor of the quality of implicit enumeration algorithms, e.g. branch-and-bound.
  - Very efficient problem-specific algorithms can be used in solving subproblems.

Decomposition approaches are usually applied in NP-hard problems. A MIP decomposition model (which linear relaxation can be solved by CG), has an exponential number of variables. Even when the set of variables is restricted, the restricted problem usually remains NP-hard (it is a general MIP). Therefore, the definition of a combinatorial model to explore a search space in an approximate manner is an attractive solution approach.

When using SearchCol, there is an implicit combinatorial model which is the base for MH search. In this model a solution is represented by a set of subproblem solutions, one from each subproblem. The rationale for using this combinatorial *decomposition* model relies on the following three aspects:

- The decomposition naturally defines a search space in a higher level which is easier to explore. In the combinatorial decomposition model, subproblems solutions are combined to form a solution to the overall problem. These subproblems solutions can be seen as higher level components when compared to subproblem variables. Note that the feasibility in the subproblems constraints is assured in subproblem solutions.
- If the linear relaxation of the MIP decomposition model associated with the combinatorial decomposition model provides high quality lower bounds, it is

reasonable to expect that the search space is of better quality (i.e. poor solutions are excluded) than search spaces based on other models.
- The decomposition of the structure of a solution naturally leads to the definition of operators for MH search.

11.1.2 Literature Review

SearchCol combines decomposition approaches mainly based on mathematical programming and MHs. We now give a necessarily brief overview of relevant works on the first type, on the second type, and also on the combination of both types of approaches.

For providing approximate solutions (as in the so called Lagrangean heuristics) or optimal solutions (as in branch-and-price), decomposition approaches based on Dantzig-Wolfe decomposition [19] or Lagrangean relaxation [30] for MIP and CO have been a major topic of research in the last decades [41]. For example, two of the most influential ten papers in one of the most relevant journals of the Institute for Operations Research and the Management Sciences [39] are papers in decomposition approaches: CG [27] and Lagrangean relaxation [24].

The roots of Lagrangean relaxation in MIP and CO lie in [37, 38, 30, 24]. More recent references can be found in [10, 25]. The roots of Dantzig-Wolfe decomposition and CG lie in [27, 19, 31, 32, 21]. A modern perspective on CG for integer programming is given in the surveys [9, 65, 46, 64] and in the book [20].

MHs are the most successfull approaches for addressing large instances of many CO problems. There is a vast literature since the 1980s where significant research on this area begun (recent books are [33, 63, 29]). A comprehensive survey paper is [14]. A landmark book is [34].

In recent years, hybrid MHs (taken as the combination of different MHs or taken as the combination of MHs with other approaches) emerged as an important area of research. In [62], a taxonomy of hybrid MHs is proposed and an annotated bibliography classifying several MHs according to the taxonomy is provided. Recent surveys on hybrid MHs are [55, 13]. Surveys and applications of hybrid MHs are in the book [11].

Among the hybrid MHs, the combination of heuristics and mathematical programming , resulting in the so-called *Matheuristics*, has received an increasing interest (see, for example, [56, 12, 47, 7] and sections 4 and 5 of [13]). SearchCol belongs to this research stream.

SearchCol is a framework which combines a decomposition method (CG) and MH search. The next two references also propose frameworks for this combination.

In [18], a framework for combining branch-and-price with local search is proposed. In that work, local search is applied in the nodes of the branch-and-price tree for attempting the improvement of the incumbent solution and generating new columns. The resulting hybrid approach is applied to a vehicle routing problem where the master problem is a set partitioning model. A central difference between this framework and Searchcol is that the latter intentionally avoids an implicit enumeration strategy, i.e. (heuristic) branch-and-price.

A framework based on Dantzig-Wolfe decomposition is proposed in [15] in which the primal and dual solutions are used to construct a problem solution in a problem-specific manner. In the same reference a framework for Lagrangean heuristics is also provided. The central step (building a solution from the subproblem solutions and from the dual variables values) is also problem dependent. In SearchCol, the subproblems are also solved in different dual points (as in the frameworks of the reference) given by the restricted master problems (RMPs) of CG (note that a subgradient method could also be used), but then the solution is obtained in a problem independent manner. Furthermore, this solution can be used to perturb CG leading to the consideration of new subproblem solutions. This perturbation is also problem independent.

Other examples and references on the combination of CG based approaches and heuristics are provided in Section 11.4 after the description of the MIP decomposition model used in SearchCol.

We now classify SearchCol according to the classification of [54]. The kind of algorithms that are hybridized define a first differentiation criterion in that classification scheme. In SearchCol, at least two types of algorithms are combined: a linear programming algorithm for solving the RMP of CG and a (hybrid) MH for the search. A problem-specific algorithm for solving the subproblem is also usual. The second differentiation criterion is the strength of combination. In SearchCol, CG influences a MH and the reverse is also true. As the algorithms retain their own identities, there is a low level of hybridization. Note that this weak coupling is a consequence of the generality of the approach. The order of execution is interleaved and the combination is collaborative: the CG and the MH exchange information but none is strongly subordinated to the other.

11.1.3 Contributions

Decomposition models are powerful modelling tools, but efficient solution methods are required for their success in problem solving. The main contribution of this work is the proposal of a framework, SearchCol, which has the purpose of solving efficiently a wide range of decomposition models. The key issue is that SearchCol explores the combinatorial structure of a problem which has been mostly tackled by mathematical programming. Furthermore, the combinatorial perspective is rather natural as we are given a number of sets, each one with a large number of elements, and we are interested in selecting one element from each such that a function is minimized.

There are two main strenghts in this framework. The first one is its ability to deal with complex problems, since the subproblems of CG can be approached with any optimization technique and their complexity is hidden from the search which is conducted in a higher level. The second one is its generality, since the problem-specific components of a SearchCol algorithm are limited to a subproblem solver and its interaction with a linear programming model. In fact, if the model results from a Dantzig-Wolfe decomposition and the subproblems are solved with a general purpose solver, SearchCol is totally problem-independent.

The type of models which may be addressed by SearchCol are the ones where the linear relaxation may be solved by CG. In this sense, SearchCol is an alternative to branch-and-price, not based on enumeration. Rigorously, we consider the case where the coefficients of a column in the rows of the master problem are a function of binary variables of the subproblem (the subproblem may have additional binary/integer/continuous variables but they are not used for this purpose). This is the most usual case in CG based approaches.

Besides the usual variables associated with the subproblem solutions, our framework also admits binary / general integer / continuous variables in the master problem, which provides flexibility when modelling a problem through decomposition.

Two related contributions of this work is the combinatorial model used in SearchCol and the proposal of a large number of algorithmic components for the definition of MHs to address it. These components can be combined originating particular MHs. We provide the examples of (advanced) multi-start local search and variable neighborhood search. Additional algorithmic components, which are easily conceivable, allow the extension to other approaches, in particular, population-based MHs.

11.1.4 Chapter Structure

This Chapter is organized as follows. In Section 2 we introduce the decomposition MIP model and give examples for different problems. In Section 3 we describe CG and perturbed CG which are base ingredients of SearchCol. CG is used to solve the linear relaxation of the MIP model. In Section 4 we review solution methods for the decomposition MIP model. In Section 5 we propose a different perspective on the problem being solved which is based on a combinatorial decomposition model. We discuss the solution representation, the search space, and how solutions are evaluated. We provide a description of Searchcol in Section 6. In Section 7, the MH search phase is detailed and examples of actual SearchCol algorithms are provided. In Section 8 we discuss the main conclusions of the Chapter.

11.2 A Decomposition Mixed Integer Programming Model

We consider problems where a set of decisions, each one from a finite but large set of alternatives, must be taken. The alternatives are modelled through binary variables and a cost value might be associated with each one. In the most general case, the relation between the decisions can be modelled by binary, general integer and continuous variables, linear constraints, and may imply additional costs. We intend to minimize the total cost.

In subsection 11.2.1 a general decomposition model for these problems is introduced. This model relies on associating a subproblem solution with each decision. In the same subsection, we define the notation used and state the assumptions made. The decomposition model is introduced independently of any solution method or previous model, what we believe can be interesting as a starting point for a fresh look to decomposition approaches.

In subsection 11.2.2 we provide several examples of decomposition models.

11.2.1 Model Statement

We consider the following MIP model (D), named decomposition model, which may be derived directly for a problem or may be the result of the application of a Dantzig-Wolfe decomposition or of a Lagrangean relaxation.

$$Min \sum_{t \in T} c_t y_t + \sum_{k \in K} \sum_{s \in S^k} c_s^k y_s^k \qquad (D)$$

subject to

$$\sum_{s \in S^k} y_s^k = 1 \qquad k \in K \qquad (11.1)$$

$$\sum_{t \in T} a_{it} y_t + \sum_{k \in K} \sum_{s \in S^k} a_{is}^k y_s^k \{\leq, =, \geq\} b_i \qquad i \in I \qquad (11.2)$$

$$y_t \in \{0,1\} \qquad t \in T^b \qquad (11.3)$$

$$y_t \geq 0 \text{ and integer} \qquad t \in T^i \qquad (11.4)$$

$$y_t \geq 0 \qquad t \in T^c$$

$$y_s^k \in \{0,1\} \qquad k \in K, s \in S^k \qquad (11.5)$$

The decomposition model (D) has two sets of decision variables. Variables y_t, which we name *static* variables, are not present in the first set of constraints and are grouped by type: binary ($t \in T^b$), general integer ($t \in T^i$), and continuous ($t \in T^c$). The second set of decision variables, y_s^k, $k \in K$, $s \in S^k$, are the *selection* variables and are partitioned in $|K|$ subsets, indexed by k. Their general definition is the following.

$$y_s^k = \begin{cases} 1 \text{ if the } s\text{-th element of the subset } k \text{ is selected} \\ 0 \text{ otherwise} \end{cases} \qquad k \in K, s \in S^k$$

Parameters c_t, c_s^k, a_{it}, a_{is}^k, and b_i are coefficients of the variables in the objective function, coefficients of the variables in some of the constraints and the right-hand sides of some of the constraints, as can be seen directly in (D).

Constraints (11.1) state that one element from each subset k must be selected, therefore they are named *selection* constraints. Constraints (11.2) may include any decision variable. We name them *global* constraints.

We are interested in the case where the number of selection variables is so large that addressing model (D) directly, for reasonable size instances, is out of the question. The growth of $|S^k|$ with respect to the size of the data defining the problem to be solved is *not* bounded by a polynomial. On the contrary, the number of static variables and constraints is polynomial.

Furthermore, we consider that each partition k is associated with a subproblem, which allows treating the selection variables implicitly as they become associated with solutions of the subproblem. The subproblem is defined such that each subproblem solution, x_s^k, is associated with one selection variable, y_s^k. Each subproblem k has a set J^k of binary decision variables represented by the vector x^k with components x_j^k. We represent the s-th feasible solution by x_s^k. The subproblem may have auxiliary variables, w_r^k not necessarily binary. Representing the set of indexes of the auxiliary binary variables of subproblem k by R^{bk}, the set of indexes of general integer variables by R^{ik}, and the set of indexes of continuous variables by R^{ck}, the feasible region of the subproblem, which we represent by Q^k, is defined by

$$
\begin{aligned}
&(x^k, w^k) \in W^k && (Q^k)\\
&x^k \in \{0,1\}^{|J^k|}\\
&w_r^k \in \{0,1\}, r \in R^{bk}\\
&w_r^k \geq 0 \text{ and integer}, r \in R^{ik}\\
&w_r^k \geq 0, r \in R^{ck}
\end{aligned}
$$

where W^k is the set of possible values for the decision variables x^k and w^k.

We represent the determination of the coefficients of the selection variables in (D) for each subproblem solution x_s^k, $k \in K$, $s \in S^k$, by the following functions:

$$
\begin{aligned}
&c_s^k = f_0^k(x_s^k, w^k)\\
&a_{is}^k = f_i^k(x_s^k) && i \in I
\end{aligned}
$$

Note that the variables w_r^k do not appear in the f_i^k functions which implies that they do not participate in the definition of the feasible region of (D). If two solutions have the same value in the x^k variables but different values in the w^k variables, then a single selection variable is associated with solution x^k, the one with a lower value (given by f_0^k).

A solution in terms of the selection variables can be translated into a subproblem solution. In order to establish this relation we define x_{js}^k as the value of the j-th variable of the subproblem k in solution s (which takes values 0 or 1).

Given a set of values for the selection variables y_s^k, the value of a subproblem variable is given by:

$$x_j^k = \sum_{s \in S^k} x_{js}^k y_s^k. \tag{11.6}$$

There are other types of models closely related to model (D) and for which all the approaches discussed and proposed in this Chapter can also be applied.

In the first type, there is only one subproblem and we want m different solutions from it. The selection constraints (11.1) are replaced by a single constraint

$$\sum_{s \in S} y_s = m$$

The second type of models are set covering, partitioning, and packing. In these models a parameter m representing an upper bound to the number of sets in an optimal solution is introduced (note that a trivial upper bound is the number of elements). The selection constraints (11.1) are again replaced by a single constraint

$$\sum_{s \in S} y_s \leq m$$

In the most general case, the selection constraints take the form:

$$y_0^k + \sum_{s \in S^k} y_s^k = m^k, k \in K$$

where m^k is the number of different solutions required for subproblem k, which may include null solutions as modelled by the slack (general integer) variables y_0^k. If the null solution of the subproblem is not feasible then $y_0^k = 0$.

In general, different decomposition models can be formulated for the same problem. As virtually all the solution approaches for solving (D) are based on linear relaxations, a first criterion for the choice of a particular decomposition model instead of another is the quality of the lower bound given by its linear relaxation. A second criterion is the easiness of dealing with the subproblem. The third criterion is the difficulty to solve the decomposition model itself: in principle, the less the number of global constraints, the better.

In the next subsection, we give examples of decomposition models for different types of problems.

11.2.2 Examples

11.2.2.1 Time Constrained Shortest Path Problem

We first consider the time constrained shortest path problem. In this problem, a directed network is given. Associated with each arc there are two parameters: its length and the time it takes to be traversed. The objective is to find the shortest path between two nodes not exceeding a given time limit. We define c_j as the length of arc j, t_j as the travel time in arc j, and T as the time limit. In order to obtain a decomposition model in the form of the general one introduced in the previous subsection, we define the following decision variables.

$$y_s = \begin{cases} 1 \text{ if path } s \text{ is the chosen path} \\ 0 \text{ otherwise} \end{cases}, s \in S$$

where S is the set of all paths. A decomposition model is:

$$Min \sum_{s \in S} c_s y_s$$

subject to

$$\sum_{s \in S} y_s = 1,$$

$$\sum_{s \in S} a_s y_s \leq T \,, j \in A$$

$$y_s \in \{0,1\} \,, s \in S$$

Note that in this model, there are no static variables and we only define one partition (we do not use the index k).

As the number of paths in a network grows exponentially with the dimension of the network, we define a subproblem where a solution corresponds to a path and the decision variables are associated with arcs (J is the set of arcs of the network). There are no auxiliary subproblem variables. The subproblem solution of path s is given by

$$x_{js} = \begin{cases} 1 \text{ if arc } j \text{ belongs to path } s \\ 0 \text{ otherwise} \end{cases} \quad k \in K, j \in J$$

The coefficients of the selection variables in the objective function and in the constraint are given by

$$c_s = f_0^k(x_s^k) = \sum_{j \in J} c_j x_{js}$$

$$a_s = f_1^k(x_s^k) = \sum_{j \in J} t_j x_{js}$$

In this model, the relation between the values of the selection variables and the subproblem variables established by (11.6) let us obtain the flows on the arcs (the values of the subproblem variables) based on the flows on paths (the values of the selection variables). This is a well known result for network flow problems [1].

We give a numerical example with an instance from [1] in Figure 11.1 where the values on each arc are the cost and time by that order. It is intended to obtain the shortest path between nodes 1 and 6 with the time limit of 14 time units.

For this very small instance, all paths can be enumerated easily. The set of all paths, S, their lenghs and times, are given in Table 11.1. Since the decomposition model is very small, it can be solved directly by a general purpose solver. The optimal solution corresponds to the path $1-3-2-4-6$ with length 13.

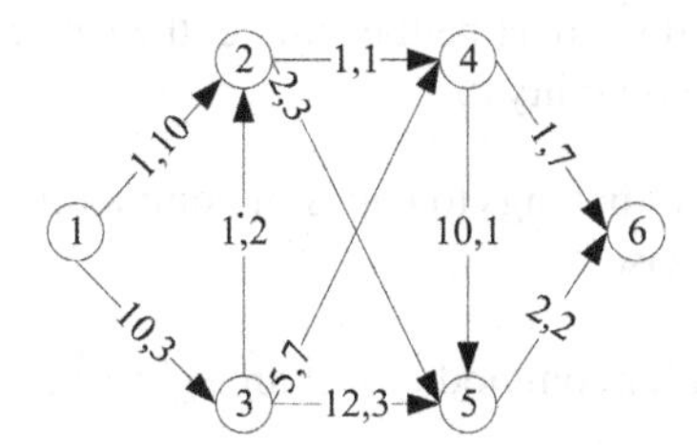

Fig. 11.1 An instance of the time constrained shortest path problem

Table 11.1 Parameters of the decomposition model for the time constrained shortest path example

s	(1,2,4,6)	(1,2,4,5,6)	(1,2,5,6)	(1,3,2,4,6)	(1,3,2,4,5,6)	(1,3,2,5,6)	(1,3,4,6)	(1,3,4,5,6)	(1,3,5,6)
c_s	3	14	5	13	24	15	16	27	24
a_s	18	14	15	13	9	10	17	13	8

11.2.2.2 Single Path Routing Problems

A single path routing problem is defined over a network in which a set of commodities, K, must be routed from their origin nodes to their destination nodes. We address the problem in which each commodity has a given demand, d^k, corresponds to an origin/destination pair, and there are capacities associated with the arcs, $b_{ij}, ij \in A$, where A is the set of arcs of the network. Furthermore, each commodity must use a unique path to route its demand. This problem is also known as the unsplittable multicommodity flow problem. In here, we consider two variants of this problem: in the first, there are costs (per unit of demand) associated with the arcs that may vary by commodity and the objective is to minimize the total cost (min cost problem); in the second, the objective is to minimize the load (i.e, the used proportion of the capacity of the arc) of the arc with the maximum load (min-max problem). We discuss the cases of directed networks but the models and solution approaches to be discussed later can be easily transformed for undirected ones. This problem have been addressed in [8, 3, 4].

Min cost problem

A decomposition model for the minimum cost single path routing problem is known as the arc-path formulation (see, for example, [1]). A selection variable, y_s^k, is associated with path s of commodity k, therefore it will take the value 1 if path s is the path selected for commodity k, and 0, otherwise. There are no static variables and there is one global constraint for each arc ij, $ij \in A$, stating that the capacity of the arc cannot be exceeded:

$$\sum_{k \in K} \sum_{s \in S^k} a^k_{(ij)s} y^k_s \leq b_{ij}$$

As in the time constrained shortest path example, the subproblem variables are associated with arcs. For commodity k:

$$x^k_{(ij)s} = \begin{cases} 1 \text{ if arc } ij \text{ belongs to path } s \text{ of commodity } k \\ 0 \text{ otherwise} \end{cases} \quad k \in K, ij \in A$$

Denoting the unitary cost for commodity k of arc ij by c^k_{ij}, then cost of the selection variable y^k_s is given by

$$c^k_s = f^k_0(x^k_{(ij)s}) = \sum_{ij \in A} c^k_{ij} x^k_{(ij)s}$$

The coefficients of the variables in the global constraints are given by:

$$a^k_{(ij)s} = f^k_{ij}(x^k_{(ij)s}) = d^k x^k_{(ij)s}$$

Min max problem

A decomposition model for the min max single path routing problem can be obtained by considering one *static* continuous variable, y_1, corresponding to the maximum load of an arc. The objective function becomes

$$Min\, y_1$$

and the global constraint for arc ij, $ij \in A$:

$$-b_{ij} y_1 + \sum_{k \in K} \sum_{s \in S^k} a^k_{(ij)s} y^k_s \leq 0$$

Note that, in this case, there are no costs associated with the subproblem solutions.

11.2.2.3 Vehicle Routing Problems

We consider the example of a decomposition model for a classical vehicle routing problem [22, 43]. We consider a network with nodes associated with customers and with the depot (one node is the origin depot and another is the destination depot). We consider that each customer has a given demand and a given time window (the demand of the client must be delivered not earlier than the beginning of the time window and not later than its end). Associated with each arc there is a cost and a time. We consider a fleet of homogeneous vehicles - all the vehicles have the same capacity. The problem is to define the route of each vehicle such that all customers are served within the time windows prescribed and minimizing the total distance travelled by the fleet. The decomposition model has no static variables, each selection variable is associated with a specific route (a path starting at the origin depot and ending at the destination depot), and the global constraints state that each customer must be visited:

$$\sum_{s\in S^k} a^k_{is} y^k_s = 1.$$

The subproblem variables are associated with arcs:

$$x_{(ij)s} = \begin{cases} 1 \text{ if arc } ij \text{ belongs to route } s \\ 0 \text{ otherwise} \end{cases} \quad ij \in A$$

Denoting the length of arc ij by c^k_{ij}, then the cost of the selection variable y_s is given by

$$c^k_s = f^k_0(x^k_{(ij)s}) = \sum_{ij\in A} c^k_{ij} x^k_{(ij)s}$$

The coefficients of the selection variables in the global constraints are given by:

$$a^k_{is} = f^k_i(x^k_{(ij)s}) = \sum_{ij\in A} x^k_{(ij)s}$$

A subproblem solution is a path from the origin depot to the destination depot which obeys the time windows of the customers associated with the nodes of the path. Auxiliary subproblem variables must be required to define feasible paths (paths obeying the time windows).

11.2.2.4 Machine Scheduling Problems

We consider the example of decomposition models for parallel machine scheduling where there are a given number of machines and a given number of jobs to be processed on them with some time-related objective. For each job, a processing time in each machine, a priority or weight, possibly a release date, and possibly a due date, are given. Machines can be identical or different, if they process the jobs with different speeds, or have different availabilities during the time horizon. Usual decomposition models [5, 16, 6, 45] have no static variables, each selection variable is related with a specific schedule, the selection constraints state that one schedule for each machine must be selected (non-identical machines) or that a given number of schedules must be selected (identical machines), and there is one global constraint for each job i stating that the job must be processed (in the case of identical machines, we drop the superscript k):

$$\sum_{k\in K} \sum_{s\in S^k} a^k_{is} y^k_s = 1.$$

As the number of schedules grows exponentially with the number of jobs, we define a subproblem for each machine where a solution corresponds to a schedule and the decision variables are associated with the assigment of jobs to the machine. The subproblem solution of schedule s of machine k is given by

$$x_{js}^k = \begin{cases} 1 \text{ if job } j \text{ belongs to schedule } s \text{ of machine } k \\ 0 \text{ otherwise} \end{cases} \quad k \in K, j \in J$$

The coefficients of the selection variables in the global constraint of job i are:

$$a_{is}^k = f_i^k(x_s^k) = x_{js}^k, \text{ with } i = j.$$

The determination of the objective coefficient of the selection variables involves a set of auxiliary subproblem variables which are related with the completion times of the jobs. For example, in [5, 16], the objective is to minimize the total weighted completion time and therefore, the cost of a schedule depends on the completion time of the jobs it includes. Representing the completion time of job j in schedule s of machine k by w_{rs}^k, then

$$c_s^k = f_0^k(x_s^k, w_r^k) = \sum_{j \in J} W_j w_{rs}^k x_{js}^k$$

where W_j is the weight of job j. The completion time of a job depends on which jobs are processed in the same machine. Therefore function f_0 is non linear.

11.2.2.5 Multiple Spanning Tree Routing Problems

Multiple spanning tree routing is a particular case of single path routing. This problem is defined in undirected networks and is motivated by the spanning tree routing protocol for switched Ethernet networks [40]. In multiple spanning tree routing, a set of spanning trees is defined and each commodity is to be routed over the edges of one of the spanning trees. Given the number of spanning trees that may be used, the problem is to define the specific spanning trees to be used and which commodities are assigned to each spanning tree. This problem and a related problem have been addressed in [58, 59].

The tree and path decomposition models are based on selection variables corresponding to paths and corresponding to spanning trees (other decompositions are studied in [58]). We describe the min cost problem, noting that a min max problem can be modelled in a similar way.

We divide partitions K in two subsets: $K1$ is related with paths and $K2$ is related with spanning trees. Therefore, a variable y_s^k with $k \in K1$ is associated with a path and a variable y_s^k with $k \in K2$ is associated with a spanning tree. Naturally, $K = K1 \cup K2$. For $k \in K1$, we define $x_{\{i,j\}s}^k$ equal to 1 if path s, which is associated with commodity, includes edge $\{i, j\}$ and equal to 0, otherwise. For $k \in K2$, we define $x_{\{i,j\}s}^k$ equal to 1 if spanning tree s, which is associated with k-th spanning tree, includes edge $\{i, j\}$ and equal to 0, otherwise.

As fixed variables, we define ϕ_{k1}^{k2} equal to 1 if path $k1 \in K1$ is assigned to spanning tree $k2 \in K2$.

A decomposition model based on paths and trees for the minimum cost multiple spanning tree routing problem is:

$$Min \sum_{k \in K1} \sum_{s \in S^k} c_s^k y_s^k \qquad (Dmctrp)$$

subject to

$$\sum_{s \in S^k} y_s^k = 1 \qquad k \in K \quad (11.7)$$

$$\sum_{s \in S^k 1} x_{\{i,j\}s}^{k1} y_s^{k1} \leq \sum_{s \in S^k 2} x_{\{i,j\}s}^{k2} y_s^{k2} + (1 - \phi_{k1}^{k2}) \quad \{i,j\} \in A, k1 \in K1, k2 \in K2 \quad (11.8)$$

$$\sum_{s \in S^{k2}} \phi_{k1}^{k2} = 1 \qquad k1 \in K1, k2 \in K2 \quad (11.9)$$

$$\sum_{k \in K1} \sum_{s \in S^k} a_k x_{\{i,j\}s}^k y_s^k \leq b_{\{i,j\}} \qquad \{i,j\} \in A \quad (11.10)$$

$$\phi_{k1}^{k2} \in \{0,1\} \qquad k1 \in K1, k2 \in K2$$

$$y_s^k \in \{0,1\} \qquad k \in K, s \in S^k$$

Constraints (11.7) force one path for each commodity and one spanning tree for each allowed spanning tree to be selected. Constraints (11.8) state that if a commodity uses edge $\{i,j\}$ and is assigned to the t-th spanning tree then this spanning tree must include edge $\{i,j\}$. Constraints (11.9) state that each commodity is assigned to one spanning tree. Constraints (11.10) are the capacity constraints. Note that in this problem the capacities are given by edge.

11.3 Column Generation

Column generation (CG) has been successfully applied in solving linear programming problems with a huge number of variables (many of those being relaxations of integer models). In this Section, we discuss how the linear relaxation of the decomposition model introduced in the previous section can be solved by CG.

In subection 11.3.1 we provide a general overview of CG. In subsection 11.3.2 we address the subproblem in detail. In subsections 11.3.3 and 11.3.4 we provide a CG algorithm and an example of its application. In subsection 11.3.5 we present how subproblem variables can be fixed, which is an important topic to the solution methods to be discussed in the remainder of this Chapter.

11.3.1 Overview

CG is a method that allows obtaining an optimal solution to the linear relaxation of the decomposition model (D). The linear relaxation of (D) (LRD) is obtained by replacing the integrality constraints, (11.3), (11.4), and (11.5) by:

$$0 \le y_t \le 1 \qquad t \in T^b$$
$$y_t \ge 0 \qquad t \in T^i$$
$$y_s^k \ge 0 \qquad k \in K, s \in S^k$$

Note that each y_s^k cannot exceed 1 because of the selection constraints (11.1). In many CG applications the (relaxed) static variables do not appear. However, their presence extends the ability to model relevant problems with decomposition models, as illustrated by some of the examples of subsection 11.2.2.

CG relies on the definition of restricted master problems (RMPs) which are linear programming problems where not all the variables of the overall problem are considered. In the case of LRD, in a RMP, not all the selection variables are considered. Consider the case where LRD has only one subproblem, which is easily extendable for $|K|$ subproblems as will be done later. A CG algorithm begins with the definition of a first RMP, which may include artificial variables for assuring feasibility. In each iteration, the current RMP is optimized and its optimal dual solution is used in evaluating the reduced costs of the selection variables outside the RMP. This is done by solving the subproblem where the objective function is related with the reduced costs of the selection variables. The smallest reduced cost is obtained and, according to linear programming theory, if it is non negative the RMP solution is optimal for the overall problem LRD and the algorithm stops. Otherwise, the selection variable associated with the subproblem solution (the one with the most negative reduced cost) is introduced in the RMP (by using functions f_0 and f_i, $i \in I$) and a new iteration begins.

11.3.2 Subproblem

11.3.2.1 General Considerations

In order to identify selection variables with negative reduced costs, a subproblem for each k, $k \in K$, is solved. The subproblem k resolution aims at identifying a selection variable y_s^k with negative reduced cost.

By definition, representing the dual variables of constraints (11.1) by π^k, $k \in K$, and the dual variables of constraints (11.2) by ω_i, $i \in I$, the reduced cost of variable y_s^k is given by

$$\bar{c}_s^k = c_s^k - \pi^k - \sum_{i \in I} a_{is}^k \omega_i.$$

The relation established between the selection variables and subproblem solutions (given by the functions f_0 and f_i, $i \in I$), allows to obtain the selection variable y_s^k with the most negative reduced cost by solving an optimization subproblem for each partition k:

$$Min - \pi^k + f_0^k(x^k, w^k) - \sum_{i \in I} \omega_i f_i^k(x^k)$$

subject to

$$(x^k, w^k) \in Q^k$$

CG is of practical interest if there is an efficient exact algorithm able to solve this (sub)problem. Typically, this algorithm is problem-specific as we are interested in exploring the common structure of the selection variables of the same partition. In many cases, such as the machine scheduling and vehicle routing models discussed in 11.2.2, those algorithms are based on dynamic programming which allows dealing with non linear objective functions and non linear constraints. Examples of subproblems solved by constraint programming can be found in [52, 23]. For the network routing examples given in the previous subsection, shortest path algorithms (e.g. Dijkstra's) and minimum spanning tree algorithms (e.g. Kruskal's or Primm's) can be used. Other usual subproblems are binary knapsacks (for example, [60]).

We now discuss two usual subproblem particular cases.

11.3.2.2 The Linear Case

In the linear case, the functions relating the subproblem solutions and the selection variables f_0^k and f_i^k, $i \in I$, $k \in K$, are linear. We define them, for $k \in K$, as

$$f_0^k(x^k, w^k) = \sum_{j \in J^k} d_{0j}^k x_j^k + \sum_{r \in R^k} d_r^k w_r^k$$

$$f_i^k(x^k) = \sum_{j \in J^k} d_{ij}^k x_j^k, i \in I$$

where d_{0j}^k, d_{ij}^k, $i \in I, j \in J^k$, $d_r, r \in R$ are scalars.

Using the example of the time constrained shortest path problem, d_{0j} corresponds to the cost of arc j and d_{1j} corresponds to the time of arc j (there is only one global constraint). For min cost single path routing, d_{0j}^k is the cost of arc j for commodity k and d_{ij}^k is the demand of commodity k if $i = j$ (the arc of the global constraint i is the arc of the subproblem j) and is 0 otherwise. X^k is set the of all paths between the origin and the destination of k.

Given the duals of the global constraints of the current RMP, π^k, $k \in K$, and ω_i, $i \in I$, the subproblem k is:

$$\bar{c}^k = min \sum_{j \in J^k} d_{0j}^k x_j^k - \pi^k - \sum_{i \in I} \omega_i \sum_{j \in J^k} d_{ij}^k x_j^k + \sum_{r \in R} d_r^k w_r^k$$

$$subject to$$

$$(x^k, w^k) \in Q^k$$

or, the objective function can be rearranged to

$$\bar{c}^k = min - \pi^k + \sum_{j \in J^k} (d_{0j}^k - \sum_{i \in I} \omega_i d_{ij}^k) x_j^k + \sum_{r \in R} d_r^k w_r^k$$

These decomposition models based on linear functions can be directly constructed or may result from a Dantzig-Wolfe decomposition. When resulting from a Dantzig-Wolfe decomposition, the subproblem variables are the original variables (the ones of the compact model before the decomposition is applied) and the parameters d_{0j}^k and d_{ij}^k, $i \in I, k \in K, j \in J^k$, correspond to the coefficients of the original variables in the compact model (in the objective function and in the constraints that are kept in the master problem when the decomposition is applied). The auxiliary subproblem variables correspond to original variables which do not appear in the linking constraints but only in constraints which were sent to the subproblem. Static variables correspond to original variables which appear in the linking constraints but not in constraints which were sent to the subproblem. In this context, it is relevant to note, we are restricting ourselves to the case where the subproblem variables are binary (the auxiliary variables may be of any type as they are not relevant for the definition of the feasible region of (D)).

11.3.2.3 Fixed Costs for Non-null Subproblem Solutions

We also consider the particular case where a non null subproblem solution has an additional cost. Its value may depend on the subproblem but is equal to all the non null solutions of the same subproblem. We represent it by e^k. As an example, let us consider a decompositiom model for a vehicle routing problem where the decision variable y_s^k is associated with route s for vehicle k and the null solution for k corresponds to not using vehicle k. If we want to model the cost of using a vehicle, it makes sense to consider it independently of the distance travelled by the vehicle, as long as the vehicle is used (i.e., y_s^k is not associated with the null solution). This cost is represented by e^k. Another component of the cost of route s may be a linear cost associated with the arcs included in route s.

Another example is for bin packing problems where a subproblem is defined for each bin and we intend to minimize the number of bins which are used to pack a set of items. The decision variable y_s^k is associated with packing pattern s for bin k and the null solution for k corresponds to not using bin k. In this case, $e^k = 1$, $k \in K$. In general, for $k \in K$, the function for computing the objective coefficient of the selection variable y_s^k becomes:

$$f_0^{k'}(x^k) = \begin{cases} 0 \text{ if } x^k = 0 \\ e^k + f_0^k(x^k, w^k) \text{ if } x^k \neq 0 \end{cases}$$

11.3.3 Algorithm

We now introduce the notation used in the presentation of a CG algorithm.

Given that π is constant, the subproblem k, which we represent by SP^k is

$$\begin{aligned} z_{SP^k} = & Min\, f_0^k(x^k, w^k) - \sum_{i \in I} \omega_i f_i^k(x^k) \\ & \text{subject to} \\ & (x^k, w^k) \in Q^k \end{aligned} \tag{11.11}$$

If the optimal solution of SP^k is the null solution, then

$$\bar{c}^k = -\pi^k + z_{SP^k}$$

otherwise it is

$$\bar{c}^k = e^k - \pi^k + z_{SP^k}$$

Of course, $\bar{c}^k < 0$ means the current dual solution is not optimal to the overall problem LRD and the selection variable corresponding to the solution obtained when solving the subproblem must be inserted in the RMP and this problem must be re-optimized.

Initialize the *RMP*, possibly with artificial variables for assuring feasibility
Initialize the SPs SP^k, $k \in K$
If the null solution is feasible for SP^k, insert the corresponding selection variable in the *RMP*
repeat
  Optimize the *RMP*
  $\pi, \omega \leftarrow$ optimal duals from the *RMP*
  $end \leftarrow true$
  for all $k \in K$ **do**
    Modify the objective function of SP^k to $f_0^k(x^k, w^k) - \sum_{i \in I} \omega_i f_i^k(x^k)$
    Optimize SP SP^k
    $z_{SP^k} \leftarrow$ optimal value of SP^k
    if $e^k - \pi^k - z_{SP^k} < 0$ **then**
      $s \leftarrow$ optimal solution of SP^k
      Obtain the RMP column associated with s by calculating the corresponding coefficients $c_s^k = e^k + f_0^k(x^k, w^k)$ and $a_{is}^k = f_i^k(x^k), \forall i \in I$
      Update the RMP by adding that column
      $end \leftarrow false$
    end if
  end for
until $end = true$

Fig. 11.2 Column generation algorithm

The CG algorithm is given in Figure 11.2. After the initializations, all the selection variables associated with null subproblem solutions are inserted in the RMP (if they are feasible). This excludes the necessity during the rest of the algorithm to test if the solution returned by the subproblem is the null solution or not. Then the main cycle begins: the RMP is optimized, all the subproblems are solved and for each of them if the smaller reduced cost, given by $e^k - \pi^k - z_{SP^k}$, is negative the coefficients of the selection variable are calculated and the RMP updated. If that happens for any subproblem more iterations are required (*end* is false).

11.3.4 Example

We give an example of a CG algorithm for the linear relaxation of the instance of the time constrained shortest path problem introduced in Subsection 11.2.2.1.

The subproblem is a shortest path problem in a network with modified costs. This problem can be solved efficiently by a specific algorithm, e.g. Dijkstra, but here we consider the linear programming model to turn explicit the relation between the selection variables and the subproblem variables. A subproblem variable x_{ij} is equal to one if the arc ij is included in the path, and equal to 0 otherwise. The subproblem is:

$$
\begin{aligned}
& Min \; -\pi + \sum_{ij \in A} (c_{ij} - \omega t_{ij}) x_{ij} \\
& subject\,to \\
& x_{12} + x_{13} = 1 \\
& x_{12} + x_{32} = x_{24} + x_{25} \\
& x_{13} = x_{32} + x_{34} + x_{35} \\
& x_{24} + x_{34} = x_{45} + x_{46} \\
& x_{25} + x_{35} + x_{45} = x_{56} \\
& x_{46} + x_{56} = 1 \\
& x_{ij} \in \{0,1\}, ij \in A
\end{aligned}
$$

In order to initialize the RMP we consider path $1-3-5-6$ (with index $s = 1$). The first RMP is then:

$$
\begin{aligned}
& Min\, 24y_1 \\
& subject\,to \\
& y_1 = 1 \\
& 8y_1 \leq 14 \\
& y_1 \geq 0
\end{aligned}
$$

The optimal primal solution of the RMP is $y_1 = 1$ and an optimal dual solution of the RMP is $\pi = 24$ and $\omega = 0$. Using this dual solution in the subproblem, its optimal solution is path $1-2-4-6$ with reduced cost -21. Given that the reduced cost is negative, this path ($s = 2$) is inserted in the RMP and a new iteration begins. The optimal primal solution of the RMP is $y_1 = 0.4, y_2 = 0.6$ and an optimal dual solution of the RMP is $\pi = 40.8$ and $\omega = -2.1$. Using this dual solution in the subproblem, its optimal solution is path $1-3-2-5-6$ with reduced cost -4.8. Given that the reduced cost is negative, this path $(s = 3)$ is inserted in the RMP and a new iteration begins. The optimal primal solution of the RMP is $y_1 = 0, y_2 = 0.5, y_3 = 0.5$ and an optimal dual solution of the RMP is $\pi = 30$ and $\omega = -1.5$. Using this dual solution in the subproblem, its optimal solution is path $1-2-5-6$ with reduced cost -2.5. Given that the reduced cost is negative, this path $(s = 4)$ is inserted in the RMP and a new iteration begins.

The current RMP is:

$$\begin{aligned}
&Min\, 24y_1 + 3y_2 + 15y_3 + 5y_4 \\
&subject\, to \\
&y_1 + y_2 + y_3 + y_4 = 1 \\
&8y_1 + 18y_2 + 10y_3 + 15y_4 \leq 14 \\
&y_s \geq 0, s = 1,2,3,4
\end{aligned}$$

The optimal primal solution is $y_1 = y_2 = 0, y_3 = 0.2, y_4 = 0.8$. An optimal dual solution is $\pi = 35$ and $\omega = -2$. An optimal solution of the subproblem is path $1-3-2-5-6$ with reduced cost 0. As the reduced cost is non negative the optimal solution of the current RMP is an optimal solution to the overall (relaxed) problem.

11.3.5 Perturbed Column Generation

We now discuss how variables can be fixed without disrupting the CG algorithm, which is a major issue in SearchCol and in branch-and-price algorithms.

We define *perturbation* as the fixing of a subproblem variable to 0 or 1. When CG is applied to a model with perturbations, we use the term *perturbed CG*. The implementation of perturbed CG is based on relation (11.6), which allows imposing values on the subproblem variables without disrupting the CG method.

We implement perturbed CG by adding constraints to the RMP and modifying accordingly the objective coefficients of the subproblem variables in the subproblems. Although this is not the only way of doing it, it has the advantage that it does modify the subproblem feasible region, being less demanding for the subproblem than other types of perturbations.

In this type of perturbed CG, constraints that fix the desired subproblem variables are added to the RMP (named perturbation constraints). The subproblem is not modified except for taking into account the duals of the perturbation constraints in the objective coefficients of the subproblem variables.

We consider a *perturbed* CG RMP which differs from the regular RMP because it has a set P of additional constraints (indexed by p). Each additional constraint forces a subproblem variable to be 1 or 0:

$$x_j^k = b,$$

where b is 0 or 1, and appears in the RMP as

$$\sum_{s\in S^k} x^k_{js} y^k_s = b.$$

We associate a dual variable σ_p to each of these constraints, $p \in P$. The general objective function of the subproblem becomes:

$$Min - \pi^k + f^k_0(x^k, w^k) - \sum_{i\in I} \omega_i f^k_i(x^k) - \sum_{p\in P} \sigma_p x^k_{j(p)}$$

where $j(p)$ represents the index of the subproblem variable in the additional constraint p.

As an example, consider that we want a solution to the linear relaxation of the time constrained shortest path in which the arc $1-2$ is not used and the arc $2-5$ is used. Adding the corresponding constraints to the last RMP, we obtain:

$$\begin{aligned}
&Min\, 24y_1 + 3y_2 + 15y_3 + 5y_4\\
&subject\,to\\
&y_1 + y_2 + y_3 + y_4 = 1\\
&8y_1 + 18y_2 + 10y_3 + 15y_4 \leq 14\\
&y_2 + y_4 = 0\\
&y_3 + y_4 = 1\\
&y_s \geq 0, s = 1,2,3,4
\end{aligned}$$

The objective function of the subproblem is now:

$$Min - \pi + \sum_{ij\in A} (c_{ij} - \omega t_{ij})x_{ij} - \sigma_1 x_{12} - \sigma_2 x_{25}$$

The optimal primal solution is $y_1 = y_2 = 0, y_3 = 1, y_4 = 0$. An optimal dual solution is $\pi = 13$, $\omega = 0$, $\sigma_1 = -10$, and $\sigma_2 = 2$. An optimal solution of the subproblem is path $1-3-2-5-6$ with reduced cost 0. As the reduced cost is non negative the optimal solution of the perturbed RMP is an optimal solution to the overall perturbed problem.

Although this was not the case in this example, when perturbations are considered, additional solutions, generated by the subproblem, may be required to achieve an optimal solution.

11.4 Solving the MIP Decomposition Model

In this Section we provide an overview of solution methods to the decomposition MIP model. We first mention methods not based on column generation (CG) (subsection 11.4.1), then a usual heuristic which is based on applying CG once and then obtaining an integer solution based on the columns of the last restricted master problem (RMP) solved (subsection 11.4.2). Then we describe the branch-and-price method (subsection 11.4.3), branch-and-price heuristics (subsection 11.4.4), and, finally, Lagrangean heuristics (subsection 11.4.5). From all these approaches to solve the decomposition MIP model of Section 11.2, only branch-and-price is an exact method.

11.4.1 Static Approaches

So far, we have discussed the decomposition MIP model and CG methods to solve its linear relaxation. We now describe an approach, which we name static approach, where the subproblem is used to generate a subset of selection variables, $\bar{S}$, $|\bar{S}| << |S|$ which defines a restricted problem - the problem restricted to the selection variables of the subset $\bar{S} \in S$. This problem, or its linear relaxation, is then optimized. If the integer restricted problem is considered then its optimal solution is the one given by the static approach. If the linear relaxation is solved then an additional procedure is needed to attempt an integer solution, for example, a rounding procedure. Of course, in any case, the obtained solution is an approximate solution to the overall problem (D).

A natural way of generating the subset of selection variables in a static approach is the application of heuristics. For example, in [50], each selection variable is associated with a subset of rectangular items packed in a rectangular bin. In order to generate the subset of selection variables, greedy procedures and fast constructive heuristic algorithms from the literature are applied. Another example where this type of approach was applied successfully is in airline crew scheduling problems [65]. In these problems, the subproblem is responsible for generating rotations (series of flights) which must obey several types of legal restrictions and have complex cost structures what makes difficult to turn it into an optimization problem with the dual variables of the RMP as parameters. However, more recent approaches usually model the subproblem through a network where the optimization is possible [42].

In [7], partial CG is described. This is a general concept that includes solving the subproblem such that only "reasonable" columns are generated. It also includes the case where columns are generated by fast heuristics executed multiple times (with some different input).

11.4.2 MIP Based CG Heuristic

Instead of using a static set of selection variables, CG may be applied to define the set of selection variables - the ones present in an *optimal* RMP (a RMP for which its optimal solution is optimal for the overall *linear* problem). An *integer* solution can then be found by solving the *integer* RMP or by a problem-specific procedure (again, rounding is the simplest example). We name the former as MIP based CG heuristic.

11.4.3 Branch-and-Price

So far, we have discussed approximate methods to solve the decomposition model (D). The previous methods are approximate because a selection variable with value 1 in an optimal solution to (D) may not be present in the set of selection variables generated by a static procedure or by CG (as CG solves the linear relaxation of the problem).

Branch-and-price is the combination of CG and branch-and-bound for obtaining an optimal solution to (D). In branch-and-price, each node of the branch-and-bound tree is solved by CG.

The main issue in a branch-and-price algorithm is how the branching is performed. It is well known that branching strategies based on the subproblem variables are preferable to branching strategies based on the selection variables.

A branching strategy based on the subproblem variables relies on the fact that the integrality of the selection variables may be imposed through the integrality of the subproblem variables. After a node of the (branch-and-price) search tree is optimized, values for the selection variables, y_s^k, are known. If they are not all integers, then a solution in the subproblem variables can be obtained through (11.6). Note that if all x_j^k are integer then the solution y_s^k was integer in the first place, because no duplicate selection variables are allowed. Note also that if y_s^k were all integers, then the subproblem solution is also integer. If there is a fractional variable y_s^k, there is at least one fractional variable x_j^k which may be used for creating the branches $x_j^k = 0$ and $x_j^k = 1$. Perturbed CG is used to solve the nodes of the search tree and the search is performed as in standard branch-and-bound.

11.4.4 Branch-and-Price Heuristics

Assuming a branch-and-price algorithm exists for problem (D) which is based on branching on the subproblem variables, the extension of branch-and-bound heuristics to branch-and-price heuristics is straightforward.

For example, the only difference in dive-and-fix (for example, [66]), relax-and-fix (for example, [66]), fix-and-relax [17], and beam search [51] heuristics, consists in how the solution of the linear programming model with some variables fixed is obtained - in a branch-and-price version of these heuristics, perturbed CG is used (with the perturbations fixing subproblem variables). For other examples of heuristics based on branch-and-bound that may be modified in a straightforward manner to accomodate CG, see [7]. In the same reference, iterative CG is described. In this approach the linear relaxation of a MIP is first solved by CG, then branch-and-bound is applied but additional columns are generated only when an heuristic criterion says so (the example given is that the value of the node is too far from the optimum of the linear relaxation). Combinations of decomposition approaches and MHs are proposed in [53], where a review on integer programming and MHs hybrids is done and hybrid approaches based on Lagrangean relaxation and CG are applied to the knapsack contrained maximum spanning tree problem and to a periodic vehicle routing problem.

11.4.5 Lagrangean Heuristics

It is well known that Dantzig-Wolfe decomposition and Lagrangean relaxation can be seen as dual decomposition techniques [28]. Therefore, for each problem in which Lagrangean relaxation has been applied (and in which the variables present in the subproblem and present in the dualized constraints are all binary), a decomposition model (D) can be defined. Inversely, the linear relaxation of model (D) can be approached by Lagrangean methods (e.g. subgradient methods) and problem (D) by Lagrangean heuristics.

Lagrangean heuristics have been applied widely. In these approaches, the decomposition model (D) is not considered explicitly. Subproblem solutions are generated at different dual points (for example, by a subgradient method) and it is attempted to modify them in such a way that the global constraints become feasible. This is done in a problem-specific manner (for references on these approaches, see [7], for a general description, see [15]).

11.5 A Combinatorial Decomposition Model

In this Section we introduce the combinatorial oprimization (CO) decomposition model which is the base for the metaheuristic (MH) search conducted in SearchCol. In subsection 11.5.1 we describe the CO decomposition model and the associated concepts of how a solution is represented and what is the search space. In 11.5.2 we describe how solutions are evaluated.

11.5.1 Solution Representation and Search Space

The general problem being addressed may be seen as a CO problem consisting in selecting one element from each partition, such that the sum of the cost associated with each element is minimum and the global constraints are satisfied. If this problem can be modelled through the decomposition MIP model (D) (see Section 11.2), it can also be modelled through the following CO decomposition model (C):

$$
\begin{aligned}
& Min\, f(s) + \lambda g(s) \qquad\qquad\qquad\qquad (C) \\
& subject\, to \\
& s(k) \in S^k, k \in K
\end{aligned}
$$

where $s(k)$, $k \in K$, represents a subproblem solution of partition k. An overall solution is represented by s, which is a vector $s = (s(1), s(2), ..., s(|K|))$, where each component is a solution from a subproblem. A feasible solution s belongs to $S = S^1 \times S^2 \times ... \times S^{|k|}$. The function $f(s)$ is the cost of the solution with the most favorable values for the static variables. Function $g(s)$ provides a measure of the infeasibility of solution s with respect to the global constraints (11.2). The parameter λ takes a value such that feasible solutions have a lower value than infeasible solutions. Note that a solution s where $g(s) > 0$ is infeasible for the problem but is feasible for this model and, in fact, can even be an incumbent solution at some point in the execution of an algorithm to solve (C). We give the details on how solutions are evaluted in the next subsection.

An important issue in (C) is that the representation of solutions is problem independent. For example, for a problem with four subproblems, a solution $s = (2,4,1,1)$ means that the second subproblem solution is chosen for the first subproblem, the fourth subproblem solution is chosen for the second subproblem, the first subproblem solution is chosen for the third subproblem, and the first subproblem solution is chosen for the fourth subproblem. These subproblems solutions can correspond to paths, trees, machine schedule, routes, ot others, depending on the problem.

Three fundamental characteristics on the design of a MH [63] are assured by this representation: completeness (all solutions can be represented), connexity (it is possible to go from any solution to any other - see neighborhood structures in Section 11.7), and efficiency of the search operators. A drawback is that when there are static variables (in particular, binary or general integer) the evaluation (see next subsection) may become too heavy.

A feasible solution in model (D) can be translated into a solution in model (C) and *vice-versa*. A variable $y_s^k = 1$ in a solution of (D) implies the component $s(k) = s$ in a solution of (C). A component $s(k) = s$ in (C) implies that $y_s^k = 1$, if $s = s(k)$, and $y_s^k = 0$, otherwise, in (D).

The number of solutions in the search phase is $n_1 \times n_2 \times ... \times n_{|K|}$ where n_k is the number of subproblem solutions associated with subproblem k, $k \in K$. The number of solutions of the combinatorial model is exponential. Therefore, model (C) is useful only if the search space S can be restricted, not excluding "good" subproblem solutions (the ones present in optimal or near optimal solutions). In SearchCol, CG is used to restrict the search space of model (C). The search is conducted only over subproblem solutions generated with (perturbed) CG.

We now extend the solution representation for the cases mentioned in the last paragraphs of subsection 11.2.1. When the subproblems are identical, a solution is represented by a set of subproblem solutions. For example, $s = \{7,4,1,3\}$. If the number of subproblem solutions have an upper bound, empty positions / null subproblem solutions are considered. For example, for an upper bound of 4, a solution made of two subproblem solutions is $s = \{0,4,0,3\}$, where 0 is the index for null subproblem solutions.

11.5.2 Evaluating Solutions and Moves

In this subsection, we describe how solutions (including partial solutions) and moves are evaluated in SearchCol. A move corresponds to the addition or removal of a component of a solution. We describe several alternatives which are controlled through parameters which are detailed in subsection 11.6.2. All evaluations comprise two values: the feasibility value and the infeasibility value. A solution/move with a lower value of infeasibility is always better than a solution/move with a larger value. If the infeasibility values of two solutions/moves are equal, the feasibility value is used to compare them. In the next subsubsections, we detail what are the alternatives for determining the feasibility and infeasibility values of a solutions and moves.

11.5.2.1 Models without Static Variables

Full Solutions

When there are no static variables, we evaluate the feasibility of a solution by using one of three functions:

- $evalNumberViolated(s)$ returns the number of violated constraints in solution s;
- $evalAmountViolation(s)$ returns the total amount of violation of the constraints in solution s;
- $evalWeightedViolation(s) = 1000 evalNumberViolated(s) + evalAmountViolation(s)$.

The amount of violation of a constraint is calculated based on the value of the slack which is defined as the right-hand side value minus the left-hand side value. If it is an equality constraint, the amount of violation is the absolute value of the slack. If it is a less han or equal to constraint, if the slack is negative than the amount of violation is the symmetric of the slack, otherwise it is zero. If it is a greater-than or equal to constraint, if the slack is positive than the amount of violation is the slack, otherwise it is zero.

The feasibility value of a solution s is given by

$$evalOriginalCosts(s) = \sum_{k \in K} c^k_{s(k)}.$$

Partial solutions

In a partial solution, the subproblem solutions of some subproblems are known (we represent the set of these subproblems by K^1) and the subproblem solutions of the other subproblems are unknown (we represent the set of these subproblems by K^0). Partial solutions are the building blocks of constructive heuristics.

Partial solutions can be evaluated by the four previous functions by considering only the components that belong to the partial solution (replacing the K by K^1). Additionally, the feasibility value of a partial solution may also be obtained by

$$evalReducedCosts(s) = \sum_{k \in K^1} \bar{c}^k_{s(k)}$$

where $\bar{c}^k_{s(k)}$ is the reduced cost of the selection variable $y^k_{s(k)}$.

The rationale behind the use of reduced costs is that when evaluating partial solutions, a solution with a lower actual objective function but higher reduced cost may be worse than one with higher actual objective function but lower reduced cost, because the reduced cost takes into account the influence of the variable in the constraints. An evaluation based on reduced cost is less greedy than one based on the actual cost. This idea was proposed in [26] for set covering and set partitioning problems.

Moves

There are two elementary moves in SearchCol: adding a subproblem solution to a partial solution and removing a subproblem solution from a partial or full solution. The add move is represented by the function $s'' = addMove(s', s(k))$. Assuming s' has not a subproblem solution associated with k. The remove move is represented by the function $s'' = dropMove(s', s(k))$. Assuming s' has a subproblem solution associated with k.

In both moves, the feasibility value of the solution can be obtained by summing (subtracting) the cost or reduced cost of the subproblem solution being added (removed). The update of the infeasibilities value requires, for each global constraint where the subproblem solution has a non zero coefficient, the calculation of the new slack, if the constraint is violated, and its amount of violation.

Based on these two elementary moves another one can be defined which replaces the solution of a subproblem by another:

$$s'' = replaceMove(s, s(k)) = addMove(DropMove(s, s'(k)), s(k))$$

Where $s'(k)$ is the solution of subproblem k in solution s. If $s(k) = s'(k)$ the same solution s is obtained, i.e. $s'' = s$.

11.5.2.2 Models with Static Variables

Full solutions

In the most general decomposition model, a solution s is evaluated by solving the MIP (E):

$$Min \sum_{t\in T} c_t y_t + \sum_{k\in K} c^k_{s(k)} + M \sum_{i\in I^=} (a_i^+ + a_i^-) + M \sum_{i\in I^{\leq}} a_i^- + M \sum_{i\in I^{\geq}} a_i^+ \qquad (E)$$

subject to

$$a_i^+ - a_i^- + \sum_{t\in T} a_{it} y_t = b_i - \sum_{k\in K} a^k_{is(k)} \qquad i \in I^=$$

$$-a_i^- + \sum_{t\in T} a_{it} y_t \leq b_i - \sum_{k\in K} a^k_{is(k)} \qquad i \in I^{\leq}$$

$$a_i^+ + \sum_{t\in T} a_{it} y_t \geq b_i - \sum_{k\in K} a^k_{is(k)} \qquad i \in I^{\geq}$$

$$y_t \in \{0,1\} \qquad t \in T^b$$

$$y_t \geq 0 \text{and integer} \qquad t \in T^i$$

$$y_t \geq 0 \qquad t \in T^c$$

$$a_i^+, a_i^- \geq 0 \qquad i \in I^=$$

$$a_i^- \geq 0 \qquad i \in I^{\leq}$$

$$a_i^+ \geq 0 \qquad i \in I^{\geq}$$

Model (E) includes the artificial variables, represented by a_i^+ and a_i^-, for equality constraints (set $I^=$), less than or equal to constraints (set $I^{\leq}$), and greater than or equal to constraints (set $I^{\geq}$). Artificial variables have a large objective function coefficient, M. This allows still evaluating solution s even if it is not feasible in (D) (if there are no values for the static variables such that all constraints are obeyed). Value M must be such that that an optimal solution to (E) has all artificials with value zero, unless there are no feasible solutions with all artificials with value zero.

The infeasibility value of a solution is given by the sum of the values of the artificial variables. The feasibility value is defined only when the sum of the value of the artificial variables is zero and is equal to the optimal value of (E).

Note that model (E) is much easier to solve than model (D) since it does not have selection variables: $\sum_{k\in K} c^k_{s(k)}$ and $\sum_{k\in K} a^k_{is(k)}$ are constants. Model (E) is a linear programming model if all static variables are continuous.

When there are static variables, it may be possible to evaluate a solution in a more efficient manner if problem-specific characteristics are used. For example, in the min-max problem for single path routing, given the set of paths (one for each commodity) the maximum load of an arc can be obtained much more efficiently by computing the highest value among all arc loads than by solving a linear programming model.

Partial solutions

The evaluation of full solutions is easily extendable for partial solutions. In this case, the selection variables associated with the subproblems belonging to K^0 are set to 0.

Moves

The two types of elementary moves, *addMove* and *dropMove*, defined for the case when there are no static variables are not efficient for the case where there are static variables. In this case, a MIP (or LP) must be solved for each move. To alleviate this issue, we define the concept of optimal move. An optimal move is associated with one subproblem k and results from the optimization of the MIP problem where the selection variables associated with the other subproblems are fixed to 1 or 0. All the the selection variables associated with subproblem solutions in the current solution are fixed to 1 (subproblems in set K^1). All the others selection variables, except the ones associated with the subproblem defining the move, are fixed to 0 (subproblems in set K^0). After the optimization, the subproblem solution for subproblem k is obtained directly from the values of the selection variables. We represent this function by

$$s' = optimalMove(s,k)$$

An extension of the optimal move concept for defining a set of subproblems solutions is straighforward by defining the set K^* of those subproblems instead of only one subproblem.

11.6 SearchCol

In this Section, we detail the main concepts of the SearchCol framework. In subsection 11.6.1 we give an overview of SearchCol including the description of its main parameters. In subsection 11.6.2, the parameters related with evaluating solutions are introduced. In the following subsections, we detail the two components involved in the exchange of information between column generation (CG) and metaheuristic (MH) search: initial solutions (subsection 11.6.3) and perturbations (subsection 11.6.4). In the last subsection, subsection 11.6.5, we describe the stopping criteria. A fundamental component of SearchCol is (perturbed) CG which was already detailed in Section 11.3. The other fundamental component is the MH search which will be detailed in Section 11.7.

11.6.1 Overview

The starting point for SearchCol is a MIP model with an exponential number of decision variables, each of them associated with a solution of a subproblem, i.e. the MIP decomposition model (D) of Section 11.2. The first step in a SearchCol algorithm is solving the linear relaxation of the MIP model by CG. In the second step, the subproblem solutions generated by CG are seen as the components of an overall solution in the combinatorial decomposition model (C). A MH search based on the representation of a solution as being made of one solution from each subproblem is conducted. This representation, and all the algorithmic components defined for the MH search, are problem independent. In the next step, a perturbation for CG is defined. A perturbation consists in a set of additional constraints in the restricted master problem (RMP) of CG, each one forcing one decision variable of the subproblem to have the value 1 or 0. This is accomplished by representing the subproblem variable through variables in the RMP as detailed in subsection 11.3.5. The aim of the perturbated CG is the generation of new subproblem solutions, that, hopefully, will improve the incumbent solution. After perturbed CG, a new search begins and then another iteration starts with the definition of a perturbation for CG. This cycle is repeated until one of the stopping criteria is met, as can be seen in Figure 11.3.

```
1: Column generation
2: Search
3: repeat
4:    Define column generation perturbation
5:    Optimize perturbed column generation
6:    Search
7: until Stopping criterion fullfilled
```

Fig. 11.3 SearchCol algorithm

In each iteration of SearchCol, a different search space is defined. The first search space corresponds to the subproblem solutions obtained during the linear relaxation. In each iteration, this search space is enlarged by the subproblem solutions generated through perturbed CG. After a perturbed CG step, the subproblem solutions with higher reduced costs (in the linear relaxation and not in the current iteration) may be removed from CG and from the search space. As many aspects of SearchCol, the number of columns that triggers the removal of subproblem solutions is given by a parameter, *PARDECmaxnumcols*. The dimension of the basis of the linear relaxation of (D), $|K|+|I|$ (the number of selection constraints plus the number of global constraints), is used as the unit for this parameter. If the number of subproblem solutions exceeds *PARDECmaxnumcols* $\times (|K|+|I|)$, a purge of columns is performed. Parameter *PARDECmaxnumcols* must be ≥ 1. Note that if the reduced cost of a variable (in CG with no perturbations) is greater than the difference between the value of the incumbent solution and the lower bound provided by CG, any solution with the corresponding subproblem solution will always be worst than the incumbent (by definition of reduced cost). Removal of these type of subproblem solutions can only improve the efficiency of the search and never worsten its effectiveness.

The search space of a given iteration is influenced by perturbations. The subproblem variables with value 1 can be seen as attributes (features) of a solution that are desired or avoidable. If, in a perturbation, a subproblem variable is fixed to 1 then, in general, more subproblem solutions with the corresponding attribute will be generated by (perturbed) CG. If a subproblem variable is fixed to 0 then, in general, subproblem solutions without the corresponding attribute will be generated by (perturbed) CG. All previous perturbations are disregarded when defining the perturbed CG for an iteration.

In order to conduct the search in the overall search space, there are several alternatives for defining perturbations. When the incumbent solution is infeasible, we use perturbations based on subproblem variables which contribute to the constraints being violated. When the incumbent solution is feasible there are the following alternatives for perturbations: (i) some subproblem variables in the constraints with higher (absolute) duals, (ii) some subproblem variables with values close to 1 in the last (perturbed) CG step, (iii) some subproblem variables in the incumbent, (iv) some subproblem variables based on the memory of the search.

To be used in memory perturbations, the search step keeps two types of memory. The recency memory value of a subproblem variable is incremented each time the solution variable is present in a good quality solution (e.g. a local optimum). The frequency memory of a subproblem variable is incremented each time the solution variable is present in a representative solution (e.g. a current solution in local search).

One way of achieving intensification and diversification is by applying different perturbations at different iterations. We define two type of iterations: in a *plus* iteration (related to intensification) the incumbent was improved in the previous iteration and in a *minus* iteration (related to diversification) that did not happen. In a plus iteration, good attributes are reinforced (for example, subproblem variables with value 1 in the incumbent, or with value 1 in many local optima are fixed to 1). In a minus iteration, attributes already explored are forbidden (roughly speaking, solutions with that attribute were already explored and the incumbent was not improved, therefore the corresponding subproblem variable is set to 0). Perturbations are described in detail in subsection 11.6.4.

Perturbations may be seen as an input for CG provided by the search. Information for defining an initial solution may be seen as an input for search provided by CG (of course, another input is the search space itself). In SearchCol, there are several alternatives for defining how the initial solution is obtained: (i) randomly, (ii) based on the values of the primal variables associated with the last RMP solved (deterministically or randomly), (iii) based on when the subproblem solutions were generated, (iv) through constructive greedy heuristics (deterministic or randomized). The alternatives for obtaining initial solutions are described in detail in subsection 11.6.3.

SearchCol defines four stopping criteria: (i) a given time limit is reached, (ii) a given number of iterations without incumbent improvement is reached, (iii) a desired quality of the incumbent is achieved, (iv) a given number of search steps is reached. Stopping criteria are described in detail in subsection 11.6.5.

Note that CG provides a lower bound (in minimization problems) that may be used to calculate the quality of the incumbent solution to be used in the third stopping criterion. This lower bound also allows the assessment of the quality of the solution obtained, which is not usually possible with usual MHs.

In Table 11.2, the main parameters of SearchCol and their purpose are provided. The algorithm with a detail in which these parameters are present is given in Figure 11.4.

We remind that CG was already addressed in Section 11.3 and search (shaded in Figure 11.4) will be detailed in the next Section. All the main parameters and other parameters are detailed in the following subsections.

Table 11.2 Main parameters of SearchCol

Parameter	Purpose
PARSCinfeasobj	Evaluation of infeasible solutions
PARSCfeasobj	Evaluation of feasible partial solutions
PARSCinitialfirst	Determination of an initial solution the first time a search is conducted
PARSCinitialother	Determination of an initial solution *not* in the first time a search is conducted
PARSCmakefeasible	Specification of the perturbation to use when the incumbent is infeasible
PARSCperturbationplus	Specification of the perturbation to use when there was an improvement in the previous iteration
PARSCperturbationminus	Specification of the perturbation to use when there was no improvement in the previous iteration

11.6.2 Evaluating Solutions

In Table 11.3 and in Table 11.4 the possible values and corresponding functions, introduced in subsection 11.5.2, used for the two parameters related with evaluating solutions are given, parameters *PARSCinfeasobj* and *PARSCfeasobj*. These parameters define the evaluation of solutions only for models *without* static variables. The parameter *PARSCfeasobj* only influence the evaluation of partial solutions (in particular in greedy procedures). For full solutions, the evaluation is based on the real costs because it is the only meaningful value since it gives the true objective function value. For models with static variables, the evaluation is done by solving a MIP and a parameter *PARSCevalmipmaxtime*(>0) sets the maximum time allowed for its optimization.

Table 11.3 Possible values and associated functions for SearchCol parameter *PARSCinfeasobj*

Value	Function
2	evalNumberViolated(s)
3	evalAmountViolation(s)
4	evalWeightedViolation(s)

Table 11.4 Possible values and associated functions for SearchCol parameter *PARSCfeasobj*

Value	Function
2	evalOriginalCosts(s)
3	evalReducedCosts(s)

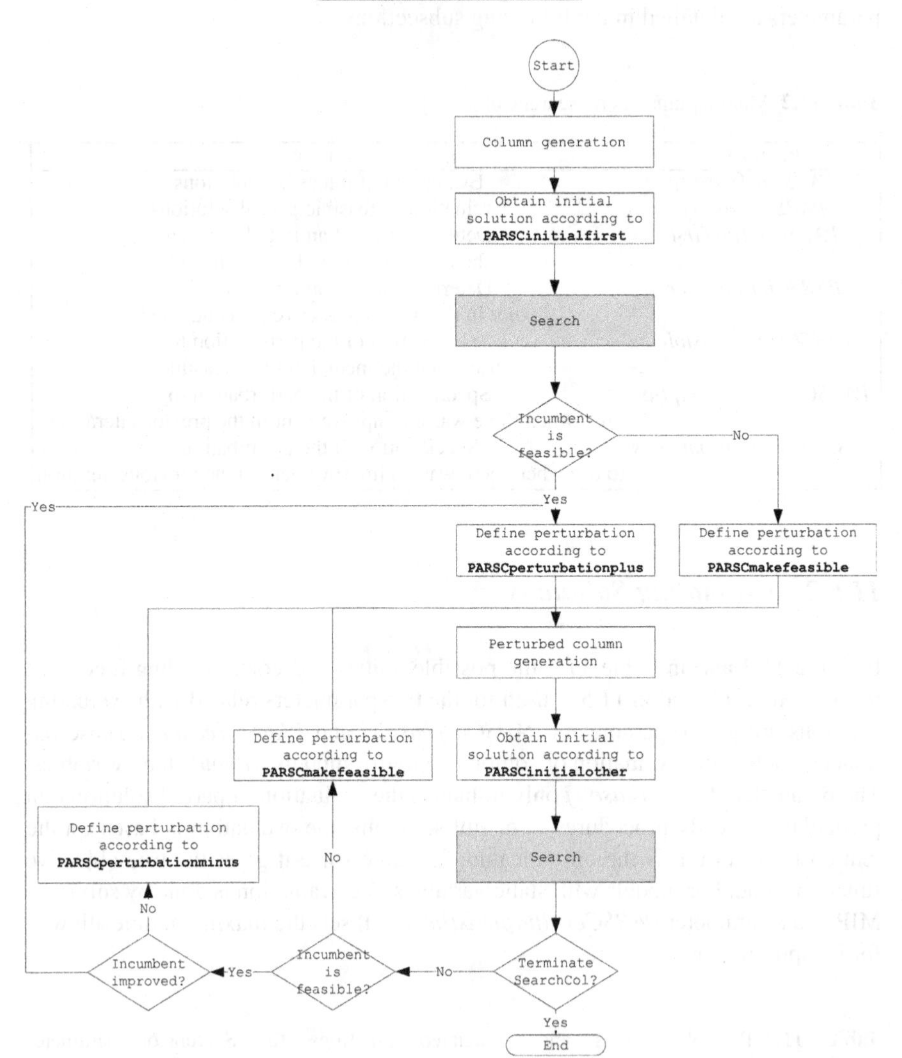

Fig. 11.4 Detailed algorithm for SearchCol

11.6.3 Initial Solutions

11.6.3.1 Uniform Random

In this alternative, a solution is defined randomly, based on the uniform distribution. In this case, each subproblem picks a solution randomly with all the subproblem solutions having the same probability of being chosen. We represent the associated function by $genSolutionRandomUniform()$.

11.6.3.2 Based on the CG Optimal Solution

In the first alternative, the subproblem solution selected is the one associated with the selection variable with higher solution value in the last RMP solved. We represent the associated function by $genSolutionRounding()$. In the second alternative, a solution is defined randomly, but now the probability of a subproblem solution being chosen is given by its solution value in the last RMP solved. We represent the associated function by $genSolutionRandomBiased()$.

11.6.3.3 Based on the CG History

In the first alternative, the selected subproblem solution is the last one that was generated by the subproblem. Note that this subproblem solution may have a value 0 at the last RMP. Since after this subproblem solution no more solutions from the subproblem were generated, the solution can be viewed as the one which stabilized the subproblem. We represent the associated function by $getLastCreatedCGSolution()$. In the second alternative, the selected subproblem solution is the first one that was generated by the subproblem. In general, this subproblem solution corresponds to an optimal solution of the subproblem if the global constraints were removed from the model. In that sense, is the best solution that the subproblem may provide. We represent the associated function by $getFirstCreatedCGSolution()$. In the third alternative, the selected subproblem solution is the one obtained when the subproblem was solved for the last time, i.e. when the dual solution was optimal. For that reason, most Lagrangean heuristics start from these subproblem solutions. We represent the associated function by $getCGSolution()$. Note that the first and third alternatives are different because the last subproblem solution generated by the subproblem is not necessarily the one obtained in the last iteration of the RMP. In fact, in the last iteration of CG no subproblem solutions are generated, otherwise at least one more iteration was required.

11.6.3.4 Greedy Construction

A solution can be constructed based on greedy procedures. In the first alternative, a deterministic greedy procedure is used. In each step, the subproblem solution which conducts to a better (partial) solution is chosen. Of course, solutions from subproblems for which there are already a solution are excluded from this candidate list. We represent the associated function by *constructDeterministicGreedy*(). The other alternative is the randomized constructive phase of GRASP [57], where in each iteration a subproblem solution is randomly chosen from a restricted candidate list which size is controlled by a parameter $\alpha \in [0,1]$. We represent the associated function by *constructRandomGreedy*(α). For $\alpha = 1$ the randomized greedy construction is the same as *genSolutionRandomUniform*() and for $\alpha = 0$, the the randomized greedy construction is the same as *constructDeterministicGreedy*().

For models without static variables, the constructive step is straightforward through the use of the functions *addMove* (for evaluating potential components and perform the move) and *dropMove* (for restoring the current partial solution after an add move was done for evaluation). For models with static variables, the elements of the restricted candidate list are the best solution from each subproblem (with no solution in the current iteration). The best solution from each subproblem is obtained by applying the optimal move described in subsubsection 11.5.2.2.

11.6.3.5 Initial Solution Parameters

We make a distinction between how an initial solution is obtained for the first search step and for the other search steps inside the loop (see Figure 11.4). For example, *getCGSolution*() can be used in the first situation and *constructDeterministicGreedy*() in the second. We define the parameters *PARSCinitialfirst* and *PARSCinitialother* for each of these situations. For *PARSCinitialother* an adiditional value, 8, is available, corresponding to use the incumbent solution as the initial solution. In this case, CG was used in the last iteration only for modifying the search space. Their possible values and corresponding functions are given in Table 11.5.

Table 11.5 Possible values and associated functions for SearchCol parameters *PARSCinitialfirst* and *PARSCinitialother*

Value	Function
0	genSolutionRounding()
1	genSolutionRandomUniform()
2	genSolutionRandomBiased()
3	getLastCreatedCGSolution()
4	getFirstCreatedCGSolution()
5	getCGSolution()
6	constructDeterministicGreedy()
7	constructRandomGreedy(*PARSCalpha*)

In Table 11.5, *PARSCalpha* belongs to $[0,1]$ and corresponds to the α parameter of the constructive phase of GRASP.

11.6.4 *Defining Perturbations*

As defined in subsection 11.3.5, a perturbation is a constraint inserted in the RMP that fixes a subproblem variable to 0 or to 1. We define different alternatives for defining perturbations depending on the feasibility of the incumbent solution.

11.6.4.1 Infeasible Incumbent

When the incumbent solution is infeasible, we define two alternatives for defining the set of perturbations. In both of them, we identify the set of subproblem variables that imply a non null coefficient in the constraints that are being violated. Then we use Table 11.6 to decide if the subproblem variable is fixed to 0 or is fixed to 1. As an example, consider the second and third lines of that table which correspond to less than or equal to violated constraints. Since the constraint is violated, the slacks are negative (by definition, the value of the slack is right-hand side minus the left-hand). In the first line of the table, a subproblem variable has a positive coefficient (more rigorously, the selection variables associated with subproblem solutions where the variable has value 1 have a positive coefficient) in the constraint. Therefore, it has a positive contribution to the violation of the constraint, which we attempt to avoid by fixing its value to zero. In the second line of the table, a subproblem variable has a negative coefficient in the constraint. Therefore, it has a negative contribution to the violation of the constraint, which we attempt to reward by fixing its value to one.

Table 11.6 Determining if a subproblem variable is fixed to 0 or 1

Sense of the constraint	Signal of the slack	Sign of the coefficient of the subproblem variable	Fixed to
$\leq$	-	+	0
$\leq$	-	-	1
$\geq$	+	+	1
$\geq$	+	-	0
=	-	+	0
=	-	-	1
=	+	+	1
=	+	-	0

If a subproblem variable is forced to 0 and simultaneously to 1 by two different perturbations, no perturbations on that variable are considered. In many models, this situation does not occur because global constraints have some type of structure. For example, in the min cost single path routing, all the constraints are less than or equal to constraints and the contributions of the variables are always positive.

The subproblem variables considered can be the ones of the incumbent solution or all the subproblem variables. The functions *makeFeasibleInc*() and

$makeFeasibleAll()$ represent these two alternatives. For example, if the capacity of an arc is being violated in a min cost single path routing problem, and the second perturbation is used, the arc will be forbidden for all the commodities. If the perturbation based on the incumbent is used, the arc is forbidden only for the commodities that use the arc in the incumbent solution.

When the incumbent is infeasible the perturbation to used is given by the parameter $PARSCmakefeasible$ with the possible values and associated functions given in Table 11.7.

Table 11.7 Possible values and associated functions for SearchCol parameter $PARSCmakefeasible$

Value	Function
0	makeFeasibleInc()
1	makeFeasibleAll()

11.6.4.2 Feasible Incumbent

If the current solution is feasible, we define the following perturbations.

Perturbation based on the duals

A parameter $p \in [0,1]$ defines the proportion of global rows that will be considered for the perturbation. The ones with higher duals in absolute value are selected. The subproblem variables appearing on those rows and belonging to the incumbent are fixed to 1 or 0 depending on argument b. The rationale behind the choice of the rows with higher values of the duals is using the dual values to estimate the importance of the rows, in the sense of the sensitivity analysis of linear programming. For example, in the min cost single path routing problem, the dual of a row corresponds to the marginal value of the capacity associated with the arc (of course in the linear relaxation of the problem). Arcs with higher duals are more attractive arcs for the commodities than the others, as the arcs may be seen as scarce resources. In the limit, a row with a dual variable equal to zero could be removed from the model without modifying the (linear relaxation) optimal solution. Function $perturbBasedDuals(p,b)$ represent this perturbation.

Perturbations based on column generation

In this type of perturbation the optimal solution of the last CG solved is used as well as a real parameter $p \in [0,1]$. The subproblem variables with a value greater than or equal to $1-p$ in the fractional solution given by the last CG are fixed to b, which may take value 1 or value 0. Function *perturbBasedCG*(p,b) represent this perturbation.

Perturbation based on the incumbent

In this type of perturbation the incumbent solution is used. Some subproblem variables with value 1 in the incumbent solution are fixed to b, which may take value 1 or value 0. Function *perturbBasedIncumbent*(p,b) represents this perturbation where p stands for the proportion of variables, randomly selected, to be fixed.

Perturbation based on memory

The memory for defining perturbations is based on subproblem variables. The recency memory value of a subproblem variable is incremented each time the solution variable is present in a good quality solution (e.g. a local optimum). The frequency memory of a subproblem variable is incremented each time the solution variable is present in a representative solution (e.g. a current solution in local search). The recency memory is used for fixing to b, which may take the value 1 or the value 0, all the variables with a recency value higher than a threshold (given as a proportion by a parameter p). When using $b = 1$, we intend to preserve good quality attributes of the subproblem solutions. The frequency memory is used for fixing to b, which may take the value 1 or the value 0, all the variables with a frequency value higher than a threshold (given as a proportion by a parameter p). When using $b = 0$, we intend to avoid attributes of subproblem solutions that were already present in many portions of the search space explored so far. We represent these perturbations by functions by *perturbBasedRecency*(p,b) and *perturbBasedFrequency*(p,b)

Parameters for perturbations

The actual perturbation used in an iteration of SearchCol depends on two parameters. If the there was an improvement in the last iteration, the perturbation is defined

Table 11.8 Possible values, associated functions and arguments for SearchCol parameters *PARSCperturbationplus* and *PARSCperturbationminus*

Value	Function	p
1	*perturbBasedDuals*$(p,1)$	PARSCproportionrows1
2	*perturbBasedDuals*$(p,0)$	PARSCproportionrows0
3	*perturbBasedCG*$(p,1)$	PARSCbinarythreshold1
4	*perturbBasedCG*$(p,0)$	PARSCbinarythreshold0
5	*perturbBasedIncumbent*$(p,1)$	PARSCproportionvariables1
6	*perturbBasedIncumbent*$(p,0)$	PARSCproportionvariables0
7	*perturbBasedRecency*$(p,1)$	PARSCrecencythreshold1
8	*perturbBasedRecency*$(p,0)$	PARSCrecencythreshold0
9	*perturbBasedFrequency*$(p,1)$	PARSCfrequencythreshold1
10	*perturbBasedFrequency*$(p,0)$	PARSCfrequencythreshold0

by the value of *PARSCperturbationplus*. Otherwise, the perturbation is defined by *PARSCperturbationminus*. For example, $perturbBasedMemory(0.5,1)$ can used in the first situation and $perturbBasedCG(0.1,0)$ in the second. The possible values, corresponding functions and arguments (which are themselves SearchCol parameters) are given in Table 11.8. The parameter p is a real belonging to the $[0,1]$ interval.

11.6.5 Termination

SearchCol has four stopping criteria: if one of them is verified, the algorithm stops and the best solution found so far is returned. The first criterion is a time limit. The second criterion is a maximum number of iterations without improvement of the incumbent solution. The third criterion is the maximum number of iterations. The last criterion is the relative gap being less than a given parameter.

$$|z_{inc} - z_{LR}| / |z_{inc}|$$

where z_{inc} is the value of the incumbent solution and z_{LR} is the value of the linear relaxation given by CG (without perturbations).

In Table 11.9 the parameters used for implementing the stopping criteria and their possible values are presented.

Table 11.9 Parameters and possible values for SearchCol stopping criteria

Parameter	Possible values
PARSCmaxtime	>0
PARSCmaxnumiterwithoutimprov	>0
PARSCmaxnumtotaliter	>0
PARSCrelgap	$[0,1]$

11.7 Metaheuristic Search in SearchCol

In this Section we detail how the search steps of SearchCol are performed and provide examples of actual SearchCol algorithms. In subsection 11.7.1 we introduce additional algorithmic components that are used, together with the ones presented in the last section, for the definition of (hybrid) metaheuristics (MHs) for the search steps. Subsections 11.7.2, 11.7.3, and 11.7.4 are devoted to the description of Search algorithms with three different search approaches: multi-start local search (MSLS), variable neighborhood search (VNS) and MIP.

11.7.1 Additional Algorithmic Components

11.7.1.1 Local Search

We define the k-neighborhood of a solution as the set of solutions which are obtained by changing the subproblem solution of k or less subproblems. For example, in single path routing, a 1-neighbor is a solution which is obtained by changing the path of one commodity and a 2-neighbor is a solution which is obtained by changing the paths of one commodity or the path of two commodities.

The size of the 1-neighborhood is $\sum_{k \in K}(n_k - 1)$ where n_k is the number of subproblem solutions for subproblem k. For 1-neighborhood we define the function $neighbor1(s,d)$ which returns the best solution in the neighborhood or the first solution better than the current one found. The parameter d specifies the chosen alternative.

The size of the 2-neighborhood is $\sum_{k1 \in K}(n_{k1} - 1)\sum_{k2 \in K:k2>k1}(n_{k2} - 1)$. For 2-neighborhood we define the function $neighbor2(s,d)$ which has a similar interpretation to $neighbor1(s,d)$.

Based on the functions $neighbor1(s,d)$ and $neighbor2(s,d)$ four different local search algorithms can be defined according to the type of neighborhood (1-neighborhood or 2-neighborhood) and the descent strategy (best or first improvement). In Table 11.10 the value 0 to *PARSCdescentstrat* corresponds to best improvement and 1 to first improvement.

Table 11.10 Parameters and possible values for SearchCol local search functions

Parameter	Possible values
PARSCneigh	1 and 2
PARSCdescentstrat	0 and 1

11.7.1.2 Initial Solution Based on Other Solution

In this subsubsection, we introduce how a solution can be obtained based on other solution.

Random solution from the k-neigborhood

The first function of this type returns a randomly generated k-neigborhood solution. A proportion of $1 - p$ subproblem solutions are kept from the base solution, where $p \in]0,1[$. The solutions for the remaining subproblems are randomly chosen. We represent this function by $perturbRandomly(s,p)$.

Solutions based on memory

The MH framework incorporates medium- and long-term memories based on subproblem solutions (note that this memory is different from the memory used in defining perturbations, which is based in subproblem *variables*). Both are based on having one integer (for each type of memory) associated with each subproblem solution counting how many times it appeared in representative solutions of the desired type of memory. For medium-term memory, we are interested in recording good quality solutions. An example of a representative set of solutions is a set of local optima solutions. For long-term memory, we are interested in recording solutions representative of portions of the search space already explored. An example of a representative set of solutions is the set of solutions that were current solutions at some time in a local search algorithm. Functions $perturbBasedRecencySols(s,p)$ and $perturbBasedFrequencySols(s,p)$ represent these two alternatives of getting a solution based on another one. Both functions sort the subproblem solutions by non increasing order of recency/frequency. For recency, the first $p \in]0,1[$ times the number of subproblems are kept. For frequency, the first $p \in]0,1[$ times the number of subproblems are changed. In both cases, the solutions to be changed are replaced randomly.

Parameters

In Table 11.11 the functions used for creating a solution based on other solution and its associated parameters are provided. Parameter p is a real belonging to the interval $[0,1]$.

Table 11.11 Functions and possible values for parameters for creating a solution based on other solution

Alternative	Function	p
8	$perturbRandomly(s,p)$	$PARSCperturbintensityrand$
9	$perturbBasedRecencySols(s,p)$	$PARSCperturbintensityrecency$
10	$perturbBasedFrequencySols(s,p)$	$PARSCperturbintensityfrequency$

11.7.1.3 Path Relinking

Path relinking is a procedure that attempts to find a better solution in the path between two good quality solutions (a starting solution and a target solution) (see, for example, [63]). A solution, we name it path solution, is initialized to be equal to the starting solution. All the components of the path solution with a different value in the target solution are evaluated in a greedy manner. The best one is fixed in the path solution and the procedure is repeated until all components of the path solution are fixed. In the case of the Searchcol, a component of a solution is a subproblem solution and the evaluation is based on the functions discussed in subsection 11.5.2. We represent the path relinking function by $pathrelinking(s,s')$. If parameter *PARSCpathrelinking* is 1 path relinking is used, otherwise path relinking is not used. Note that, as in the case of the other algorithmic components of this subsubsection, it is up to the actual MH to define where and how path relinking is used.

11.7.2 SearchCol with Multi-start Local Search

In a MSLS algorithm, local search is applied from different initial solutions. In Figure 11.5, an algorithm for SearchCol with MSLS is displayed. The shaded blocks correspond to the MSLS and replace the search blocks in 11.4.

The local search can be performed by one of the four procedures described in 11.7.1.1.

As in SearchCol main cycle, we distinguish between two types of iterations in MSLS: iterations where the MSLS incumbent was improved in the previous iteration and the others. Parameters *PARMSLSinitialplus* and *PARMSLSinitialminus* define how the initial solution is obtained in those two types of iterations. Each of these parameters can take values corresponding to the alternative 1, 2, or 7 presented in subsubsection 11.6.3.5 (the ones involving randomness) or one of three alternatives presented in subsubsection 11.7.1.2. The rationale behind the use of two parameters is exploring intensification and diversification concepts with the available algorithmic components.

The stopping criterion of MSLS is a maximum number of iterations without improvement, represented by *PARSCMSLSmaxiterwithoutimprov*. At the end of each local search step, a path relinking between the local optimum and the incumbent solution is attempted (if $PARSCpathrelinking = 1$). This additional procedure is not represented in Figure 11.5 for clarity.

According to [48], multi-start methods can be classified based on three key elements: memory, randomization, and degree of rebuild (related to the number of solutions components that remain fixed from one iteration to the next start). MSLS based on the available algorithmic components of SearchCol allows the implementation of an actual algorithms in all the spectrum of each of these elements. Different settings of parameters for SearchCol with MSLS result in different algorithms. For example, GRASP with path relinking is obtained by setting *PARMSLSinitialplus* and *PARMSLSinitialminus* to alternative 7 presented in subsubsection 11.6.3.5. MSLS also can be seen as an adaptive memory programming technique [61]. In the main cycle of this unifying perspective for MHs, a solution is generated using data in memory, then this solution is improved and the memory is udpated.

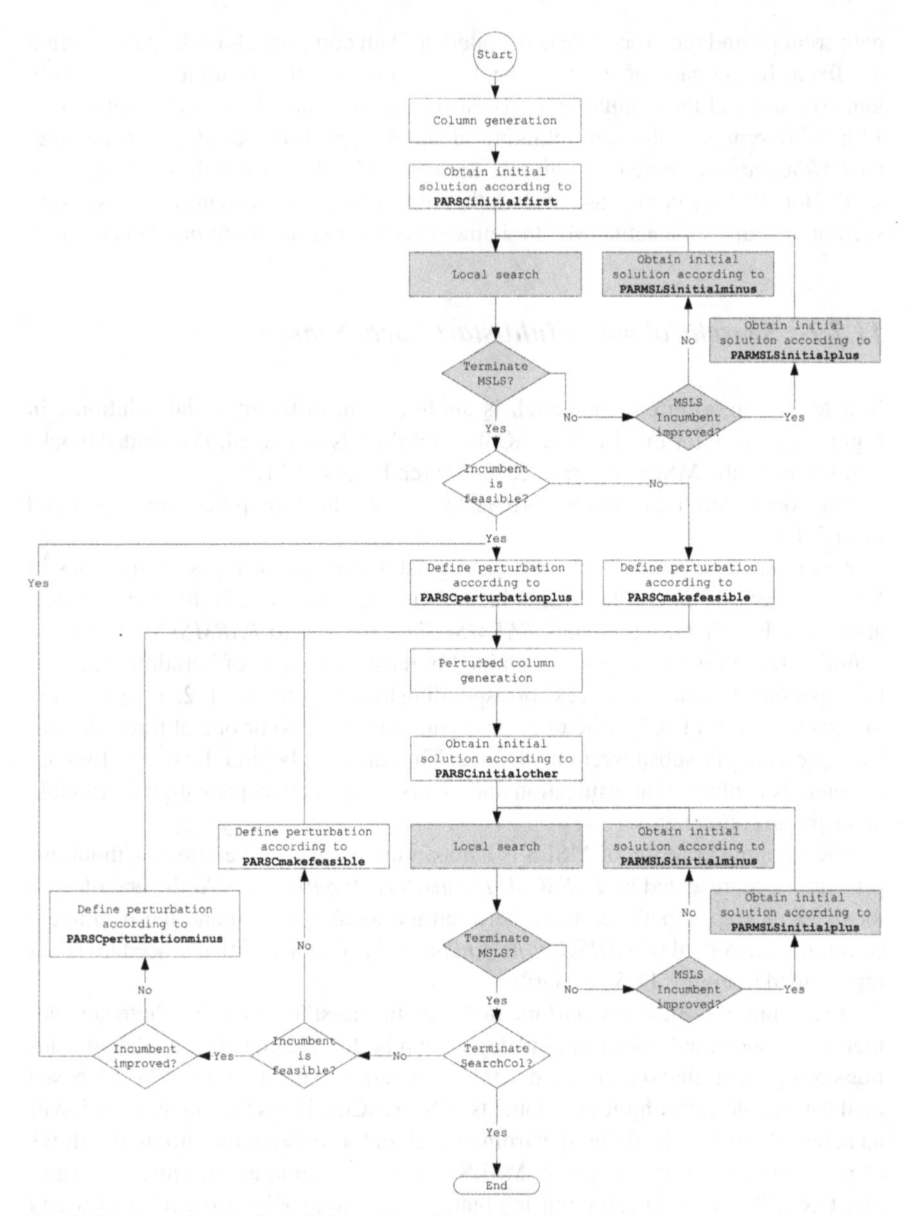

Fig. 11.5 Algorithm for SearchCol with multi-start local search

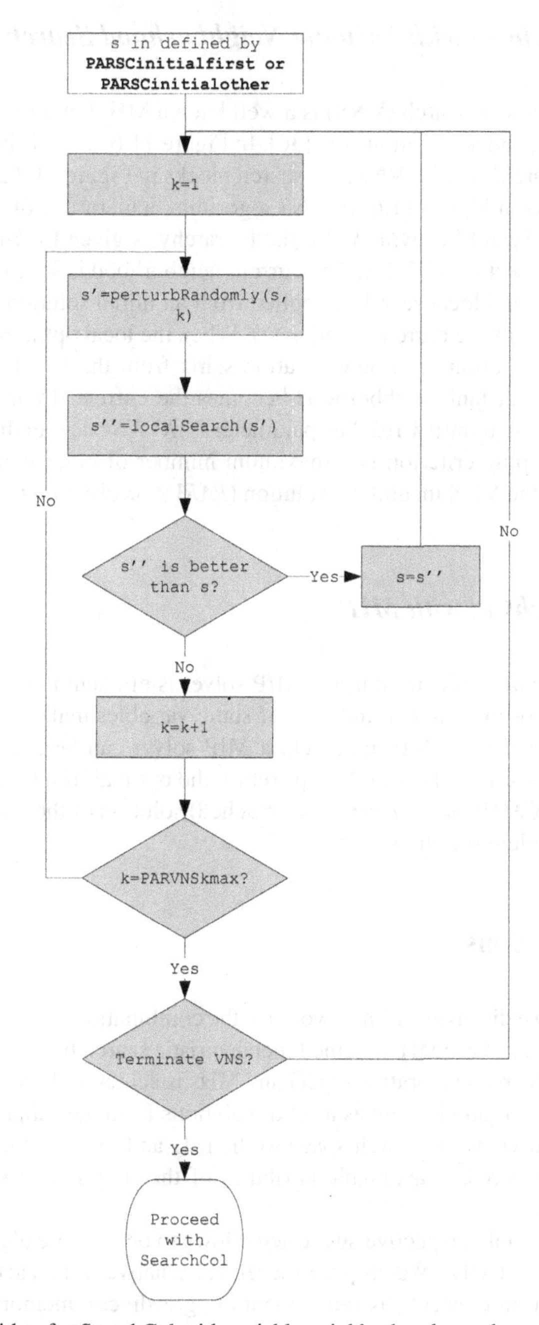

Fig. 11.6 Algorithm for SearchCol with variable neighborhood search

11.7.3 SearchCol with Variable Neighborhood Search

Variable neighborhood search (VNS) is a well known MH. For a recent description of VNS, variants, and applications see [36]. In Figure 11.6, an algorithm for VNS is displayed. In SearchCol with VNS, the search blocks in Figure 11.4 are replaced by the shaded blocks in Figure 11.6. In VNS algorithm, a hierarchy of neighborhoods are defined. For SearchCol with VNS, the hierarchy is given by k-neighborhoods defined in subsubsection 11.7.1.1. The current neigborhood is set to the first of the hierarchy ($k = 1$) and local search is applied from an initial solution (s') in the current neighborhood of the current solution (s). When the local optimum (s'') is better than the current solution (s) a new iteration starts from the $k = 1$ neighborhood, otherwise a more distant neighborhood becomes the current (by incrementing k). This cycle is repeated until k reaches parameter *PARVNSkmax*, ending an outer iteration. The stopping criterion is a maximum number of outer iterations without improvement of the VNS incumbent solution (*PARVNSiterwithoutimprov*).

11.7.4 SearchCol with MIP

For problems where a general-purpose MIP solver is efficient in solving restricted versions of (D), or the number and type of static variables make the evaluation of solutions too heavy for a MH approach, a MIP solver can be used in the search phase, i.e. in Figure 11.3 the search steps return the optimal (if a time limit set by a parameter *PARSCMIPmaxtimemip* is not reached) solution of the problem with the available subproblem solutions.

11.8 Conclusions

In this Chapter, we discussed a framework for the combination of column generation (CG) and metaheuristics (MHs), named metaheuristic search by column generation (SearchCol). A deep collaboration of CG and MHs is achieved through representing a solution to the overall problem as a set of solutions from the subproblems of CG. Basically, CG provides the search space to the MH and, in turn, the MH provides information on desired or avoidable attributes of the subproblem solutions to be generated by CG.

The combinatorial perspective suggested allows to define core algorithmic components for different MHs. We proposed several alternatives for an actual SearchCol algorithm, based on concepts as randomization, greediness, memory, duality, and neighborhoods. We also detailed three examples of actual SearchCol algorithms: SearchCol with multi-start local search, SearchCol with VNS, and SearchCol with MIP.

Although the uselfulness of heuristics based on decomposition models, as the ones resulting from Lagrangean relaxation or the ones using CG, has been proved for several decades in specific problems, their combination in a more general setting seldom has been tried. In this Chapter, we discussed such a general framework for combining CG and metaheuristics and highlighted its potential.

Acknowledgements. This work have been partially funded by FCT (Fundação para a Ciência e a Tecnologia - Portugal) through project "SearchCol: Metaheuristic search by column generation" (PTDC/EIA-EIA/100645/2008) and through the post-doc grant SFRH/BPD/41581/2007 of D. Santos.

References

1. Ahuja, R.K., Magnanti, T.L., Orlin, J.B.: Network flows: theory, algorithms, and applications. Prentice Hall, Englewood Cliffs (1993)
2. Alvelos, F., de Sousa, A., Santos, D.: SearchCol: Metaheuristic Search by Column Generation. In: Blesa, M.J., Blum, C., Raidl, G., Roli, A., Sampels, M. (eds.) HM 2010. LNCS, vol. 6373, pp. 190–205. Springer, Heidelberg (2010)
3. Alvelos, F., Valério de Carvalho, J.M.: Comparing branch-and-price algorithms for the unsplittable multicommodity flow problem. In: Ben-Ameur, W., Petrowski, A. (eds.) Proceedings of the International Network Optimization Conference, INOC 2003, Evry/Paris, pp. 7–12 (October 2003)
4. Alvelos, F., Valério de Carvalho, J.M.: A Local Search Heuristic based on Column Generation Applied to the Binary Multicommodity Flow Problem. In: Proceedings of International Network Optimization Conference, INOC 2007, Spa, Belgium, p. 6 (April 2007)
5. Akker, J.M., van den Hoogeveen, J.A., van de Velde, S.L.: Parallel machine scheduling by column generation. Operations Research 47, 862–872 (1999)
6. Akker, J.M., van den Hoogeveen, H., van de Velde, S.L.: Appplying column generation to machine scheduling. In: Desaulniers, G., Desrosiers, J., Solomon, M.M. (eds.) Column Generation, ch. 11, Springer (2005)
7. Ball, M.O.: Heuristics based on mathematical programming. Surveys in Operations Research and Management Science 16, 21–38 (2006)
8. Barnhart, C., Hane, C.A., Vance, P.H.: Using branch-and-price-and-cut to solve origin-destination integer multicommodity flow problems. Operations Research 48, 318–326 (2000)
9. Barnhart, C., Johnson, E.L., Nemhauser, G.L., Savelsbergh, M.W.P., Vance, P.H.: Branch-and-price: column generation for solving huge integer programs. Operations Research 46, 316–329 (1998)
10. Beasley, J.E.: Lagrangian relaxation. In: Reeves, C.R. (ed.) Modern Heuristic Techniques for Combinatorial Problems. John Wiley and Sons (1993)
11. Blum, C., Aguilera, M.J.B., Roli, A., Sampels, M. (eds.): Hybrid metaheuristics: An emerging approach to optimization. Springer (2008)
12. Blum, C., Cotta, C., Fernandez, A.J., Gallardo, J.E., Mastrolilli, M.: Hybridizations of metaheuristics with branch-and-bound derivatives. In: Blum, C., Aguilera, M.J.B., Roli, A., Sampels, M. (eds.) Hybrid Metaheuristics: An Emerging Approach to Optimization. Springer (2008)
13. Blum, C., Puchinger, J., Raidl, G.R., Roli, A.: Hybrid metaheuristics in combinatorial optimization: A survey. Applied Soft Computing 11, 4135–4151 (2011)

14. Blum, C., Roli, A.: Metaheuristics in Combinatorial Optimization: Overview and Conceptual Comparison. ACM Computing Surveys 35, 268–308 (2011)
15. Boschetti, M., Maniezzo, V., Roffilli, M.: Decomposition Techniques as Metaheuristic Frameworks. In: Maniezzo, V., Stützle, T., Voß, S. (eds.) Matheuristics - Hybridizing Metaheuristics and Mathematical Programming, Annals of Information Systems, vol. 10, ch. 5. Springer (2009)
16. Chen, Z.-L., Powell, W.B.: Solving Parallel Machine Scheduling Problems by Column Generation. INFORMS Journal on Computing 11, 78–94 (1999)
17. Dillenberger, C., Escudero, L.F., Wollensak, A., Zhang, W.: On Practical Resource Allocation for Production Planning and Scheduling with Period Overlapping Setups. European Journal of Operational Research 75, 275–286 (1994)
18. Danna, E., Pape, C.L.: Branch-and-Price Heuristics: A Case Study on the Vehicle Routing Problem with Time Windows. In: Desaulniers, G., Desrosiers, J., Solomon, M.M. (eds.) Column Generation, ch. 4. Springer (2005)
19. Dantzig, G.B., Wolfe, P.: Decomposition principle for linear programs. Operations Research 8, 101–111 (1960)
20. Desaulniers, G., Desrosiers, J., Solomon, M.M. (eds.): Column Generation. Springer, New York (2005)
21. Desrosiers, J., Soumis, F., Desrochers, M.: Routing with time windows by column generation. Networks 14, 545–565 (1984)
22. Desrosiers, J., Dumas, Y., Solomon, M.M., Soumis, F.: Time Constrained Routing and Scheduling. In: Ball, M.O., Magnanti, T.L., Monma, C.L., Nemhauser, G.L. (eds.) Network Routing, Handbooks in OR & MS, vol. 8, ch. 2. Elsevier Science B.V. (1995)
23. Fahle, T., Junker, U., Karisch, S.E., Kohl, N., Sellmann, M., Vaaben, B.: Constraint Programming Based Column Generation for Crew Assignment. Journal of Heuristics 18, 59–81 (2002)
24. Fisher, M.L.: The Lagrangian relaxation method for solving integer programming problems. Management Science 27, 1–18 (1981)
25. Fisher, M.L.: The Lagrangian relaxation method for solving integer programming problems. Management Science 50, 1872–1874 (2004)
26. Fisher, M.L., Kedia, P.: Optimal solutions of set covering/partitioning problems using dual heuristics. Management Science 36, 674–688 (1990)
27. Ford, L.R., Fulkerson, D.R.: A suggested computation for maximal multicommodity network flows. Management Science 5, 97–101 (1958)
28. Frangioni, A.: About Lagrangian Methods in Integer Optimization. Annals of Operations Research 139, 163–193 (2005)
29. Gendreau, M., Potvin, J.-Y. (eds.): Handbook of metaheuristics. Springer (2010)
30. Geoffrion, A.M.: Lagrangean relaxation for integer programming. Mathematical Programming Study 2, 82–114 (1974)
31. Gilmore, P.C., Gomory, R.E.: A linear programming approach to the cutting stock problem. Operations Research 9, 849–859 (1961)
32. Gilmore, P.C., Gomory, R.E.: A linear programming approach to the cutting stock problem - Part II. Operations Research 11, 863–888 (1963)
33. Glover, F., Kochenberger, G. (eds.): Handbook of metaheuristics. Kluwer (2003)
34. Glover, F., Laguna, M.: Tabu Search. Kluwer (1997)
35. Gualandi, S., Malucelli, F.: Constraint programming-based column generation. A Quarterly Journal of Operations 7, 113–137 (2009)
36. Hansen, P., Mladenovic, N., Brimberg, J., Perez, J.A.M.: Variable neighborhood search. In: Gendreau, M., Potvin, J.-Y. (eds.) Handbook of Metaheuristics. Springer (2010)
37. Held, M., Karp, R.M.: The traveling-salesman problem and minimum spanning trees. Operations Research 18, 1138–1167 (1970)
38. Held, M., Karp, R.M.: The traveling-salesman problem and minimum spanning trees: Part II. Mathematical Programming 1, 6–25 (1971)
39. Hopp, W.J. (Editor-in-Chief): Ten Most Influential Titles of "Management Science's" First Fifty Years. Management Science 50 (2004)

40. IEEE Standard 802.1s: Virtual Bridged Local Area Networks - Amendment 3: Multiple Spanning Trees (2002)
41. Jünger, M., Liebling, T.M., Naddef, D., Nemhauser, G.L., Pulleyblank, W.R., Reinelt, G., Rinaldi, G., Wolsey, L.A. (eds.): 50 Years of Integer Programming 1958-2008, From the Early Years to the State-of-the-Art. Springer (2010)
42. Klabjan, D.: Large-scale models in the airline industry. In: Desaulniers, G., Desrosiers, J., Solomon, M.M. (eds.) Column Generation, ch. 3, Springer (2005)
43. Kallehauge, B., Larsen, J., Madsen, O.B.G.: Vehicle Routing with Time Windows. In: Desaulniers, G., Desrosiers, J., Solomon, M.M. (eds.) Column Generation, ch. 3. Springer (2005)
44. Kelley, J.E.: The cutting-plane method for solving convex programs. Journal of the SIAM 8, 703–712 (1960)
45. Lopes, M.J.P., Valério de Carvalho, J.M.: A branch-and-price algorithm for scheduling parallel machines with sequence dependent setup times. European Journal of Operational Research 176, 1508–1527 (2007)
46. Lübbecke, M.E., Desrosiers, J.: Selected topics in column generation. Operations Research 53, 1007–1023 (2005)
47. Maniezzo, V., Stutzle, T., Voss, S. (eds.): Matheuristics, hybridizing metaheuristics and mathematical programming. Springer (2009)
48. Marti, R., Moreno-Vega, J.M., Duarte, A.: Advanced multi-start methods. In: Gendreau, M., Potvin, J.-Y. (eds.) Handbook of Metaheuristics. Springer (2010)
49. Martin, R.K.: Large Scale Linear and Integer Optimization, A Unified Approach. Kluwer Academic Publishers (1999)
50. Monaci, M., Paolo, T.: A Set-Covering-Based Heuristic Approach for Bin-Packing Problems. INFORMS Journal on Computing 18, 71–85 (2006)
51. Ow, P.S., Morton, T.E.: Filtered beam search in scheduling. International Journal of Production Research 26, 35–62 (1988)
52. Pisinger, D., Sigurd, M.: Using decomposition techniques and constraint programming for solving the two-dimensional bin-packing problem. INFORMS Journal on Computing 19, 1007–1023 (2007)
53. Puchinger, J., Raidl, G.R., Pirkwieser, S.: MetaBoosting: enhancing integer programming techniques by metaheuristics. In: Maniezzo, V., Stutzle, T., Voss, S. (eds.) Matheuristics, Hybridizing Metaheuristics and Mathematical Programming. Springer (2009)
54. Raidl, G.R.: A Unified View on Hybrid Metaheuristics. In: Almeida, F., Aguilera, M.J., Blum, C., Moreno Vega, J.M., Perez, M., Roli, A., Sampels, M. (eds.) Hybrid Metaheuristics. Springer (2006)
55. Raidl, G.R., Puchinger, J., Blum, C.: Metaheuristic Hybrids. In: Gendreau, M., Potvin, J.-Y. (eds.) Handbook of Metaheuristics. Springer (2010)
56. Raidl, G.R., Puchinger, J.: Combining (integer) linear programming techniques and metaheuristics for combinatorial optimization. In: Blum, C., Aguilera, M.J.B., Roli, A., Sampels, M. (eds.) Hybrid Metaheuristics: An Emerging Approach to Optimization. Springer (2008)
57. Resende, M., Ribeiro, C.: Greedy randomized adaptive search procedures: advances, hybridizations, and applications. In: Gendreau, M., Potvin, J.-Y. (eds.) Handbook of Metaheuristics, 2nd edn., ch. 10. Springer (2010)
58. Santos, D., de Sousa, A., Alvelos, F.: Traffic Engineering of Telecommunication Networks Based on Multiple Spanning Tree Routing. In: Valadas, R., Salvador, P. (eds.) FITraMEn 2008. LNCS, vol. 5464, pp. 114–129. Springer, Heidelberg (2009)
59. Santos, D., Sousa, A.F., Alvelos, F., Dzida, M., Pióro, M.: Optimization of link load balancing in multiple spanning tree routing networks. Telecommunication Systems 48, 109–124 (2011)
60. Savelsbergh, M.: A branch-and-price algorithm for the generalized assignment problem. Operations Research 45, 831–841 (2007)

61. Taillard, E., Gambardella, L., Gendreau, M., Potvin, J.-Y.: Adaptive memory programming: A unified view of metaheuristics. European Journal of Operational Research 135, 1–16 (2001)
62. Talbi, E.-G.: A taxonomy of hybrid metaheuristics. Journal of Heuristics 8, 541–564 (2002)
63. Talbi, E.-G.: Metaheuristics. John Wiley and Sons (2009)
64. Vanderbeck, F.: Implementing Mixed Integer Column Generation. In: Desaulniers, G., Desrosiers, J., Solomon, M.M. (eds.) Column Generation, ch. 12, Springer (2005)
65. Wilhelm, W.E.: A technical review of column generation in integer programming. Optimization and Engineering 2, 159–200 (2001)
66. Wolsey, L.A.: Integer Programming. John Wiley and Sons (1998)

Chapter 12
Application of Large Neighborhood Search to Strategic Supply Chain Management in the Chemical Industry

Pedro J. Copado-Méndez, Christian Blum,
Gonzalo Guillén-Gosálbez, and Laureano Jiménez

Abstract. Large neighborhood search is a popular hybrid metaheuristic which results from the use of a complete technique—such as dynamic programming, constraint programming or MIP solvers—for finding the best neighbor within a large neighborhood of the incumbent solution. In this work we present an application of large neighborhood search to a strategic supply chain management problem from the Chemical industry, namely the configuration of a three-echelon hydrogen network for vehicle use with the goal of minimizing the total cost. Traditionally, these large-scale combinatorial optimization problems have been solved by means of mathematical programming techniques. Our experimental results show that large neighborhood search has the potential to be a viable alternative, especially when the complexity of the problem grows.

12.1 Introduction

Supply chain management (SCM) problems [18, 15] can be classified into strategic, tactical and operational according to the temporal and spatial scales considered in the analysis [7]. In this work we will focus on the strategic level, which deals with decisions that have a long lasting effect on the firm, such as those related with the establishment of new facilities and transportation links between the supply

---

Pedro J. Copado-Méndez · Gonzalo Guillén-Gosálbez · Laureano Jiménez
Departament d'Enginyeria Quimica, Universitat Rovira Virgili, Tarragona, Spain
e-mail: {pedrojesus.copado,gonzalo.guillen}@urv.cat
laureano.jimenez@urv.cat

Christian Blum
ALBCOM Research Group, Universitat Politècnica de Catalunya, Barcelona, Spain
e-mail: cblum@lsi.upc.edu

E.-G. Talbi (Ed.): Hybrid Metaheuristics, SCI 434, pp. 335–352.
springerlink.com

chain entities. Spatially explicit models have recently gained wider interest in SCM. These formulations are particularly suited for strategic SCM problems in which the supply chain performance shows a strong geographical dependence. They give rise to large-scale MILP models with three types of variables: (1) integers representing the number of facilities opened in a given location, (2) binary variables denoting the existence of transportation links between two sub-regions, and (3) continuous variables that quantify the materials flows and inventory levels. In this work we will deal with a spatially explicit model that concerns the strategic planning of hydrogen supply chains for vehicle use [1, 9, 10, 12, 19].

In spatially explicit SCM models a trade-off exists between modelling accuracy and computational burden. Realistic models require the definition of a large number of discrete variables. Mathematical programming is probably the prevalent approach for solving SCM problems. Hereby, decomposition strategies that exploit the mathematical structure of the problem are sometimes used to make the problem tractable. A general review on the application of mathematical programming techniques in SCM can be found in [14], whereas more specific reviews devoted to process industries have been presented in [8, 16]. Apart from mathematical programming, metheuristics have also been applied so strategic SCM problems. In [21], for example, a method to solve the vehicle routing problem (VRP) is proposed that combines genetic algorithms with mathematical programming. The authors of [5] examine the open vehicle routing problem with time windows (OVRPTW) using tabu search. Several evolutionary algorithms for the application fo SCM models have been proposed in [2], while in [6] the authors employed genetic algorithms for solving the coordinated scheduling of production and air transportation. Other applications can be found in [24, 3].

The goal of this work is the application of a popular algorithm from the field of hybrid metaheuristics to the above mentioned SCM problem. Hybrid metaheuristics [4] are algorithms for optimization that combine metaheuristics with components of other techniques for optimization. Examples are combinations of metaheuristics with dynamic programming, contraint programming, and branch & bound. The specific algorithm that is applied in this work is known as *large neighborhood search* (LNS) [17]. The characteristic feature of LNS algorithms is the use of complete techniques for searching large neighborhoods within a metaheuristic framework. Our method, as shown by means of numerical examples, produces near optimal solutions in a fraction of the computational time required by stand-alone deterministic branch and cut techniques applied to the original full-space MILP. The same approach can be easily extended to tackle similar engineering problems with large numbers of discrete decisions, expediting current solution approaches for a certain class of process systems engineering models.

The remainder of this chapter is organized as follows. In Section 12.2, we provide a generic formulation of spatially explicit supply chain models. The full description of the mathematical model of the hydrogen supply chains for vehicle use is given in Appendix A. In Section 12.3 we describe the proposed LNS approach, whereas in

Section 13.5 the experimental results are outlined in detail. Finnally, the conclusions of the work are presented in Section 12.5.

12.2 Spatially Explicit Supply Chain Models

As mentioned before, we address the solution of MILPs resulting from the formulation of spatially explicit models used in SCM. The problem under study can be formally stated as follows (see also Figure 12.1). Given are a set of available production, storage and transportation technologies that can be adopted in different locations of a region in order to fulfill the demand of a product of interest. We are also given economic and environmental data associated with the establishment and operation of these facilities. The goal of the analysis is to determine the optimal supply chain configuration, including the type of technologies selected, the capacity expansions over time, and their optimal location, along with the associated planning decisions that optimize a predefined objective function.

The strategic planning problem presented above can be described in mathematical terms as an MILP of the following form:

$$\min_{x,Y,N} f(x,Y,N)$$

such that

$$h(x,Y,N) = 0$$
$$g(x,Y,N) \leq 0$$
$$x \subset \mathbb{R}, \quad Y \subset \{0,1\}, \quad N \subset \mathbb{Z}^+$$

This generic formulation includes three types of variables: continuous variables x, denoting capacity expansions, production rates, inventory levels and materials flows; discrete variables N, representing the number of transportation units and production and storage facilities opened in a given region; and binary variables Y employed for modelling the establishment of transportation links between two potential locations within the overall region of interest. The inequality and equality constraints, denoted by $g(x,Y,N)$ and $h(x,Y,N)$ respectively, represent mass balances, capacity limitations and objective function calculations. In this work, without loss of generality, we address the solution of a spatially explicit SCM model that was introduced in [13, 11, 9, 20]. The solution of this multi-period model provides the optimal supply chain structure along with the capacity expansions over time required to follow a given demand pattern.

For the sake of brevity, a detailed description of the mathematical model for the hydrogen supply chains problem for vehicle use can be found in Appendix A. Moreover, further details on the complete MILP formulation can be found in the original works. From now on, we will refer to this model as HYDROGEN.

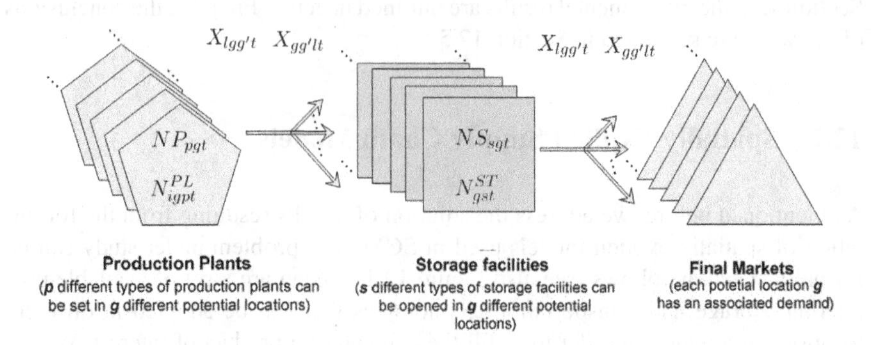

Fig. 12.1 Main sets of decision variables involved in spacially explicit supply chain management

12.3 The Proposed LNS Algorithm

The difficulty in solving model exposed in the previous section is highly dependent on the number of integer and binary variables since they are responsible for the combinatorial complexity of the problem. The number of discrete variables required increases with the number of time periods and sub-regions considered in the model. The MILP can be solved via standard branch-and-cut techniques implemented in software packages such as CPLEX. Models accounting for a large number of time periods and/or sub-regions may lead to branch-and-bound trees with a prohibitive number of nodes, thus making the MILP computationally intractable. We next present a hybrid method, LNS, that combines local search with standard branch and cut for the efficient solution of the tackled problem. LNS was first introduced by [22]. In LNS, an initial solution is gradually improved by alternately destroying and repairing it. This approach combines components from different search techniques, and has many potential applications in the fields of operations research and artificial intelligence. Classifications, taxonomies and overviews on the subject can be found in the work by [4, 23].

All LSN algorithms are based on the observation that searching a large neighbourhood results in finding local optima of high quality. Specifically, LNS decomposes the original problem into a number of smaller sub-problems that are solved in a sequential way. Each sub-problem emerges from a partial solution, in which some decision variables are fixed and others released. A partial solution defines a neighbourhood of solutions that can be explored rather fast by either tailored (e.g., another heuristic or meta-heuristic) or general purpose algorithms (e.g., branch and cut MIP solvers). LNS is a general framework that must be adapted to the particularities of the problem under study. Hence, the definition of the large neighbourhood is highly dependent on the problem of interest. In the simplest case, an appropriate portion of the decision variables is fixed to the values that they have in the current solution, and only the remaining "free" variables are considered by the optimization algorithm (typically, a MIP-solver). If the MIP-solver finds an improved solution, it becomes the new current solution, a new large neighbourhood is defined around it, and the process is repeated in subsequent iterations. Obviously, the selection of the variables that remain fixed and the ones that are subject to optimization, respectively, plays a crucial role in the performance of the algorithm. Particularly, the number of free variables directly defines the size of the neighbourhood. Too restricted neighbourhoods—that is, sub-problems—are unlikely to yield improved solutions, while too large neighbourhoods might result in excessive running times for solving the sub-problems by the MIP-solver. Therefore, a strategy for dynamically adapting the number of free variables is sometimes used. Furthermore, the variables to be optimized might be selected either purely at random or in a more sophisticated guided way by considering the variables with largest potential impact on the objective function and their relatedness. The section that follows describes the main features of our algorithm.

12.3.1 Algorithm

In this section we describe the LNS implementation for our particular problem. The algorithm requires the following input data:

- t_{max}: a maximum execution time of the algorithm;
- n_{max}: a maximum number of variables to be released;
- m_{max}: a maximum number of attempts (the meaning of this parameter is described below).

The algorithm works as follows (see Algorithm 11). First, the initial solution is generated in function *generate_initial_solution*(). The HYDOGENE model includes three main types of discrete variables that are relevant for our algorithm:

- **Integer variables** N_{igpt}^{PL}**:** Number of facilities producing hydrogen in form i using technology p established in location g at period t.
- **Integer variables** N_{gst}^{ST}**:** Number of storage facilities of type s opened in location g at period t.
- **Binary variables** $X_{gg'lt}$**:** Equals 1 if there is a link between g and g' using transportation mode l in period t and 0 otherwise.

The initial solution is generated by solving the HYDROGENE model with the variables NP_{igpt}^{PL}, N_{gst}^{ST} and $X_{gg'lt}$ fixed to the values obtained from a reduced-space model that considers a single time period with a demand equal to the average demands over all the time periods. We have used CPLEX for this purpose. The pseudo-code of this procedure is given in Algorithm 12).

After the generation of the initial solution, the main loop of the algorithm starts. While the maximum computation time limit is not reached, in each trial m the following is done. First, a set V of n variables that are to be released is chosen in

Algorithm 11. LNS for solving the HYDROGENE model

Require: The model HYDROGENE to be solved
AND $t_{max} > 0$ **AND** $m_{max} > 0$ **AND** $n_{max} > 0$
Ensure: s
1: $s :=$ generate_initial_solution()
2: **while** computation time limit t_{max} not reached **do**
3: $n := 1$
4: *improved* := **FALSE**
5: **while** $n \leq n_{max}$ **AND NOT** *improved* **do**
6: $m := 1$;
7: **while** $m \leq m_{max}$ **AND NOT** *improved* **do**
8: $V := choose\_variables\_to\_be\_released(n)$
9: $s' := release\_variables(s, V)$
10: $s'' := MIP\_solve(s')$
11: **if** $f(s'' < f(s)$ **then**
12: $s := s''$
13: *improved* := **TRUE**
14: **end if**
15: $n := n + 1$;
16: **end while**
17: $m := m + 1$;
18: **end while**
19: **end while**

Algorithm 12. Generating the initial solution

for all g **do**
$D_g := \frac{\sum_t^T D_{gt}}{T}$
end for
Solve HYDROGENE considering one period ($t = 1$) with demand D_g
Solve HYDROGENE for all the time periods fixing
$\langle N^{PL}_{igpt}, N^{ST}_{gst}, X_{gg'lt} \rangle := \langle N^{PL}_{igp1}, N^{ST}_{gs1}, X_{gg'l1} \rangle$

function $choose\_variables\_to\_be\_released(n)$. Second, solution s is copied, resulting in solution s'. Next, the n variables from V are released in s'. Third, the CPLEX solver is invoked. The solver determines the best solution that can be obtained on the basis of the partial solution s'. In case $f(s' < f(s)$—where $f(\cdot)$ refers to the value of the objective function—variable *improved* is assigned the value true.

12.4 Experimental Evaluation

In the following subsection we present numerical results that illustrate the performance of LNS as compared to the commercial full-space branch and cut code implemented in CPLEX. We have selected different instances of the HYDROGENE model concerning the number of time periods. More specifically, we tested $t \in \{2,4,6,8,10,12,14,16\}$. As computation time limits for the resulting eight models of HYDROGENE we chose $\{1000,2000,3000,4000,5000,6000,7000,8000\}$

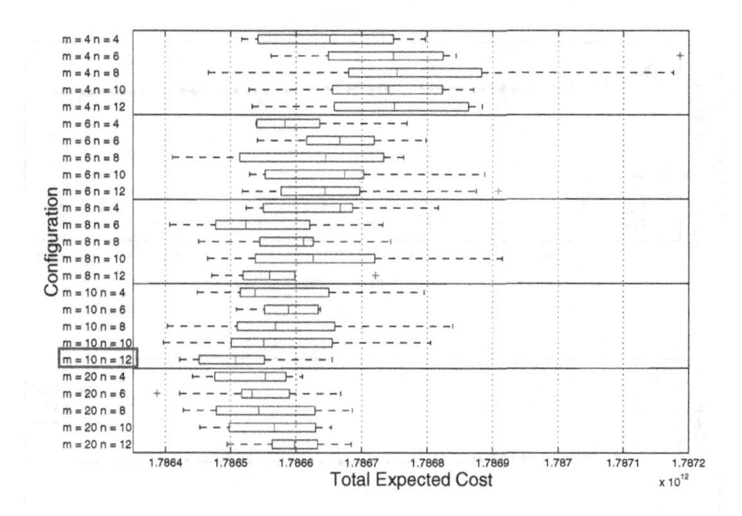

Fig. 12.2 Tuning of algorithm LNS. Note that n and m refer to parameters n_{max} and m_{max}.

seconds. All experiments were performed on a PC Intel (R) Core (TM) Quad CPU Q9550@2.83 GHz and 3 GB of RAM.

12.4.1 Algorithm Tuning

In order to obtain reasonable values for parameters n_{max} and m_{max}, we applied LNS for each combination of n_{max} and m_{max} 10 times to each of the eight different models (resulting from eight different time periods). The values considered for n_{max} are taken from $\{4,6,8,10,20\}$, while the values considered for m_{max} are taken from $\{4,6,8,10,12\}$. The results are shown for each combination of n_{max} and m_{max} in the form of boxplots in Figure 12.2. This is a standard and convenient way of graphically depicting sets of numerical data through their five-number summaries: the smallest observation (sample minimum), lower quartile (Q1), median (Q2), upper quartile (Q3), and largest observation (sample maximum). A boxplot also indicates which observations, if any, are to be considered as outliers. When observing these results, the general impression is that the results become better when m_{max} grows. Concerning n_{max}, no conclusions can be drawn. The final setting that we chose based on these results is marked by a box. In particular, we chose the setting of $n_{max} = 10$ and $m_{max} = 12$.

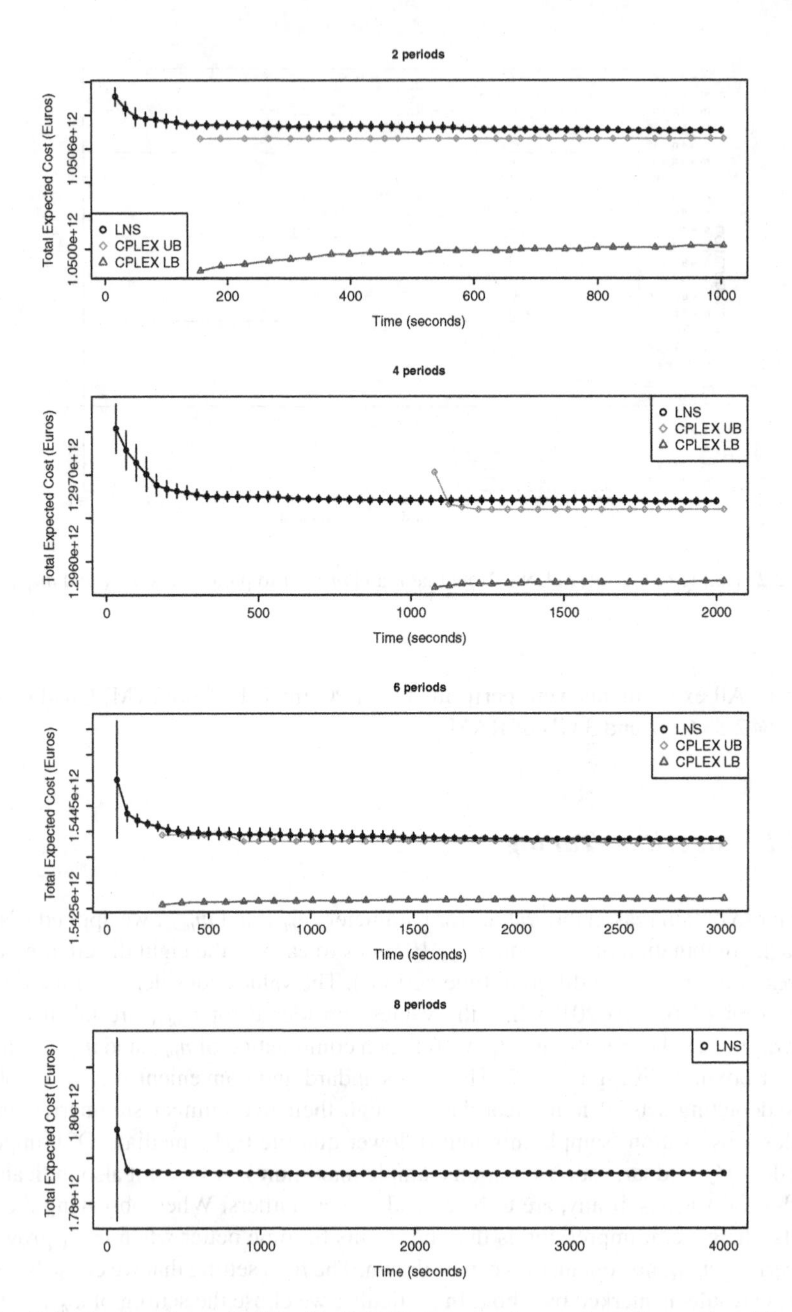

Fig. 12.3 Commparison of LNS with CPLEX (lower and upper bound) over time. The four graphics show the results for a different number of periods (value of t). From top to down, t takes values $\{2,4,6,8\}$). The vertical bars show the standard deviation of LNS over 10 runs. In the cases in which CPLEX results are missing, CPLEX was not able to find any solution within the given time.

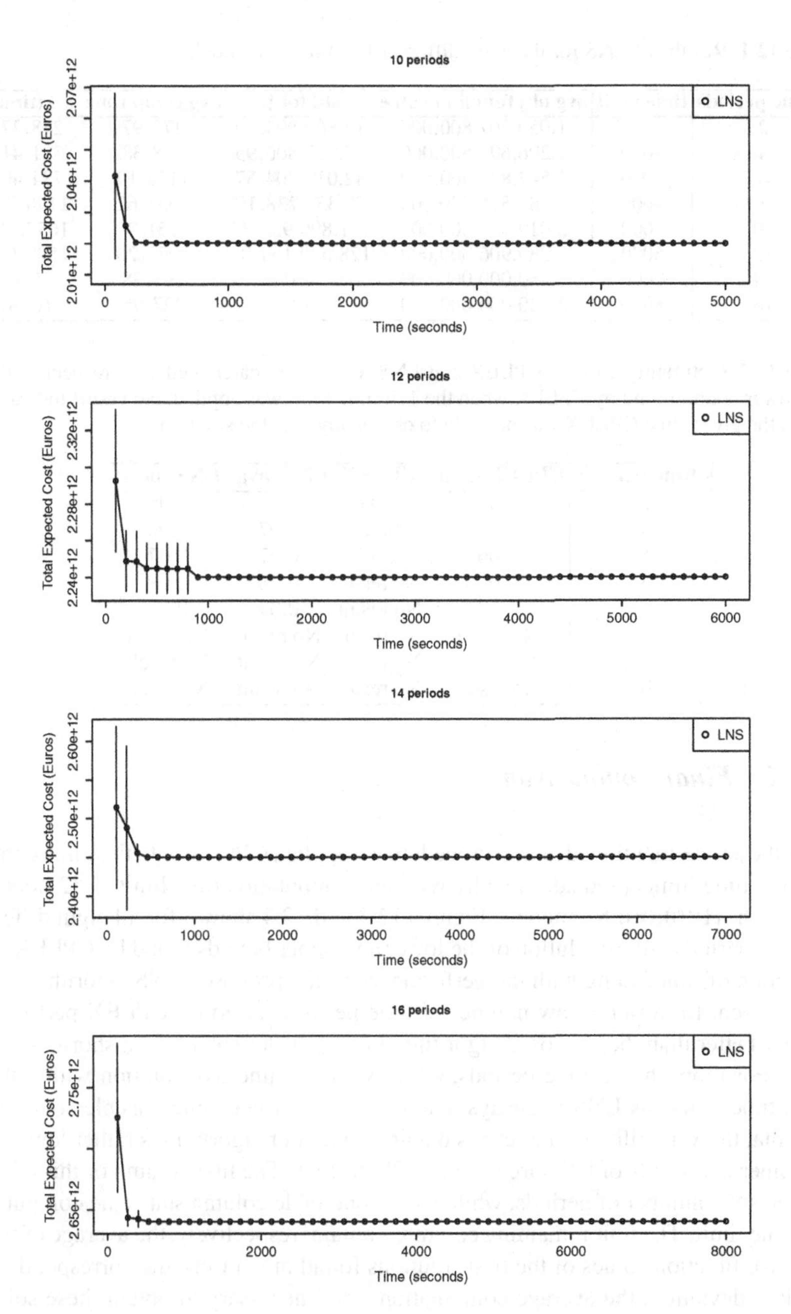

Fig. 12.4 Commparison of LNS with CPLEX (lower and upper bound) over time. The four graphics show the results for a different number of periods (value of t). From top to down, t takes values $\{10, 12, 14, 16\}$). The vertical bars show the standard deviation of LNS over 10 runs. In the cases in which CPLEX results are missing, CPLEX was not able to find any solution within the given time.

Table 12.1 Results of LNS for the eight different HYDROGEN models

# time periods	time limit	avg obj function value	std (obj)	avg comp time	std (time)
2	1000	1.050.707.800.000	18.860.894,29	477,97	248,27
4	2000	1.296.693.500.000	37.997.806,95	748,38	581,41
6	3000	1.543.836.900.000	42.019.704,37	1156,19	784,40
8	4000	1.786.516.900.000	58.333.238,10	1651,62	1196,73
10	5000	2.019.793.600.000	111.899.955,32	2751,97	1657,26
12	6000	2.239.906.300.000	128.670.164,03	2973,20	1508,65
14	7000	2.450.000.000.000	0,0	195,28	125,80
16	8000	2.640.000.000.000	0,0	227,95	116,70

Table 12.2 Optimality gaps of CPLEX and LNS. GAP's are calculated with respect to the best lower bound found by CPLEX when the 12h time limit was applied. **No result** indicates that in the given time CPLEX was not able to obtain any feasible solution.

# time periods	CPLEX (12h)	CPLEX	LNS (avg)	LNS (best)
2	0.05	0.05	0.05	0.05
4	0.05	0.06	0.07	0.06
6	0.06	0.07	0.07	0.07
8	0.08	No result	0.08	0.08
10	0.10	No result	0.09	0.09
12	No result	No result	No result	No result
14	No result	No result	No result	No result
16	No result	No result	No result	No result

12.4.2 Final Comparison

After the above-mentioned tuning procedure we applied CPLEX with the same computation time limits (and additionally with the computation time limit of 12 hours) to all eight HYDROGENE models. Figures 12.3 and 12.4 show—for all eight different time periods—the evolution of the lower and upper bounds found by CPLEX as a function of time, along with the performance of the proposed LNS algorithm. As can be seen, for a rather low number of time periods (up to 6), CPLEX performs slightly better than the proposed algorithm, finding better solutions in shorter CPU times. For more than 6 time periods, CPLEX cannot find any solution within the given time, whereas LNS is always able to provide at least one feasible solution. Note that the variability of the results obtained with our algorithm is rather low.

Numerical results of LNS are shown in Table 12.1. The first column of this table indicates the number of periods, while the second table column states the computation time limit. The four remaining columns contain, respectively, the average of the objective function values of the best solutions found in ten runs, the corresponding standard deviation, the average computation times necessary to obtain these solutions, and, again, the corresponding standard deviation.

Finally, Table 12.2 displays the optimality gaps obtained by the following algorithms: the best solution calculated by CPLEX after 12 hours of CPU time (column labelled **CPLEX (12h)**) and after the same CPU time limit applied to LNS (column labelled **CPLEX**), the average solution quality obtained by LNS (column labelled **LNS (avg)**), and the value of the best solution found by LNS (column labelled **LNS (best)**). The optimality gap is determined with respect to the best lower bound calculated by CPLEX within 12 hours. Note that in some cases (12, 14, and 16 periods), CPLEX is unable to provide any bound even after 12 hours.

Summarizing, we can say that LNS appears to be a useful alternative to pure mathematical programming for what concerns the application to large-scale models from the Chemical industries.

12.5 Conclusions

In this work we have introduced an efficient hybrid algorithm for a spatially explicit supply chain management model. Our algorithm combines mathematical programming techniques with local search, and is known as large neighborhood search in the literature. The capabilities of the proposed method were illustrated through its application to the strategic planning of infrastructures for hydrogen production. Our algorithm was shown to outperform the stand-alone branch and cut method implemented in CPLEX especially for large-scale problems. Numerical examples have demonstrated that our method is particularly suited for tackling large scale problems with a high number of time periods and potential locations (and, therefore, a high number of integer and binary variables).

Future work will particularly focus on investigating how to incorporate the information obtained after solving sub-problems of the mathematical program into the original model in order to expedite the solution of the full space formulation.

Acknowledgements. Pedro J. Copado wishes to acknowledge support of this research work from the Spanish Ministry of Education and Science (DPI2008-04099/DPI). The authors also wish to acknowledge support from the CON- 844 ICET (Argentina), the Spanish Ministry of Education and Science (DPI2008-04099/DPI, CTQ2009-14420 and ENE2008-06687-C02-01), and the Spanish Ministry of External Affairs (projects A/8502/07, A/023551/09 and HS2007-0006). Finally, Christian Blum acknowledges support from grant TIN2007-66523 (FORMALISM) and from the *Ramón y Cajal* program of the Spanish Government.

References

1. Almansoori, A., Shah, N.: Design and Operation of a Future Hydrogen Supply Chain: Snapshot Model. Chemical Engineering Research and Design 84(6), 423–438 (2006)
2. Amodeo, L., Prins, C., Sánchez, D.R.: Comparison of Metaheuristic Approaches for Multi-objective Simulation-Based Optimization in Supply Chain Inventory Management. In: Giacobini, M., Brabazon, A., Cagnoni, S., Di Caro, G.A., Ekárt, A., Esparcia-Alcázar, A.I., Farooq, M., Fink, A., Machado, P. (eds.) EvoWorkshops 2009. LNCS, vol. 5484, pp. 798–807. Springer, Heidelberg (2009)
3. Baykasoglu, A., Gocken, T.: Multi-objective aggregate production planning with fuzzy parameters. Advances in Engineering Software 41(9), 1124–1131 (2010)
4. Blum, C., Puchinger, J., Raidl, G.R., Roli, A.: Hybrid metaheuristics in combinatorial optimization: A survey. Applied Soft Computing Journal 11(6), 4135–4151 (2011)

5. Chiang, W., Russell, R., Xu, X., Zepeda, D.: A simulation/metaheuristic approach to newspaper production and distribution supply chain problems. International Journal of Production Economics 121(2), 752–767 (2009)
6. Delavar, M.R., Hajiaghaei-Keshteli, M., Molla-Alizadeh-Zavardehi, S.: Genetic algorithms for coordinated scheduling of production and air transportation. Expert Systems with Applications 37(12) (2010)
7. Fox, M.S., Barbuceanu, M., Teigen, R.: Agent-oriented supply-chain management. International Journal of Flexible Manufacturing Systems 12(2), 165–188 (2000)
8. Grossmann, I.: Enterprise-wide optimization: A new frontier in process systems engineering. AICHE Journal 51(7), 1846–1857 (2005)
9. Guillén-Gosálbez, G., Mele, F.D., Grossmann, I.E.: A bi-criterion optimization approach for the design and planning of hydrogen supply chains for vehicle use. AICHE Journal 56(3), 650–667 (2010)
10. Kim, J., Lee, Y., Moon, I.: Optimization of a hydrogen supply chain under demand uncertainty. International Journal of Hydrogen Energy 33(18), 4715–4729 (2008)
11. Kostin, A.M., Guillén-Gosálbez, G., Mele, F.D., Bagajewicz, M.J., Jiménez, L.: A novel rolling horizon strategy for the strategic planning of supply chains. Application to the sugar cane industry of argentina. Computers & Chemical Engineering (2010)
12. Li, Z., Gao, D., Chang, L., Liu, P., Pistikopoulos, E.N.: Hydrogen infrastructure design and optimization: A case study of china. International Journal of Hydrogen Energy 33(20), 5275–5286 (2008)
13. Mele, F.D., Kostin, A.M., Guillen-Gosalbez, G., Jimenez, L.: Multiobjective model for more sustainable fuel supply chains. A case study of the sugar cane industry in argentina. Industrial & Engineering Chemistry Research (2011)
14. Mula, J., Peidro, D., Diaz-Madronero, M., Vicens, E.: Mathematical programming models for supply chain production and transport planning. European Journal of Operational Research 204(3), 377–390 (2010)
15. Naraharisetti, P.K., Adhitya, A., Karimi, I.A., Srinivasan, R.: From pse to pse2-decision support for resilient enterprises. Computers and Chemical Engineering 33(12), 1939–1949 (2009)
16. Papageorgiou, L.G.: Supply chain optimisation for the process industries: Advances and opportunities. Computers and Chemical Engineering 33(12), 1931–1938 (2009)
17. Pisinger, D., Ropke, S.: Large Neighborhood Search. International Series in Operations Research & Management Science, vol. 146. Springer, US (2010)
18. Puigjaner, L., Guillén-Gosálbez, G.: Towards an integrated framework for supply chain management in the batch chemical process industry. Computers & Chemical Engineering 32(4-5), 650–670 (2008)
19. Sabio, N., Gadalla, M., Jimnez, L., Guillén-Gosálbez, G.: Risk management on the design and planning of a hydrogen supply chain for vehicle use under uncertainty in production prices: A case study of spain (2009)
20. Sabio, N., Gadalla, M., Guillén-Gosalbéz, G., Jimenez, L.: Strategic planning with risk control of hydrogen supply chains for vehicle use under uncertainty in operating costs: A case study of spain. International Journal of Hydrogen Energy 35(13), 6836–6852 (2010)
21. Sadjadi, S.J., Jafari, M., Amini, T.: A new mathematical modeling and a genetic algorithm search for milk run problem (an auto industry supply chain case study). International Journal of Advanced Manufacturing Technology 44(1-2), 194–200 (2009)
22. Shaw, P.: Using Constraint Programming and Local Search Methods to Solve Vehicle Routing Problems. In: Maher, M.J., Puget, J.-F. (eds.) CP 1998. LNCS, vol. 1520, pp. 417–431. Springer, Heidelberg (1998)
23. Talbi, E.-G.: A Taxonomy of Hybrid Metaheuristics. Journal of Heuristics 8(5), 541–564 (2002)
24. Warren Liao, T., Chang, P.C.: Impacts of forecast, inventory policy, and lead time on supply chain inventorya numerical study. International Journal of Production Economics 128(2), 527–537 (2010)

Appendix A

In this appendix we provide the complete mathematical model of the three-echelon hydrogen network problem for vehicle use with the goal of minimizing the total cost. Moreover, we give a brief description of its components. Further details may be found in [20, 9, 11].

Notation

Indices

e	scenarios
i	hydrogen form
g	potential locations
l	transportation mode
p	manufacturing technologies
s	storage technologies
t	time period

Sets

$IL(l)$	set of hydrogen forms that can be transported via transportation mode l
$IS(s)$	set of hydrogen forms that can be stored via technology s
$LI(i)$	set of transportation modes that can transport hydrogen form i
$SI(i)$	set of storage technologies that can store hydrogen form i

Parameters

av_l	availability of transportation mode l
cc_{lt}	capital cost of transport mode l in period t
cud_{lt}	maintenance cost of transportation mode l in period t per unit of distance traveled
$\overline{D_{gt}}$	total demand of hydrogen in location g in period t
$distance_{gg'}$	average distance traveled between locations g and g'
$dsat$	demand satisfaction level to be fulfilled
$fuelc_l$	fuel consumption of transportation mode l
$fuelp_{lt}$	price of the fuel consumed by transportation mode l in period t
ge_{lt}	general expenses of transportation mode l in period t
ir	interest rate
$lutime_l$	loading/unloading time of transportation mode l
$\overline{PC_p^{PL}}$	upper bound on the capacity expansion of manufacturing technology p

$\underline{PC_p^{PL}}$ lower bound on the capacity expansion of manufacturing technology p

$\overline{QC_{gg'l}}$ upper bound on the flow of materials between locations g and g' via transportation model l

$\underline{QC_{gg'l}}$ lower bound on the flow of materials between locations g and g' via transportation model l

$\overline{SC_s^{ST}}$ upper bound on the capacity expansion of storage technology s

$\underline{SC_s^{ST}}$ lower bound on the capacity expansion of storage technology s

$speed_l$ average speed of transportat mode l

$tcap_l$ capacity of transport mode l

upc_{igpte} mean value of unit production cost of hydrogen form i produced via technology p in location g in period t in scenario e

$Vupc_{igpte}$ Variance associated to the probability distribution of upc_{igpte}

usc_{igst} unit storage cost of hydrogen form i stored via technology s in location g in period t

$wage_{lt}$ driver wage of transportation mode l in period t

α_{gpt}^{PL} fixed investment term associated with manufacturing technology p installed in location g in period t

α_{gst}^{ST} fixed investment term associated with storage technology s installed in location g in period t

β_{gpt}^{PL} variable investment term associated with manufacturing technology p installed in location g in period t

β_{gst}^{ST} variable investment term associated with storage technology s installed in location g in period t

θ average storage period

τ minimum desired percentage of the capacity that must be used

$prob_e$ occurrence probability of scenario e

Variables

C_{gpt}^{PL} capacity of manufacturing technology p in location g in period t

C_{gst}^{ST} capacity of storage technology s in location g in period t

CE_{gpt}^{PL} capacity expansion of manufacturing technology p in location g in period t

CE_{gst}^{ST} capacity expansion of storage technology s in location g in period t

D_{igt} amount of hydrogen form i distributed in location g in period t

FC_t fuel cost in period t

FCC_t facility capital cost in period t

FOC_{te} facility operating cost in period t in scenario e

GC_t general cost in period t

LC_t labor cost in period t

MC_t maintenance cost in period t

$TPIC$ capital cost of pipelines establishment (euros/km)

UTP unit transportation cost of pipelines (euros/kg day)

$UTCB$ unit transportation cost of ship rental (euros/h kg)

$PICC_t$	pipeline capital cost (euros/yr)
PIC_t	pipeline operating cost (euros/yr)
$TOCB_t$	ship operating cost
N^{PL}_{gpt}	number of plants of type p installed in location g in period t (integer variable)
N^{ST}_{gst}	number of storage facilities of type s installed in location g in period t (integer variable)
N^{TR}_{lt}	number of transportation units of type l purchased in period t (integer variable)
PR_{igpt}	production of hydrogen mode i via technology p in period t in location g
$Q_{igg'lt}$	flow of hydrogen mode i via transportation mode l between locations g and g' in period t
S_{igst}	amount of hydrogen in physical form i stored via technology s in location g in period t
TC_{te}	total amount of money spent in period t for scenario e
TCC_t	total transportation capital cost in period t
TDC_e	total discounted cost for scenario e
TOC_t	transportation operating cost in period t
$X_{gg'lt}$	binary variable (1 if a link between locations g and g' using transportation technology l is established, 0 otherwise)

Equation 12.1 defines the mass balance for the grids considered in the analysis, whereas Equation 12.2 forces the model to fulfill a minimum demand satisfaction level. Equation 12.3 limits the production capacity between lower and upper bounds. Equation 12.4 determines the production capacity in a time period from the previous one plus the expansion in capacity executed in the same period. Equation 12.5 limits the capacity expansions within lower and upper bounds given by the number of facilities opened.

$$\begin{aligned}\sum_{s\in SI(i)} S_{igst-1} + \sum_{p} PR_{igpt} + \sum_{g'\neq g}\sum_{l} Q_{ilg'glt} \\ = \sum_{s\in SI(i)} S_{igst} + D_{igt} + \sum_{g'\neq g}\sum_{l} Q_{ilgg'lt} \forall i,g,t\end{aligned} \tag{12.1}$$

$$\overline{D_{gt}}dsat \leq \sum_{i} D_{igt} \leq \overline{D_{gt}} \forall g,t \tag{12.2}$$

$$\tau C^{PL}_{gpt} \leq \sum_{i} PR_{igpt} \leq C^{PL}_{gpt} \forall g,p,t \tag{12.3}$$

$$C^{PL}_{gpt} = C^{PL}_{gpt-1} + CE^{PL}_{gpt} \quad \forall g,p,t \tag{12.4}$$

$$\underline{PC^{PL}_{p}} N^{PL}_{gpt} \leq CE^{PL}_{gpt} \leq \overline{PC^{PL}_{p}} N^{PL}_{gpt} \forall g,p,t \tag{12.5}$$

Equations 12.6 to 12.9 are equivalent to equations 12.3 to 12.5, but apply to warehouses. Particularly, equation 12.6 limits the amount of materials stored to be lower than the existing capacity. Equation 12.7 forces the average inventory level, which is determined from the demand and turnover ratio, to be lower than the existing capacity. Equation 12.8 provides the storage capacity in a time period from the previous one and the expansion in capacity in the previous period, whereas equation 12.9 limits the expansion in capacity between lower and upper limits given by the number of storage facilities installed.

$$\sum_{i \in IS(s)} S_{igst} \leq C^{ST}_{gst} \forall g,s,t \tag{12.6}$$

$$2(\theta D_{igt}) \leq \sum_{s \in SI(i)} C^{ST}_{gst} \forall i,g,t \tag{12.7}$$

$$C^{ST}_{gst} = C^{ST}_{gst-1} + CE^{ST}_{gst} \forall g,s,t \tag{12.8}$$

$$\underline{SC^{ST}_{s}} N^{ST}_{gst} \leq CE^{ST}_{gst} \leq \overline{SC^{ST}_{s}} N^{ST}_{gst} \forall g,s,t \tag{12.9}$$

Equation 12.10 limits the transportation links between lower and upper bounds provided the link is finally established. Equations 12.11 and 12.12 are defined for the construction of pipelines. Equation 12.13 is a logic constraint that makes the formulation tighter. Equations 12.14 and 12.15 avoid the transportation between certain maritime grids, whereas equation 12.16 is a symetric cut. Finally, equations 12.17 to 12.31 allow to determine the cost of the network.

$$\begin{aligned} &\underline{QC_{lgg'}} X_{gg'lt} \leq \sum_{i} Q_{ilgg't} \leq \overline{QC_{lgg'}} X_{gg'lt} \\ &\forall g,g'(g \neq g'), l \in LI(i) \cup NPL, t \end{aligned} \tag{12.10}$$

$$\begin{aligned} &\sum_{t' \leq t+1} \underline{QC_{lgg'}} X_{gg'lt'} \leq \sum_{i} Q_{ilgg't} \leq \sum_{t' \leq t+1} \overline{QC_{lgg'}} X_{gg'lt} \\ &\forall g,g'(g \neq g'), l = pipeline, t \end{aligned} \tag{12.11}$$

$$\sum_{t' \leq t+1} X_{gg'lt'} \leq 1 \qquad \forall g,g'(g \neq g'), l = pipeline, t \tag{12.12}$$

$$X_{gg'lt} + X_{g'glt} \leq 1 \qquad \forall g,g'(g \neq g'), l \in LI(i,t \tag{12.13}$$

$$\begin{aligned} &X_{lgg't} = 0 \qquad \forall l,g,g' \in LG' \\ &LG' = \{l,g,g' : (l = ship) \wedge ((g,g') \notin SGG(gg'))\} \end{aligned} \tag{12.14}$$

$$\begin{aligned} &X_{lgg't} = 0 \qquad \forall l,g,g' \in LG \\ &LG = \{l,g,g' : (l \neq ship) \wedge ((g,g') \in SGG'(gg'))\} \end{aligned} \tag{12.15}$$

$$X_{lgg't} = 0 \qquad \forall l, g = g' \tag{12.16}$$

$$TDC = \sum_t \frac{TC}{(1+ir)^{t-1}} \tag{12.17}$$

$$TC_t = FCC_t + TCC_t + FOC_t + TOC_t \qquad \forall t \tag{12.18}$$

$$\begin{aligned} FOC_t = & \sum_i \sum_g \sum_p upc_{igpt} PR_{igpt} \\ & + \sum_i \sum_g \sum_s \in SI(i) usc_{igst} \left(\theta D_{igt}\right) \qquad \forall t \end{aligned} \tag{12.19}$$

$$\begin{aligned} FCC_t = & \sum_g \sum_p \left(\alpha^{PL}_{gpt} N^{PL}_{gpt} + \beta^{PL}_{gpt} CE^{PL}_{gpt}\right) \\ & + \sum_g \sum_s \left(\alpha^{ST}_{gst} N^{ST}_{gst} + \beta^{ST}_{gst} CE^{ST}_{gst}\right) \qquad \forall t \end{aligned} \tag{12.20}$$

$$TCC_t = \sum_{l \neq ship, pipeline} N^{TR}_{lt} \cdot cc_{lt} + PCC_t \tag{12.21}$$

$$PCC(t) = \sum_g \sum_{g' \neq g} \sum_{l \in LI(i)} upcc_t X_{lgg't} distance_{gg'} \qquad \forall t \tag{12.22}$$

$$\begin{aligned} & \sum_{t' \leq t+1} N^{TR}_{lt'} \geq \sum_{i \in IL(l)} \sum_g \sum_{g' \neq g} \sum_t \frac{Q_{igg'lt}}{av_l tcap_l} \left(\frac{2distance_{gg'}}{speed_l} + lutime_l\right) \\ & \forall l \neq ship, pipeline \end{aligned} \tag{12.23}$$

$$TOC_t = ROC_t + POC_t + SOC_t \qquad \forall t \tag{12.24}$$

$$ROC_t = FC_t + LC_t + MC_t + GC_t \qquad \forall t \tag{12.25}$$

$$FC_t = \sum_i \sum_g \sum_{g' \neq g} \sum_{l \in LI(i)} fuelp_{lt} \frac{2distance_{gg'} Q_{ilgg't}}{fuelc_l tcap_l} \qquad \forall t \tag{12.26}$$

$$\begin{aligned} LC_t = & \sum_i \sum_g \sum_{g' \neq g} \sum_{l \in LI(i)} wage_{lt} \\ & \times \left[\frac{Q_{ilgg't}}{tcap_l} \left(\frac{2distance_{gg'}}{speed_l} + lutime_l\right)\right] \qquad \forall t \end{aligned} \tag{12.27}$$

$$MC_t = \sum_i \sum_g \sum_{g' \neq g} \sum_{l \in LI(i)} cud_l \frac{2distance_{gg'} Q_{ilgg't}}{tcap_l} \qquad \forall t \tag{12.28}$$

$$GC_t = \sum_l \sum_{t' \leq t} g_{lt} N^{TR}_{lt'} \qquad \forall t \tag{12.29}$$

$$POC(t) = \sum_{i} \sum_{g} \sum_{g' \neq g} \sum_{l \in LI(i)} upoc_t Q_{ilgg't} \qquad \forall t \tag{12.30}$$

$$SOC_t = \sum_{i} \sum_{g} \sum_{g' \neq g} \sum_{l \in LI(i)} usoc_t \left(\frac{distance_{gg'}}{speed_l} \right) Q_{ilgg't} \qquad \forall t \tag{12.31}$$

Chapter 13
A VNS-Based Heuristic for Feature Selection in Data Mining

A. Mucherino and L. Liberti

Abstract. The selection of features that describe samples in sets of data is a typical problem in data mining. A crucial issue is to select a maximal set of pertinent features, because the scarce knowledge of the problem under study often leads to consider features which do not provide a good description of the corresponding samples. The concept of consistent biclustering of a set of data has been introduced to identify such a maximal set. The problem can be modeled as a 0–1 linear fractional program, which is NP-hard. We reformulate this optimization problem as a bilevel program, and we prove that solutions to the original problem can be found by solving the reformulated problem. We also propose a heuristic for the solution of the bilevel program, that is based on the meta-heuristic Variable Neighborhood Search (VNS). Computational experiments show that the proposed heuristic outperforms previously proposed heuristics for feature selection by consistent biclustering.

13.1 Introduction

Nowadays technologies are able to produce a large quantity of data which needs to be analyzed. Data mining is a well-established field whose aim is to discover hidden patterns in the data for acquiring novel knowledge. A classic example is given by the huge quantity of data that is contained in DNA molecules of living beings. The relationships among the different genes of a DNA molecule, under different conditions, can provide important information regarding diseases and the functioning of life.

Data can be collected from different resources. *Samples* represent a single measurement of what is under study, and *features* are employed for describing the

A. Mucherino
IRISA, University of Rennes, Rennes, France
e-mail: antonio.mucherino@irisa.fr

L. Liberti
LIX, École Polytechnique, Palaiseau, France
e-mail: liberti@lix.polytechnique.fr

E.-G. Talbi (Ed.): Hybrid Metaheuristics, SCI 434, pp. 353–368.
springerlink.com

samples. In the example of the DNA molecule, a sample can represent the patients' condition, such as "healthy" and "sick", which is monitored through the expression levels of each gene in his DNA. In other words, each feature represents the expression level of a gene, and a list of feature measurements represents a sample. In many applications, the number of samples is scarce (only a few measurements are available), while the number of features is usually large (many factors are involved in the phenomenon under study).

In this context, *feature selection* is the problem of extracting only important and pertinent features from a set of data. Some considered features, indeed, may not be adequate for describing the samples, and, in such a case, they should be removed from the set of data. This brings two important consequences. First, if only pertinent features are used and all the others are rejected, the memory space necessary for storing this set in databases is optimized. Secondly, a strict relationship between samples and features may be identified, which could be exploited for discovering important information.

If a set of data contains n samples which are described by m features, then the whole set can be represented by a $m \times n$ matrix A, where the samples are organized column by column, and the features are organized row by row. In this context, we refer to a *bicluster* of A as a submatrix of A, whose elements are a subset of samples and features. Equivalently, a bicluster can be seen as a pair of subsets (S_r, F_r), where S_r is a class (or cluster) of samples, and F_r is a class (or cluster) of features.

Definition 13.1. *A biclustering is a partition of A in k biclusters:*

$$\mathbb{B} = \{(S_1, F_1), (S_2, F_2), \ldots, (S_k, F_k)\},$$

such that the following conditions are satisfied:

$$\bigcup_{r=1}^{k} S_r = A, \qquad S_\zeta \cap S_\xi = \emptyset \quad 1 \le \zeta \ne \xi \le k,$$

$$\bigcup_{r=1}^{k} F_r = A, \qquad F_\zeta \cap F_\xi = \emptyset \quad 1 \le \zeta \ne \xi \le k,$$

where $k \le \min(n,m)$ is the number of biclusters [2, 6].

If a classification for the samples of A is available, as well as a classification for its features, a biclustering $\mathbb{B}$ can be trivially constructed. Inversely, classifications of samples and features can be extracted from $\mathbb{B}$.

In some data mining applications, there exist sets of data for which a classification of its samples is already given. In the example of the DNA molecules, samples may be taken from patients affected by different diseases, so that their classification is already known. In this case, the set A is named *training set*. However, the classification of the features used for describing the samples is not known, or, equivalently, there is no biclustering $\mathbb{B}$ associated to A. Therefore, we have no a priori information about possible relationships between samples and features.

A way to obtain a classification for the features from a training set A is to assign each feature to the class where it is "mostly expressed" (see Section 13.2 for more details). Then, once a classification for the features is also available, a biclustering $\mathbb{B}$ for A can be obtained by simply applying Definition 13.1. If the found biclustering is *consistent* (in the sense stated in Section 13.2), then the selected features are most likely the ones that better describe the samples.

The feature selection problem related to consistent biclustering can be formulated as a 0–1 linear fractional optimization problem, which is NP-hard [9]. In this paper, we propose a new heuristic for solving the feature selection problem, that is based on a bilevel reformulation of the 0–1 linear fractional optimization problem. The proposed heuristic is based on the meta-heuristic Variable Neighborhood Search (VNS) [5, 10] and on the idea of solving exactly, at each iteration, the inner problem of the bilevel program, which is linear. Preliminary studies regarding the proposed heuristic for features selection have been previously presented in [12].

The rest of the paper is organized as follows. In Section 13.2, we develop the concept of consistent biclustering in more details, and we present the corresponding feature selection problem. In Section 13.3, we reformulate this feature selection problem as a bilevel optimization problem and we formally prove that solutions to the original problem can be found by solving this bilevel program. In Section 13.4, we introduce a new VNS-based heuristic for an efficient solution of the bilevel program. Computational experiments on real-life sets of data are presented in Section 13.5, as well as a comparison to the heuristic presented in [17]. Conclusions are given in Section 13.6.

13.2 Consistent Biclustering

Let $A = (a_{ij}) \in \Re^{m \times n}$ be a matrix representing a certain set of data, where samples a^j are organized column by column, and their features a_i are organized row by row. In the following, k is the number of biclusters (known a priori) forming the biclustering, and the index $r \in \{1,2,\ldots,k\}$ will refer to the generic class of samples or features.

If the set of data A is a training set, then the classification of its samples in k classes is known:

$$B_S = \{S_1, S_2, \ldots, S_k\}.$$

Let s_{ir} be a binary vector which indicates if the ith sample belongs to the class S_r of samples ($s_{ir} = 1$) or not ($s_{ir} = 0$). Since A is a training set, the vector s_{ir} is known a priori. From the classification B_S, we can use the following procedure to construct a classification of the features in k classes:

$$B_F = \{F_1, F_2, \ldots, F_k\}.$$

The basic idea is to assign each feature to the class $F_{\hat{r}}$ (with $\hat{r} \in \{1,2,\ldots,k\}$) such that it is mostly expressed (i.e. *it has higher value*), in average, in the class of

Algorithm 13. Procedure for constructing B_F from B_S.

1: **for** (each feature i, $i \in \{1,2,\dots,n\}$) **do**
2: let $\hat{r} = \arg\max_r \left(\frac{\sum_{j=1}^m a_{ij} s_{ir}}{\sum_{j=1}^m s_{ir}} \right)$;
3: **for** each class r, $r \in \{1,2,\dots,k\}$ **do**
4: let $f_{ir} = 0$;
5: **end for**
6: let $f_{i\hat{r}} = 1$;
7: **end for**

samples $S_{\hat{r}}$. Let f_{ir} be a binary vector which indicates if the ith feature belongs to the class F_r of features ($f_{ir} = 1$) or not ($f_{ir} = 0$), which is not known a priori. In order to define it and hence to give a classification B_F to the features in A, we can employ Algorithm 13 [2, 17].

We remark that the same procedure can be used for finding a classification of the samples from a known classification of its features. Let

$$\hat{B}_S = \{\hat{S}_1, \hat{S}_2, \dots, \hat{S}_k\}$$

be the classification of samples obtained from B_F. A biclustering $\mathbb{B}$ for A can be defined by combining the two classifications B_S and B_F (see Definition 13.3). Moreover, if the classifications of samples B_S and $\hat{B}_S$ are equivalent, then the biclustering $\mathbb{B}$ has a particular property that we call *consistency*.

Definition 13.2. *Let A be a training set with classification of samples B_S. Let B_F be the classification of its features obtained by Algorithm 13 from B_S, and let $\hat{B}_S$ the classification of samples obtained by Algorithm 13 from B_F. If $B_S = \hat{B}_S$, then the biclustering $\mathbb{B}$ of A obtained by combining B_S and B_F is consistent [2].*

By definition, when a biclustering is consistent, the classification of the samples can be correctly reconstructed from the classification of its features, and vice versa. Therefore, the features are all able to describe accurately the samples of the set of data.

If a consistent biclustering exists for a certain set of data A, then A is said to be *biclustering-admitting*. However, sets of data admitting consistent biclusterings are very rare in real-life applications. In other words, the situation $B_S \equiv \hat{B}_S$ is very difficult to be verified in practice, because some of the features used for describing the samples may not be actually pertinent. As a consequence, non-pertinent features should be removed from the set of data with the aim of finding a consistent biclustering for submatrices of A in which some rows have been removed [2]. Note that it is very important to remove the least number of features, in order to preserve the information in the set of data.

Let us suppose that only a subset of features is considered: let

$$x = (x_1, x_2, \dots, x_m)$$

be a binary vector of variables, where x_i is 1 if the ith feature is selected, and it is 0 otherwise. Let $A[x]$ be the submatrix of A obtained by removing all the rows a_i for which $x_i = 0$. We give the following definition.

Definition 13.3. *A biclustering for $A[x]$ is consistent if and only if, $\forall \hat{r}, \xi \in \{1,2,\ldots,k\}$, $\hat{r} \neq \xi, j \in S_{\hat{r}}$, the following inequality is satisfied [2]:*

$$\frac{\sum_{i=1}^{m} a_{ij} f_{i\hat{r}} x_i}{\sum_{i=1}^{m} f_{i\hat{r}} x_i} > \frac{\sum_{i=1}^{m} a_{ij} f_{i\xi} x_i}{\sum_{i=1}^{m} f_{i\xi} x_i}. \tag{13.1}$$

Note that the two fractions in (13.1) are used for computing the *centroids* of the considered biclusters (for each sample in $S_{\hat{r}}$, the average over the features belonging to same class is computed). On the left hand side of (13.1), the jth component of the centroid of the bicluster $(S_{\hat{r}}, F_{\hat{r}})$ is computed. On the right hand side of (13.1), the jth component of the centroid of the bicluster $(S_{\hat{r}}, F_{\xi})$ is computed. In order to have a consistent biclustering for $A[x]$ (i.e. $B_S \equiv \hat{B}_S$), all components of the centroid of $(S_{\hat{r}}, F_{\hat{r}})$ must have a value that is larger than any other. This condition on the classification B_F of features allows Alg. 13 to generate a classification of samples $\hat{B}_S$ that is equivalent to the original classification B_S.

In order to overcome issues related to sets of data containing noisy data and errors, the concepts of α-consistent biclustering and β-consistent biclustering have been introduced in [17]. The basic idea is to artificially increase the margin between the centroids of the different biclusters in the constraints (13.1). In this way, small variations due to noisy data and errors should not be able to spoil the feature selection.

Definition 13.4. *Given a real parameter $\alpha > 0$, a biclustering for $A[x]$ is α-consistent if and only if, $\forall \hat{r}, \xi \in \{1,2,\ldots,k\}, \hat{r} \neq \xi, j \in S_{\hat{r}}$, the following inequality is satisfied [17]:*

$$\frac{\sum_{i=1}^{m} a_{ij} f_{i\hat{r}} x_i}{\sum_{i=1}^{m} f_{i\hat{r}} x_i} > \alpha + \frac{\sum_{i=1}^{m} a_{ij} f_{i\xi} x_i}{\sum_{i=1}^{m} f_{i\xi} x_i}. \tag{13.2}$$

The additive parameter $\alpha > 0$ is used to guarantee that the margin between the centroid of $(S_{\hat{r}}, F_{\hat{r}})$ and any other bicluster concerning $S_{\hat{r}}$ is at least greater than α, independently from the considered data. Similarly, in the case of β-consistent biclustering, a multiplicative parameter β is employed.

Definition 13.5. *Given a real parameter $\beta > 1$, a biclustering for $A[x]$ is β-consistent if and only if, $\forall \hat{r}, \xi \in \{1,2,\ldots,k\}, \hat{r} \neq \xi, j \in S_{\hat{r}}$, the following condition is satisfied [11]:*

$$\begin{cases} \dfrac{\sum_{i=1}^{m} a_{ij} f_{i\hat{r}} x_i}{\sum_{i=1}^{m} f_{i\hat{r}} x_i} > \beta \dfrac{\sum_{i=1}^{m} a_{ij} f_{i\xi} x_i}{\sum_{i=1}^{m} f_{i\xi} x_i} & \text{if} \, c > 0 \\ \dfrac{\sum_{i=1}^{m} a_{ij} f_{i\hat{r}} x_i}{\sum_{i=1}^{m} f_{i\hat{r}} x_i} > (2-\beta) \dfrac{\sum_{i=1}^{m} a_{ij} f_{i\xi} x_i}{\sum_{i=1}^{m} f_{i\xi} x_i} & \text{if} \, c < 0 \end{cases} \tag{13.3}$$

where

$$c = \frac{\sum_{i=1}^{m} a_{ij} f_{i\xi} x_i}{\sum_{i=1}^{m} f_{i\xi} x_i}.$$

We remark that the concept of β-consistent biclustering was firstly introduced in [17], but the given definition was only suitable for sets of data containing non-negative entries. In general, different values for the parameters α and β could be used for each j in Definitions 13.4 and 13.5. Usually, however, only one value is set up for all components of the centroids.

In real-life applications, there are usually no biclusterings which are consistent, α-consistent or β-consistent if all features are selected (this situation corresponds to a binary vector x with all its components equal to 1). As already mentioned before, this happens because some of the considered features may actually be inadequate. Such features must therefore be removed from the set of data, while the total number of considered features must be maximized in order to preserve as much information as possible. The following combinatorial optimization problem is therefore considered:

$$\max_{x} \left(f(x) = \sum_{i=1}^{m} x_i \right), \tag{13.4}$$

subject to constraints (13.1), (13.2) or (13.3) depending on the fact that a consistent, α-consistent or β-consistent biclustering, respectively, is searched. The three problems are linear with fractional constraints and binary variables. The solution of this kind of optimization problems could be attempted by general-purpose solvers, such as `Baron` [19, 20] or `Couenne` [1], but the large size of real-life sets of data can make their converge very slow and the computational experiments too expensive. The three optimization problems are in fact all NP-hard [9]. In [2] and [17], two heuristics have been proposed. The heuristic we propose in this paper is able to provide better solutions with respect to the ones previously obtained.

13.3 A Bilevel Reformulation

In the following discussion, only the optimization problem (13.1)-(13.4) will be considered, because similar observations can be made for the other two problems. The computational experiments reported in Section 13.5, however, will be related to all three optimization problems.

We propose a reformulation of the problem (13.1)-(13.4) as a bilevel optimization problem. To this aim, we substitute the denominators in the constraints (13.1) with new continuous variables y_r, $r = 1,2,\ldots,k$, where each y_r is related to the bicluster (S_r, F_r). We can rewrite the constraints (13.1) as follows:

$$\frac{1}{y_{\hat{r}}}\sum_{i=1}^{m} a_{ij} f_{i\hat{r}} x_i > \frac{1}{y_{\xi}}\sum_{i=1}^{m} a_{ij} f_{i\xi} x_i, \tag{13.5}$$

where $y_{\hat{r}}$ and y_{ξ} replace the original fractional parts. The constraints (13.5) must be satisfied $\forall \hat{r}, \xi \in \{1,2,\ldots,k\}$, $\hat{r} \neq \xi$ and $j \in S_{\hat{r}}$, in order to have a consistent biclustering.

Let us consider $\bar{y}_r = \delta y_r$, where $\delta > 0$. It is easy to see that, given certain values for the variables x_i, the constraints (13.5) are satisfied with $\bar{y}_r$ if and only if they are satisfied with y_r. As an example, if $k = 3$ and there is a consistent biclustering in which 20, 30 and 50 features are selected in the k biclusters, then the constraints (13.5) are also satisfied if 0.20, 0.30 and 0.50, respectively, replace the actual number of features (in this example, the proportional factor δ is 0.01). For this reason, the variables y_r can be used for representing the *proportions* among the cardinalities of the classes of features. In the previous example, 20% of the selected features are in the first bicluster, 30% of the features in the second one, and 50% in the last one. As a consequence, the variables y_r can be bound in the real interval $[0,1]$, so that we can consider the following constraint:

$$\sum_{r=1}^{k} y_r \leq 1.$$

A percentage of features is not selected when this sum is smaller than 1.

We introduce the function:

$$c(x,\hat{r},\xi) = \sum_{j\in S_{\hat{r}}} \left| \frac{\sum_{i=1}^{m} a_{ij} f_{i\xi} x_i}{\sum_{i=1}^{m} f_{i\xi} x_i} - \frac{\sum_{i=1}^{m} a_{ij} f_{i\hat{r}} x_i}{\sum_{i=1}^{m} f_{i\hat{r}} x_i} \right|_{+},$$

where $x = (x_1, x_2, \ldots, x_m)$ and $\hat{r}, \xi \in \{1,2,\ldots,k\}$, with $\hat{r} \neq \xi$, and where the symbol $|\cdot|_+$ represents the function which returns its argument if it is positive, and it returns 0 otherwise. As a consequence, the value of $c(x,\hat{r},\xi)$ is positive if and only if at least one constraint (13.1) is not satisfied.

We reformulate the optimization problem (13.1)-(13.4) as the following bilevel optimization problem:

$$\min_y \left(g(x,y) = \sum_{r=1}^{k} \left[(1-y_r) + \sum_{\xi=1:\xi\neq r}^{k} c(x,r,\xi) \right] \right) \tag{13.6}$$

subject to:

$$\begin{aligned} x = \arg\max_x & \left(f(x) = \sum_{i=1}^{m} x_i \right) \\ \text{subject to} & \begin{cases} \sum_{i=1}^{m} f_{ir} x_i = \lfloor y_r \sum_{i=1}^{m} f_{ir} \rfloor & \forall r \in \{1,\ldots,k\} \\ \text{constraint (5)} \end{cases} \end{aligned} \tag{13.7}$$

$$\sum_{r=1}^{k} y_r \leq 1.$$

The objective function g of the outer problem depends on both variables x_i, with $i \in \{1,2,\ldots,m\}$, and y_r, with $r \in \{1,2,\ldots,k\}$. For each class S_r, the generic term of g is the sum of two parts, one depending on the vector y and the other one depending on the vector x. The first part is simply the difference $(1-y_r)$, that must be minimized in order to maximize the value for y_r, which represents the percentage of selected features in the class F_r (recall that $y_r \leq 1$). The second part is the sum over all the other classes S_ξ, with $\xi \neq r$, of the function $c(x,r,\xi)$ (when its value is positive). The minimization of this second part allows to minimize the number of constraints (13.1) that are not satisfied.

The bilevel program is subject to two constraints. The first one is based on the solution of another optimization problem, to which we refer as inner problem. The inner problem can be seen as a linear simplification of the original problem (13.1)-(13.4), where the fractional parts have been substituted by the variables y_r, which indicate the percentage of features to be selected in each bicluster. The solution of the inner problem provides a set of values for the variables x_i from the variables y_r. Therefore, whatever method is employed for the solution of the outer optimization problem, the search can be reduced to the variables y_r only, because the corresponding values for the variables x_i can be obtained by solving the inner problem. The inner problem is subject to two constraints: the constraints (13.5), as well as another constraint that forces the number of selected features in each bicluster to respect the percentages given by the variables y_r. The second constraint of the outer problem requires that the sum of all variables y_r must be smaller or equal to 1 (no more than 100% of features can be selected in total). We formally prove that solutions to the proposed bilevel optimization problem are also solutions to the original problem (13.1)-(13.4).

Proposition 1 *If* $(\hat{x},\hat{y})$ *is solution for (13.6)-(13.7), then* $\hat{x}$ *is solution for (13.1)-(13.4).*

Proof. By contradiction, let us suppose that there is a solution $\bar{x}$ such that $f(\bar{x}) > f(\hat{x})$ and constraints (13.1) are satisfied. Let

$$\bar{y}_r = \frac{\sum_{i=1}^{n} f_{ir}\bar{x}_i}{\sum_{i=1}^{n} f_{ir}} \qquad \forall r \in \{1,2,\dots,k\}.$$

Since the constraints (13.1) are satisfied,

$$g(\bar{x},\bar{y}) = \sum_{r=1}^{k}(1-\bar{y}_r) = k - \sum_{r=1}^{k}\bar{y}_r.$$

Then,

$$f(\bar{x}) > f(\hat{x}) \Longrightarrow \sum_{r=1}^{k}\bar{y}_r > \sum_{r=1}^{k}\hat{y}_r \Longrightarrow g(\bar{x},\bar{y}) < g(\hat{x},\hat{y}),$$

which brings to a contradiction. □

13.4 A VNS-Based Heuristic

The heuristic we propose for the solution of the bilevel program presented in Section 13.3 is based on the meta-heuristic Variable Neighborhood Search (VNS) [5, 10], which is one of the most successful heuristics for global optimization. The VNS is based on the idea of exploring small neighbors of currently known solutions, which are increased in size when no better solutions can be found. At each iteration of the VNS, a local search algorithm is often employed, so that a path of local optima can be defined, that may lead to the global optimum of the considered problem. The local search can however be replaced by another VNS, which is nested in the main one.

The proposed heuristic actually implements a VNS in two main steps with an adaptive value for the percentage of unselected features *unsel*, which is small at the beginning (*unsel* $\simeq 0$), and then it increases when no better solutions can be found in the current neighbor. In this way, the algorithm firstly tries to find solutions where the number of selected features is high. Afterwards, solutions where fewer features are selected are considered. For each neighbor of the first step of VNS, there is a full execution of another step. The neighbors of the second step of VNS are generated so that the set of variables y_r can be slightly perturbed at the beginning (*range* = *starting_range*), and larger perturbations can be performed only when no better solutions can be found by considering the current neighbor.

Algorithm 14 is a sketch of our heuristic for feature selection by consistent biclustering. At the beginning, the variables x_i are all set to 1, and the variables y_r are set so that they represent the distribution of all m features among the k classes.

Algorithm 14. A VNS-based heuristic for feature selection.

1: let $iter = 0$;
2: let $x_i = 1, \forall i \in \{1,2,\ldots,m\}$;
3: let $y_r = \sum_i f_{ir}/m, \forall r \in \{1,2,\ldots,k\}$;
4: let $y_r^{best} = y_r, \forall r \in \{1,2,\ldots,k\}$;
5: let $range = starting\_range$;
6: let $unsel = 0$;
7: **while** (constraints (13.1) unsatisfied and $unsel \leq max\_unsel$) **do**
8:   **while** (constraints (13.1) unsatisfied and $range \leq max\_range$) **do**
9:     let $iter = iter + 1$;
10:     solve inner optimization problem (linear & cont.);
11:     **if** (constraints (13.1) unsatisfied) **then**
12:       increase *range*;
13:       **if** (g has improved) **then**
14:         let $y_r^{best} = y_r, \forall r \in \{1,2,\ldots,k\}$;
15:         let $range = starting\_range$;
16:       **end if**
17:       let $y_r = y_r^{best}, \forall r \in \{1,2,\ldots,k\}$;
18:       let r' = random in $\{1,2,\ldots,k\}$;
19:       choose randomly $y_{r'}$ in $[y_{r'} - range, y_{r'} + range]$;
20:       let r'' = random in $\{1,2,\ldots,k\}\ :\ r' \neq r''$;
21:       set $y_{r''}$ so that $1 - unsel \leq \sum_r y_r \leq 1$;
22:     **end if**
23:   **end while**
24:   **if** (constraints (13.1) unsatisfied) **then**
25:     increase *unsel*;
26:   **end if**
27: **end while**

If the biclustering is already consistent, then all features can be selected, and the algorithm stops.

For each neighbor defined by the second VNS step, the variables y_r are randomly modified. $y_{r'}$ and $y_{r''}$ are chosen randomly so that $r' \neq r''$. Then, $y_{r'}$ is perturbed, and its value is chosen randomly in the interval centered in $y_{r'}^{best}$ and with length $2 \times range$. Then, a random value for $y_{r''}$ is chosen so that $1 - unsel \leq \sum_r y_r \leq 1$. In this way, the new set of values for y_r falls in the two current neighbors defined by the VNS.

The inner optimization problem is solved for each random choice for the variables y_r. It is a linear 0–1 optimization problem, and we consider its continuous relaxation, i.e. we allow the variables x_i to take any real value in the interval $[0,1]$. Therefore, after a solution has been obtained, we substitute the fractional values of x_i with 0 if $x_i \leq 1/2$, or with 1 if $x_i > 1/2$. In our experiments, the equality of the first constraint of the inner problem is relaxed to an inequality:

$$\sum_{i=1}^{m} f_{ir} x_i \leq \lfloor y_r \sum_{i=1}^{m} f_{ir} \rfloor \quad \forall r \in \{1,\ldots,k\}.$$

The strict inequality of the constraints (13.5) is also relaxed, so that the domains defined by the constraints are closed domains. Under these hypotheses, the inner problem can be solved by commonly used solvers for mixed integer linear programming (MILP), e.g. CPLEX [7].

After the solution of the inner problem, the original set of constraints (13.1) is checked. If the obtained values for the variables x_i, along with the used values for the variables y_r, define a consistent biclustering, then the algorithm stops. Otherwise, some of the variables y_r are modified and a new iteration of the algorithm is performed.

We point out that heuristics offer no guarantee of optimality. One way to enhance the algorithm is to restart it and to allow only values for the variables y_r corresponding to a larger number of selected features. Moreover, since the algorithm can provide different solutions if it is executed more than once (with different seeds for the generator of random numbers), it can be executed a certain number of times. The best obtained solution is then taken into consideration.

13.5 Computational Experiments

We implemented the presented heuristic for feature selection in AMPL [4], from which the ILOG CPLEX11 solver [7] is invoked for the solution of the inner optimization problem. We also implemented in AMPL the heuristic previously proposed in [17] (for more details about this heuristic, the reader is referred to the reference paper). Experiments are carried out on an Intel Core 2 CPU 6400 @ 2.13 GHz with 4GB RAM, running Linux.

The following four subsections are devoted to four different training sets from different real-life applications for which we selected a subset of pertinent features. They are ordered by the increasing number of features originally contained in the training set. We will briefly describe each considered training set and then we will focus our attention on the presented experiments. The interested reader can find more information about these sets of data in the provided references. The comparison of the two algorithms will be carried out by comparing the quality of the found solutions. The heuristic in [17] is in general faster to converge (or to get stuck in non-optimal solutions, see experiments), whereas our heuristic is generally able to find better-quality solutions. CPU times range from a few seconds (wine fermentations) to about half an hour (ovarian cancer).

13.5.1 Wine Fermentations

Problems occurring during the fermentation process of wine can impact the productivity of wine-related industries and also the quality of wine [13, 14]. The fermentation process of wine can be too slow or it can even become stagnant. Predicting

Table 13.1 Wine fermentations. Total features: 450.

	VNS-based heuristic							Heuristic in [17]						
α	0	0.20	0.40	0.60	0.80	1.00	1.20	0	0.20	0.40	0.60	0.80	1.00	1.20
$f(x)$	431	430	427	427	424	421	415	425	424	424	420	stuck	stuck	stuck
β	1	1.01	1.02	1.04	1.06	1.08	1.10	1	1.01	1.02	1.04	1.06	1.08	1.10
$f(x)$	431	430	429	425	422	411	401	425	424	423	420	415	400	386

how good the fermentation process is going to be may help enologists who can then take suitable steps to make corrections when necessary and to ensure that the fermentation process concludes smoothly and successfully.

We present some analysis performed on a set of data obtained from a winery in Chile's Maipo Valley, which is the result of 24 measurements of industrial vinifications of *Cabernet sauvignon* [22, 23]. The data are related to the harvest of 2002. The level of 30 compounds are analyzed during time: the whole set of data consists of approximately 22000 data points. In this paper, the considered set of data contains 24 fermentations described by $15 \times 30 = 450$ features: the first class contains *normal* fermentations (9 in total), whereas the second class contains *problematic* fermentations (15 in total).

Table 13.1 shows some computational experiments. Note that α-consistent biclusterings with $\alpha = 0$ and β-consistent biclusterings with $\beta = 1$ correspond to consistent biclusterings (see Definition 13.3). We executed Algorithm 14 with different choices for the two parameters α and β. In the table, the number of selected features $f(x)$ is given in correspondence with each experiment. We can remark that the number of selected features decreases as the values of α or β increases. This was expected, because fewer features should be selected when the required margin between the centroids of the biclusters is enlarged. Only the features that better describe the samples in the set of data should be contained in the biclustering we found that contain fewer features (in particular, the α-consistent biclustering with $\alpha = 1.20$ and the β-consistent biclustering with $\beta = 1.10$).

Table 13.1 also shows some results obtained by using the heuristic presented in [17] (the reader is referred to the reference paper for a sketch of the algorithm). The comparison with the VNS-based heuristic proposed in this paper shows that our heuristic was able to find better solutions for all experiments, i.e. it was able to find biclusterings having the desired consistency property where a larger number of features are selected. Moreover, in some experiments, the heuristic in [17] got stuck and was not able to provide any solution. This heuristic is based on the solution of a sequence of linear optimization problems, where some parameters are modified on the basis of partial found solutions. The algorithm got stuck when such parameters stopped changing iteration after iteration.

By using the found biclusterings, we were able to identify a subset of compounds that are most likely the cause of problematic wine fermentations [15]. For example, among the organic acids, the features related to lactic, malic, succinic, and tartaric acids are always preserved during the feature selection. Moreover, all the features

Table 13.2 Colon cancer, set I. Total features: 2000.

	VNS-based heuristic						Heuristic in [17]					
α	0	1	2	3	5	8	0	1	2	3	5	8
$f(x)$	1700	1698	1617	1596	1583	1352	1700	1531	1590	1590	1530	1211
β	1	1.02	1.03	1.05	1.08	1.10	1	1.02	1.03	1.05	1.08	1.10
$f(x)$	1700	1593	1577	1566	1432	1269	1700	1578	1509	1108	1082	stuck

related to each of these organic acids are assigned to only one bicluster, showing that they can play a very important role for the classification of the fermentations. Features related to the same compound can also be always discarded, or they can show some regular patterns. The study of all these features in the biclusterings can give some insights on fermentation process of wine. Moreover, the found biclusterings can also be exploited for performing supervised predictions of new fermentations from which the selected compounds have been monitored [16].

13.5.2 Colon Cancer – Set I

This set of data contains 62 samples collected from colon-cancer patients [21]. Among them, 40 *tumor biopsies* are from tumors and 22 *normal biopsies* are from healthy parts of the colons of the same patients. 2000 out of around 6500 genes were selected based on the confidence in the measured expression levels. This set of data, along with the known classification of its samples, is available on the Kent Ridge Database [8].

Table 13.2 shows the results of some experiments performed with the aim of finding consistent, α-consistent and β-consistent biclusterings of this set of data. As in the previous experiments, the two algorithms selected a smaller number of features when the values for the parameters α or β were larger. In these experiments, the two heuristics found two consistent biclusterings with the same number of features only once (1700 out of 2000 features), when $\alpha = 0$ and $\beta = 1$. In all other cases, our VNS-based heuristic was able to provide better solutions. The heuristic in [17] got stuck when β was set to 1.10. Moreover, in the experiments regarding α-consistent biclustering, $f(x)$ does not decrease regurarly with larger α values, showing that the heuristic in [17] was not able to find the optimal solution.

13.5.3 Colon Cancer – Set II

The third set of data that we consider is a set of gene expressions related to human tissues from sick patients (affected by colon cancer) and healthy patients [18]. This set of data is available on the web site of the Princeton University (see the reference

Table 13.3 Colon cancer, set II. Total features: 7457.

	VNS-based heuristic					Heuristic in [17]				
α	0	1	2	5	10	0	1	2	5	10
$f(x)$	7450	7448	7444	7413	7261	7450	7430	7291	stuck	stuck
β	1	1.10	1.50	2.00	3.00	1	1.10	1.50	2.00	3.00
$f(x)$	7450	7420	7107	6267	5365	7450	7349	7099	6054	5252

Table 13.4 Ovarian cancer. Total features: 15154.

	VNS-based heuristic			
α	0	0.001	0.005	0.009
$f(x)$	12701	12471	12198	11027
β	1	1.001	1.005	1.009
$f(x)$	12701	12519	12392	12233

for the web link). It contains 36 samples classified as *normal* or *cancer*, and each sample is described through 7457 features.

Table 13.3 shows some computational experiments. Even in these experiments, there is the tendency to select a smaller number of features when α or β are increased in value. The number of features that are selected by the heuristic in [17] is always smaller than the number of features selected by the VNS-based heuristic.

13.5.4 Ovarian Cancer

This set contains data collected from experiments performed with the aim of identifying gene patterns that can distinguish ovarian cancer from non-cancer [3]. As the authors of the reference paper remark, this study is significant to women who have a high risk of ovarian cancer due to family or personal history of cancer. The set of data includes 91 samples classified as *normal* and 162 samples classified as *ovarian cancer*. The total number of considered features is 15154. After the experiments, the intensity values of the raw data were normalized so that each intensity value can fall within the interval $[0,1]$. More details on these experiments can be found in [3]. The set of data can be downloaded from [8]: it is the largest set of data ever considered for feature selection by consistent biclustering. Our VNS-based heuristic was able to identify some consistent, α-consistent and β-consistent biclusterings by selecting a subset of pertinent features. Table 13.4 shows some computational experiments.

13.6 Conclusions

We considered a problem of great interest in data mining, that is the one of the identification of consistent biclusterings of sets of data, which are used to identify pertinent features describing the samples of a set of data. We presented a reformulation of the problem, originally modeled as a 0–1 linear fractional optimization problem, as a bilevel program and we proposed a new heuristic for its solution. This heuristic is based on the meta-heuristic Variable Neighborhood Search. Computational experiments on various sets of data available in the literature show that the proposed heuristic outperforms previously proposed ones and is promising for the solution of large instances.

Data are nowadays obtained from many resources and they need to be efficiently analyzed. The VNS-based heuristic we proposed represents a good step forward a satisfactory solution of feature selection problems. However, a wider test analysis of the algorithm on other training sets is needed in order to study possible improvements. To this aim, we plan to implement the heuristic in C/C++, so that the CPLEX solver can be invoked more efficiently and the overall execution can be optimized.

Acknowledgements. The authors are thankful to Sonia Cafieri for the fruitful discussions on the bilevel reformulation presented in the paper.

References

1. Belotti, P.: Couenne: a user's manual. Technical report, Lehigh University (2009)
2. Busygin, S., Prokopyev, O.A., Pardalos, P.M.: Feature selection for consistent biclustering via fractional 0–1 programming. Journal of Combinatorial Optimization 10, 7–21 (2005)
3. Hitt, B.A., Levine, P.J., Fusaro, V.A., Steinberg, S.M., Mills, G.B., Simone, C., Fishman, D.A., Kohn, E.C., Liotta, L.A., Petricoin III, E.F., Ardekani, A.M.: Use of proteomic patterns in serum to identify ovarian cancer. The Lancet 359, 572–577 (2002)
4. Fourer, R., Gay, D.M., Kernighan, B.W.: AMPL: A Modeling Language for Mathematical Programming. Brooks/Cole Publishing Company, Cengage Learning (2002)
5. Hansen, P., Mladenovic, N.: Variable neighborhood search: Principles and applications. European Journal of Operational Research 130(3), 449–467 (2001)
6. Hartigan, J.: Clustering Algorithms. John Wiles & Sons, New York (1975)
7. Ilog cplex solver, `http://www.ilog.com/products/cplex/`
8. Kent ridge database, `http://datam.i2r.a-star.edu.sg/datasets/krbd/`
9. Kundakcioglu, O.E., Pardalos, P.M.: The complexity of feature selection for consistent biclustering. In: Butenko, S., Pardalos, P.M., Chaovalitwongse, W.A. (eds.) Clustering Challenges in Biological Networks. World Scientific Publishing (2009)
10. Mladenovic, M., Hansen, P.: Variable neighborhood search. Computers and Operations Research 24, 1097–1100 (1997)
11. Mucherino, A.: Extending the definition of β-consistent biclustering for feature selection. In: Proceedings of the Federated Conference on Computer Science and Information Systems, FedCSIS 2011. IEEE (2011)
12. Mucherino, A., Cafieri, S.: A new heuristic for feature selection by consistent biclustering. Technical Report arXiv:1003.3279v1 (March 2010)
13. Mucherino, A., Papajorgji, P., Pardalos, P.M.: Data Mining in Agriculture. Springer (2009)
14. Mucherino, A., Papajorgji, P., Pardalos, P.M.: A survey of data mining techniques applied to agriculture. Operational Research: An International Journal 9(2), 121–140 (2009)

15. Mucherino, A., Urtubia, A.: Consistent biclustering and applications to agriculture. In: Proceedings of the Industrial Conference on Data Mining, ICDM 2010, Workshop on Data Mining and Agriculture DMA 2010, IbaI Conference Proceedings, pp. 105–113. Springer, Berlin (2010)
16. Mucherino, A., Urtubia, A.: Feature selection for datasets of wine fermentations. In: Proceedings of the 10th International Conference on Modeling and Applied Simulation, MAS 2011. I3A (2011)
17. Nahapatyan, A., Busygin, S., Pardalos, P.M.: An improved heuristic for consistent biclustering problems, vol. 102, pp. 185–198. Springer
18. Notterman, D.A., Alon, U., Sierk, A.J., Levine, A.J.: Transcriptional gene expression profiles of colorectal adenoma, adenocarcinoma, and normal tissue examined by oligonucleotide arrays. Cancer Research 61, 3124–3130 (2001)
19. Sahinidis, N.V., Tawarmalani, M.: BARON 9.0.4: Global Optimization of Mixed-Integer Nonlinear Programs. User's Manual (2010)
20. Tawarmalani, M., Sahinidis, N.V.: A polyhedral branch-and-cut approach to global optimization. Mathematical Programming 103, 225–249 (2005)
21. Notterman, D.A., Gish, K., Ybarra, S., Mack, D., Levine, A.J., Alon, U., Barkai, N.: Broad patterns of gene expression revealed by clustering analysis of tumor and normal colon tissues probed by oligonucleotide arrays. PNAS 96, 6745–6750 (1999)
22. Urtubia, A., Perez-Correa, J.R., Meurens, M., Agosin, E.: Monitoring large scale wine fermentations with infrared spectroscopy. Talanta 64(3), 778–784 (2004)
23. Urtubia, A., Perez-Correa, J.R., Soto, A., Pszczolkowski, P.: Using data mining techniques to predict industrial wine problem fermentations. Food Control 18, 1512–1517 (2007)

Chapter 14
Scheduling English Football Fixtures: Consideration of Two Conflicting Objectives

Graham Kendall, Barry McCollum, Frederico R.B. Cruz,
Paul McMullan, and Lyndon While

Abstract. In previous work the distance travelled by UK football clubs, and their supporters, over the Christmas/New Year period was minimised. This is important as it is not only a holiday season but, often, there is bad weather at this time of the year. Whilst searching for good quality solutions for this problem, various constraints have to be respected. One of these relates to *clashes*, which measures how many *paired* teams play at home on the same day. Whilst the supporters have an interest in minimising the distance they travel, the police also have an interest in having as few pair clashes as possible. This is due to the fact that these fixtures are more expensive, and difficult, to police. However, these two objectives (minimise distance and minimise pair clashes) conflict with one another in that a decrease in one intuitively leads to an increase in the other. This chapter explores this question and shows that there are compromise solutions which allow fewer pair clashes but does not statistically increase the distance travelled. We present a detailed set of computational experiments, on datasets covering seven seasons. We conclude that it

---

Graham Kendall
School of Computer Science, University of Nottingham, Malaysia Campus, Jalan Broga, 43500 Semenyih, Selangor Darul Ehsan, Malaysia
e-mail: graham.kendall@nottingham.edu.my

Barry McCollum · Paul McMullan
School of Electronics, Electrical Engineering and Computer Science, Queen's University Belfast, BT7 1NN, UK
e-mail: {B.McCollum,p.p.mcmullan}@qub.ac.uk

Frederico Cruz
Departamento de Estatística - ICEx - UFMG, Av. Antônio Carlos, 6627, 31270-901 - Belo Horizonte - MG, Brazil
e-mail: fcruz@est.ufmg.br

Lyndon While
School of Computer Science & Software Engineering, The University of Western Australia, Perth WA 6009, Australia
e-mail: lyndon@csse.uwa.edu.au

E.-G. Talbi (Ed.): Hybrid Metaheuristics, SCI 434, pp. 369–385.
springerlink.com

is sometimes possible to reduce the number of pair clashes whilst not significantly increasing the overall distance that is travelled.

14.1 Introduction

The English Premier League is one of the most high profile, and successful, football (soccer in the USA) leagues in the world. It comprises 20 teams which have to play each other both home and away (i.e. a double round robin tournament), resulting in 380 fixtures that have to be scheduled. The other three main divisions in England (the Championship, League One and League Two) each have 24 teams, resulting in 552 fixtures having to be scheduled for each division. Therefore, for the four main divisions in England 2036 fixtures have to be scheduled every season. The divisions operate a system of promotion and relegation such that the teams in each division changes each year so it is not possible to simply use the same schedule every time.

Of particular interest are the schedules that need to be generated for the Christmas/New Year period. At this time of the year it is a requirement that every team plays two fixtures, one on Boxing Day (26th December) and one on New Years Day (1st January). Whilst scheduling these two sets of fixtures the overriding aim is to minimise the total distance that has to be travelled by the supporters. An analysis of the fixtures that were actually used, and also following discussions with the football authorities, confirm that this is a real world requirement and that the distances travelled by the supporters are the minimum when compared against other fixtures when all teams play. In addition, there are various other constraints that have to be respected, which are described in sections 14.3 and 14.4.

The problem we address in this chapter is to attempt to minimise two competing objectives to ascertain if there is a good trade off between them. The objectives we minimise are the distances travelled by the supporters and the number of *pair clashes*. Pairing matches two (or more) teams and dictates that these clubs should not play at home on the same day. If they do, this is termed a pair clash. In fact, a certain number of pair clashes are allowed. The exact number is taken from the number that were present in the published fixtures for a given season. Importantly, paired teams cannot play each other on the two days in question. This is treated as a hard constraint. It is this constraint that causes a problem. If we allow Liverpool and Everton (for example) to play each other, one set of supporters would only travel four miles. If these teams are paired (as they are) then they cannot play each other so the distances are likely to increase as either Liverpool or Everton would have to travel more than four miles. As pair clashes usually involve teams which are geographically close this gives rise to the conflicting objectives.

In [19], an initial study of the problem considered the 2003-2004 football season, suggesting that it may be possible to minimise both of these competing objectives but still produce results which are acceptable to both the supporters (who are interested in minimising the amount they travel) and the police (who are interested in having fewer pair clashes). In this chapter, we carry out a more in depth study by considering more seasons and carrying out statistical analysis of the results in order to draw stronger conclusions.

14.2 Related Work

Producing a double round robin tournament is relatively easy in that the algorithms are well known, with the polygon construction method being amongst the most popular [9]. The problem with utlising such an algorithm is that the fixtures it generates, although being a valid round robin tournament, will not adhere to all the additional constraints for a particular problem. Moreover, every problem instance will be subtly different and, often, a bespoke algorithm is required for each instance. This is even the case when faced with seemingly the same problem. For example, the English Football League consists of four divisions and 92 teams. It would be easy to assume that once an algorithm has been developed it can be used every season. This may indeed be the case but due to the promotion/relegation system the problem changes year on year and, perhaps, there are additional features/constraints in one season that were not previously present. Rasmussen and Trick [21] provide an excellent overview of the issues, methods and theoretical results for scheduling round robin tournaments.

The Travelling Tournament Problem (TTP) [11] is probably the most widely used test bed in sports scheduling. The problem was inspired by work carried out for Major League Baseball [11]. The aim of the TTP is to generate a double round robin tournament, while minimising the overall distance travelled by all teams. Unlike the problem studied in this paper, it is possible to minmise the overall travel distance as teams go on *road trips* so, with a suitable schedule, the length of these trips can be reduced. The TTP is further complicated by the introduction of two constraints. The first says that no team can play more than three consecutive home or away games. The second stipulates that if team i plays team j in round, r, then team j cannot play team i in round $r+1$. These constraints add sufficient complexity to the problem so as to make it challenging, but it still does not reflect all the constraints that are present in the real world problem.

The TTP has received significant research attention. Some of the important papers being [12, 2, 8, 22, 25]. A recent annotated bibliography of TTP papers can be found in [18]. An up to date list of the best known solutions, as well as details of all the instances, can be found at the web site maintained by Michael Trick [23].

With respect to minimising travel costs/distances, previous studies have considered a variety of sports. Campbell and Chen [6] and Ball and Webster [3] both studied basketball, attempting to minimise the distance travelled. Bean and Birge [4] also studied basketball, attempting to minimise airline travel costs. Minimising travel costs was also the focus of [5], for baseball. Minimising travel distances for hockey [16] and umpires for baseball [15] have also been studied. Wright [28], as one part of the evaluation function, considered travel between fixtures for English cricket clubs. Costa [7] considered the National Hockey League, where minimisation of the distance travelled by the teams was just one factor in the objective function.

Urrutia and Ribeiro [24] have shown that minimising distance and maximising breaks (two consecutive home games (home break) or two consecutive away games (away break)) is equivalent. This followed previous work by de Werra [26, 27] and Elf et al. [14] who showed how to construct schedules with the minimum number of breaks.

The scheduling problem that we are considering in this chapter is minimising the distance travelled for two complete fixtures (a complete fixture is defined as a set of fixtures when every team plays) while, at the same time, minimising the number of pair clashes. These two complete fixtures can then be used over the Christmas holiday period when, for a variety of reasons, teams wish to limit the amount of travelling undertaken. Note, that this is a different problem to the Travelling Tournament Problem as the TTP assumes that teams go on road trips, and so the total distance travelled over a season can be minimised. In English football, there is no concept of road trips. Therefore, over the course of a season, the distance cannot be minimised. However, we can minimise the distance on particular days. Kendall [17] adopted a two-phase approach to produce two complete fixtures for this problem. A depth first search was used to produce fixtures for one day, for each division. A further depth first search created another set of fixtures for the second day. This process produced eight separate fixtures (two sets of fixtures for each division) which adhered to some of the constraints (e.g. a team plays at home on one day and away on the other) but had not yet addressed the constraints with regards to pair clashes (see [17] for a detailed description). The fixture lists from the depth first searches were input to a local search procedure which aimed to satisfy the remaining constraints, whilst attempting to minimise the overall distance travelled. The output of the local search, and a post-process operation to ensure feasibility, produced the results presented in the paper.

Overviews of sports scheduling can be found in [13, 9, 10, 21, 29, 20, 18].

14.3 Problem Definition

In previous work [17] the only objective was to minimise the total distance travelled by the teams/supporters. The aim of that study was to investigate if we were able to generate better quality solutions than those used by the football league. We demonstrated that it was possible. As stated in the Introduction, the police also have an interest in the fixtures that are played at this time of the year. If we are able to generate acceptable schedules, with fewer pair clashes then the policing costs would be reduced.

The purpose of this chapter is to investigate if there is an acceptable trade off between the minimisation of distance and the minimisation of pair clashes. In order to do this we will utilise a multi-objective methodology.

14.4 Experimental Setup

We use a two stage algorithm. In [17] a depth first search (DFS) was used, followed by a local search . DFS was used as we wanted to carry out a preliminary study just to see if this area was worthy of further study. As we were able to produce superior solutions to the published fixtures we have now decided to utilise more sophisticated methods, due to the large execution times of DFS which were typically a few hours for each division. In this work we utilise CPLEX as a replacement for DFS and simulated annealing [1] as a replacement for the local search. This reduces the overall execution time from tens of hours to a few minutes.

14.4.1 Phase 1: CPLEX

The first phase uses CPLEX to produce an optimal solution to a relaxed version of the problem. In generating *relaxed* optimal solutions we respect the following constraints, whilst minimizing the overall distance.

1. Each of the 92 teams has to play on two separate days (i.e. 46 fixtures will be scheduled on each day).
2. Each team has to play at home on one day and away on the other.
3. Teams are not allowed to play each other on both days.
4. A team is not allowed to play itself.

The CPLEX model is executed four times. Each run returns the Boxing Day and New Years Day fixtures for a particular division. Each run takes less than 10 seconds.

In solving the CPLEX model we do not take into account many of the constraints that ultimately have to be respected. For example, pair clashes, geographical constraints such as the number of London or Manchester clubs playing at home on the same day etc. (see [17] for details).

14.4.2 Phase 2: Simulated Annealing

The schedules from CPLEX are input to the second phase, where we utilise simulated annealing . This operates across all the divisions in order to resolve any hard constraint violations whilst still attempting to minimise the distance.

The simulated annealing parameters are as follows:

Start_Temperature = 1000 The same value is used across all seven datasets and was found by experimentation. We could have used different values for each dataset but we felt that it was beneficial to be consistent across all the datasets.

Stop_Temperature The algorithm continues while the temperature is > 0.1.

Cooling Schedule $CurTemp = CurTemp * 0.95$.
Number of Iterations 2000 iterations are carried out at each temperature.

14.4.3 Evaluation Function

The evaluation function we use for simulated annealing is dynamic in that the hard constraint violations are more heavily penalised as the search progresses. This enables more exploration at the start of the search, which gets tighter as the temperature is reduced. The objective function is formulated as follows:

$$f(x) = d\_fb + d\_fy + w \times penalty \tag{14.1}$$

where:

d_fb = total distance travelled by teams on Boxing Day.
d_fy = total distance travelled by teams on New Years Day.
w = is a weight for the penalty (see below). It is given by (*Start_Temperature - CurTemp*). *Start_Temperature* is the maximum temperature for the simulated annealing algorithm. *CurTemp* is the current temperature of the simulated annealing algorithm. As the simulated annealing algorithm progresses, the weight of the penalty gradually increases, driving the search towards feasible solutions, but allowing it to search the infeasible region at the start of the search.
penalty = This is given by a summation of the following terms (the limits referred to are available in [17] and represent the values found by analyzing the published fixtures):

ReverseFixtures The number of *reverse* fixtures (the same teams cannot meet on both days).
Boxing Day Local Derby Clashes The number of paired teams playing each other on Boxing Day.
New Years Day Local Derby Clashes The number of paired teams playing each other on New Years Day.
Boxing Day London Clashes The number of London clubs playing at home on Boxing Day, which exceed a given limit.
New Years Day London Clashes The number of London clubs playing at home on New Years Day, which exceed a given limit.
Boxing Day Greater Manchester Clashes The number of Greater Manchester based clubs playing at home on Boxing Day, which exceed a given limit.
New Years Day Greater Manchester Clashes The number of Greater Manchester based clubs playing at home on New Years Day, which exceed a given limit.
Boxing Day London Premier Clashes The number of Premiership London clubs playing at home on Boxing Day, which exceed a given limit.
New Years Day London Premier Clashes The number of Premiership London clubs playing at home on New Years Day, which exceed a given limit.

Boxing Day Clashes The number of Boxing Day clashes greater than an allowable limit.

New Years Day Clashes The number of New Years Day clashes greater than an allowable limit.

14.4.4 Perturbation Operators

Simulated annealing often has a single neighborhood operator but we have defined sixteen operators in order to match the hard constraints within the model. The operators are as follows:

1. Examines the Boxing Day fixtures and if the number of clashes exceeds an upper limit, randomly select one of the clashing fixtures and swap the home and away teams.
2. Same as 1 expect that it considers New Years Day fixtures.
3. Examines the Boxing Day fixtures and if the number of London based clubs exceeds an upper limit, randomly select one of the fixtures that has a London based club playing at home and swap the home and away teams.
4. Same as 3 except that it considers Greater Manchester based clubs.
5. Same as 3 except that it considers London based premiership clubs.
6. Same as 3 except that it considers the New Years Day fixtures.
7. Same as 4 except that it considers the New Years Day fixtures.
8. Same as 5 except that it considers the New Years Day fixtures.
9. Examines the Boxing Day and New Years Day fixture lists, returning the number of reverse fixtures (where team i plays team j and team j plays team i). While there are reverse fixtures, one of the reverse fixtures on Boxing Day is chosen and the home team is swapped with a randomly selected home team, with the condition that the swaps must be made between teams in the same division. This operator iterates until all reverse fixtures have been removed from the fixture list.
10. Same as 9 except the swaps are made in the New Years Day fixtures.
11. This operator examines the Boxing Day and New Years Day fixture lists, returning the number fixtures where paired teams are playing each other. While this is the case, one of the Boxing Day fixtures is chosen and the home team is swapped with a randomly selected home team in the Boxing Day fixtures, with the condition that the swaps must be made between teams in the same division. This operator iterates until all local pair clashes have been removed from the fixture lists.
12. Same as 11 except the swaps are made in the New Years Day fixtures.
13. This operator chooses a random fixture from a candidate list (we use a candidate list size of 250) which represents the potential fixtures that have the shortest distances. Swaps are carried out in the Boxing Day fixtures in order to allow the two teams from the selected item in the candidate list to play each

other. The necessary swaps are also done in the New Years Day fixture to ensure feasibility.
14. Same as 13 except that it considers the New Years Day fixtures.
15. Selects a random fixture in the Boxing Day fixture list and swaps the home and away teams.
16. Same as 15, but swaps a random fixture in the New Years Day fixture list.

At each iteration, one of the sixteen operators is chosen at random. *Start_Temperature* is initially set to enable infeasible solutions during the early stages of the algorithm, but they are more heavily penalised at lower temperatures (eq. 14.1), ensuring that the final solution is feasible.

14.4.5 Experimental Methodology

We are investigating this problem from a multi-objective perspective but rather than using a multi-objective algorithm we run the same algorithm a number of times, adjusting the parameters for each run. As an example, for the 2002-2003 season the number of pair clashes, in the published fixtures, was 10 and 8 for Boxing Day and New Years Day respectively. We denote this as 10-8 in the tables below. Therefore, the first experiment fixes the values as 10 and 8 as the number of pair clashes that cannot be exceeded. In this respect, these values represent hard constraints. The next experiment reduces one of these values so that the next experiment uses 10-6. We then reduce the other value to run a further experiment using 8-8. There are two points worthy of note. Firstly, we reduce the value by two as a pair clash of, say, Everton and Liverpool actually counts as two pair clashes as both teams are considered to be clashing. Secondly, we do not reduce the total number of pair clashes below 16.

14.5 Results

Tables 14.1 thru 14.7 shows the results of each of the seven seasons that we use. The *Clashes* column shows the number of pair clashes (see section 14.4.5 for the notation that we use). *Min* represents the best solution found. *Max* is worst solution found and *Average* and *Std Dev* are self-explanatory. All experiments were runs 30 times.

In tables 14.8 and 14.9 we analyse the results from table 14.1. Table 14.8 shows the results of independent two-tailed t-tests (at the 95% confidence level) to compare the means of each experiment against every other experiment for that season. Where two experiments are statistically significant the relevant cell shows "Yes", otherwise the cell is empty. As an example, if we compare 10-8 (column) with 10-6 (row) in table 14.8 we see that the means (i.e. the travel distances from 30 independent runs) are statistically different. By comparing the means in table 14.1, 5630 and 6183

Table 14.1 2002-2003: Summary of results from 30 runs

Clashes	Min	Max	Average	Std Dev
10-8	5243	6786	5630	288.46
10-6	5674	7222	6183	410.71
8-8	5562	6797	6070	309.50

Table 14.2 2003-2004: Summary of results from 30 runs

Clashes	Min	Max	Average	Std Dev
8-14	5464	6173	5698	165.46
8-12	5412	6519	5827	228.66
8-10	5511	7093	6053	417.00
8-8	5887	7674	6535	433.83
6-14	5550	6334	5805	176.02
6-12	5559	6587	6036	289.75
6-10	5898	7416	6454	395.37
4-14	5592	6911	6059	274.61
4-12	5886	7848	6635	484.59
2-14	6028	7704	6704	448.87

Table 14.3 2004-2005: Summary of results from 30 runs

Clashes	Min	Max	Average	Std Dev
10-10	5365	6986	5644	318.33
10-8	5345	6348	5727	259.17
10-6	5812	7714	6431	421.63
8-10	5443	6982	5923	469.01
8-8	5645	7612	6428	550.67
6-10	5810	7824	6486	487.26

respectively, we conclude that reducing the number of pair clashes from 18 (10-8) to 16 (8-8) the travel distances for the clubs/supporters increases by a significant amount. Looking at 10-6 and 8-8, there is no statistical difference. However, as both of these experiments represent 16 pair clashes it is, perhaps, not surprising that the average distance travelled over the 30 runs is (statistically) the same.

Table 14.9 summarises the results from table 14.8 by only showing those experiments where there are statistical differences, AND when the total number of pair clashes is different (i.e. it will ignore 10-6 and 8-8). We can see from table 14.9 that there are no experiments where we can reduce the number of pair clashes that leads to no statistical difference in the distance travelled.

Table 14.4 2005-2006: Summary of results from 30 runs

Clashes	Min	Max	Average	Std Dev
12-14	5234	6046	5575	184.74
12-12	5335	6002	5596	153.90
12-10	5240	6511	5641	238.58
12-8	5334	6423	5754	231.81
12-6	5481	6958	6010	339.63
12-4	6041	6989	6468	271.99
10-14	5171	6683	5606	304.33
10-12	5308	6322	5610	204.96
10-10	5460	6674	5846	359.65
10-8	5595	6380	5872	216.82
10-6	6027	7561	6660	421.25
8-14	5335	6674	5680	286.00
8-12	5334	6133	5722	211.02
8-10	5608	7078	5979	356.15
8-8	6146	7277	6587	302.48
6-14	5500	6694	5843	254.23
6-12	5528	6655	5951	233.54
6-10	5884	7291	6529	382.80
4-14	5713	7391	6161	331.25
4-12	6032	7904	6662	434.72
2-14	6084	7551	6682	399.34

Table 14.5 2006-2007: Summary of results from 30 runs

Clashes	Min	Max	Average	Std Dev
14-8	5713	7040	6077	300.71
14-6	5735	7065	6117	270.59
14-4	5872	7000	6259	227.84
14-2	6110	7778	6741	402.35
12-8	5721	6784	6084	244.28
12-6	5714	6894	6234	326.99
12-4	6195	7546	6791	405.86
10-8	5762	7671	6209	411.02
10-6	5894	7376	6618	423.94
8-8	6071	6958	6513	251.33

Table 14.6 2007-2008: Summary of results from 30 runs

Clashes	Min	Max	Average	Std Dev
14-10	5366	5902	5595	145.26
14-8	5403	5975	5674	152.93
14-6	5425	7172	5870	372.17
14-4	5690	6995	6172	364.78
14-2	5905	7856	6698	435.98
12-10	5370	6506	5736	294.88
12-8	5321	7139	5850	338.15
12-6	5625	7394	6084	365.93
12-4	5961	7580	6575	411.41
10-10	5340	6552	5754	228.71
10-8	5616	6365	5944	183.52
10-6	6101	7468	6619	369.10
8-10	5536	7081	6056	369.47
8-8	6091	7884	6725	402.08
6-10	5951	7709	6647	381.12

Table 14.7 2008-2009: Summary of results from 30 runs

Clashes	Min	Max	Average	Std Dev
10-10	5564	6806	5833	246.11
10-8	5574	6235	5829	140.52
10-6	5736	6523	6106	208.78
8-10	5581	6817	5936	281.83
8-8	5790	6900	6148	230.42
6-10	5809	7194	6208	274.67

Table 14.8 2002-2003: Are the Results Statistically Different?

Clashes	10-8	10-6	8-8
10-8	X	Yes	Yes
10-6		X	
8-8			X

Table 14.9 2002-2003: Are different total clashes significantly different?

Clashes	10-8	10-6	8-8
10-8	X		
10-6		X	
8-8			X

Table 14.10 2003-2004: Are the Results Statistically Different?

Clashes	8-14	8-12	8-10	8-8	6-14	6-12	6-10	4-14	4-12	2-14
8-14	X	Yes	Yes	Yes	Yes	Yes	Yes	Yes	Yes	Yes
8-12		X	Yes	Yes		Yes	Yes	Yes	Yes	Yes
8-10			X	Yes	Yes		Yes		Yes	Yes
8-8				X	Yes	Yes		Yes		
6-14					X	Yes	Yes	Yes	Yes	Yes
6-12						X	Yes		Yes	Yes
6-10							X	Yes		Yes
4-14								X	Yes	Yes
4-12									X	
2-14										X

Table 14.11 2003-2004: Are different total clashes significantly different?

Clashes	8-14	8-12	8-10	8-8	6-14	6-12	6-10	4-14	4-12	2-14
8-14	X									
8-12		X								
8-10			X							
8-8				X						
6-14					X					
6-12						X				
6-10							X			
4-14								X		
4-12									X	
2-14										X

Table 14.12 2004-2005: Are the Results Statistically Different?

Clashes	10-10	10-8	10-6	8-10	8-8	6-10
10-10	X		Yes	Yes	Yes	Yes
10-8		X	Yes		Yes	Yes
10-6			X	Yes		
8-10				X	Yes	Yes
8-8					X	
6-10						X

Table 14.13 2004-2005: Are different total clashes significantly different?

Clashes	10-10	10-8	10-6	8-10	8-8	6-10
10-10	X	Yes				
10-8		X				
10-6			X			
8-10				X		
8-8					X	
6-10						X

Table 14.14 2005-2006: Are different total clashes significantly different?

Clashes	12-14	12-12	12-10	12-8	12-6	12-4	10-14	10-12	10-10	10-8	10-6	8-14	8-12	8-10	8-8	6-14	6-12	6-10	4-14	4-12	2-14
12-14	X	Yes	Yes				Yes	Yes				Yes									
12-12		X	Yes					Yes				Yes									
12-10			X	Yes			Yes						Yes								
12-8				X								Yes									
12-6					X				Yes												
12-4						X															
10-14							X	Yes				Yes	Yes								
10-12								X													
10-10									X	Yes		Yes		Yes			Yes				
10-8										X						Yes					
10-6											X										
8-14												X	Yes								
8-12													X								
8-10														X		Yes					
8-8															X						
6-14																X	Yes				
6-12																	X				
6-10																		X			
4-14																			X		
4-12																				X	
2-14																					X

Table 14.15 2006-2007: Are different total clashes significantly different?

Clashes	14-8	14-6	14-4	14-2	12-8	12-6	12-4	10-8	10-6	8-8
14-8	X	Yes			Yes	Yes		Yes		
14-6		X				Yes		Yes		
14-4			X							
14-2				X						
12-8					X			Yes		
12-6						X				
12-4							X			
10-8								X		
10-6									X	
8-8										X

Table 14.16 2007-2008: Are different total clashes significantly different?

Clashes	14-10	14-8	14-6	14-4	14-2	12-10	12-8	12-6	12-4	10-10	10-8	10-6	8-10	8-8	6-10
14-10	X														
14-8		X								Yes					
14-6			X			Yes					Yes		Yes		
14-4				X											
14-2					X										
12-10						X	Yes			Yes					
12-8							X				Yes				
12-6								X							
12-4									X						
10-10										X					
10-8											X				
10-6												X			
8-10													X		
8-8														X	
6-10															X

Table 14.17 2008-2009: Are different total clashes significantly different?

Clashes	10-10	10-8	10-6	8-10	8-8	6-10
10-10	X	Yes		Yes		
10-8		X				
10-6			X			
8-10				X		
8-8					X	
6-10						

Tables 14.10 and 14.11 show similar analysis for the 2003-204 season. Again, it is not possible to reduce the number of pair clashes without an (statistically) increase in the distance travelled.

Tables 14.12 and 14.13 are more interesting. Table 14.12 shows that there is no statistical difference between the 10-10 (20 pair clashes) experiment and the 10-8 (18 pair clashes) experiment. Removing all the *noise* from the table (see table 14.13) we can see that it is possible to reduce the number of pair clashes from 20 to 18 without a significant rise in the distance travelled (the respective means from table 14.3 are 5644 and 5727).

For the remaining four seasons, we only present the summary tables. Where a"Yes" appears in these tables (tables 14.14 thru 14.17) it indicates that it is possible to reduce the number of pair clashes and not have an (statistical) increase in travel distance. The tables show that there are a number of opportunities to reduce policing costs. We are probably most interested in the top rows as they represent the fixtures that were actually used.

14.6 Conclusion

We have demonstrated that it is sometimes possible to reduce the number of pair clashes without a statistical difference to the distance that has to be travelled by the club/supporters. This provides the police with the ability to reduce their costs for these two days, which might have included paying overtime. We hope that we are able to discuss these results with the football authorities and the police in order for them to validate our work and to provide us with potential future research directions. We already recognise that some pair clashes might provide the police with more *problems* than others and it might be worth prioritising certain clashes so that these can be removed, rather than removing less high profile fixtures. As a longer term research aim, we would like to include in our model details about public transport as some routes might be more difficult than other routes, even if they are shorter. We also plan to run our algorithms for every future season, as well as for previous seasons. Executing the algorithm is not the main issue. Data collection provides the real challenge due to the distance data that has to be collected. To date, this has been carried out manually by using motoring organisation's web sites but we have recently started experimenting with services such as Google Maps™and Multimap which will speed up the data collection.

References

1. Aarts, E., Korst, J., Michels, W.: Simulated annealing. In: Burke, E.K., Kendall, G. (eds.) Search Methodologies: Introductory Tutorials in Optimization and Decision Support Methodologies, 1st edn., ch. 7, pp. 97–125. Springer (2005)
2. Anagnostopoulos, A., Michel, L., Van Hentenryck, P., Vergados, Y.: A simulated annealing approach to the traveling tournament problem. Journal of Scheduling 9, 177–193 (2006)

3. Ball, B.C., Webster, D.B.: Optimal scheduling for even-numbered team athletic conferences. AIIE Transactions 9, 161–169 (1977)
4. Bean, J.C., Birge, J.R.: Reducing travelling costs and player fatigue in the national basketball association. Interfaces 10, 98–102 (1980)
5. Cain, W.O.: The computer-aided heuristic approach used to schedule the major league baseball clubs. In: Ladany, S.P., Machol, R.E. (eds.) Optimal Strategies in Sports, pp. 33–41. North Holland, Amsterdam (1977)
6. Campbell, R.T., Chen, D.S.: A minimum distance basketball scheduling problem. In: Machol, R.E., Ladany, S.P., Morrison, D.G. (eds.) Management Science in Sports. Studies in the Management Sciences, vol. 4, pp. 15–25. North-Holland, Amsterdam (1976)
7. Costa, D.: An evolutionary tabu search algorithm and the NHL scheduling problem. INFOR 33, 161–178 (1995)
8. Di Gaspero, L., Schaerf, A.: A composite-neighborhood tabu search approach to the traveling tournament problem. Journal of Heuristics 13, 189–207 (2007)
9. Dinitz, J.H., Fronček, D., Lamken, E.R., Wallis, W.D.: Scheduling a tournament. In: Colbourn, C.J., Dinitz, J.H. (eds.) Handbook of Combinatorial Designs, 2nd edn., pp. 591–606. CRC Press (2006)
10. Drexl, A., Knust, S.: Sports league scheduling: Graph- and resource-based models. Omega 35, 465–471 (2007)
11. Easton, K., Nemhauser, G.L., Trick, M.A.: The Traveling Tournament Problem Description and Benchmarks. In: Walsh, T. (ed.) CP 2001. LNCS, vol. 2239, pp. 580–585. Springer, Heidelberg (2001)
12. Easton, K., Nemhauser, G.L., Trick, M.A.: Solving the Travelling Tournament Problem: A Combined Integer Programming and Constraint Programming Approach. In: Burke, E., De Causmaecker, P. (eds.) PATAT 2002. LNCS, vol. 2740, pp. 100–109. Springer, Heidelberg (2003)
13. Easton, K., Nemhauser, G.L., Trick, M.A.: Sports scheduling. In: Leung, J.T. (ed.) Handbook of Scheduling, pp. 52.1–52.19. CRC Press (2004)
14. Elf, M., Jünger, M., Rinaldi, G.: Minimizing breaks by maximizing cuts. Operations Research Letters 31(3), 343–349 (2003)
15. Evans, J.R.: A microcomputer-based decision support system for scheduling umpires in the American Baseball League. Interfaces 18, 42–51 (1988)
16. Ferland, J.A., Fleurent, C.: Computer aided scheduling for a sport league. INFOR 29, 14–25 (1991)
17. Kendall, G.: Scheduling English football fixtures over holiday periods. Journal of the Operational Research Society 59, 743–755 (2008)
18. Kendall, G., Knust, S., Ribeiro, C.C., Urrutia, S.: Scheduling in sports: An annotated bibliography. Computers & Operations Research 37, 1–19 (2010)
19. Kendall, G., While, L., McCollum, B., Cruz, F.: A multiobjective approach for UK football scheduling. In: Burke, E.K., Gendreau, M. (eds.) Proceedings of the 7th International Conference on the Practice and Theory of Automated Timetabling (2008)
20. Knust, S.: Classification of literature on sports scheduling (2010), `http://www.inf.uos.de/knust/sportssched/sportlit_class/` (last visited July 15, 2010)
21. Rasmussen, R.V., Trick, M.A.: Round robin scheduling – A survey. European Journal of Operational Research 188, 617–636 (2008)
22. Ribeiro, C.C., Urrutia, S.: Heuristics for the mirrored traveling tournament problem. European Journal of Operational Research 179, 775–787 (2007)
23. Trick, M.: Traveling tournament problem instances (2010), `http://mat.gsia.cmu.edu/TOURN/` (last accessed July 15, 2010)
24. Urrutia, S., Ribeiro, C.: Minimizing travels by maximizing breaks in round robin tournament schedules. Electronic Notes in Discrete Mathematics 18-C, 227–233 (2004)

25. Urrutia, S., Ribeiro, C.C., Melo, R.A.: A new lower bound to the traveling tournament problem. In: Proceedings of the IEEE Symposium on Computational Intelligence in Scheduling, pp. 15–18. IEEE, Honolulu (2007)
26. de Werra, D.: Scheduling in sports. In: Hansen, P. (ed.) Studies on Graphs and Discrete Programming, pp. 381–395. North Holland, Amsterdam (1981)
27. de Werra, D.: Some models of graphs for scheduling sports competitions. Discrete Applied Mathematics 21, 47–65 (1988)
28. Wright, M.: Timetabling county cricket fixtures using a form of tabu search. Journal of the Operational Research Society 45, 758–770 (1994)
29. Wright, M.: 50 years of OR in sport. Journal of the Operational Research Society 60, S161–S168 (2009)

Part IV
Combining Metaheuristics with Constraint Programming Approaches

Chapter 15
A Multi-paradigm Tool for Large Neighborhood Search

Raffaele Cipriano, Luca Di Gaspero, and Agostino Dovier

Abstract. We present a general tool for encoding and solving optimization problems. Problems can be modeled using several paradigms and/or languages such as: Prolog, MiniZinc, and GECODE. Other paradigms can be included. Solution search is performed by a hybrid solver that exploits the potentiality of the Constraint Programming environment GECODE and of the Local Search framework EasyLocal++ for Large Neighborhood Search . The user can modify a set of parameters for guiding the hybrid search. In order to test the tool, we show the development phase of hybrid solvers on some benchmark problems. Moreover, we compare these solvers with other approaches, namely a pure Local Search, a pure constraint programming search, and with a state-of-the-art solver for constraint-based Local Search.

15.1 Introduction

The number of known approaches for dealing with constraint satisfaction problems (CSP) and constrained optimization problems (COP) is as huge as the difficulty of these problems. They range from mathematical approaches (i.e., methods from Operations Research, such as Integer Linear Programming, Column Generation, ...) to Artificial Intelligence approaches (such as Constraint Programming, Evolutionary Algorithms, SAT-based techniques, Local Search, just to name a few). It is also well-known (and formally proved) that methods that are adequate for a given set of

Raffaele Cipriano · Agostino Dovier
Dipartimento di Matematica e Informatica, Università degli Studi di Udine
via delle Scienze 208, I-33100, Udine, Italy
e-mail: {cipriano,dovier}@dimi.uniud.it

Luca Di Gaspero
Dipartimento di Ingegneria Elettrica, Gestionale e Meccanica, Università degli Studi di Udine
via delle Scienze 208, I-33100, Udine, Italy
e-mail: l.digaspero@uniud.it

E.-G. Talbi (Ed.): Hybrid Metaheuristics, SCI 434, pp. 389–414.
springerlink.com

problem instances are often useless for others [20]. Notwithstanding, there is agreement on the importance of developing tools for challenging these kind of problems in an easy and effective way. In this paper we go in that direction providing a multi-paradigm hybrid-search tool called GELATO (Gecode + Easy Local = A Tool for Optimization).

The tool comprises three main components, each of them dealing with one specific aspect of the problem solution phase. In the *modeling component* the user defines the problem at a high-level, and chooses the strategy for solving it. The *translation component* deals with the compilation of the model and the abstract algorithm defined by the user into the solver frameworks supported by the tool. Finally, the *solving component* runs the compiled program on the problem instance at hand, interleaving the solution strategies decided at the modeling stage.

GELATO currently supports a number of programming paradigms/languages for modeling. In particular, a declarative logic programming approach can be used for modeling, supporting Prolog and MiniZinc, and exploiting the front-end translator of Prolog presented in [4] and the current support of MiniZinc in GECODE. Moreover, an object-oriented approach can be used through the GECODE framework. Additional paradigm/languages can be easily added to the system, for example extending the front-end developed for the Haskell language [21] or other front-ends to GECODE.

The solving phase makes use of pure Constraint Programming tree search as well as a combination of Constraint Programming with Local Search. In particular, for this combination we have implemented a parametric schema for Large Neighborhood Search [14]. Large Neighborhood Search is a particular Local Search heuristic that relies on a constraint solver for blending the inference capabilities of Constraint Programming with the effectiveness of the Local Search techniques.

GELATO is based on two state-of-the-art, existing systems: the GECODE Constraint Programming environment [15] and the Local Search framework EasyLocal++ [6]. Both these systems are free and open C++ systems with a growing community of users. The main contribution of this paper is to blend together these two streams of works so as to generate a comprehensive tool, that allows to exploit the multi-paradigm/multi-language modeling and the multi-technique solving.

We provide some examples of the effectiveness of this approach by testing the solvers on three classes of benchmark problems (namely the Asymmetric Traveling Salesman Problem, the Minimum Energy Broadcast, and the Course Timetabling Problem) and comparing them against a pure Local Search approach, a pure Constraint Programming search, and an implementation of the same models in Comet [11], another language for hybrid systems.

We have defined a small set of few parameters that can be tuned by the user *against* a particular problem. In the paper we also provide some tuning of these parameters and show their default values.

A relevant aspect of our system is that the very same GECODEmodel (either directly written in GECODE or obtained by translation from Prolog or Minizinc) can be used either for pure Constraint Programming search or for LNS search (and by degenerating LNS with small neighborhoods, for Local Search). The resulting system is available from `http://www.dimi.uniud.it/GELATO`.

We compare the solvers obtained from GECODE with other approaches, namely a pure Local Search, a pure Constraint Programming search, and with Comet, a state-of-the-art solver for constraint-based Local Search. The results of the experimentation show that GECODE Large Neighborhood Search solver is able to outperform the Local Search and the Constraint Programming search on the set of benchmark problems and it achieves the same performances as Comet.

15.2 Preliminaries

A *Constraint Satisfaction Problem* (CSP) (see, e.g., [2]) $\mathscr{P} = \langle \mathscr{X}, \mathscr{D}, \mathscr{C} \rangle$ is modeled by a set $\mathscr{X} = \{x_1, \ldots, x_k\}$ of *variables*, a set $\mathscr{D} = \{D_1, \ldots, D_k\}$ of *domains* associated to the variables (i.e., if $x_i = d_i$ then $d_i \in D_i$), and a set $\mathscr{C}$ of *constraints* (i.e., relations) over $\mathbf{dom} = D_1 \times \cdots \times D_k$.[1] A tuple $\langle d_1, \ldots, d_k \rangle \in \mathbf{dom}$ *satisfies* a constraint $C \in \mathscr{C}$ if and only if $\langle d_1, \ldots, d_k \rangle \in C$. $\mathbf{d} = \langle d_1, \ldots, d_k \rangle \in \mathbf{dom}$ is a *solution* of a CSP $\mathscr{P}$ if $\mathbf{d}$ satisfies every constraint $C \in \mathscr{C}$. The set of the solutions of $\mathscr{P}$ is denoted by $\mathbf{sol}(\mathscr{P})$. $\mathscr{P}$ is said to be *consistent* if and only if If $\mathbf{sol}(\mathscr{P}) \neq \emptyset$.

A *Constrained Optimization Problem* (COP) $\mathscr{O} = \langle \mathscr{X}, \mathscr{D}, \mathscr{C}, f \rangle$ is a CSP $\mathscr{P} = \langle \mathscr{X}, \mathscr{D}, \mathscr{C} \rangle$ with an associated function $f : \mathbf{sol}(\mathscr{P}) \to E$ where $\langle E, \leq \rangle$ is a well-ordered set (typical instances of E are $\mathbb{N}$, $\mathbb{Z}$, or $\mathbb{R}$). A *feasible solution* for $\mathscr{O}$ is any $\mathbf{d} \in \mathbf{sol}(\mathscr{P})$. When clear from the context, we use $\mathbf{sol}(\mathscr{O})$ for $\mathbf{sol}(\mathscr{P})$. A tuple $\mathbf{e} \in \mathbf{sol}(\mathscr{O})$ is a *solution* of the COP $\mathscr{O}$ if it minimizes the cost function f, namely if it holds that $\forall \mathbf{d} \in \mathbf{sol}(\mathscr{O})\ f(\mathbf{e}) \leq f(\mathbf{d})$.

Constraint Programming (CP) solves a CSP $\mathscr{P}$ by alternating the following two phases:

- a deterministic *constraint propagation* stage that reduces domains preserving $\mathbf{sol}(\mathscr{P})$, typically based on a *local* analysis of each constraint, one at a time;
- a non-deterministic *variable assignment*, in which one variable is selected together with one value in its current domain.

The process is repeated until a solution is found or a domain becomes empty. In the last case, the computation proceed by backtracking assignments, until possible. A COP $\mathscr{O}$ is solved by exploring the set $\mathbf{sol}(\mathscr{O})$ obtained in the previous way and storing the best value of the function f found during the search. However, a constraint analysis based on a partial assignment and on the best value already computed, might allow to sensibly prune the search tree. This complete search heuristics is called (with a slight of ambiguity with respect to Operations Research terminology) *branch and bound*.

A *Local Search* (LS) algorithm (e.g. [1]) for a COP $\mathscr{O}$ is given by defining a *search space* $\mathbf{sol}(\mathscr{O})$, a *neighborhood relation* $\mathscr{N}$, a *cost function* f, and a *stop*

[1] Constraints do not necessarily relate all the variables. For instance, $x_1 + x_2 = 3$ is a binary constraint on the variables x_1 and x_2. This binary constraint, however, has a k-ary constraint counterpart $C = \{\langle d_1, d_2, \ldots, d_k \rangle : d_i \in D_i, d_1 + d_2 = 3\}$.

criterion. The neighborhood relation $\mathcal{N}$, i.e. a set $\mathcal{N}(\mathbf{d}) \subseteq \mathbf{sol}(\mathcal{O})$, is defined for each element $\mathbf{d} \in \mathbf{sol}(\mathcal{O})$. The set $\mathcal{N}(\mathbf{d})$ is called the *neighborhood* of $\mathbf{d}$ and each $\mathbf{d}' \in \mathcal{N}(\mathbf{d})$ is called a *neighbor* of $\mathbf{d}$. Commonly, $\mathcal{N}(\mathbf{d})$ is implicitly defined by referring to a set of possible *moves*, which are transitions between feasible solutions in the form of perturbations to be applied. Usually these perturbations insist on a small part of the problem, involving only few variables.

Starting from an initial solution $\mathbf{s}_0 \in \mathbf{sol}(\mathcal{O})$, a LS *algorithm* iteratively navigates the space $\mathbf{sol}(\mathcal{O})$ by stepping from one solution $\mathbf{s}_i$ to a neighboring one $\mathbf{s}_{i+1} \in \mathcal{N}(\mathbf{s}_i)$, choosing s_{i+1} using a rule $\mathcal{M}$. The selection of the neighbor by $\mathcal{M}$ might depend on the whole computation that leaded $\mathbf{s}_0$ to $\mathbf{s}_i$ and on the values of f on $\mathcal{N}(\mathbf{s}_i)$. Moreover, it depends on the specific LS technique considered.

The stop criterion depends on the technique at hand, but it is typically based on stagnation detection (e.g., a maximum number of consecutive non improving moves, called *idle moves*) or a timeout. Several techniques for effectively exploring the search space have been presented in the literature (e.g., Montecarlo methods, Simulated Annealing, Hill Climbing, Steepest descent, Tabu Search, just to name a few).

LS algorithms can also be used to solve CSPs by relaxing some constraints and using f as a *distance to feasibility*, which accounts for the number of relaxed constraints that are violated.

Large Neighborhood Search (LNS) (e.g., [14]) is a LS method that relies on a particular definition of the neighborhood relation and on the strategy to explore the neighborhood. Differently from classical Local Search moves, which are small perturbations of the current solution, in LNS a large part of the solution is perturbed and searched for improvments. This part can be represented by a set $\mathbf{FV} \subseteq \mathcal{X}$ of variables, called *free variables*, that determines the neighborhood relation $\mathcal{N}$. More precisely, given a solution $\mathbf{s} = \langle d_1, \ldots, d_k \rangle$ and a set $\mathbf{FV} \subseteq \{X_1, \ldots, X_k\}$ of free variables

$$\mathcal{N}(s, \mathbf{FV}) = \{\langle e_1, \ldots, e_k \rangle \in \mathbf{sol}(\mathcal{O}) : (X_i \notin \mathbf{FV}) \to (e_i = d_i)\}$$

Given $\mathbf{FV}$, the neighborhood exploration can be performed with any searching technique, ranging from solution enumeration, to CP or Operations Research methods, and so on. In this work we focus on CP techniques for this exploration. In this case, the search technique is known as Constraint Based Local Search (CBLS) [11]. The following aspects are crucial for the performance of this technique: *which variables* have to be selected (i.e., the definition of $\mathbf{FV}$), and *how to perform the exploration* on these variables.

Once $\mathbf{FV}$ has been defined, the exploration of the neighborhood can be be made searching for:

- the *best neighbor*: namely, given a solution $\mathbf{s}$ and a set $\mathbf{FV} \subseteq \mathcal{X}$, look for a tuple $\mathbf{e} \in \mathcal{N}(s, \mathbf{FV})$ such that $(\forall \mathbf{d} \in \mathcal{N}(s, \mathbf{FV}))(f(\mathbf{e}) \leq f(\mathbf{d}))$.
- the best neighbor within a certain exploration timeout: namely the point in $\mathcal{N}(s, \mathbf{FV})$ that minimizes the value f found within a fixed timeout
- the first found neighbor point improving the value $f(\mathbf{s})$

- the first found neighbor point improving the objective function of at least a given value (e.g., below the 95% of $f(\mathbf{s})$).

Deciding *how many variables* will be free (i.e., $|\mathbf{FV}|$) affects the time spent on every large neighborhood exploration and the improvement of the objective function for each exploration. A small **FV** will lead to very efficient and fast search but, possibly at the price of very little improvement of the objective function. Otherwise, a big **FV** can lead to big improvements at each step, but every single exploration can take a lot of time. This trade-off should be investigated experimentally, looking at a size of **FV** which is a compromise between fast enough explorations and good improvements. Obviously, the choice of $|\mathbf{FV}|$ is strictly related to the search technique chosen (e.g., an efficient technique can manage more variables than a naïve one) and the fact that a timeout is used or not. The choice of *which variables* will be included in **FV** is strictly related to the problem we are solving: for simple and not too structured problems we can select the variables in a naïve way (randomly, or iterating between given sets of them); for complex and well-structured problems, we should define **FV** cleverly, selecting the variables which are most likely to give an improvement to the solution.

During the search of the new solution a portion of the neighborhood that does not fit the requirements (e.g., that do not improve the last known solution) is visited. Each of these neighbors is counted as a *failure*. The time spent on each exploration and the corresponding improvement of the cost function depends on $|\mathbf{FV}|$ (and, of course, on the structure of the problem) and a local stop criterion can be given either as a timeout or by setting a maximum number of failures (briefly, `maxF`) for the improving stage.

Experimental tests must be made for choosing values of $|\mathbf{FV}|$ and `maxF` that lead to quick explorations and good improvements. This choice might depend from other parameter choices. For instance, the choice of the order in which variables have to be instantiated in the CP stage can be naïve (e.g., choose the leftmost one) or clever (e.g., the one involved in more constraints and in case of equality break a tie choosing the one with smallest domain first). Usually, a clever technique allows to manage more variables, but this is not a general rule (it can be the case that the extra time for choosing a variable causes to exceed the timeout allowed).

15.2.1 A Working Example: The Simple Course Timetabling Problem

Let us give the definition as a COP of a basic timetabling problem that we will use as a working example in the rest of the paper. The problem is referred as the SCTT problem in the rest of the paper.

Given a set of courses $S = \{c_1,\ldots,c_n\}$, each course c_i requires a number of weekly lectures $l(c_i)$ where $l : S \rightarrow \mathbb{N}$, and is taught by a teacher $t(c_i)$ where $t : S \rightarrow T = \{t_1,\ldots,t_g\}$. Five teaching days (from Monday to Friday) and two time slots for each day (Morning and Afternoon) are allowed, thus having a set

$P = \{p_1, \ldots, p_{10}\}$ of ten possible time periods. Each teacher t_j can be unavailable for some periods $u(t_j)$ where $u : T \rightarrow 2^P$.

The problem consists in finding a schedule for all the courses, so that:

(C_1) All the required lectures for each course are given.

(C_2) Lectures of courses taught by the same teacher are scheduled in distinct periods.

(C_3) Teacher un-availabilities are taken into account.

Moreover, a cost function is defined on the basis of the following criterion: the lectures of each course c_i should be at least spread into a given minimum number of days $\delta(c_i)$, where $\delta : S \rightarrow \{1, \ldots, 5\}$.

Let us consider the following toy instance, just to fix the ideas:

- $S = \{OS, PL, AI\}$ (Operating Systems, Programming Languages, Artificial Intelligence)
- $l(OS) = 3$, $l(PL) = 4$, $l(AI) = 3$
- $t(OS) = Schroeder$, $t(PL) = Schroeder$, $t(AI) = Linus$
- $P = \{MO_m, MO_a, TU_m, TU_a, WE_m, WE_a, TH_m, TH_a, FR_m, FR_a\}$
- $\delta(OS) = 3$, $\delta(PL) = 4$, $\delta(AI) = 3$
- $u(Schroeder) = \{TU_m, TU_a\}$, $u(Linus) = \{TH_m, TH_a, FR_m, FR_a\}$

A possible model for this problem can be defined as follows:

- $\mathscr{X}$ consists of $|S| \cdot |P|$ boolean variables $x_{c,p}$. The variable $x_{c,p} = 1$ if and only if course c is scheduled at period p (otherwise it is 0).
- The required number of lectures is assigned to each course:

$$\sum_{p \in P} x_{c,p} = l(c) \quad \forall c \in S \qquad (C_1)$$

- The constraints stating that lectures of courses taught by the same teacher are scheduled in distinct periods can be modeled as:[2]

$$x_{c,p} \cdot x_{c',p} = 0 \quad \forall c, c' \in S \text{ s.t. } (c \neq c' \wedge t(c) = t(c')), \forall p \in P \qquad (C_2)$$

- The "un-available" constraints are modeled as follows:

$$x_{c,p} = 0 \quad \forall c \in S \forall p \in u(t(c)) \qquad (C_3)$$

- The objective function f is defined by summing, for each course c, the difference between $\delta(c)$ and the actual number of days occupied by the lectures of c in a feasible solution:

$$f(\mathscr{X}) = \sum_{c \in S} \max(0, \delta(c) - |\{d(p) : p \in P \wedge x_{c,p} > 0\}|) \qquad (15.1)$$

[2] A linear constraint such as: $(x_{c,p} + x_{c',p} \leq 1)$ can be used here. We have employed a non-linear constraint to point out the flexibility of Constraint Programming.

where $d : P \to \{MO, TU, WE, TH, FR\}$ is a function assigning a period to a day.

In Figure 15.1 we report an assignment that is not a solution (on Monday morning Prof. *Schroeder* has two different lectures in the same time slot and (moreover) Prof. *Linus* gives a lecture on Thursday, when he is unavailable), a solution which is not optimal, and an optimal solution. With γ, in Figure 15.1, we denote $\delta(c) - |\{d(p) : p \in P \wedge x_{c,p} > 0\}|$, where c is OS, PL, or AI.

$P \Rightarrow$	MO		TU		WE		TH		FR		f
$S \Downarrow$	m	a	m	a	m	a	m	a	m	a	γ
OS	$\underline{1}$	0	0	0	1	0	1	0	0	0	3-3
PL	$\underline{1}$	0	0	0	0	1	0	1	0	1	4-4
AI	0	0	0	0	1	1	$\underline{1}$	0	0	0	3-2
(1)							Total f:				1

MO		TU		WE		TH		FR		f
m	a	m	a	m	a	m	a	m	a	γ
1	0	0	0	1	0	1	0	0	0	3-3
0	0	0	0	0	1	0	1	1	1	4-3
1	1	1	0	0	0	0	0	0	0	3-2
(2)						Total f:				2

MO		TU		WE		TH		FR		f
m	a	m	a	m	a	m	a	m	a	γ
1	0	0	0	1	0	1	0	0	0	3-3
0	1	0	0	0	1	0	1	0	1	4-4
1	0	1	0	1	0	0	0	0	0	3-3
(3)						Total f:				0

Fig. 15.1 An assignment which is not a solution (1), a solution (2), and an optimum solution (3) for the given toy instance of SCTT

15.3 The Existing CP and LS Systems Used

The core of GELATO is based on two programming environments for CSPs/COPs, namely GECODE and EasyLocal++. We briefly introduce them in this section (for more details, see [15] and [6], respectively).

15.3.1 Gecode Overview

GECODE (GEneric COnstraint Development Environment) [15] is a programming environment for developing constraint-based systems and applications developed by a group led by Christian Schulte, and including Mikael Lagerkvist, and Guido Tack. It is a modular and extensible C++ constraint programming toolkit, whose development started in 2002 and that counts 28 improving versions.[3]

GECODE is *open source* and is distributed under the MIT license. All of its parts (source code, example, documentation...) are available for download from `http://www.gecode.org` and anyone can modify the implementation of its classes, extend or improve their functionalities, or interface GECODE to other systems.

Being completely implemented in C++ and adhering to the language standards, GECODE is *portable*, and it can be compiled and run on most current platforms. Moreover, GECODE is *documented* in a tutorial [17] that explains its architecture and some programming tricks, and in an on-line reference documentation [16] that gives the technical specifications of each module/class/function implemented in the framework.

[3] At the time of writing, September 2011, the current version is 3.7.0.

GECODE has excellent *performance* with respect to both runtime and memory usage, it won severals competitions, such as the MiniZinc Challenge, in 2008–2011 and it can be considered a state-of-the-art constraint solver.

GECODE also implements *parallel* search by exploiting the multiple cores of today's hardware, giving to an already efficient base system an additional edge.

The discussion of GECODE architecture is out of the scope of this paper, we refer the to [16] for the specific details.

15.3.2 Gecode Modeling

GECODE comes with extensive modeling support, that allows the user to encode his/her problem using higher-level language facilities rather than rough low-level statements. The modeling support includes: regular expressions for extensional constraints, arithmetic expressions, matrices, sets, and boolean constraints.

A GECODE model is implemented using *spaces*: a *space* is the repository for variables, constraints, objective function, and searching options. Modeling exploits the C++ notion of *inheritance*: a model must implement the class *Space*, and the subclass constructor implements the actual model. In addition to the constructor, a model must implement some other functions (e.g., those performing a copy of the space, or returning the objective function, ...). A GECODE space can be asked to perform the propagation of the constraints, to find the first solution (or the next one), exploring the search tree, or to find the best solution in the whole search space.

We survey the description of GECODE modeling using the SCTT example defined in Section 15.2.1. We recall that a problem encoded in GECODEcan be solved by GELATO as well.

Every GECODE model is defined into a class that inherits from a GECODE `Space` superclass. The superclasses available are `Space` for CSPs and `MinimizeSpace` or `MaximizeSpace` for COPs. In our working example, reported in Listing 15.1 we declare the class `Timetabling` to be a subclass of `MinimizeSpace` (line 1).

Listing 15.1 SCTT encoding: model heading definition

```
1  class Timetabling : MinimizeSpace {
2    IntVarArray x;
3    IntVar      fobj;
4    Timetabling(const Faculty& in ):
5      x(*this, in.Courses() * in.Periods(), 0, 1),
6      fobj(*this, 0, Int::Limits::max)
7    { ...
```

On line 2 the array `x` for the variables $\mathscr{X}$ of the problem is declared, and on line 3 we declare the variable `fobj` that will store the value of the objective function. `IntVarArray` and `IntVar` are built-in GECODE data structures. Line 4 starts the constructor method. It takes as input an object of the class `Faculty`, that contains all the information specific to the instance at hand. All these input information are stored in the variable `in`. Line 5 sets variables and domains: the array `x` is constructed so to contain a number `in.Periods()*in.Courses()` of boolean variables. At line 6, the interval between 0 and the largest allowed integer value is set as the domain of the variable `fobj`.

Moving to modeling constraints, GECODE provides a `Matrix` support class for accessing an array as a two dimensional matrix (matrices are, of course, very common in modeling). Going back to our working example, in Listing 15.2, line 10 sets up a matrix interface to access the vector x, with the number of columns equal to the number of time periods and the number of rows equal to the number of courses.

Listing 15.2 SCTT modeling: main constraints

```
7      unsigned int cols =  in.Periods() ;
8      unsigned int rows = in.Courses();

10     Matrix<IntVarArgs> mat(x, cols, rows);

12      for (int r = 0; r < rows; r++)
13        linear(*this,  mat.row(r), IRT_EQ, in.
              NumberOfLectures(r));

15      for (int p = 0; p < cols; p++)
16        for (int r1 = 0; r1 < rows - 1; r1++ )
17          for (int r2 = r1+1; r2 < rows; r2++ )
18            if (in.SameTeacher(r1,r2))
19              rel(*this, mat(c,r1) * mat(c,r2) == 0);

21        for (int p = 0; period < in.Periods(); p++)
22          for (int c = 0; c < in.Courses() ; c++ )
23            if (!in.Available(c,p))
24              rel(*this, mat(p,c) == 0);
```

The matrix has the same structure as those reported in Figure 15.1. Lines 12–13 post the constraint (C_1) (see, Section 15.2.1) on each row of the matrix. The constraint `linear` is a built-in arithmetical operator of GECODE, and line 12 corresponds to the formula

$$\sum_{\texttt{col}} \texttt{mat.row}(r)[\texttt{col}] = \texttt{in.NumberOfLectures}(r)$$

where `in.NumberOfLectures(r)` contains the value of function l (number of weekly lectures) for each course (i.e., for each row). `IRT_EQ` is the built-in predicate for equality between finite domain constraints.[4]

[4] A similar syntax is used for $\neq, <, \leq$, etc.

Lines 15–19 post the constraint (C_2). Line 18 corresponds to the constraint $x(c,r1) \cdot x(c,r2) = 0$, applied to the variables of the same time periods sharing the same teacher.

Lines 21–24 post the constraint (C_3). In particular, at line 24, whenever $x(p,c)$ is a time slot unavailable for the given professor, the constraint $x(p,c) = 0$ is posted.

Objective Function and Branching

The formula computing the objective function is reported in Listing 15.3.

Listing 15.3 SCTT modeling: cost function

```
24      IntVarArgs vectorWD(*this, in.Courses(), 0, Int::
           Limits::max);
25      IntVarArgs sumWD(*this, in.Courses(), 0, Int::Limits::
           max);
26      IntVarArgs diff(*this, in.Courses(), Int::Limits::min,
            Int::Limits::max);

28      for (int course = 0 ; course < in.Courses(); course++)
29      {
30        IntVarArgs workingDays(in.Days(), 0, 1);

32        for (int day = 0 ; day < in.Days(); day++)
33        {
34          IntVarArgs Day( in.PeriodsPerDay(), 0,1 );
35          for (unsigned int slot = 0; slot < in.
               PeriodsPerDay(); slot++)
36          {
37            int period = day*in.PeriodsPerDay() + slot;
38            expr(*this, (mat(period, course) != 0) == (Day[
                 slot] == 1));
39          }
40          IntVar nLecturesPerDay(*this, 0, in.PeriodsPerDay
               ());
41          linear(*this, Day, IRT_EQ, nLecturesPerDay);
42          expr(*this, (nLecturesPerDay != 0) == (workingDays
               [day] == 1));
43        }
44        linear(*this, workingDays, IRT_EQ, sumWD[course]);
45        diff[course] = post(*this,in.CourseVector(course).
             MinWD()-sumWD[course]);
46        Gecode::max(*this, ZERO, diff[course], vectorWD[
             course]);
47      }

49      linear(*this, vectorWD, IRT_EQ, fobj);

51      branch(*this,x,tiebreak(INT_VAR_DEGREE_MAX,
           INT_VAR_SIZE_MAX),INT_VAL_MED);
```

As for the objective function, some auxiliary arrays of temporary variables are introduced. They are declared as `IntVarArgs`, a GECODE built-in data type for array of temporary variables. At the end of this piece of code, the formula is bound to the `fobj` variable.

During the search, a variable is selected and a value of its domain is attempted. This operation is called *branching*, and the branching strategy is defined at line 51. The variables to branch are those in the array `x` and the variable selection strategy is `tiebreak(INT_VAR_DEGREE_MAX,INT_VAR_SIZE_MAX)`, i.e., the variable with the highest number of constraint on it is selected, breaking ties choosing the variable with largest domain size. The values selection strategy is `INT_VAL_MED`, i.e., the greatest value not greater than the median is selected. Other choices are, of course, possible.

In order to use the model defined above in a constraint based branch and bound search engine (see Section 15.2) the following functions need to be defined into the `Timetabling` class (see Listing 15.4):

- The function `cost`, that returns the variable representing the cost of a solution, i.e., the objective function of the model. This function is defined at lines 54–55 and simply returns the variable `fobj`.
- A branch and bound search engine needs to know what constraint to add every time a new solution is found, in order to drive the search to better solutions and cut the search space. The function `constrain` (lines 57–61) defines the desired constraint to add: it takes in input a `Space` object, (i.e., a solution of the model), and posts a constraint (line 60) stating that the value of the variable `fobj` has to be less than the `fobj` value of the solution `sol`. The GECODE branch and bound search engine calls this method, every time it finds a new solution, passing the solution found as parameter.

Listing 15.4 SCTT modeling: branch and bound

```
54  IntVar Timetabling::cost(void) const
55  { return fobj; }

57  void Timetabling::constrain(const Space& sol)
58  {
59    const Timetabling& s = static_cast<const Timetabling&>(
        sol);
60    rel(*this, fobj, IRT_LT, s.fobj.val());
61  }
```

15.3.3 EasyLocal++

EasyLocal++ [6] is an object-oriented framework that allows programmers to design, implement and test LS algorithms in an easy, fast and flexible way. The underlying idea is to capture the essential features of most LS meta-heuristics, and their possible compositions, allowing the user to address the design and implementation issues of new LS heuristics in a more principled way. EasyLocal++ has been entirely developed in C++ with wide use of object-oriented patterns and currently it is at its 2.0. release.

Modeling a problem using EasyLocal++ requires that the C++ classes representing the problem specific layers of the EasyLocal++ hierarchy to be defined. On the other hand, the framework provides the full control structures for the invariant part of the LS algorithms. Consequently, the user is required only to supply the problem specific details by defining concrete classes and implementing concrete methods.

Going into the details of the EasyLocal++ development process is out of the scope of this paper, therefore we refer the interested reader to [5].

15.4 GELATO: Gecode + Easy Local = A Tool for Optimization

The aim of GELATO is to allow the programmer to easily model a CSP/COP using one of the three modeling languages: Prolog, MiniZinc, and GECODE, define or select the meta-heuristics using a tiny meta-heuristics modeling language that will allow to program search heuristics in a wide range.

15.4.1 High Level Modeling and Translation

The system is currently able to handle CSPs and COPs modeled in Prolog exploiting the front-end translators of Prolog to GECODE presented in [4], and in MiniZinc, exploiting the (two steps) translator of MiniZinc in GECODE available in the GECODE distribution. Moreover, the modeling capabilities can benefit from the front-end developed for the Haskell language [21] and, in general, from other front-ends to GECODE, usually listed in the GECODE web-site [15].

We do not enter here in the translation details, but we just say that the translations are based on the low-level modeling language FlatZinc. FlatZinc models are a list of simple constraints, without other programming constructs. Basically, a model is the unfolded version of a MiniZinc model, which can be interpreted directly by GECODE.

In what follows, instead, we will show how to encode the SCTT problem (Section 15.2.1) in Prolog and in MiniZinc.

Prolog and Constraint Logic Programming.

According to [8], Logic Programming was the first community embedding constraint programming giving raise to the so-called Constraint Logic Programming (CLP) paradigm. Nowadays all available Prolog system comes equipped with a constraints solver on finite domains (and on other domains, such as booleans, sets, rational numbers, ...). Due to the slight differences in the syntax of some primitives, we are focusing on one system, namely SICStus Prolog [13].

We are going to show the encoding of the SCTT problem in SICStus Prolog. First, let us focus on the input format. `L` is a list storing the function ℓ where courses are assumed ordered as `OS`, `PL`, `AI`. `T` is a list storing the function t, and `Unav` is a list of lists of unavailable time periods, given as numbers (precisely, Monday morning is 1, ..., Friday afternoon is 10), where the order of the teachers must be the same in the two lists (Schroeder, then Linus). `D` is a list storing the minimum desired duration of the courses. Our toy instance would be therefore represented by:

```
L    = [3,4,3]                    T = [[os,pl],[ai]]
Unav = [[3,4],[7,8,9,10]]         D = [3,4,3]
```

In Listing 15.5 we report the complete encoding of the problem in Prolog. If, on the one side, one might appreciate the compactness of the code, on the other side recursion is employed to implement the various "for loops" needed. Actually, most Prolog interpreters now implements a `foreach` iterator. However, this is not (yet) the standard way of encoding in Prolog. If the reader wants to run it, the libraries `lists` and `clpfd` must be included.

The sketch of the code is as follows: at line 2 the matrix is generated row by row and at line 3 the domains are assigned to its variables. Then the three predicates adding the corresponding constraints are called (line 4) and the objective function is defined, using constraints, by predicate `build_fobj` at line 7. Search is called at line 6 by the built-in predicate `labeling` with the option of minimizing the function `FOBJ`.

The predicates `c1`, `c2`, and `c3` called at line 4, implement the constraints (C_1), (C_2), and (C_3). They are defined at lines 10–13, 15–29, and 31–43, respectively. The function `fobj` is defined by the code at lines 45–58.

Even though the code is mostly self-contained, we just focus on some points. At line 11 the built-in `sum` predicate is used; this is similar to the use of `linear` in Listing 15.2. At lines 40–41 the variables corresponding to an unavailable period are set to 0, while the "iff" definition of line 56 states that the flag variable TD is set to 1 if and only if a class is held in that day (either in the morning or in the afternoon).

Listing 15.5 SICStus Prolog encoding of the SCTT problem

```
1  sctt(L, T, Unav, D, Mat):-
2    length(OS, 10), length(PL, 10), length(AI, 10), Mat=[OS,
         PL,AI],
3    append([OS,PL,AI],Vars), domain(Vars, 0, 1),
4    c1(Mat, L),
5    c2(Mat, T),
6    c3(Mat, Unav, T),
7    build_fobj(Mat,D,FOBJ),
8    labeling([minimize(FOBJ)], Vars).

10 c1([Row|Tr],[L|Tl]):-
```

```
11     sum( Row, #=, L ),
12     c1(Tr,Tl).
13   c1([],[]).

15   c2(MAT,[]).
16   c2(MAT,[[_]|T]) :- c2(MAT, T).
17   c2([OS,PL,AI],[[A,B|R]|T]) :-
18     c2_aux([OS,PL,AI], [A,B]),
19     c2([OS,PL,AI], [[A|R]|T]),
20     c2([OS,PL,AI], [[B|R]|T]).
21   c2_aux([OS,PL,AI],[A,B]):-
22     (A=os,B=pl) -> c2_post(OS,PL);
23     (A=os,B=ai) -> c2_post(OS,AI);
24     (A=pl,B=ai) -> c2_post(PL,AI).

26   c2_post([X1|T1],[X2|T2]):-
27     X1 * X2 #= 0,
28     c2_post(T1, T2).
29   c2_post([], []).

31   c3(MAT, [], []).
32   c3(MAT, [_|UnAvailable], [[]|T]) :-
33     c3(MAT,  UnAvailable, T).
34   c3([OS,PL,AI], [Un|Available], [[A|R]|T]) :-
35     (A = os -> c3_aux(OS,Un);
36     A = pl -> c3_aux(PL,Un);
37     A = ai -> c3_aux(AI,Un)),
38     c3([OS,PL,AI], [Un|Available], [R|T]).
39   c3_aux(Row, [Un|Tu]):-
40     nth1(Un, Row, X),
41     X#=0,
42     c3_aux(Row, Tu).
43   c3_aux(_, []).

45   build_fobj([Row|M], [D|Ds], FOBJ):-
46     build_fobj( M, Ds, FObjT),
47     single_contribution(Row, D, FObjR),
48     FObjT + FObjR #= FOBJ.
49   build_fobj([], _, 0).
50   single_contribution(Row, D, FObjR):-
51     number_of_days(Row, Days),
52     Days+Diff  #= D,
53     max(Diff, 0 ) #= FObjR.
54   number_of_days([M,P|Row], TotalDays):-
55     number_of_days(Row, Days),
56     TD #<=> (M+P #> 0),
57     TD + Days #= TotalDays.
58   number_of_days([], 0).
```

MiniZinc.

MiniZinc is a medium-level declarative constraint modeling language developed by NICTA (National ICT Australia) designed for specifying constrained optimization and decision problems over integers and real numbers [12]. MiniZinc is designed so as to be easily interfaced to backend solvers, via compilation in a low level language called FlatZinc. We refer to the MiniZinc manual [9] for details. We just report here the encoding of the SCTT problem (Listing 15.6).

Listing 15.6 The SCTT problem encoded in MiniZinc.

```
1  int: courses  = 3;
2  int: teachers = 2;
3  int: periods = 10;
4  int: weekdays = 5;
5  array[1..courses]  of 1..periods: l = [3,4,3];
6  array[1..courses]  of 1..teachers: t = [1,1,2];
7  array[1..teachers] of set of 1..periods: u =
       [{3,4},{7,8,9,10}];
8  array[1..courses]  of int: delta = [3,4,3];

10 array[1..courses,1..periods] of var 0..1: x;
11 var int: fobj;

13 constraint forall(c in 1..courses)
14   (sum(p in 1..periods)(x[c,p]) = l[c]);
15 constraint forall(ci, cj in 1..courses where ci<cj /\ t[ci
       ] = t[cj])
16   (forall (p in 1..periods) (x[ci,p] * x[cj,p] = 0));
17 constraint forall(c in 1..courses)
18   (forall (p in u[t[c]]) (x[c,p] = 0));

20 constraint fobj = sum(c in 1..courses)
21   (max(0,delta[c] - sum(d in 1..weekdays) (max(x[c,2*d],x[
         c,2*d-1]))));

23 solve minimize fobj;

25 output ["x = ",show(x),"\n", "Fobj = ", show(fobj), "\n"];
```

Let us focus on the lines 1–8, which store the input data. Constants for the number of courses, teachers, periods, and days are initialized and used in arrays (in particular at line 7 an array of sets is initialized). Line 10 defines the boolean matrix and line 11 defines the objective function `fobj`. Constraints (C_1), (C_2), and (C_3) are posted in a very natural way at lines 13–18, while the objective function is initialized at lines 20–21. The predicate `max` (line 21, second occurrence) between the two periods per day returns 1 if that day is used, 0 otherwise. The solver is colled at line 23 with the `minimize` option (whereas `solve satisfy` is used for a CSP) while line 25 states how to output the result.

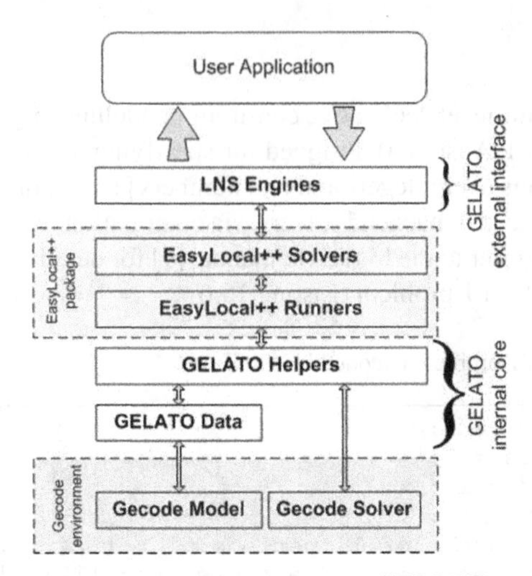

Fig. 15.2 Structure of GELATO and interactions with GECODE and EasyLocal++

15.4.2 *The Core of the GELATO Hybrid Solver*

The core of the tool is constituted by a hybrid solver whose preliminary implementation has been presented in [3].

GELATO functionalities are divided into two main parts: an *internal core*, that merges GECODE and EasyLocal++ together, and an *external interface*, that provides an easy way for the user to interact with the system. Figure 15.2 shows the general architecture of the tool.

The *internal core* is a *middle layer* that defines the interaction between EasyLocal++ and GECODE. In particular, these classes include the encodings of GECODE models and CP search engines.

The *external interface* provides high level functions that can easily be called by the user to access the inner functionalities, thus hiding all the internal GECODE and EasyLocal++ calls. This external layer act as a *Façade* object-oriented pattern (see [7]): the aim of the Façade pattern is to provide a simplified interface to a larger body of code, making the subsystem easier to use. Thus, this external layer transforms the overall GELATO tool into a black-box and allows the user to interact with it just by passing the input (i.e., the problem model and the solving metaheuristic) and retrieving the output (i.e., the solution obtained), without caring about complicated object interactions in the GELATO internal core.

In this scenario, the user is asked to:

1. *model the CSP or COP* with a modeling language supported by GELATO
2. *specify the instance* of the problem in a different file, since GELATO requires the concepts of *problem* and *instance* to be separated;

3. *select the meta-heuristic* to be used to find the solution (e.g., Hill Climbing, Tabu Search, Large Neighborhood Search) and specify the desired parameters for its execution.

It is important to note that GELATO does not require any modification of GECODE and EasyLocal++. In this way GELATO does not affect the single development of GECODE or EasyLocal++ and every improvement of the two basic tools, such as new functionalities or performance improvements, is immediately inherited by GELATO.

15.5 Benchmarks

In this section we briefly describe the three benchmark problems chosen to test our tool. They are well-known COPs with several available sets of input instances.

15.5.1 Asymmetric Traveling Salesman Problem

This problem is taken from the TSPLib [18] and it is defined as follows.

Definition 15.1 (Asymmetric Traveling Salesman Problem (ATSP)). Given a directed graph $G = (V,E)$ and a function c that assigns a cost to each directed edge (i, j), find a roundtrip of minimal total cost visiting each node exactly once.

Let us observe that the edge costs might be asymmetric, in the sense that $c(i, j)$ and $c(j,i)$ can be different. The problem is therefore a generalization of the well-known Traveling Salesman Problem problem.

The problem is modeled as follows: let $\mathscr{X} = \{x_1,\ldots,x_n\}, n = |V|$ be the set of variables with domains $D_1 = D_2 = \ldots = D_n = \{1,\ldots,n\}$. The value of x_i represents the *successor* of the vertex i in the tour. We exploit the global constraint `circuit`$([x_1,\ldots,x_n])$ available in most CP frameworks, to constrain the solutions that represent a tour. The cost function is defined as $f = \sum_{i=1}^{n} c(i,x_i)$.

15.5.2 Minimum Energy Broadcast

This problem is drawn from CSPLIB (available from `http://www.csblib.org`, problem number 48) and it is an optimization problem for the design of Wireless Networks. Further information can be found in [19].

Definition 15.2 (MEB). Given a set of n nodes $V = \{1,\ldots,n\}$ forming a complete graph K_n, a source node $s \in V$, and and a cost function $p : V \times V \to \mathbb{R}$, representing the transmission cost between two nodes, the problem consists in finding a (directed) spanning tree rooted at s that minimizes a cost function f that measures the energy needed by a node for broadcasting information, and defined as follows. Assume a tree τ is given, and let us denote by $\texttt{children}(i)$ the set of children nodes of the node i in the tree τ (namely, $\texttt{children}(i) = \{j \in V \mid (i,j) \in \tau\}$). Then $f(\tau) = \sum_{i=1}^{n} \max\{p(i,j) : j \in \texttt{children}(i)\}$.

We model the problem using a set $\mathscr{X} = \{x_{i,j} \in \{0,1\} \mid i,j \in V\}$ of n^2 boolean variables, with the meaning that $x_{i,j} = 1$ if and only if the edge (i,j) is in the solution τ. Let us define the function $\delta : V \to V$ as follows: $\delta(s) = s$ (s is the source node) and $\delta(j) = i$ if $j \neq s$ and $x_{i,j} = 1$ (the parent node in the tree—the function is well-defined as long as we know that for all $j \neq s$ exists exactly one i such that $x_{i,j} = 1$). Then, for $k \in \mathbb{N}$, $\delta^k(i)$ is recursively defined as follows: $\delta^0(j) = j$ and $\delta^{k+1}(j) = \delta(\delta^k(j))$.

The constraints added to the problem to ensure the tree structure of the output are the following:

$$\sum_{i=1}^{n} x_{i,j} = 1 \qquad j = 1,\dots,n, j \neq s \qquad (C_{1.1})$$

$$\sum_{i=1}^{n} x_{i,s} = 0 \qquad (C_{1.2})$$

$$\sum_{i=1}^{n}\sum_{j=1}^{n} x_{i,j} = n-1 \qquad (C_2)$$

$$\delta^n(j) = s \qquad j = 1,\dots,n \qquad (C_3)$$

and the objective function is: $f = \sum_{i=1}^{n} \max_{j=1}^{n}(x_{i,j}p(i,j))$.

As a minor implementation remark, we have multiplied the input data by 100 so as to use (finite) integer values for p instead of the real values stored in the problem instances used.

15.5.3 Course Timetabling

This problem has been introduced as Track 3 of the second International Timetabling Competition held in 2007 [10]. It consists in the weekly scheduling of the lectures of a set of university courses on the basis of a set of predefined curricula published by the University.

Definition 15.3 (CTT). Given a set of courses $C = \{c_1,\dots,c_n\}$, each course c_i consists of a set of lectures $L_i = \{l_{i_1},\dots,l_{i_a}\} \in \mathscr{L}$, is taught by a teacher $t : C \to T = \{t_1,\dots,t_g\}$, it is attended by a number of students $s : C \to \mathbb{N}$, and belongs to one or more curricula $Q = \{q_1,\dots,q_b\}$, where $q_i \subseteq C, i = 1,\dots,b$, which are subset of courses that have students in common. Moreover it is given a set of periods $P = \{1,\dots,p\}$, each period belonging to a single teaching day $d : P \to \{1,\dots,h\}$, and a set of rooms $R = \{r_1, \ldots, r_m\}$, each with a capacity $w : R \to \mathbb{N}$. Each teacher t_i can be unavailable for some periods $u : T \to 2^P$.

The problem consists in finding an assignment $\tau : \mathcal{L} \to R \times P$ of a room and period to each lecture of a course so that:

(H_1) all lectures of courses are assigned;
(H_2) for each (room, period) pair, only one lecture is assigned;
($H_{3.1}$) lectures of courses in the same curriculum are scheduled in distinct periods;
($H_{3.2}$) lectures of courses taught by the same teacher are scheduled in distinct periods;
(H_4) teacher unavailabilities are taken into account.

Moreover, a cost is defined for the following criteria (soft constraints):

(S_1) each lecture should be scheduled in a room large enough for containing all its students;
(S_2) the lectures of each course should be spread into a given minimum number of days $\delta : C \to \{1, \ldots, h\}$;
(S_3) lectures belonging to a curriculum should be adjacent to each other;
(S_4) all lectures of a course would be preferably assigned to the same room.

We model the CTT problem by defining the set $\mathcal{X}$ of $C \cdot P$ variables $x_{c,p} \in \{0, \ldots, r\}$, with the intuitive meaning that $x_{c,p} = r > 0$ if and only if course c is scheduled at period p in room r, and $x_{c,p} = 0$ if course c is not scheduled at period p. The constraints are modeled as follows:

$$\sum_{j \in P} (x_{c_i,j} = 0) = p - |L_i| \qquad i = 1, \ldots, n \qquad (H_1)$$

$$\sum_{i=1}^{n} (x_{c_i,j} = 0) \geq n - m \qquad j = 1, \ldots, p \qquad (H_2)$$

$$x_{c,j} \cdot x_{c',j} = 0 \qquad c, c' \in C \text{ s.t. } t(c) = t(c'), j = 1, \ldots, p \qquad (H_{3.1})$$

$$\sum_{c \in q_i} x_{c,j} \leq 1 \qquad i = 1, \ldots, b, j = 1, \ldots, p \qquad (H_{3.2})$$

$$x_{c,j} = 0 \qquad j \in u(t(c)), c \in C \qquad (H_4)$$

The objective function f is the sum of the following four components:

$$s_1 = \sum_{i=1}^{n} \sum_{j=1}^{p} \max(0, s(c_i) - w(x_{c_i,j})) \qquad (S_1)$$

$$s_2 = 5 \cdot \sum_{i=1}^{n} \max(0, \delta(c_i) - |\{d(j) : x_{c_i,j} > 0\}|) \qquad (S_2)$$

$$s_3 = 2 \cdot \sum_{i=1}^{b} \left|\{k \in P : \mathsf{start}(k) \wedge \sum_{c \in q_i} x_{c,k} \neq 0 \wedge \sum_{c \in q_i} x_{c,k+1} = 0\}\right| + \left|\{k \in P : \mathsf{end}(k) \wedge \sum_{c \in q_i} x_{c,k} \neq 0 \wedge \sum_{c \in q_i} x_{c,k-1} = 0\}\right| + \left|\{k \in P : \mathsf{mid}(k) \wedge \sum_{c \in q_i} x_{c,k} \neq 0 \wedge \sum_{c \in q_i} x_{c,k-1} = 0 \wedge \sum_{c \in q_i} x_{c,k+1} = 0\}\right| \quad (S_3)$$

$$s_4 = \sum_{i=1}^{n} (|\{r : \exists p(x_{c_i,p} = r)\}| - 1) \quad (S_4)$$

where start, end, and mid are three boolean functions that state if a period is initial, ending, or in the middle of a day, respectively.

15.6 Experiments and Comparison

We briefly explain how experiments drove us in tuning search parameters and show that the performances of GELATO are comparable to those of an LNS solver implemented in Comet. All computations were run on an AMD Opteron 2.2 GHz Linux Machine. We used GECODE 3.1.0, EasyLocal++ 2.0, and Comet 2.0.1. Additional tests have shown that GELATO is fully compatible with GECODE 3.7.0, the latest release at the time of writing.

15.6.1 Solving Techniques

The problem instances have been solved using:

1. a pure constraint programming approach in GECODE
2. a pure Local Search approach in EasyLocal++, and
3. LNS meta-heuristics encoded in GELATO.

We use the same model, the same meta-heuristics, and the same parameters used for (3) in the Comet system to guarantee a fair comparison with that system.

We tested the ATSP on the following instances of growing size, taken from the TSPLib [18]: br17 (that we call *instance 0*, with $|V| = 17$), ftv33 (*inst. 1*, $|V| = 34$), ftv55 (*inst. 2*, $|V| = 56$), ftv70 (*inst. 3*, $|V| = 71$), kro124p (*inst. 4*, $|V| = 100$), and ftv170 (*inst. 5*, $|V| = 171$). For the MEB problem we selected the following six instances of size $|V| = 20$ (hence, $|\mathscr{X}| = 400$) from the set used in [19]: p20.02/03/08/14/24/29. For the CTT problem we selected from the website `http://tabu.diegm.uniud.it/ctt/` six instances with different features (instance size, average number of conflicts, average teacher availability, average room occupation, ...): comp01/04/07/09/11/14.

Pure CP Approach: GECODE.

The pure CP approach in GECODE exploits the straightforward encoding of the models described in Sect. 15.5. For the sake of showing the possibility of using GELATO starting from already available CP models in the case of ATSP we used the model reported in the set of GECODE examples, slightly adapted for its integration. For MEB and CTT, instead, we encoded the GECODE models from scratch. We tried different search strategies for each problem, and choose the ones with best performances: for ATSP the variable with smallest domain is selected, and the values are chosen in increasing order; for MEB the variable with smallest domain is selected and the values are chosen randomly; for CTT the most constrained variable is selected, breaking ties randomly, and values are chosen randomly.

Pure LS Approach: EasyLocal++.

The pure LS approaches in EL are based on elementary move definitions. For ATSP we use a *swap-move*: given a tour, two nodes are randomly selected and swapped. For MEB we define a *change-parent-move*: given a directed rooted tree, two nodes A and B are selected, so that B is not an ancestor of A; then A becomes the parent of B. For CTT we use a *time-and-room* exchange move: given a timetable, we select randomly a lesson scheduled at period p in room r and move it to another period p_1 into another room r_1, chosen from the empty ones.

In order to compare the algorithms based on pure LS with LNS on a fair base, we decided to drive the Local Search by means of a Randomized Hill Climbing scheme. Indeed, LNS actually perform a "large" hill climbing in a wider neighborhood, we call *mountain* climbing. Developing LS strategies with more complex move definitions and more elaborated meta-heuristics is out of the scope of the present work.

It is worth noticing that for LNS implementation we can directly reuse the existing GECODE models, whereas using EasyLocal++ we had to implement from scratch the basic LS classes for the problem.

As a final note, all LS algorithms are stopped after 1000 iterations without improvement.

Large Neighborhood Search Approach: GELATO (and Comet).

The LNS meta-heuristic is composed by a deterministic CP search for the first solution and then a LNS exploration based on a mountain climbing algorithm with a maximum number of idle iteration.

Given a COP $\mathscr{O} = \langle \mathscr{X}, \mathscr{D}, \mathscr{C}, f \rangle$, the first solution is obtained by a CP search using GECODE without any pre-assigned value of the variables, and without any timeout. The Large Neighborhood definition we have chosen is the following: given a number $N < |\mathscr{X}|$ and given a solution in $\mathbf{sol}(\mathscr{O})$ we randomly choose a set $\mathbf{FV} \subseteq \mathscr{X}$ of free variables, so that $|\mathbf{FV}| = N$. Therefore, the exploration of the neighborhood consists in the possible assignments to the constrained variables $\mathbf{FV}$ and it is regulated by the value of N and by the number `maxF` of maximum number of failures in an improving stage. At each stage variables and values selection are the same used for the first solution.

15.6.2 Parameters Tuning and Data Analysis

For all instances of each problem we perform just one run of GECODE with a timeout of one hour. Let us observe that even in MEB and CTT some random choices for tie breaking are used, GECODE is in fact deterministic.

We tried several LNS approaches, that differs on the parameters N and `maxF`. N is determined as a proportion of $|\mathscr{X}|$ (10%, 20%, 30%, etc.). For relating the number `maxF` to the exponential growth of the search space w.r.t. the number N of free variables, we calculate $\texttt{maxF} = 2^{\sqrt{N*\texttt{Mult}}}$, and allow to fix the values for the parameter `Mult`. Reasonable values for `Mult` range from 0.01 to 2. Of course, these values cannot be chosen independently of the size of the problem instance and of the problem difficulty. These values are given by optional arguments of the command-line call to the execution.

The range of parameter we have tested is the following:

ATSP: $N \in \{20\%, 25\%, 30\%, 35\%, 40\%, 45\%\}$ and $\texttt{Mult} \in \{1, 1.5, 2\}$
MEB: $N \in \{35\%, 40\%, 45\%, 50\%\}$ and $\texttt{Mult} \in \{0.5, 0.75, 1\}$
CTT: $N \in \{2\%, 3\%, 4\%, 5\%, 10\%, 15\%\}$ and $\texttt{Mult} \in \{0.1, 0.5, 1\}$

Choosing a range to analyze can be done by few preliminary runs setting $\texttt{Mult} = 1$ and start by small N (e.g. 1% and then double it iteratively) in which the number k of consecutive idle iterations is kept low (e.g., 20).

With small values of N the algorithm performs little improvements at each step (and frequently a number of steps without improvements). Execution stops soon at high values of the objective function.

With increasing values of N the running time become slower, but one might notice the computation of better optima. At a certain point one notice that the algorithm is slower and slower, but, even worse, the optimum is not improved. In the next subsection we will come back on this training stage.

Once an interval of "good" behavior is found, one might enlarge the number k of consecutive idle iterations allowed before forcing the termination (in all our tests $k = 50$) and starting tests with different values of `Mult`.

Since LNS and pure LS computations are stochastic in nature, we collect the results of 20 runs in order to allow for basic statistical analyses. During each run, we stored the values of the objective function that correspond to improvements of the current solution, together with the running time spent. These data have been aggregated in order to analyze the average behavior of the different LNS and pure LS strategies on each sample. To this aim we performed a discretization of the data in regular time intervals; subsequently, for each discrete interval we computed the average value of the objective function.

We experimentally determine that the best parameter combinations are: $N = 35\%$ and $\texttt{Mult} = 2$ for ATSP; $N = 50\%$ and $\texttt{Mult} = 0.5$ for MEB; $N = 5\%$ and $\texttt{Mult} = 0.5$ for CTT. We show the plot representing the behavior of the various parameters in the CTT problems (see also Figure 15.3). Diagrams for other problems are similar. Further LNS experiments (either in GELATO or Comet) have been run with this parameters setting.

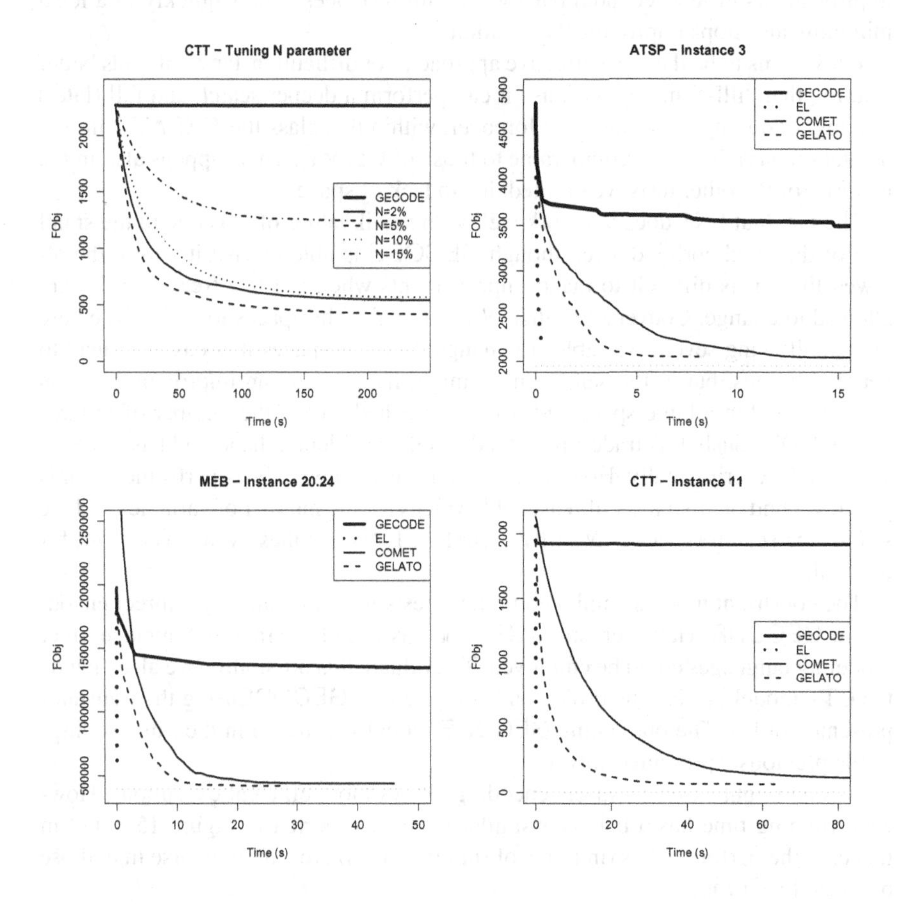

Fig. 15.3 Parameters and methods comparison (FObj is f)

15.6.3 Comparison

Let us briefly analyze the results of the comparison of GECODE, hill-climbing EasyLocal++ , GELATO, and Comet. In Figure 15.3 we show an excerpt of these results.[5] In those pictures, the horizontal dotted line represents the best known solution for the instance considered.

From the results, we can make some observations. First, it is clear that pure CP (GECODE) is unsuitable to improve the objective function f and to reach a good solution in reasonable time, while it is very useful to provide a good starting point for LS/LNS approaches. Conversely, hill-climbing is very fast (providing large cost improvements in few seconds) but for difficult instances it falls quickly in a local minimum and stops improving the solution.

LNS seems to be the most effective approach: for difficult instances it finds better solution than hill-climbing, because it can perform a deeper search and fall (later) into higher quality local-optima. Moreover, within this class the GELATO implementation has a behavior comparable to those of Comet[6]. This happens also in the majority of the other tests we omitted due to lack of space.

The fact that LNS does not work well with small values of N is due to the small size of the neighborhood: even though GECODE is able to visit it within the allowed time, it is difficult to make improvements when too few free variables are allowed to change. Conversely, when N is large also the spaces to be analyzed are bigger, allowing several variables to change. In these spaces it is easier to lead to better solutions, but at the same time it might happen that an improving solution is hidden within a huge space and it is not reached within the number of failures allowed. Obviously this trade-off depends on the problem at hand and has to be investigated experimentally. However, on the basis of the analysis performed in this paper, we find out that a set of reasonable values for the number of variables and the size of the search spaces is $N = 10\%, \mathtt{mafF} = 1$. We set these value as default for our tool.

The constraint models employed in these tests were obtained by a direct encoding in GECODE. However, since GELATO is a multi-paradigm language other modeling languages could be employed. To evaluate this possibility, we also encode the CTT model in SICStus Prolog and compile it to GECODEusing the translator presented in [4]. The outcoming GECODE model was tested in the same settings as the previously presented solvers.

As one might expect, a direct encoding allows more efficient executions. However, running time has a behavior similar to those reported in Figure 15.3, but in this case the performances (in terms of running times) are slightly worse than those obtained by Comet.

15.7 Conclusions and Future Developments

We have presented a multi-language tool for combinatorial optimization, called GELATO, which is able to deal with CSP/COP models expressed in different

[5] An exhaustive comparison can be found at `http://www.dimi.uniud.it/GELATO`.

[6] Somehow slightly better than Comet.

modeling languages and to use a combination of Local Search and Constraint Programming techniques for solving them. Being built upon GECODE the user might exploit the emerging literature of problems already encoded in GECODE and use GELATO for speeding-up the search. Being also build upon EasyLocal++, the user can inherit various Local Search techniques from EL, without the need of reformulating the model.

The immediate future work is the first release of the tool that will be made available in the near future. We will improve the quality of the GECODE models obtained through the translation of the Prolog/MiniZinc models and to simplify the compilation task (currently demanded to many routines written in different languages).

We also plan to develop a more clever technique for determining good candidates for the values of the tool parameters so as to allow the user to exploit the solver as a black-box. However, being completely written in C++ the skilled programmer can use it as a free fully configurable search engine.

Acknowledgements. This work is partially supported by MUR-PRIN project "*Innovative and multi-disciplinary approaches for constraint and preference reasoning*". We would like to thank Andrea Schaerf, Tom Schrijvers, and Guido Tack for their precious advises in several phases of the developing of the tool.

References

1. Aarts, E., Lenstra, J.K.: Local Search in Combinatorial Optimization. John Wiley & Sons, Chichester (1997)
2. Krzysztof, R.: Apt. In: Principles of Constraint Programming. Cambridge University Press, Cambridge (2003)
3. Cipriano, R., Di Gaspero, L., Dovier, A.: A Hybrid Solver for Large Neighborhood Search: Mixing Gecode and EASYLOCAL++. In: Blesa, M.J., Blum, C., Di Gaspero, L., Roli, A., Sampels, M., Schaerf, A. (eds.) HM 2009. LNCS, vol. 5818, pp. 141–155. Springer, Heidelberg (2009)
4. Cipriano, R., Dovier, A., Mauro, J.: Compiling and Executing Declarative Modeling Languages to Gecode. In: Garcia de la Banda, M., Pontelli, E. (eds.) ICLP 2008. LNCS, vol. 5366, pp. 744–748. Springer, Heidelberg (2008)
5. Di Gaspero, L., Schaerf, A.: Writing local search algorithms using EASYLOCAL++. In: Voß, S., Woodruff, D.L. (eds.) Optimization Software Class Libraries, OR/CS, Kluwer Academic Publisher, Boston (2002)
6. Di Gaspero, L., Schaerf, A.: EASYLOCAL++: An object-oriented framework for flexible design of local search algorithms. Software — Practice & Experience 33(8), 733–765 (2003)
7. Gamma, E., Helm, R., Johnson, R., Vlissides, J.: Design patterns: elements of reusable object-oriented software. Addison-Wesley Professional (1995)
8. Jaffar, J., Maher, M.J.: Constraint logic programming: A survey. Journal of Logic Programming 19/20, 503–581 (1994)
9. Marriot, K., Stuckey, P.J., De Koninck, L., Samulowitz, H.: An introduction to minizinc, http://www.g12.cs.mu.oz.au/minizinc/downloads/doc-1.3/minizinc-tute.pdf

10. McCollum, B., Schaerf, A., Paechter, B., McMullan, P., Lewis, R., Parkes, A.J., Di Gaspero, L., Qu, R., Burke, E.K.: Setting the research agenda in automated timetabling: The second international timetabling competition. INFORMS Journal on Computing 22(1), 120–130 (2010)
11. Michel, L., Van Hentenryck, P.: The comet programming language and system. In: van Beek, P. (ed.) CP 2005. LNCS, vol. 3709, pp. 881–881. Springer, Heidelberg (2005) doi: 10.1007/11564751/119
12. Nethercote, N., Stuckey, P.J., Becket, R., Brand, S., Duck, G.J., Tack, G.: MiniZinc: Towards a Standard CP Modelling Language. In: Bessière, C. (ed.) CP 2007. LNCS, vol. 4741, pp. 529–543. Springer, Heidelberg (2007)
13. Swedish Institute of Computer Science. Sicstus prolog, http://www.sics.se/sicstus.html
14. Shaw, P.: Using Constraint Programming and Local Search Methods to Solve Vehicle Routing Problems. In: Maher, M.J., Puget, J.-F. (eds.) CP 1998. LNCS, vol. 1520, pp. 417–431. Springer, Heidelberg (1998)
15. Gecode Team. Gecode: Generic constraint development environment, http://www.gecode.org
16. Gecode Team. Gecode reference documentation, http://www.gecode.org/doc-latest/reference/index.html
17. Gecode Team. Modeling and programming with gecode, http://www.gecode.org/doc-latest/MPG.pdf
18. Universität Heidelberg, Institut für Informatik. TSPLIB, http://www.iwr.uni-heidelberg.de/groups/comopt/software/TSPLIB95/
19. Wolf, S., Merz, P.: Evolutionary Local Search for the Minimum Energy Broadcast Problem. In: van Hemert, J., Cotta, C. (eds.) EvoCOP 2008. LNCS, vol. 4972, pp. 61–72. Springer, Heidelberg (2008)
20. Wolpert, D.H., Macready, W.G.: No free lunch theorems for optimization. IEEE Transactions on Evolutionary Computation 1(1), 67–82 (1997)
21. Pieter Wuille and Tom Schrijvers. Monadic Constraint Programming with Gecode. In: Proc. of ModRef 2009, Eighth International Workshop on Constraint Modelling and Reformulation, Lisbon, Portugal (September 2009)

Part V
Combining Metaheuristics with Machine Learning and Data Mining Techniques

Chapter 16
Predicting Metaheuristic Performance on Graph Coloring Problems Using Data Mining

Kate Smith-Miles, Brendan Wreford, Leo Lopes, and Nur Insani

Abstract. This chapter illustrates the benefits of using data mining methods to gain greater understanding of the strengths and weaknesses of a metaheuristic across the whole of instance space. Using graph coloring as a case study, we demonstrate how the relationships between the features of instances and the performance of algorithms can be learned and visualized. The instance space (in this case, the set of all graph coloring instances) is characterized as a high-dimensional feature space, with each instance summarized by a set of metrics selected as indicative of instance hardness. We show how different instance generators produce instances with various properties, and how the performance of algorithms depends on these properties. Based on a set of tested instances, we reveal the generalized boundary in instance space where an algorithm can be expected to perform well. This boundary is called the algorithm footprint in instance space. We show how data mining methods can be used to visualize the footprint and relate its boundary to properties of the instances. In this manner, we can begin to develop a good understanding of the strengths and weaknesses of a set of algorithms, and identify opportunities to develop new hybrid approaches that exploit the combined strength and improve the performance across a broad instance space.

16.1 Introduction

The quest to develop a powerful heuristic that performs well on all optimization problems and classes of instances may appear quixotic, especially in light of the No Free Lunch (NFL) Theorems [1, 2]. Without providing an algorithm with sufficient knowledge of the properties of the problem and instance classes, there will always be

Kate Smith-Miles · Brendan Wreford · Leo Lopes · Nur Insani
School of Mathematical Sciences, Monash University, Victoria 3800, Australia
e-mail: {kate.smith-miles,brendan.wreford,leo.lopes}@monash.edu
nur.insani@monash.edu

E.-G. Talbi (Ed.): Hybrid Metaheuristics, SCI 434, pp. 417–432.
springerlink.com

some instances where its performance will suffer compared to tailored heuristics designed to exploit problem-dependent knowledge. Hybrid metaheuristics [3, 4] hold much potential to integrate problem-dependent knowledge with broadly applicable search strategies. One successful example is the hyper-heuristic approach [5], which automates the process of selecting or generating heuristics based on a portfolio of heuristic components. Another approach is the algorithm portfolio concept [6, 7, 8], whereby the algorithm best suited to a particular instance is predicted based on a regression model that has already been trained based on past experience of the relationship between measurable properties of the instances and the performance of each algorithm in a portfolio. These are powerful ideas that help to circumvent the NFL Theorems somewhat by enabling us to use some knowledge of the problem and instance properties to ensure we adopt a search strategy well suited to the task.

Some of the key challenges though are to develop a greater understanding of the true strengths and weaknesses of a given algorithm, or component of an algorithm. Specifically, across the space of all possible instances that could be generated, where do we expect an algorithm to perform well, and where is it likely to fail? Typically, we see the performance of optimisation algorithms reported only on a chosen set of benchmark test instances, with little discussion about how the performance of the algorithm could be expected to generalize across a broader untested instance space [9]. Our objective is to develop the tools to understand the boundaries of algorithm performance across instance space so we will better placed to develop hybrid metaheuristics that exploit this knowledge to provide effective solution across a useful breadth of problems and instances.

Corne and Reynolds [10] have recently introduced the idea of a *footprint* in instance space as a means to visualise the generalized region where the algorithm could statistically be expected to perform well, and noted that "understanding these footprints, how they vary between algorithms and across instance space dimensions, may lead to a future platform for wiser algorithm-choice decisions" [10]. The definition of an instance space is critical to this approach, and Corne and Reynolds chose two features that define an instance of an optimization problem, for ease of visualization. For the vehicle routing problem they defined the instance space according to a pair of time windows for regular and non-regular customers; for the single machine tardiness problem the instance space was defined by the due dates and processing times. All generated instances can be plotted in this two dimensional instance space, and then the footprint of an algorithm can be depicted in the same two-dimensional instance space to show where the algorithm performance is most effective.

More recently, we have generalized these ideas to consider a higher-dimensional instance space, generated by knowledge of what makes an optimization problem hard [11]. The question of what makes an optimization problem (or even a particular instance of an optimization problem) hard for an algorithm was originally posed by Macready and Wolpert in 1996 [12], and the answer depends on both the algorithm and the particular instance properties. For example, a travelling salesman problem (TSP) involving cities arranged in a perfect circle will be trivial for all algorithms to solve, yet if the cities were tightly clustered, some algorithms may struggle more than others [13, 14, 15]. A concrete measure for the hardness of an instance for a particular algorithm has been proposed [12] as the fraction of the search space that corresponds to a better solution than the algorithm was able to find. Other measures of hardness have compared algorithm performance on the basis of the optimisation precision reached after a certain number of iterations, or the number of iterations taken to reach the best solution [13]. However, defining the characteristics of the instance that affect these hardness metrics has been much more complicated. While generic properties of the landscape of the problem instance have been proposed, such as autocorrelation structures and distributions of local minima [16, 17], many of the characteristics that greatly affect hardness have been shown to be problem-specific. Over many decades, numerous studies have reported on the effect of characteristics or features of a particular problem, such as the cluster ratio in the TSP [13] or the edge density in graph colouring [18], and demonstrated the contribution of such features to problem hardness. These features frequently serve as phase transition parameters [19], whereby some critical value of a feature serves as the boundary between easy and hard instances for a particular algorithm. Our recent survey paper [11] provides a review of metrics that have been shown to relate to instance difficulty for a variety of common combinatorial optimization problems including TSP, assignment, bin-packing and knapsack problems, timetabling and graph problems.

Once we have suitable metrics to characterize the hardness of optimization problem instances, we are able to define an instance space, but this will typically be of high-dimensionality requiring the use of dimension reduction methods to visualise the algorithm footprint in the instance space. We have demonstrated the effectiveness of this approach in a series of papers focused on the TSP [14, 15], scheduling [20], quadratic assignment problem [21], and university course timetabling [22]. It has become clear though, that the quality of the instance space, and the degree to which conclusive footprints can be determined to indicate the strengths and weaknesses of an algorithm depends greatly upon the chosen instances. Randomly generated instances often fail to generate instances of sufficient diversity, and consequently they provide little assistance in providing a discriminating boundary of where one algorithm outperforms another [14, 15]. Likewise, benchmark instances reported in the literature are often lacking diversity in the spectrum of difficulty [23].

In this chapter we explore further these ideas of footprints and generalisation of algorithm performance in high dimensional instance spaces. Using a case study of graph coloring, we demonstrate how this view of the instance space is useful for visualizing and defining the footprint of an algorithm, but is also revealing of the relationship between the instance generation method and the resulting features of the instances. We consider the performance of two heuristics across five classes of instance generators, and reveal the footprint of each heuristic, defining its boundary using data mining methods.

The remainder of this chapter is as follows: In Section 16.2 we describe the graph coloring problem and the meta-data for our case study. In particular, we describe the five classes of instance generators, we provide a comprehensive list of the graph properties that we use to define the instance space, we discuss the two heuristics in our algorithm portfolio, and how we measure the performance of the heuristics on the instances. Once our meta-data has been defined in this way, the analysis of the relationship between features of the instances and algorithm performance can begin. In Section 16.3 we present some data mining approaches to understanding the relationships in the meta-data. We begin with a self-organising feature map to visualise the high-dimensional feature space as a two-dimensional map of the instance space. Each instance generator creates instances with restricted properties, and these are revealed, along with the footprints of each algorithm in instance space. We also partition the instance space using a decision tree to provide rules describing the differences between the classes of instances, and the differences between the performance of the two heuristics in terms of properties of the instances. In Section 16.4 we discuss these findings and draw conclusions.

16.2 Graph Coloring and Experimental Meta-data

A graph $G = (V,E)$ comprises a set of vertices or nodes V and a set of edges E that connect certain pairs of vertices. The graph coloring problem is to assign colors to the vertices, minimizing the number of colors used, subject to the constraint that two vertices connected by an edge (called adjacent vertrices) do not share the same color. The optimal (minimal) number of colors needed to to solve this NP-complete problem is called the chromatic number of the graph. Graph coloring finds important applications in problems such as timetabling, where events to be scheduled are represented as vertices, with edges representing conflicts between events, and the color representing the time period.

16.2.1 Algorithms

Due to the NP-completeness of this problem, many heuristics have been designed [24, 25]. One of the earliest heuristics was DSATUR [26], which was shown to be exact for bipartite graphs. The saturation degree of a vertex is defined to be the number of different colors to which it is adjacent. The DSATUR (degree saturation) heuristic is a simple approach shown in Algorithm 15.

Algorithm 15. DSATUR

Step 1: Arrange the vertices by decreasing order of degrees.
Step 2: Color a vertex of maximal degree with color 1.
Step 3: Choose a vertex with a maximal saturation degree. If there is an equality, choose any vertex of maximal degree in the uncolored subgraph.
Step 4: Color the chosen vertex with the least possible (lowest numbered) color.
Step 5: If all the vertices are colored, stop. Otherwise, return to 3

For non-bipartite graphs though, the performance of DSATUR is not optimal, and many more sophisticated search strategies have been employed to provide effective heuristics for general graphs, including tabu search [27], simulated annealing [28], iterated local search [29], scatter search [30], genetic algorithms [31], and hybrid approaches [32, 33]. Just as the performance of DSATUR depends on the bipartivity of the graph, we should also expect the performance of any method to depend on various properties of the graph, but the relationship between the properties of the graph and the performance of various heuristics is not well understood. Empirical analysis of algorithms is needed [34], particularly to understand the complex relationship between graph properties, instance generators, and algorithm performance.

Culberson's web resources for graph coloring [35] provide a valuable starting point for this kind of empirical analysis, providing a number of heuristics (such as DSATUR and TABU) as well as a set of graph instance generators. In our empirical analysis, we will compare the performance of DSATUR to the TABU algorithm provided on Culberson's webpage [35], with no attempt made to fine-tune the performance via parameter tuning. In that sense, we aim to understand how the properties of the graphs generated using a variety of instance generators affect the performance of two basic heuristics: DSATUR and TABU. The implementation of tabu search is shown in Algorithm.

Algorithm 16. TABU Algorithm

Step 1: Estimate the number of colors (x) required to solve the problem.
Step 2: Generate a random solution using x color partitions.
Step 3: Move a conflicting vertex to a different color partition.
Step 4: if No conflicts, move vertex in smallest partition, to another partition that won't cause a conflict.
Step 5: Update tabu list, based on vertex moved.
Step 6: if After some maximum number of iterations there are still conflicts, $x = x + 1$
Step 7: else if No improvements found after maximum number of iterations, end algorithm.
Step 8: else Go to step 3.

16.2.2 Instances

Culberson states about his resources for graph coloring, "My intention is to provide several graph generators that will support empirical research into the characteristics of various coloring algorithms. Thus, I want generators that will exhibit variations of various characteristics of graphs, such as degree (expectation and variation), hidden colorings, girth, edge distributions etc." [35]. The five instance generators available from Culberson's website are described as [36]:

1. Uniform or IID: edges are assigned to vertex pairs with a fixed probability p
2. Girth and degree inhibited: Each graph is assigned a probability p, girth limit g; and a degree limit δ. The girth limit indicates that no cycle will be created with

girth less than g. Hence if an edge (v,w) is being considered as a new edge, every pair of vertices (x,y) which will have a distance of less than g after the addition is blocked, and will never be selected as a possible new edge. p is the probability that a possible edge will be used. δ is a hard limit on the difference between the average node degree and the maximum degree of any vertex.
3. Geometric: These graphs are generated by choosing a radius r and uniformly distributing n pairs of numbers (x,y) in the range of $0 \leq x,y < 1$. Vertices in the graph correspond to the points (x,y) in the plane, with edges included if the distance between a pair of vertices is less than the radius r.
4. Weight-biased Graphs: These graphs contain cliques limited to a given size. Each clique is generated by randomly creating h color partitions, then randomly selecting one of the vertices in each partition and joining every pair by an edge. Each clique is generated independently. Such restrictions on the size of the cliques reduces large variations in structure that might aid an algorithm like DSATUR.
5. Cycle Driven Graphs: Similar to the weight-biased graphs, this generator creates cycles of a specified length. h color partitions are created, the algorithm then generates a cycle by randomly generating a path with each vertex from a different color partition than the last.

Empirical studies have already shown how the performance of some algorithms depends on the source of the instances. Observations include the fact that DSATUR seems to outperform other methods such as tabu search and simulated annealing on geometric graphs [28]. This was conjectured to be due to the easiness of coloring graphs with large cliques [36], with the girth and degree inhibited and weight-biased graphs designed to prove more difficult for DSATUR.

16.2.3 Graph Properties or Features

While some properties of the generated instances are clear from the description above, there are still many other properties that can be calculated, and whose relationship to algorithm performance may be revealing. A recent survey of what makes optimization problem instances difficult [11] shows that there are many features of a graph that can be calculated in polynomial time, that could be used to shed some light on the relationships between graph instances, instance generators, and algorithm performance. In this study we consider the following features of a graph $G = (V,E)$:,mostly of which were generated using software available from igraph.sourceforge.net:

1. The number of nodes or vertices in a graph: $n = |V|$
2. The number of edges in a graph: $m = |E|$
3. The diameter of a graph: the greatest distance between any pair of vertices. To find the diameter we find the shortest path between each pair of vertices and take the greatest length of these paths.
4. The density of a graph: the ratio of the number of edges to the number of possible edges.

5. Average path length: the average number of steps along the shortest paths for all possible pairs of nodes. It is a measure of the efficiency of traveling between nodes.
6. The girth of a graph: the length of the shortest cycle in a graph. If a graph has a girth greater than three, it is triangle free. If a graph has a girth of three, it cannot possibly be bipartite.
7. Mean node degree: the degree of a node is the number of connections a node has to other nodes.
8. Standard deviation of node degree: the average node degree and its standard deviation can give us an idea of how connected a graph is.
9. The clustering coefficient: a measure of degree to which nodes in a graph tend to cluster together. This is a ratio of the closed triplets to the total number of triplets in a graph. A closed triplet is a triangle, while an open triplet is a triangle without one side.
10. Mean eigenvector centrality: the eigenvector of the adjacency matrix, averaged across all nodes.
11. Standard deviation of eigenvector centrality: together with the mean, the standard deviation of egenvector centrality gives us a measure of the importance of a node inside a graph.
12. Mean betweenness centrality: the fraction of all shortest paths connecting all pairs of vertices that pass through a given vertex, averaged over all nodes.
13. Standard deviation of betweenness centrality: together with the mean, the standard deviation gives us a measure of how central the nodes are in a graph. A graph that has its nodes clustered together will have a high mean betweenness centrality, while a spread out graph will have a low mean centrality.
14. Mean spectrum: the mean of the set of eigenvalues of the adjacency matrix. The spectrum is known to be symmetric for bipartite graphs.
15. Standard deviation of the set of eigenvalues of the adjacency matrix:
16. Algebraic connectivity is the second smallest eigenvalue of the Laplacian matrix [37]. This reflects how well connected a graph is. Cheeger's constant, another important graph property, is bounded by half the algebraic connectivity [38].

16.2.4 Experimental Meta-data

We are now in a position to describe the meta-data for our experimental analysis using the framework proposed in Rice [39], and adapted for the study of optimization algorithm performance by Smith-Miles [40].

- The *problem space* $\mathscr{P}$ is a set of 5000 graph coloring instances, with 1000 instances from each of the five generators described in section 16.2.2. The number of nodes was randomly selected from the range $[100, 1000]$, with all other parameters randomly selected within default settings.

- The *algorithm space* $\mathscr{A}$ comprises two heuristics: DSATUR and TABU described in Algorithms 15 and 16.
- The *performance space* $\mathscr{Y}$ is the minimum number of colors found by an algorithm after one run using default settings from the code provided by [35].
- The *feature space* $\mathscr{F}$ is defined by the 16 metrics listed in section 16.2.3.

The starting point for empirical analysis based on this meta-data is to use data mining methods to endeavour to learn the relationships between the feature space defining the instances and the performance space.

16.3 Empirical Analysis of Instances and Algorithms

Our goal is to identify the various types of instances within the high-dimensional feature space (which we will also call the instance space), and to understand the effect of instance generation method on the properties of the instances. We also seek to visualise the generalisation footprint of each algorithm's performance behaviours, and to determine the parts of feature or instance space where one algorithm dominates the other. In order to visualise the high dimensional instance space (defined by the set of 16 features discussed in section 16.2.3) we will be employing self-organising maps that produce a topologically-preserved reduction to a two-dimensional space for ease of visualization.

16.3.1 Self-organising Feature Maps

Self-Organising Feature Maps (SOFMs) are the most well known unsupervised neural network approach to clustering. Their advantage over traditional clustering techniques such as the k-means algorithm lies in the improved visualisation capabilities resulting from the two-dimensional map of the clusters. Often patterns in a high dimensional input space have a very complicated structure, but this structure is made more transparent and simple when they are clustered in a lower dimensional feature space. Kohonen [41] developed SOFMs as a way of automatically detecting strong features in large data sets. SOFMs find a mapping from the high dimensional input space to low dimensional feature space, so the clusters that form become visible in this reduced dimensionality. They can be viewed as an approximation to a nonlinear generalisation of principal component analysis.

The architecture of the SOFM is a feed-forward neural network with a single layer of neurons arranged into a rectangular array. Figure 16.1 depicts the architecture with n inputs connected via weights to a 3×3 array of 9 neurons. The number of neurons used in the output layer is determined by the user.

When an input pattern is presented to the SOFM, each neuron calculates how similar the input is to its weights. The neuron whose weights are most similar (minimal distance d in input space) is declared the winner of the competition for the

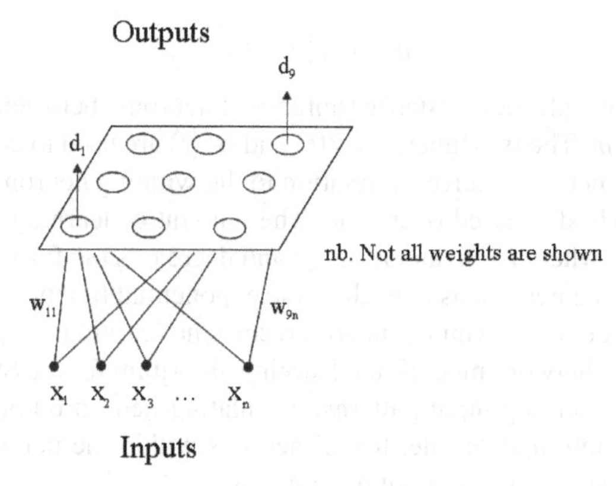

Fig. 16.1 Architecture of Self-Organising Feature Map

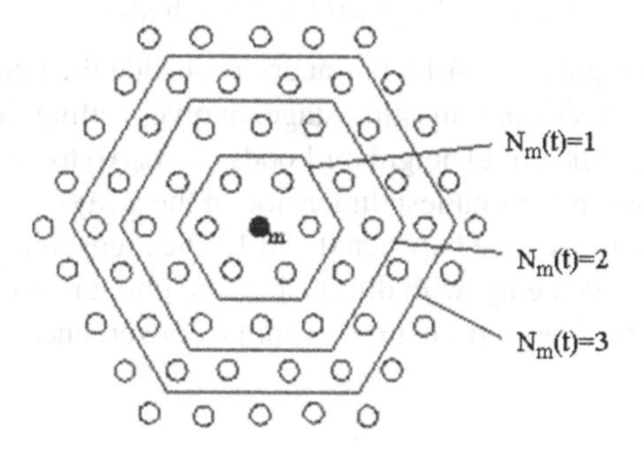

Fig. 16.2 Varying neighbourhood sizes around winning neuron *m*

input pattern, the weights of the winning neuron are strengthened to reflect the outcome, and the learning is shared with neurons in the neighbourhood of the winning neuron. This creates a process of global competition, followed by local cooperation. Figure 16.2 provides an example of how a neighbourhood N_m can be defined around a winning neuron m. Initially the neighbourhood size around a winning neuron is allowed to be quite large to encourage the regional response to inputs. As the learning proceeds however, the neighbourhood size is slowly decreased so that the response of the network becomes more localised. The localised response, which is needed to help clearly differentiate distinct input patterns, is also encouraged by varying the amount of learning received by each neuron within the winning neighbourhood. The winning neuron receives the most learning at any stage, with neighbours receiving less the further away they are from the winning neuron. If we denote the size of the neighbourhood around winning neuron m at time t by $N_{m(t)}$, then the amount of learning that every neuron i within the neighbourhood of m receives is determined by:

$$c = \alpha(t)e^{-\frac{\|r_i - r_m\|}{\sigma^2(t)}} \tag{16.1}$$

where $r_i - r_m$ is the physical distance (number of neurons) between neuron i and the winning neuron m. The two functions $\alpha(t)$ and $\sigma^2(t)$ are used to control the amount of learning each neuron receives in relation to the winning neuron. These functions are usually slowly decreased over time. The amount of learning is greatest at the winning neuron (where $i = m$ and $r_i = r_m$) and decreases the further away a neuron is from the winning neuron, as a result of the exponential function. Neurons outside the neighbourhood of the winning neuron receive no learning.

Like all neural network models, the learning algorithm for the SOFM follows the basic steps of presenting input patterns, calculating neuron outputs, and updating weights. The weight update rule, for all neurons within the neighbourhood of the winning neuron m for a given input pattern x_i is:

$$w_{ji}(t+1) = w_{ji}(t) + c[x_i - w_{ji}(t)]$$

with c as defined by equation (16.1). For neurons outside the neighbourhood of the winning neuron, $c = 0$. The initialisation stage involves setting the weights to small random values, setting the initial neighbourhood size $N_m(0)$ to be large (but less than the number of neurons in the smallest dimension of the array), and setting the values of the parameter functions to be between 0 and 1. The algorithm iterates through all of the input patterns repeatedly, with diminishing neighbourhood size and decaying functions $\alpha(t)$ and $\sigma^2(t)$ each time, until eventual convergence of the weights.

16.3.2 Visualising the Instance Space

The instance space is characterised by a set of 5000 graph coloring instances, each defined by a set of 16 features related to the graph properties. All features were normalised to the range [0,1] using variance. The software package Viscovery SOMine [42] was used to generate the SOFM, using a rectangular map of approximate ratio 100:40 based on the dimensions of the plane spanned by the two largest eigenvectors of the correlation matrix of the features (i.e. the first two principal components of the correlation matrix). The final map contains 2108 neurons arranged in 31 rows and 68 columns. 54 complete presentations of all 5000 feature vectors (instances) were required to achieve convergence, with a decay factor of 0.5 applied to the functions $\alpha(t)$ and $\sigma^2(t)$. The initial neighbourhood size was 7. While these values were chosen arbitrarily based on past experience, experimentation with different values showed that the resulting maps were quite robust. The final quantization error of the map was 0.005999.

Figure 16.3 shows the location of the instances in the 16-dimensional feature space when projected onto a two-dimensional map of instance space. Instances that are close to each other in the map are similar (according to Euclidean distance in 16-dimensional feature space) to each other, and significantly different from other instances further away. It should be noted that the type of generator used to create

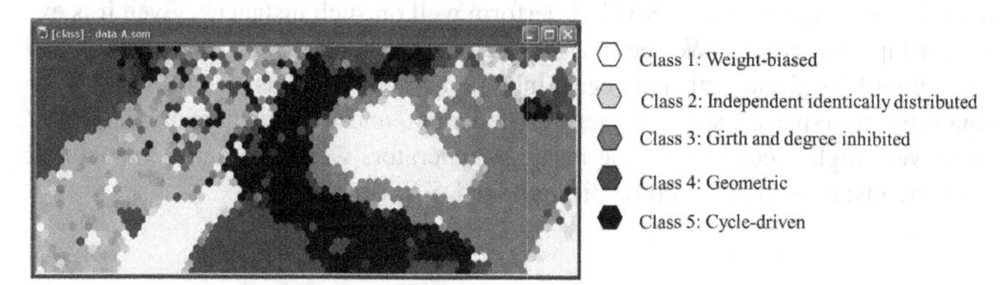

Fig. 16.3 The distribution of the five classes of instances across instance space (16-dimensional feature space projected onto 2-dimensional SOFM)

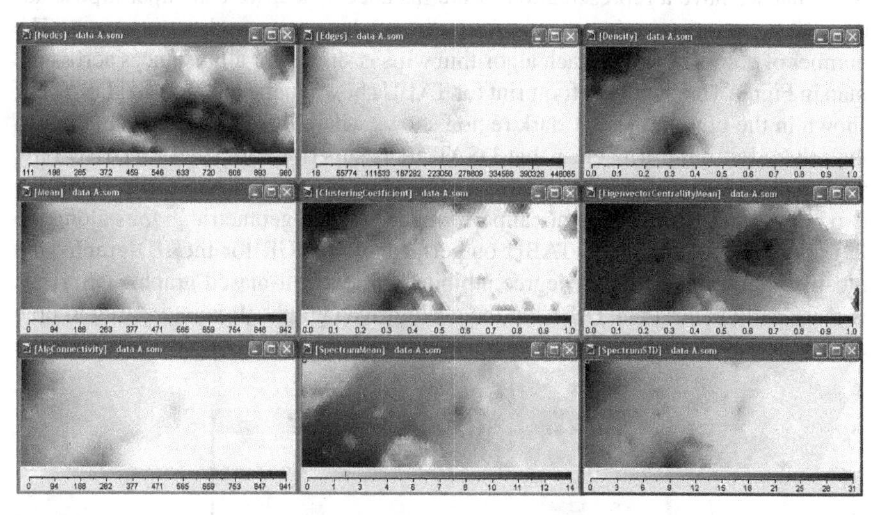

Fig. 16.4 The distribution of some of the features across instance space (white represents a minimal value of the feature and black represents a maximal value of the feature)

an instance was not used as a feature and yet, based only on the measured properties of the resulting graph instances, we can clearly see that a given generator creates a group of instances that are quite similar to each other. Some of the generators are capable of creating instances that are diverse (for example, pockets of weight-biased and geometric instances are found across the map), whereas other generators appear to create instances that are in the same region of instance space (for example, girth and degree inhibited, and IID graphs). It is also clear that the feature vector of an instance is determined to a large extent by the instance generator, with very few regions containing instances coming from multiple generators.

In order to determine which features are dominant in the various classes of instances, we can inspect the distribution of features across the map. A subset of the features are shown in Figure 16.4. We see that the region of instance space corresponding to geometric graphs contains instances with a high density, high clustering coefficient, and very low spectrum mean indicating a graph close to being bipartite. We expect to see DSATUR perform well on such instances, given it is exact for bipartite graphs. We see that the region of instance space corresponding to the weight-biased and girth and degree inhibited graphs contains instances with very small density (sparse), small clustering coefficient, and with eigenvector centrality mean very high. Recall that these instance generators were originally conjectured to create instances that would be difficult for DSATUR to solve.

16.3.3 Visualising the Footprints of Algorithm Performance

Now that we have a representation of the instance space, we can superimpose additional information such as the performance of algorithms on those instances. The number of colors by which each algorithm wins is shown for all instances across the map in Figure 16.5, with the footprint for TABU shown in the top map and DSATUR shown in the bottom map. A dark region shows where the algorithm is superior to the other algorithm. It is clear that DSATUR is superior compared to TABU only for a small subset of the Geometric graphs close to the top left corner, with the two algorithms achieving identical performance for the geometric graphs along the left corner of the top edge. TABU outperforms DSATUR for the IID graphs and mostly ties on the girth and degree inhibited and weight-biased graphs. DSATUR is marginally better than TABU on the cycle-driven graphs. It is interesting to note that, even though the SOFM was not given any algorithm performance data, only the

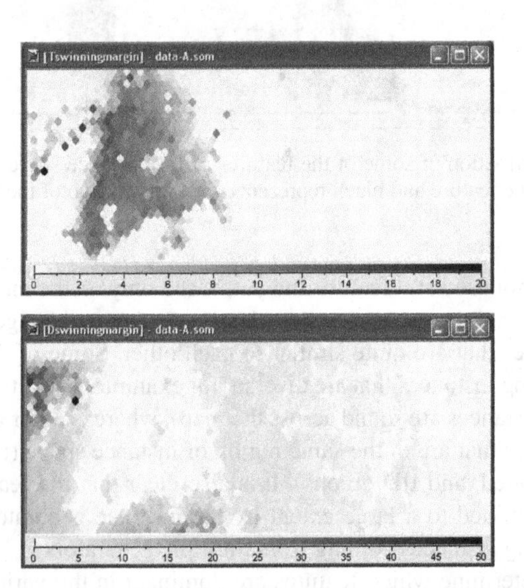

Fig. 16.5 The performance of each algorithm across instance space. The number of colors by which TABU wins is shown in the top map, and DSATUR is shown in the bottom map. White represents no difference in solution, with darker regions indicating graphs where a superior performance is obtained for the given algorithm (small number of colors needed to color the graph).

16 features, we still find clear regions in instance space where the performance of algorithms is similar. This confirms that the chosen features are well-suited to characterizing the similarities between the instances, and the effect that their properties have on algorithm performance.

16.3.4 Partitioning the Instance Space via Decision Trees

Decision trees can be very powerful tools for elucidating rules that can help explain performance differences between algorithms. We randomly extract 80% of the instances to form a training set, and use a decision tree with CHAID (chi-squared automatic interaction detector) method in SPSS to find rules describing which instances (in terms of the 16 features) are better solved with TABU. The resulting rules are then evaluated on the remaining 20% test set of instances. The accuracy of the rules, on both the training and test sets, are shown in Table 16.1. If TABU and DSATUR achieve a tied result, then the instance is labelled as TABU not best. The results enable us to claim a 93.7% accuracy in correctly identifying, based only on the measurable graph features, if TABU will outperform DSATUR. Considering that many of the instances incorrectly classified are actually tied results, this is a highly accurate prediction of algorithm performance on the basis of graph features alone.

Table 16.1 Classification accuracy of predicting when TABU will be better than DSATUR on training set (top) and test set (bottom)

TRAINING SET	predicted TABU best	predicted TABU not best	percent correct
actually TABU best	2951	133	95.7%
actually TABU not best	127	748	85.5%
	95.9%	84.9%	93.4%

TEST SET	predicted TABU best	predicted TABU not best	percent correct
actually TABU best	772	41	95.0%
actually TABU not best	22	171	88.6%
	97.2%	80.6%	93.7%

An inspection of the rules generated by the decision tree reveals that TABU is better suited to instances where the clustering coefficient is less than 0.5, and the spectrum mean is greater than 9.18 (not bipartite graphs, and not geometric graphs). In addition, TABU excels when the algebraic connectivity is between 83.7 and 246.9. These are just a few of the small pockets of the instance space that were found to describe when TABU outperforms DSATUR. While these rules may not be particularly interesting for the chosen heuristics (especially considering the default parameter setting used in TABU which made it not much more powerful than DSATUR across many instances), we hope that the case study presented here illustrates the benefits of the presented methodology for gaining insights into the relationships between features of instances and the performance of algorithms. It is a straightforward extension of these ideas to consider the effect of parameter tuning - for example, simulated annealing run with two different cooling schedules could be considered as two different algorithms, and the footprint of each algorithm (cooling schedule) can be identified and understood across instance space.

16.4 Conclusions

In this paper we have shown, through a case study of graph coloring, that data mining techniques like self-organising feature maps and decision trees can be used to explore the high-dimensional feature space that defines an instance space. Specifically, the instance space can be visualised with a view to understanding the kind of instances that different generators create, and examining the generalisation footprint of algorithm performance across a broad instance space. For our case study, we have utilised a comprehensive set of features based on properties of graphs. We have demonstrated that these features create an instance space where the differences between instance generators are readily visualised. The effectiveness of different algorithms can then be superimposed across the instance space and the footprint can be visualised. For our chosen algorithms, DSATUR and TABU, we have been able to partition the instance space to separate instances where one algorithm outperforms the other. The chosen features have proven to be sufficient for discovering the properties of instances that make TABU outperform DSATUR, and vice versa, achieving over 93% accuracy in predicting when TABU will win. It remains for future research to consider additional features of instances that could further distinguish between these two competitive algorithms, and the relationship between landscape metrics [43, 44, 45] and algorithm performance should also be considered.

The ability to generate instances that are discriminating of algorithm performance is critical for progress in understanding the strengths and weaknesses of various algorithms, so that improved hybrid algorithms can be designed. Analysis of the kind presented in this paper provides a starting point to examine the characteristics of a set of instances, and enables feedback into the instance generation process [46] to develop a meaningful set of instances to drive future research developments.

References

1. Wolpert, D.H., Macready, W.G.: No free lunch theorems for optimization. IEEE Transactions on Evolutionary Computation 1(1), 67–82 (1997)
2. Culberson, J.: On the futility of blind search: An algorithmic view of 'no free lunch'. Evolutionary Computation 6(2), 109–127 (1998)
3. Blum, C., Roli, A.: Hybrid metaheuristics: An introduction. In: Hybrid Metaheuristics, pp. 1–30 (2008)
4. Talbi, E.: A taxonomy of hybrid metaheuristics. Journal of Heuristics 8(5), 541–564 (2002)

5. Burke, E., Kendall, G., Newall, J., Hart, E., Ross, P., Schulenburg, S.: Hyper-heuristics: An emerging direction in modern search technology. International Series in Operations Research and Management Science, pp. 457–474 (2003)
6. Leyton-Brown, K., Nudelman, E., Shoham, Y.: Learning the Empirical Hardness of Optimization Problems: The Case of Combinatorial Auctions. In: Van Hentenryck, P. (ed.) CP 2002. LNCS, vol. 2470, pp. 556–572. Springer, Heidelberg (2002)
7. Leyton-Brown, K., Nudelman, E., Andrew, G., McFadden, J., Shoham, Y.: A portfolio approach to algorithm selection. In: International Joint Conference on Artificial Intelligence, vol. 18, pp. 1542–1543 (2003)
8. Xu, L., Hutter, F., Hoos, H.H., Leyton-Brown, K.: `SATzilla-07`: The Design and Analysis of an Algorithm Portfolio for SAT. In: Bessière, C. (ed.) CP 2007. LNCS, vol. 4741, pp. 712–727. Springer, Heidelberg (2007)
9. Hooker, J.: Testing heuristics: We have it all wrong. Journal of Heuristics 1(1), 33–42 (1995)
10. Corne, D.W., Reynolds, A.P.: Optimisation and Generalisation: Footprints in Instance Space. In: Schaefer, R., Cotta, C., Kołodziej, J., Rudolph, G. (eds.) PPSN XI, Part I. LNCS, vol. 6238, pp. 22–31. Springer, Heidelberg (2010)
11. Smith-Miles, K.A., Lopes, L.B.: Measuring instance difficulty for combinatorial optimization problems. Computers and Operations Research 39(5), 875–889 (2012)
12. Macready, W., Wolpert, D.: What makes an optimization problem hard. Complexity 5, 40–46 (1996)
13. van Hemert, J., Urquhart, N.: Phase Transition Properties of Clustered Travelling Salesman Problem Instances Generated with Evolutionary Computation. In: Yao, X., Burke, E.K., Lozano, J.A., Smith, J., Merelo-Guervós, J.J., Bullinaria, J.A., Rowe, J.E., Tiňo, P., Kabán, A., Schwefel, H.-P. (eds.) PPSN 2004. LNCS, vol. 3242, pp. 151–160. Springer, Heidelberg (2004)
14. Smith-Miles, K., van Hemert, J., Lim, X.Y.: Understanding TSP Difficulty by Learning from Evolved Instances. In: Blum, C., Battiti, R. (eds.) LION 4. LNCS, vol. 6073, pp. 266–280. Springer, Heidelberg (2010)
15. Smith-Miles, K., van Hemert, J.: Discovering the suitability of optimisation algorithms by learning from evolved instances. Annals of Mathematics and Artificial Intelligence (in press)
16. Merz, P., Freisleben, B.: Fitness landscape analysis and memetic algorithms for the quadratic assignment problem. IEEE Transactions on Evolutionary Computation 4(4), 337–352 (2000)
17. Merz, P.: Advanced fitness landscape analysis and the performance of memetic algorithms. Evolutionary Computation 12(3), 303–325 (2004)
18. Achlioptas, D., Naor, A., Peres, Y.: Rigorous location of phase transitions in hard optimization problems. Nature 435(7043), 759–764 (2005)
19. Cheeseman, P., Kanefsky, B., Taylor, W.: Where the really hard problems are. In: Proceedings of the 12th IJCAI, pp. 331–337 (1991)
20. Smith-Miles, K., James, R., Giffin, J., Tu, Y.: Understanding the Relationship between Scheduling Problem Structure and Heuristic Performance using Knowledge Discovery. LNCS (2009) (in press)
21. Smith-Miles, K.: Towards insightful algorithm selection for optimisation using meta-learning concepts. In: IEEE International Joint Conference on Neural Networks, IJCNN 2008 (IEEE World Congress on Computational Intelligence), pp. 4118–4124 (2008)
22. Smith-Miles, K., Lopes, L.: Generalising Algorithm Performance in Instance Space: A Timetabling Case Study. In: Coello, C.A.C. (ed.) LION 2011. LNCS, vol. 6683, pp. 524–538. Springer, Heidelberg (2011)
23. Hill, R., Reilly, C.: The effects of coefficient correlation structure in two-dimensional knapsack problems on solution procedure performance. Management Science, 302–317 (2000)
24. Pardalos, P., Mavridou, T., Xue, J.: The graph coloring problem: A bibliographic survey, vol. 2. Kluwer Academic Publishers (1998)

25. Galinier, P., Hertz, A.: A survey of local search methods for graph coloring. Computers & Operations Research 33(9), 2547–2562 (2006)
26. Brélaz, D.: New methods to color the vertices of a graph. Communications of the ACM 22(4), 251–256 (1979)
27. Hertz, A., Werra, D.: Using tabu search techniques for graph coloring. Computing 39(4), 345–351 (1987)
28. Johnson, D., Aragon, C., McGeoch, L., Schevon, C.: Optimization by simulated annealing: an experimental evaluation; part ii, graph coloring and number partitioning. Operations Research, 378–406 (1991)
29. Chiarandini, M., Stützle, T., et al.: An application of iterated local search to graph coloring problem. In: Proceedings of the Computational Symposium on Graph Coloring and its Generalizations, pp. 7–8. Citeseer (2002)
30. Hamiez, J.-P., Hao, J.-K.: Scatter Search for Graph Coloring. In: Collet, P., Fonlupt, C., Hao, J.-K., Lutton, E., Schoenauer, M. (eds.) EA 2001. LNCS, vol. 2310, pp. 168–213. Springer, Heidelberg (2002)
31. Galinier, P., Hao, J.: Hybrid evolutionary algorithms for graph coloring. Journal of Combinatorial Optimization 3(4), 379–397 (1999)
32. Fleurent, C., Ferland, J.: Genetic and hybrid algorithms for graph coloring. Annals of Operations Research 63(3), 437–461 (1996)
33. Blöchliger, I., Zufferey, N.: A graph coloring heuristic using partial solutions and a reactive tabu scheme. Computers & Operations Research 35(3), 960–975 (2008)
34. Hooker, J.: Needed: An empirical science of algorithms. Operations Research, 201–212 (1994)
35. Culberson, J.: Graph coloring page (2006), `http://www.cs.ualberta.ca/~joe/Coloring`
36. Culberson, J., Beacham, A., Papp, D.: Hiding our colors. In: CP 1995 Workshop on Studying and Solving Really Hard Problems. Citeseer (1995)
37. Mohar, B.: The laplacian spectrum of graphs. Graph Theory, Combinatorics, and Applications 2, 871–898 (1991)
38. Biggs, N.: Algebraic graph theory, vol. 67. Cambridge Univ. Pr. (1993)
39. Rice, J.: The Algorithm Selection Problem. Advances in Computers 15, 65–117 (1976)
40. Smith-Miles, K.: Cross-disciplinary perspectives on meta-learning for algorithm selection. ACM Computing Surveys 41(1) (2008)
41. Kohonen, T.: Self-organized formation of topologically correct feature maps. Biological Cybernetics 43(1), 59–69 (1982)
42. Somine, V.: Eudaptics software Gmbh
43. Knowles, J., Corne, D.: Towards landscape analyses to inform the design of a hybrid local search for the multiobjective quadratic assignment problem. Soft Computing Systems: Design, Management and Applications, 271–279 (2002)
44. Bierwirth, C., Mattfeld, D.C., Watson, J.-P.: Landscape Regularity and Random Walks for the Job-Shop Scheduling Problem. In: Gottlieb, J., Raidl, G.R. (eds.) EvoCOP 2004. LNCS, vol. 3004, pp. 21–30. Springer, Heidelberg (2004)
45. Schiavinotto, T., Stützle, T.: A review of metrics on permutations for search landscape analysis. Comput. Oper. Res. 34(10), 3143–3153 (2007)
46. Lopes, L., Smith-Miles, K.: Generating applicable synthetic instances for branch problems, under review (2011)

Chapter 17
Boosting Metaheuristic Search Using Reinforcement Learning

Tony Wauters, Katja Verbeeck, Patrick De Causmaecker, and Greet Vanden Berghe

Abstract. Many techniques that boost the speed or quality of metaheuristic search have been reported within literature. The present contribution investigates the rather rare combination of reinforcement learning and metaheuristics. Reinforcement learning techniques describe how an autonomous agent can learn from experience. Previous work has shown that a network of simple reinforcement learning devices based on learning automata can generate good heuristics for (multi) project scheduling problems. However, using reinforcement learning to generate heuristics is just one method of how reinforcement learning can strengthen metaheuristic search. Both existing and new methodologies to boost metaheuristics using reinforcement learning are presented together with experiments on actual benchmarks.

17.1 Introduction

Researchers developing search methods to solve combinatorial optimization problems are faced with a number of challenges. An important challenge is to avoid convergence to a local optimum. A second challenge is to create a method applicable to different problems of various sizes and properties, while still being able to produce good quality solutions in a short amount of time. Metaheuristics [17, 30] and the more recently introduced hyper-heuristics [6] try to address these issues. Hybrid systems and their various perspectives cope with these challenges even better. One possible hybridization, which is the main topic of this contribution, involves the inclusion of a Reinforcement Learning (RL) [19, 29] component to these meta- and hyper-heuristic methods. This idea fits in the area of intelligent optimization [2],

Tony Wauters
CODeS, KAHO Sint-Lieven, Gebroeders Desmetstraat 1, 9000 Gent, Belgium
e-mail: tony.wauters@kahosl.be

E.-G. Talbi (Ed.): Hybrid Metaheuristics, SCI 434, pp. 433–452.
springerlink.com

where some intelligent (learning) component aids the optimization method in order to obtain a better informed search. During the search process a learning component can adjust parameters or support the optimization method in making decisions. Intelligent optimization can be defined as a combination of techniques from Operations Research and Artificial Intelligence.

In what follows, several arguments for combining RL and metaheuristics are enlisted. Reinforcement learning causes the algorithm to be adaptive, and as such it minimizes the weaknesses of strongly parameterized methods. As long as the algorithm is granted enough time facilitating the learning of valuable information. Reinforcement learning offers interesting advantages. It does not require a complete model of the underlying problem. RL methods learn the model by gathering experience, often referred to as trial-and-error. Many model free RL methods exist. This is noteworthy since ordinarily no model is available for most combinatorial optimization problems. Some reinforcement learning methods can handle incomplete information, although this is obviously a much harder learning task. Reinforcement learning permits applying independent learning agents, and thus, it is applicable for fully decentralized problems. Furthermore, it is computationally cheap, i.e. often it uses only a single update formula at each step. Additionally, if only general problem features and no instance specific features are used, then the learned information can possibly be transfered to other instances of the same problem. Recently some type of RL algorithms that are well suited for this task have been introduced, these are called transfer learning [32, 31]. Lastly, one can build on theoretical properties showing that many RL methods converge to optimal state-action pairs under certain conditions (e.g. the policy for choosing the next action is ergodic) [29]. In this chapter we will show that these interesting theoretical properties show good results in practice.

The remainder of the chapter describes the combination of reinforcement learning and search in more detail. Section 17.2 gives a short introduction to reinforcement learning and some common RL algorithms. Section 17.3 discusses different opportunities for combining these two methods. Section 17.4 gives an extensive literature overview of the combination of learning and search. An example of a succesfull application of RL is presented in Sect. 17.5. A conclusion and some future prospects are described in Sect. 17.6.

17.2 Reinforcement Learning

Reinforcement Learning (RL) [19, 29] is the computational task of learning what action to take in a given situation (state) to achieve one or more goal(s). The learning process takes place through interaction with an environment (Fig. 17.1), and is therefore different from supervised learning methods which require a teacher. At each discrete time step an RL agent receives observations, i.e. an indication of the current state s. In each state s the agent can take some action a from the set of actions available in that state. An action a can cause a transition from state s to another state s', based on transition probabilities . The environment's model contains these transition probabilities. A numerical *reward signal* r is returned to the agent to inform the RL agent about the 'goodness' of its actions or the intrinsic desirability of a state. The reward signal is also a part of the model of the environment. An RL agent searches for the optimal policy. A *policy* π maps states to actions or action probabilities. $\pi(s,a)$ denotes the probability that action a is selected in state s. An RL agent wants to maximize the expected sum of future rewards. When an infinite horizon is assumed, the discount factor γ is used to discount these future rewards. As such, less importance is given to rewards further into the future.

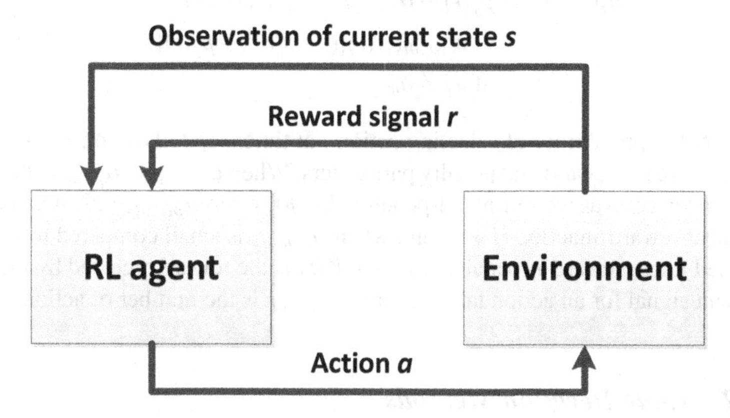

Fig. 17.1 The basic reinforcement learning model

One of the main issues in RL is balancing exploration and exploitation, i.e. whether to use the already gathered experience or to gather new experience. Another important issue is the credit assignment problem, where one have to deal with delayed rewards, and thus which action should receive credit for a given reward. The latter property is currently getting few attention in hybrid RL inspired search methods.

17.2.1 Policy Iteration Methods

When an RL method searches directly for the optimal policy in the space of policies it is called a policy iteration method. Learning Automata (LA) [24, 33] belong to this category. Other methods belonging to this type of RL algorithms are the policy gradient methods, like the REINFORCE algorithm [39]. LA are simple reinforcement learning devices that take actions in single state environments. A single learning automaton maintains an action probability distribution p, which it updates using some specific learning algorithm or reinforcement scheme. Several reinforcement schemes are available in the literature with varying convergence properties. These schemes use information from a reinforcement signal provided by the environment, and thus the LA operates with its environment in a feedback loop. Examples of linear reinforcement schemes are linear reward-penalty, linear reward-inaction and linear reward-ε-penalty. The philosophy of these schemes is to increase the probability of selecting an action in the event of success and decrease it when the response is a failure. The general update scheme is given by:

$$\begin{aligned} p_m(t+1) &= p_m(t) + \alpha_{reward}(1-\beta(t))(1-p_m(t)) \\ &\quad - \alpha_{penalty}\beta(t)p_m(t) \\ &\quad \text{if } a_m \text{ is the action taken at time } t \end{aligned} \tag{17.1}$$

$$\begin{aligned} p_j(t+1) &= p_j(t) - \alpha_{reward}(1-\beta(t))p_j(t) \\ &\quad + \alpha_{penalty}\beta(t)[(r-1)^{-1} - p_j(t)] \\ &\quad \text{if } a_j \neq a_m \end{aligned} \tag{17.2}$$

With $p_i(t)$ the probability of selecting action i at time step t. The constants α_{reward} en $\alpha_{penalty}$ are the reward and penalty parameters. When $\alpha_{reward} = \alpha_{penalty}$, the algorithm is referred to as linear reward-penalty (L_{R-P}), when $\alpha_{penalty} = 0$, it is referred to as linear reward-inaction (L_{R-I}) and when $\alpha_{penalty}$ is small compared to α_{reward}, it is called linear reward-ε-penalty ($L_{R-\varepsilon P}$). $\beta(t)$ is the reward received by the reinforcement signal for an action taken at time step t. r is the number of actions.

17.2.2 Value Iteration Methods

Value iteration methods are more common than policy iteration methods. Value iteration methods do not directly search for optimal policy, instead they are learning evaluation functions for states or state-action pairs. Evaluation functions, as in the popular Q-learning algorithm [35, 34], can be used to evaluate the quality of a state-action pair. The Q-learning algorithm maintains an action-value function called Q-values. The Q-learning update rule is defined by

$$Q(s,a) = Q(s,a) + \alpha\left[r + \gamma \max_{a'} Q(s',a') - Q(s,a)\right], \tag{17.3}$$

where s is the previous state, s' the new state, a the action taken in state s, a' a possible action in state s', r the received reward, α the learning rate or step size, and γ the discount rate which indicates the importance of future rewards. In many cases the number of states or state-action pairs is too large to store, and thus the (action)-value function must be approximated. Often an artificial neural network is used to accomplish this task, but any linear or nonlinear function approximation can be used. Q-learning is known as a Temporal Difference (TD) learning method, because the update rule uses the difference between the new estimate and the old estimate of the value function. Other common TD methods are SARSA [27, 28] and TD(λ) [35].

17.2.3 Relationships between Reinforcement Learning and Metaheuristics

The link between reinforcement learning and search algorithms is not tenuous. Given a state space, an action space and a reward function, then the reinforcement learning problem can be reduced to a search in the space of policies. Thus similar issues like exploration and exploitation are faced, often called diversification and intensification in metaheuristic literature.

Unlike the subject of the present chapter where (reinforcement) learning methods are used to support metaheuristic search, there are methods which do exactly the opposite, and thus are using metaheuristic methods to improve learning. [3] show this interplay between optimization and machine learning.

17.3 Opportunities for Learning

Reinforcement learning methods can be utilized in a variety of ways to boost metaheuristic or hyper-heuristic search. Learning may help to find good settings for various parameters or components. For example, RL methods can learn properties of good *starting solutions* or an *objective function* that guides a metaheuristic towards good quality solutions. Such an approach is adopted in [4, 5], where a function is learned that is able to guide the search to good starting solutions. Another component on which learning can be applied is the *neighborhood* or *heuristic selection*. A learning method can learn which ones are the best neighborhoods or heuristics to construct or change a solution at any time during the search, such that in the end good quality solutions can be generated. Such an approach is applied in [40] and [25]. When a classical hyper-heuristic with acceptance and selection mechanism is used learning can be applied to both mechanisms. A learning method can learn to select the low-level heuristics (e.g. in [7] and [22]), or it can ascertain when to accept a move. To summarize, the possible involved components include but are not limited to:

- starting solution,
- objective function,
- neighborhoods or heuristics selection,
- acceptance of new solutions/moves.

All these parameters or components can be updated by the RL algorithm in an adaptive way.

Alternatively RL methods can also be applied directly to solve optimization problems. In other words, they are not hybridized but are themselves used as a metaheuristic. The RL method learns and directly assigns the values of the variables. Such approaches are investigated in [16], [21], [36] and [37].

A possible indication for the inclusion of RL in a search algorithm is the presence of a random component. By replacing this random component with an RL component the algorithm develops a more intelligent decision mechanism. For example in a hyper-heuristic with a simple-random selection step, the selection can be replaced by some RL method like the one presented by [22].

Yet another opportunity for learning arises both when either the problem is intrinsically distributed or can be split into several subproblems. Examples of such include distributed scheduling and planning problems such as the decentralized resource-constrained multi-project scheduling problem (DRCMPSP) [9] and [18]. This problem considers scheduling multiple projects simultaneously, with each project having multiple jobs. A job requires local or global resources, which are available for either all jobs in the project or have to be shared among all projects respectively. Some local objectives can be optimized, for example the makespan of the individual projects, whereas global objectives, such as the average project delay or the total makespan, can be minimized. For this kind of problems multi-agent reinforcement learning methods are appropriate. When one or more global objectives need to be optimized the agents share a common goal and can thus be cooperative. The agents have to coordinate to jointly improve their decisions. Using a common reward one can simply but effectively coordinate the agents' decisions. Through sharing one reward signal the agents coordinate their decisions. This approach is applied in [36] for the DRCMPSP.

17.3.1 States and Actions

Before applying RL to a problem, the set of possible states and the set of possible actions available in each state, have to be defined. Many possible ways exist to accomplish this. First of all we can make a distinction between *search-dependent*, *problem-dependent* and *instance-dependent* state space definitions. A search-dependent state space definition uses observations of the search process itself, such as the current iteration, the number of successive non-improving iterations, or the total improvement over the initial solution. A problem-dependent setting is defined by the usage of generic problem features, like the Resource Dilation Factor for scheduling problems, as defined by [14]. An instance-dependent setting uses instance-specific features, like the number of tardy jobs in a scheduling problem, or the number of full bins in a bin-packing problem. Combinations of these three settings are also possible. When a problem-dependent or a search-dependent state space definition is used, the learned information can possibly be transfered to other instances of the same problem, or even to other problems. In many cases the properties of the solutions to the optimization problem itself cannot be used directly, due to the curse of dimensionality. Better is to use some extracted problem features. Take for example a TSP problem. If one should use the encoding of a complete tour directly as the state, then the number of states grows exponentially, i.e. $n!$ with n the number of states.

The set of possible actions in each state is defined by the set of parameters or components of the metaheuristic that one wants to learn.

17.3.2 Reward Function

Experience gathering is a prerequisite to learning, which can be achieved either online or offline. Experience in combinatorial optimization problems is scarce. Often only a single numerical value is available, indicating the quality of a complete solution. However, reward functions are very important for an RL method in order to learn some valuable information. As stated in [14], there are three requirements that a reward function should satisfy. First of all, it should give higher rewards to better solutions. Secondly, it should encourage the reinforcement learning system to find efficient search policies, i.e. search policies that involve only a few steps. Thirdly, it should be a normalized measure, in order to be transferred to new problem instances. A fourth requirement may be added; that it should be computationally efficient. When designing a reward function for a hybrid RL-metaheuristic method we do take into account these four requirements.

17.4 Literature Overview

The combination of metaheuristics or hyper-heuristics and (reinforcement) learning is relatively new. Only a limited number of papers describe a combination of the two domains. Applied problem domains include scheduling, packing and routing. Table 17.1 compares these methods by the used RL-method, metaheuristic and involved component.

One of the first papers covering the combination of learning and metaheuristic search is [40]. A reinforcement learning method is applied to learn domain-specific heuristics for the NASA space shuttle payload processing problem, which is modeled as a job shop scheduling problem. A value function is learned offline using a temporal difference algorithm TD(λ) together with a neural network. General features of the schedules (solutions) such as the percentage of the time units with a violation are used to represent a state. The possible actions are taken from a set of repair heuristics. After learning the value function on a number of small problem instances, it is used over multiple instances of the same problem. The TD algorithm is compared to an existing method for the problem, i.e. an iterative repair method with simulated annealing. The reinforcement learning based method outperforms the iterative repair method. It is noteworthy that the value functions that were learned on small problem instances also have a very good performance on larger instances. In [14] a more detailed description and application of this approach is given.

Another early contribution to the application of RL for solving combinatorial optimization problems can be found in [16]. The paper describes the Ant-Q algorithm, which combines the Ant System and the Q-Learning algorithm, and has been successfully applied to the Asymmetric Traveling Salesman Problem (ATSP). Ant System is based on the observation of ant colonies behaviour. Each ant from a colony constructs a solution for the ATSP, called a tour. The method uses a modified version of Q-values, called AQ-values. These AQ-values are updated using a Q-learning update rule. The delayed reward, which is calculated when each ant completes a tour, is based on the best tour of the current iteration or on the best tour from all past iterations, taking into account each ant. The Ant-Q algorithm shows an interesting property. It was observed that the Ant-Q agents do not make the same tour, demonstrating the explorative character of the search method.

[21] present an algorithm that combines reinforcement learning with genetic algorithms for the Asymmetric Traveling Salesman Problem (ATSP) and Quadratic Assignment Problem (QAP). For the ATSP a Q-learning [35, 34] method is used to both express and update the desirability of choosing city a after city b. A state is a city, and an action is another city following the aforementioned city in the tour. A population of RL agents with desirability values (Q-values) is formed, with each agent holding one solution. The offspring is constructed by replicating solution parts from one parent and filling in the other parts using the desirability values (updated by a q-learning update rule) of the other parent, rather than the traditional genetic crossover operators method. As a reward a weighted combination of immediate and global rewards based on the tour lengths of the new solution and the solutions of the parents is used. The QAP is solved using a simplified update rule, that does not require a particular order as opposed to the q-learning update rule. Competitive results are shown for both addressed problems.

[4] and [5] describe the STAGE algorithm, which searches for good quality solution using two alternating phases. The first phase runs a local search method, e.g. hillclimbing or simulated annealing from a starting solution until a local optimum is reached. During this phase, the search trajectory is analyzed and used for learning an evaluation function. This is achieved by training a linear or quadratic regression method using the properties or features of the visited solutions and the objective function value of the local optimum. The authors point out that in some conditions a reinforcement learning method like TD(λ) of the temporal-difference algorithms may make better use of the training data, converge faster, and use less memory during training. The second phase performs hillclimbing on the learned evaluation function to reach a new starting solution for the first phase. This phase enables the algorithm to learn to find good starting solutions for a local search method. Empirical results are provided on seven large-scale optimization domains, e.g. bin-packing, channel routing, ... This demonstrates the ability of the STAGE algorithm to perform well on many problems.

[23] combine aspects taken from the research by [40], [4] and [5]. A reinforcement learning algorithm TD(λ) is applied to learn a value function in an offline training phase, and then uses this learned value function to solve other instances of the same problem. This method also uses features of solutions for representing a state. A linear function approximation algorithm is used. The method is applied to the dial-a-ride problem, and was compared to both the STAGE algorithm, and a 2-opt and 3-opt local search method. The method performs better than 2-opt and STAGE if the same calculation time is used. It was not as performant as 3-opt, but was a lot faster.

[25] describes a non-stationary reinforcement learning method for choosing search heuristics. At each decision point weights are used to select the search heuristics via a probabilistic selection rule (softmax) or by randomly selecting among the choices with maximal value. Based on the increase/decrease of the objective function the weights of the search heuristics are updated using simple positive/negative reinforcement rules (e.g. incrementing/decrementing the weight value). Different selection and reinforcement method combinations are tested on two types of problems - the Orc Quest problem and problems from the Logistics Domain benchmark. The author concludes that a weak positive reinforcement rule combined with a strong negative reinforcement rule works best on the tested problems.

[7] present a hyper-heuristic in which the selection of low-level heuristics makes use of basic reinforcement learning principles combined with a tabu-search mechanism. The reinforcements are performed by increasing/decreasing the rank of the low-level heuristics when the objective function value improves/worsens. The hyper-heuristic was evaluated on various instances of two distinct timetabling and rostering problems and showed to be competitive with the state-of-the-art approaches. The paper states that a key ingredient in implementing a hyper-heuristic is the learning mechanism.

An interesting study on memory length in learning hyper-heuristics is performed in [1]. Utility values or weights are used to select the low-level heuristics, similar to [25] and [7]. A dsicount factor is added to this mechanism to discount rewards later on in the search process, and thus obtaining a short term memory. The results obtained on a course timetabling problem show that a short term memory can produce better results than both no memory and infinite memory.

[15] gives an extensive overview of single and multi-agent RL approaches for distributed job-shop scheduling problems. Both value function-based and policy search-based RL methods are discussed, including policy gradient RL methods and Q-learning.

[26] present a hyper-heuristic with an RL selection mechanism and a great-deluge acceptance method for the examination timetabling problem. A set of exams must be assigned a timeslot and possibly a room while respecting a number of hard and soft constraints, such as the room capacity. An RL method based on utility values with simple update rules is used, similar to what was presented in [25]. The idea is that a heuristic is selected when it results in a lot of improving moves, and thus has a higher utility value. When a heuristic i results in an improving move the utility value u_i of that heuristic is incremented, and in case of a worsening move the utility value is lowered using three different rules, namely subtractive ($u_i = u_i - 1$), divisional ($u_i = u_i/2$) and root ($u_i = \sqrt{u_i}$). Upper and lower bounds are applied to the utility values to encourage exploration in further steps. Experiments are performed with different settings for the selection of the heuristics, the upper and lower bound, and the negative utility adaptation mechanism. The setting with a maximal selection (i.e. selecting the heuristic with a maximal utility value) and subtractive negative utility adaption mechanism performed the best. The method improves the performance of a non learning simple-random great-deluge hyper-heuristic on the examination timetabling problem.

Table 17.1 Comparison of hybrid RL-metaheuristic methods

Method	RL method	Metaheuristic	Involved component	Problem(s)
[40]	TD(λ)	Constructive Method	Heuristic selection	Job scheduling
[16]	Q-learning	Ant system	Direct	ATSP
[21]	Q-learning	GA	Direct	ATSP and QAP
[4, 5]	No RL (regression)	Local search (e.g. SA)	Obj. function, starting solution	Bin packing, channel routing, SAT,...
[23]	TD(λ)	2-opt local search	Obj. function	Dial-a-ride
[25]	Utility values	CSP solver	Heuristic selection	Orc Quest and Logistics Domain
[7]	Utility values	Hyper-heuristic	Heuristic selection	Timetabling and rostering
[1]	Utility values + discount	Hyper-heuristic	Heuristic selection	Course timetabling
[22]	LA	Hyper-heuristic	Heuristic selection	TTP
[38]	LA	GA	Direct	Project scheduling (MRCPSP)
[26]	Utility values	Hyper-heuristic	Heuristic selection	Examination timetabling
[36]	LA (+Dispersion Game)	-	Direct	Project scheduling (DRCMPSP)
[37]	LA	-	Direct	Project scheduling (MRCPSP)

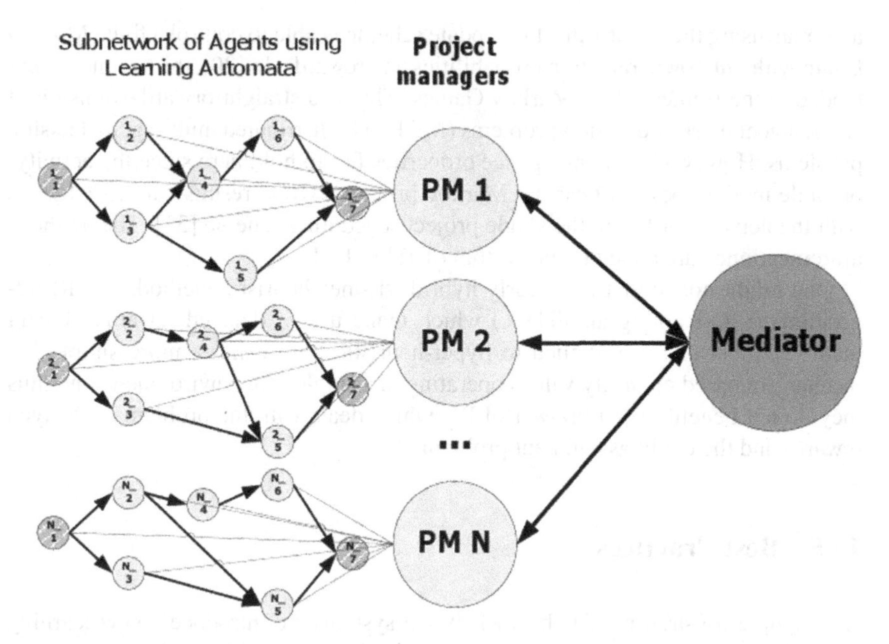

Fig. 17.2 Multi-agent system employed to solve the decentralized resource-constrained multi-project scheduling problem

Recently, learning automata have been introduced to solve combinatorial optimization problems. [22] present a heuristic selection method for hyper-heuristics, which they have applied to the traveling tournament problem. Instead of using simple reinforcement rules, a learning automaton was used for the selection. [38] describe the combination of a genetic algorithm and learning automata to solve the Multi-Mode Resource-Constrained Project Scheduling Problem (MRCPSP). The GA is applied to find good activity orders, while the LA are used to find good modes, a mode being an important decision variable of the scheduling problem. This work is extended in [37] where the GA is replaced by a network of learning automata [33]. All decision variables (i.e. activity order and modes) of the scheduling problem (MRCPSP) are now directly chosen by multiple LA. The method produces state-of-the-art results for the MRCPSP. [36] follow a very similar approach for the Decentralized Resource-Constrained Multi-Project Scheduling Problem (DRCMPSP). In the DRCMPSP multiple projects are scheduled factoring in the availability of both private and shared resources, while a global objective, i.e. the average project delay, is optimized. A network of learning automata searches for activity orders resulting in good schedules for each single project, while a dispersion game is employed to coordinate the projects. Figure 17.2 shows the multi-agent system with project managers and network of LA for solving the DRCMPSP, as applied by [36]. One motivating factor for organizing the activities in a project as learning automata is that theoretical convergence properties hold in both single and multi automata environments. One of the foundations for LA theory is that a set of decentralized learning automata using the reward-inaction update scheme is able to control a finite Markov Chain with unknown transition probabilities and rewards. In [20], this result was extended to the framework of Markov Games. That is a straightforward extension of single-agent markov decision problems (MDP's) to distributed multi-agent decision problems. However, the convergence properties fail to hold here since the activity-on-node model does not bear the Markov property. Good results can be achieved with the network of LA in the single project scheduling scheme [37]. The methods aforementioned are added to the bottom of Table 17.1.

One might notice that most early hybrid RL-metaheuristic methods use RL algorithms as Q-learning and TD(λ) which make use of delayed rewards. Recent methods, most of them applied to hyper-heuristics, are using a more simple RL mechanism based on utility values operating in a single state environment, and thus they do not benefit the full power of RL which deals with the problem of delayed rewards and the credit assignment problem.

17.5 Best Practices

As a simple illustration of hybrid RL based systems we introduce a new learning method (LA-ILTA) using RL and show how to boost an exiting acceptance mechanism (ILTA), which was recently published [22]. We discuss possible overheads, such as extra parameters belonging to the learning components and time overhead.

17.5.1 LA-ILTA

Iteration Limited threshold acceptance (ILTA) is an acceptance mechanism for meta- and hyper-heuristics, introduced by [22]. ILTA is based on the improving or equal (IE) acceptance criterion, which only accepts non-worsening moves. Like every good acceptance criterion, ILTA tries to efficiently balance intensification and diversification. In addition to IE, ILTA accepts worsening moves under certain conditions, i.e. it accepts a move if k consecutive worsening moves are generated, and the new solution's fitness is within a certain range R of the current best solution's fitness. These two parameters k and R are fixed during the complete search. We now propose a method, called LA-ILTA, which uses RL and more specifically Learning Automata to adaptively change and learn good parameter values for ILTA. We have chosen to learn the R values, based on the place in the search process (e.g. beginning, middle or end of the search). The method thus belongs to the category of methods that use a search-dependent state space representation. We first define the state and action spaces for the RL component. We divide the search process into 10 separate periods, each one lasting 10% of the search duration. The start of each period is a state. We define the R values to be the actions possibly in each state. Thus a chosen R value is used for the next 10% of the search. The chosen discrete R values (and thus actions) are $\{1.0, 1.1, 1.2, 1.3, 1.4, 1.5\}$. In total we have 10 states and 6 actions. The reward function, which expresses the learning goal, is the percentage improvement realized by using the chosen R value in the past search moment. In each state we use one learning automaton with LRI update scheme to select the R values. LA-ILTA is applied to the Patient Admission Scheduling (PAS) problem [12, 13]. Current best results are presented in [8]. Patients in a hospital have to be assigned to beds such that multiple hard and soft constraints regarding hospital regulations and patient preferences are met. A weighted objective function including a term for each soft constraint must be minimized. Thirteen problem instances are available [11].

We perform the following experiment. For each PAS problem instance we compare the learning LA-ILTA method to the static ILTA. For the LA-ILTA method we perform 1000 learning runs, each run from a different starting solution. Then we perform 1000 validation runs also starting from different starting solutions. For ILTA we perform only 1000 validation runs because ILTA does not include learning. We run ILTA for each static R setting $\{1.0, 1.1, 1.2, 1.3, 1.4, 1.5\}$. During these experiments all runs perform $500,000$ iterations and the k value was fixed to 100. The LA use a learning rate $\alpha_{reward} = 0.1$ and a linear reward-inaction update scheme. Table 17.2 shows the results of these experiments on the second problem instance of the PAS problem. A comparison is made between LA-ILTA and six static ILTA versions in terms of best, average and worst objective function value over 1000 validation runs. It is clear that the static ILTA with $R = 1.0$ outperforms the other static ILTA versions. However, the learning LA-ILTA method which learns a parameter setting that is dependent on the current search progress performs even better, and thus boosts the original ILTA method. All methods started from the same set of 1000 initial solutions. Similar results were obtained for the other problem instances.

Figure 17.3 shows the evolution of the objective function value over the 1000 validation runs for LA-ILTA on the second PAS problem instance. A moving average is also shown. The figure clearly shows that solutions with better objective values are reached when more learning runs are performed.

When we examine the learned policy of the LA-ILTA method, we notice that it favours higher R values (1.5) at the start of the search and lower R values (1.0) for the rest of the search progress. A transition from 1.5 to 1.0 is observed early in the search. In the beginning the learned policy allows for a lot of diversification, while later on it chooses to have more intensification. We can find a similar diversification/intensification strategy in the popular simulated annealing acceptance criterion. Figure 17.4 shows the evolution of the selected R values for the first 10% (state 1), middle 10% (state 5), and last 10% (state 10) of the search. The evolution in states 2, 3, 4, 6, 7 and 8 are ommitted from the figure for clarity. The R value for the first

Table 17.2 LA-ILTA compared to six static ILTA versions over 1000 validation runs on the second PAS problem instance

	LA-ILTA	ILTA-1.0	ILTA-1.1	ILTA-1.2	ILTA-1.3	ILTA-1.4	ILTA-1.5
Best obj.	**14898**	14960	24168	24656	24998	25220	25884
Average obj.	**16200.1**	16233.9	26796.9	27732.1	27746.4	27964.4	28610.4
Worst obj.	**17776**	18100	29272	30448	30268	30300	30842

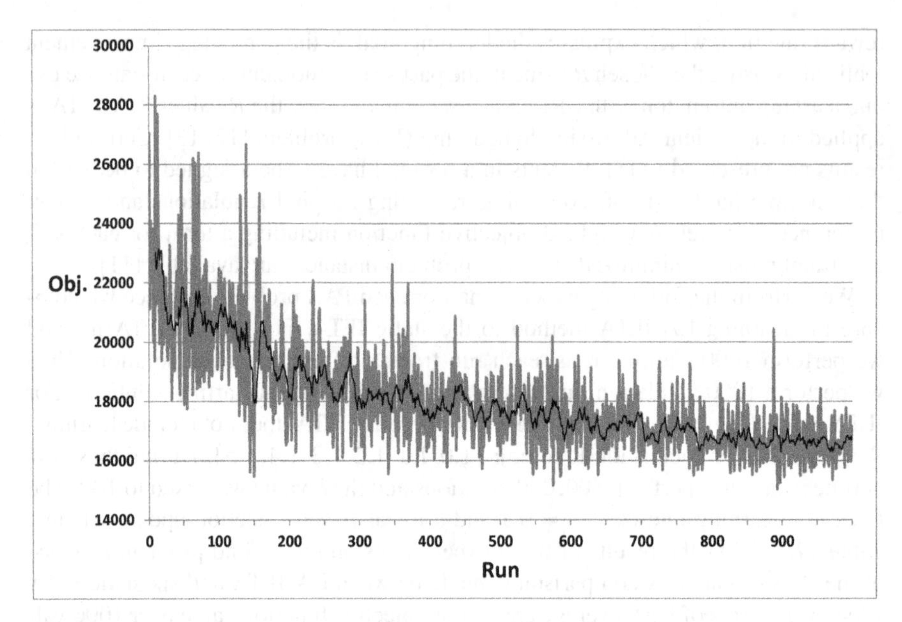

Fig. 17.3 Learning curve for LA-ILTA on the second PAS problem instance

part of the search converges rapidly to a value of 1.5, which is the highest value in the range. The middle and the last state move towards a value of 1.0, but the value for the middle state converges faster than the last state, be it slower than the first state. In general, learning in the first states goes faster than in the last states, because more information is available in the beginning of the search than at the end. This idea was also discussed in [1].

We also have applied LA-ILTA to other problems, such as the Edge Matching Puzzle (EMP) problem [10]. The problem consists of placing $n \times n$ square tiles on a board of size $n \times n$. A tile has four edges, each edge containing a pattern from a set of available patterns. All tiles must be rotated and placed on the board, such that the shared edge between neighboring tiles has a matching pattern. A special pattern (pattern 0) must only occur on the outer edges of the board. Figure 17.5 shows the results of LA-ILTA over 1000 validation runs on an Edge Matching Puzzle problem of size 10×10. Each run performing $100,000$ iterations. The LA use a learning rate $\alpha_{reward} = 0.1$ and a linear reward-inaction update scheme. Higher scores are better. The LA-ILTA method is at least as good as the best static ILTA methods. The learned policy shows similar characteristics as the learned policy on the PAS problems, i.e. high diversification in the beginning of the search process and more intensification at the end. However, unlike in the PAS experiments, the diversification does not fade away completely.

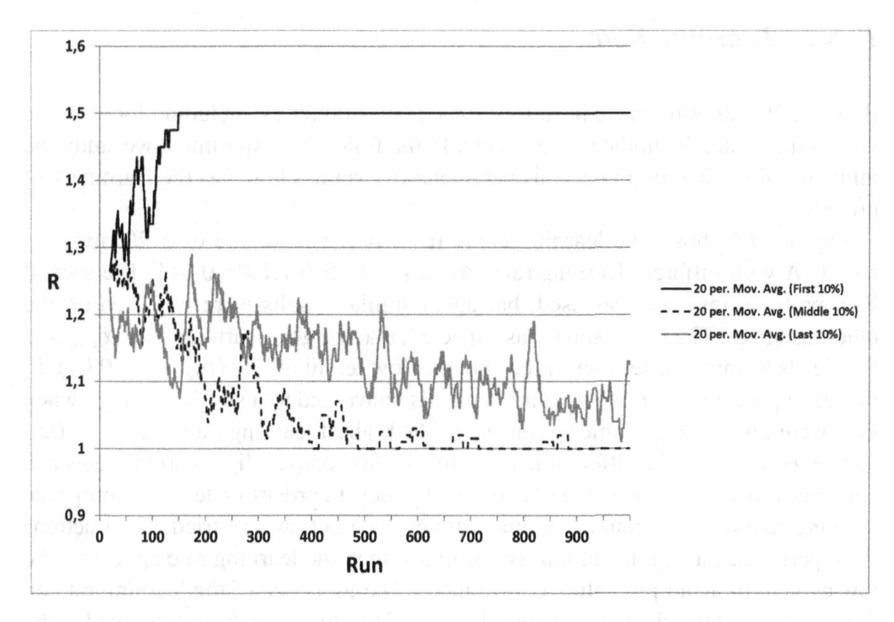

Fig. 17.4 Evolution of the selected R value setting for the first, middle and last 10% of the search on the second PAS problem instance

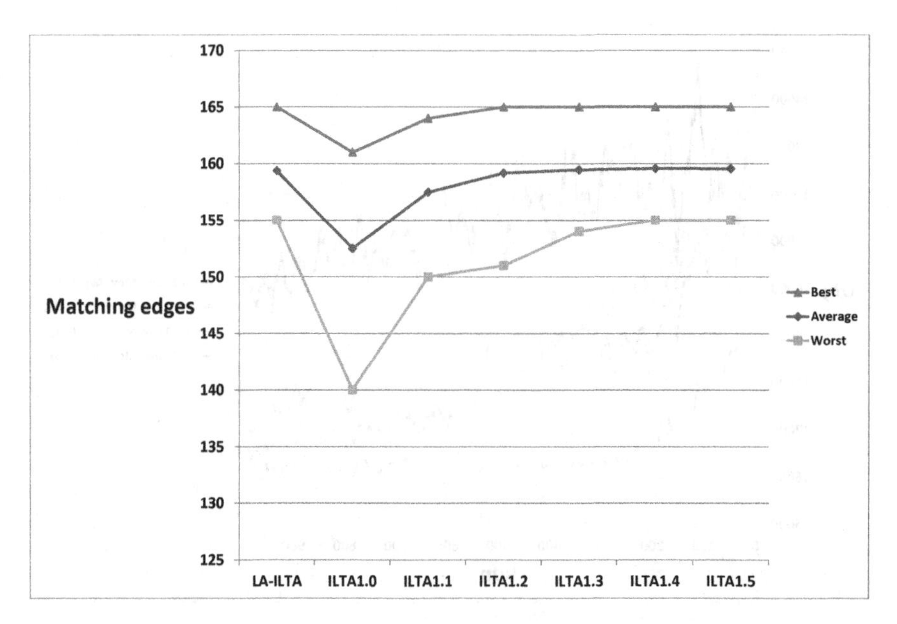

Fig. 17.5 LA-ILTA compared to six static ILTA versions over 1000 validation runs on an edge matching puzzle problem of size 10×10

17.5.2 Learning Rate

By using RL algorithms we introduce some new parameters, including for example the learning rate, the update scheme, etc. In the following experiment we study the influence of the learning rate. This parameter determines how fast the learning will proceed.

Figure 17.6 shows the learning curve (as a moving average over 20 runs) for LA-ILTA with different learning rates $\alpha_{reward} = \{0.5, 0.1, 0.05, 0.01\}$. The second PAS problem instance was used, but again similar results were observed on the other instances. The figure shows, as expected, that a higher learning rate ($\alpha_{reward} = 0.5$) leads to much faster convergence than a low learning rate ($\alpha_{reward} = 0.01$). In this example all except one learning methods converged to a similar strategy when they were given enough time to converge. The highest learning rate ($\alpha_{reward} = 0.5$) converged to a slightly different R value for the first phase. High learning rates can converge too quickly and reach a suboptimal policy. In order to select an appropriate learning rate one can count how many times each action was tried. If all actions were performed a significant number of times, then the learning rate appears to be low enough to avoid premature convergence. The influence of the learning rate on the results of a hybrid RL-metaheuristic method has to be carefully examined in the future.

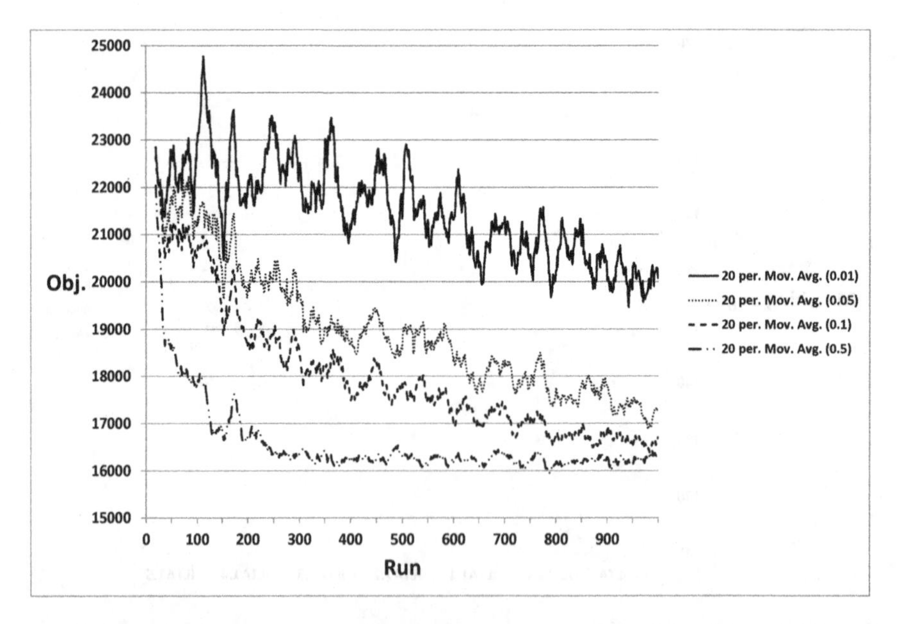

Fig. 17.6 The learning curve (moving average over 20 runs) for LA-ILTA with different learning rates $\alpha_{reward} = \{0.5, 0.1, 0.05, 0.01\}$ on the second PAS problem instance

Table 17.3 Average calculation time overhead in percentage introduced by the LA-ILTA method. Tested on 12 PAS problem instances and the eternity 2 puzzle problem.

PAS1	PAS2	PAS3	PAS4	PAS5	PAS6	PAS7	PAS8	PAS9	PAS10	PAS11	PAS12	EMP
0.41%	2.65%	1.45%	2.45%	0.38%	1.35%	4.68%	7.00%	5.40%	5.84%	9.58%	8.11%	0.31 %

17.5.3 Calculation Time Overhead

In this experiment we will investigate the calculation time overhead introduced by the reinforcement learning component. Table 17.3 shows the average calculation time overhead in % introduced by the reinforcement learning component in the LA-ILTA method on 12 PAS problem instances and an edge matching puzzle (EMP) problem instance of size 16 by 16. The LA-ILTA method uses 10 LRI learning automata, each having 6 actions. The results in the table show that the RL component never introduces more than 10% overhead to the calculation time. The overhead is hard to measure, since it is subject to various factors, such as implementation details, the metaheuristic, the RL method, the problem type and the problem size. The calculation time of the RL methods are mostly determined by the number of actions which can be selected at each decision point. Figure 17.7 shows the calculation time in milliseconds to perform 1 million selections and updates for an LRI learning automaton on a modern desktop pc with Intel Core I7-2600 3.4Ghz CPU. The calculation time grows linearly with the number of actions. For good performance, one should keep the number of actions as small as possible. The number of

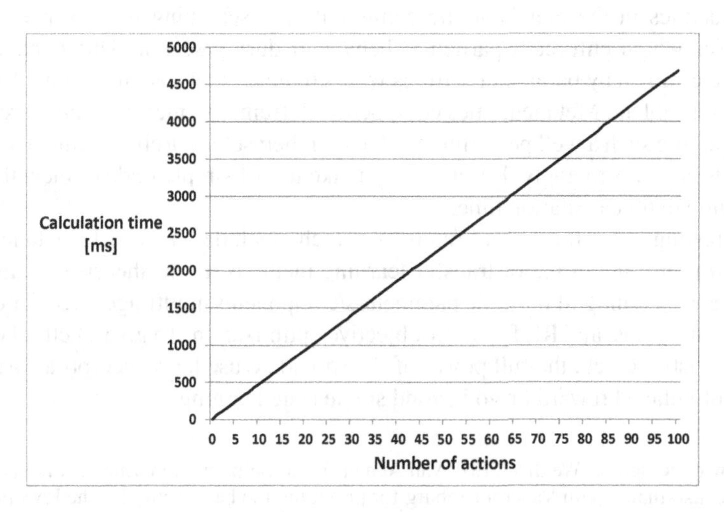

Fig. 17.7 Calculation time in ms for an LRI learning automaton to perform 1 million selections and updates, for a varying number of actions

states has no impact on the calculation time, since only one state is being updated at a time. However, the number of state transitions determines how many action selections and updates are performed, and thus affects the calculation time. The memory requirements on the other hand are determined by the product of the number of states and the number of actions.

17.6 Conclusion and Suggestions for Future Research

In the present chapter we have discussed the combination of reinforcement learning and metaheuristic search. This hybridization of two well studied research topics shows some promising methods and many opportunities for future research directions. We have discussed some main reinforcement learning components in detail, together with a literature overview on hybrid RL-metaheuristic methods. We have illustrated some examples where learning automata are able to boost the performance of metaheuristic search. An example of a new application of reinforcement learning to meta- or hyper-heuristic methods is also given, i.e. a learning acceptance method called LA-ILTA. LA-ILTA uses several learning automata to learn a search-dependent parameter value, which was used in an existent acceptance method (ILTA). This simple method was able to boost the results of the original ILTA on two tested benchmark problems, i.e. the patient admission scheduling problem and the edge matching puzzle problem. The overhead in terms of calculation time and extra parameters introduced by the RL components was studied. We have shown that simple RL methods, called learning automata, introduce little overhead.

Academics in the metaheuristic community are searching for advanced metaheuristics which enforce a particular behaviour during search. This behaviour is often determined by parameter settings which require extensive fine-tuning for each different problem. Metaheuristics equipped with (reinforcement) learning are capable of finding such a well performing behaviour themselves. Reinforcement learning methods are easy to apply. Because they make use of simple update rules, they require little extra calculation time.

Interesting directions for future research include; new RL-metaheuristic hybridizations, the usage of transfer learning methods in metaheuristic search, simultaneous learning of multiple parameters/components, multi-agent RL for decentralized problems, and RL for multi-objective optimization. To give a better boost to metaheuristic search, the full power of RL should be used, i.e. incorporate a mechanism of delayed reward or go beyond single state learning.

Acknowledgements. We thank Erik Van Achter for his help on improving the quality of this text. We also thank Wim Vancroonenburg for providing the basic setup for the PAS problem experiments.

References

1. Bai, R., Burke, E.K., Gendreau, M., Kendall, G., Mccollum, B.: Memory length in hyper-heuristics: An empirical study. In: Proceedings of the 2007 IEEE Symposium on Computational Intelligence in Scheduling, CI-Sched 2007 (2007)
2. Battiti, R., Brunato, M., Mascia, F.: Reactive Search and Intelligent Optimization. Operations research/Computer Science Interfaces, vol. 45. Springer (2008)
3. Bennett, K.P., Parrado-Hernández, E.: The interplay of optimization and machine learning research. J. Mach. Learn. Res. 7, 1265–1281 (2006)
4. Boyan, J.: Learning Evaluation Functions for Global Optimization. PhD thesis, Carnegie-Mellon University (1998)
5. Boyan, J., Moore, A.W., Kaelbling, P.: Learning evaluation functions to improve optimization by local search. Journal of Machine Learning Research 1, 2000 (2000)
6. Burke, E., Hart, E., Kendall, G., Newall, J., Ross, P., Schulenburg, S.: Hyper-heuristics: An emerging direction in modern search technology. In: Handbook of Metaheuristics, pp. 457–474. Kluwer Academic Publishers (2003)
7. Burke, E.K., Kendall, G., Soubeiga, E.: A tabu-search hyperheuristic for timetabling and rostering. Journal of Heuristics 9, 451–470 (2003)
8. Ceschia, S., Schaerf, A.: Local search and lower bounds for the patient admission scheduling problem. Computers & Operartions Research 38, 1452–1463 (2011)
9. Confessore, G., Giordani, S., Rismondo, S.: A market-based multi-agent system model for decentralized multi-project scheduling. Annals of Operational Research 150, 115–135 (2007)
10. Demaine, E.D., Demaine, M.L.: Jigsaw puzzles, edge matching, and polyomino packing: Connections and complexity. Graphs and Combinatorics 23, 195–208 (2007); Special issue on Computational Geometry and Graph Theory: The Akiyama-Chvatal Festschrift
11. Demeester. P.: Patient admission scheduling website (2009), `http://allserv.kahosl.be/~peter/pas/` (last visit August 15, 2011)
12. Demeester, P., De Causmaecker, P., Vanden Berghe, G.: Applying a local search algorithm to automatically assign patients to beds. In: Proceedings of the 22nd Conference on Quantitive Decision Making (Orbel 22), pp. 35–36 (2008)
13. Demeester, P., Souffriau, W., De Causmaecker, P., Vanden Berghe, G.: A hybrid tabu search algorithm for automatically assigning patients to beds. Artif. Intell. Med. 48, 61–70 (2010)
14. Dietterich, T.G., Zhang, W.: Solving combinatorial optimization tasks by reinforcement learning: A general methodology applied to resource-constrained scheduling. Journal of Artificial Intelligence Research (2000)
15. Gabel, T.: Multi-agent Reinforcement Learning Approaches for Distributed Job-Shop Scheduling Problems. PhD thesis, Universität Osnabrück, Deutschland (2009)
16. Gambardella, L.M., Dorigo, M.: Ant-q: A réinforcement learning approach to the traveling salesman problem, pp. 252–260. Morgan Kaufmann (1995)
17. Glover, F., Kochenberger, G.A.: Handbook of metaheuristics. Springer (2003)
18. Homberger, J.: A (μ, λ)-coordination mechanism for agent-based multi-project scheduling. OR Spectrum (2009), doi:10.1007/s00291-009-0178-3
19. Kaelbling, L.P., Littman, M.L., Moore, A.W.: Reinforcement learning: A survey. Journal of Artificial Intelligence Research 4, 237–285 (1996)
20. Littman, M.L.: Markov games as a framework for multi-agent reinforcement learning. In: Proceedings of the Eleventh International Conference on Machine Learning, pp. 157–163. Morgan Kaufmann (1994)
21. Miagkikh, V.V., Punch III, W.F.: An approach to solving combinatorial optimization problems using a population of reinforcement learning agents (1999)
22. Misir, M., Wauters, T., Verbeeck, K., Vanden Berghe, G.: A new learning hyper-heuristic for the traveling tournament problem. In: Proceedings of Metaheuristic International Conference (2009)

23. Moll, R., Barto, A.G., Perkins, T.J., Sutton, R.S.: Learning instance-independent value functions to enhance local search. In: Advances in Neural Information Processing Systems, pp. 1017–1023. MIT Press (1998)
24. Narendra, K., Thathachar, M.: Learning Automata: An Introduction. Prentice-Hall International, Inc. (1989)
25. Nareyek, A.: Choosing search heuristics by non-stationary reinforcement learning. In: Metaheuristics: Computer Decision-Making, pp. 523–544. Kluwer Academic Publishers (2001)
26. Özcan, E., Misir, M., Ochoa, G., Burke, E.K.: A reinforcement learning - great-deluge hyper-heuristic for examination timetabling. Int. J. of Applied Metaheuristic Computing, 39–59 (2010)
27. Rummery, G.A., Niranjan, M.: On-line q-learning using connectionist systems. Technical Report CUED/F-INFENG/TR 166, Engineering Department, Cambridge University (1994)
28. Richard, S., Sutton, R.S.: Generalization in reinforcement learning: Successful examples using sparse coarse coding. In: Touretzky, D.S., Mozer, M.C., Hasselmo, M.E. (eds.) Advances in Neural Information Processing Systems: Proceedings of the 1995 Conference, pp. 1038–1044. MIT Press, Cambridge (1996)
29. Sutton, R.S., Barto, A.G.: Reinforcement learning: an introduction. MIT Press (1998)
30. Talbi, E.-G.: Metaheuristics: From Design to Implementation. John Wiley and Sons (2009)
31. Taylor, M.E., Stone, P.: Transfer learning for reinforcement learning domains: A survey. J. Mach. Learn. Res. 10, 1633–1685 (2009)
32. Taylor, M.E., Stone, P., Liu, Y.: Transfer learning via inter-task mappings for temporal difference learning. Journal of Machine Learning Research 8(1), 2125–2167 (2007)
33. Thathachar, M.A.L., Sastry, P.S.: Networks of Learning Automata: Techniques for Online Stochastic Optimization. Kluwer Academic Publishers (2004)
34. Watkins, C.J.C.H., Dayan, P.: Q-learning. Machine Learning 8, 279–292 (1992)
35. Watkins, C.J.C.H.: Learning from Delayed Rewards. PhD thesis, Cambridge University (1989)
36. Wauters, T., Verbeeck, K., De Causmaecker, P., Vanden Berghe, G.: A game theoretic approach to decentralized multi-project scheduling (extended abstract). In: Proc. of 9th Int. Conf. on Autonomous Agents and Multiagent Systems, AAMAS 2010, vol. R24 (2010)
37. Wauters, T., Verbeeck, K., Vanden Berghe, G., De Causmaecker, P.: Learning agents for the multi-mode project scheduling problem. Journal of the Operational Research Society 62(2), 281–290 (2011)
38. Wauters, T., Verstichel, J., Verbeeck, K., Vanden Berghe, G.: A learning metaheuristic for the multi mode resource constrained project scheduling problem. In: Proceedings of the Third Learning and Intelligent OptimizatioN Conference, LION3 (2009)
39. Williams, R.J.: Simple statistical gradient-following algorithms for connectionist reinforcement learning. Machine Learning, 229–256 (1992)
40. Zhang, W., Dietterich, T.: A reinforcement learning approach to job-shop scheduling. In: Proceedings of the Fourteenth International Joint Conference on Artificial Intelligence, pp. 1114–1120. Morgan Kaufmann (1995)

Index

Author Index